# 深入理解Django
## 框架内幕与实现原理

沈聪　全树强　编著

電子工業出版社
Publishing House of Electronics Industry
北京•BEIJING

## 内容简介

全书共分为 8 章，分别是 Django 源码的前置知识、Django 命令原理解析、Django 内置的 ORM 框架、Django 内置的模板系统、解读 Django 核心模块的源码、Django 的视图层、Django 的中间件原理和解读 Django 中的辅助代码。本书展示了剖析 Django 框架的完整过程，这种剖析方式同样适用于其他框架，如 Flask、Ansible 和 Scrapy 等。

本书适合有一定 Python 基础的开发人员、Django 框架的初学者，以及渴望进入开源世界的读者阅读。



**图书在版编目（CIP）数据**

深入理解 Django：框架内幕与实现原理 / 沈聪，全树强编著. —北京：电子工业出版社，2021.11
ISBN 978-7-121-42188-4

Ⅰ. ①深… Ⅱ. ①沈… ②全… Ⅲ. ①软件工具－程序设计 Ⅳ. ①TP311.561

中国版本图书馆 CIP 数据核字(2021)第 203988 号

责任编辑：安　娜
印　　刷：北京天宇星印刷厂
装　　订：北京天宇星印刷厂
出版发行：电子工业出版社
　　　　　北京市海淀区万寿路 173 信箱　　　邮编：100036
开　　本：787×980　1/16　印张：37.75　　字数：900 千字
版　　次：2021 年 11 月第 1 版
印　　次：2021 年 11 月第 1 次印刷
定　　价：118.00 元

凡所购买电子工业出版社图书有缺损问题，请向购买书店调换。若书店售缺，请与本社发行部联系，联系及邮购电话：（010）88254888，88258888。

质量投诉请发邮件至 zlts@phei.com.cn，盗版侵权举报请发邮件至 dbqq@phei.com.cn。

本书咨询联系方式：（010）51260888-819，faq@phei.com.cn。

# 前言

随着深度学习的流行，Python 逐渐走到了互联网的幕前，它在多个领域使用广泛：机器学习、数据分析、Web 服务、自动化运维，等等。Django 是 Python 在 Web 框架中的代表作品。学习 Django 源码，一方面，能从中学到许多 Python 的高级用法；另一方面，能深度掌握 Django 框架，并能随时对其进行定制和改造，这对一个 Python 程序员的成长至关重要。此外，在 Django 源码中，很多函数与类均可在简单改造后直接加入开发人员手头的代码库中。

撰写本书的主要原因有两个：一是想将笔者剖析 Django 源码的过程记录下来，二是想给一些 Django 初学者开启一个新的进阶方向。

Django 源码十分适合初学者学习，因为 Python 源码与 C++源码相比难度要低许多，而且调试方便，读者可以随时打印结果进行验证。以 Django 源码为入口切入开源世界，对 Python 爱好者而言，再合适不过了。

## 本书特色

本书的主要特色有：

- ◎ 对 Django 的源码进行了合理选择，并不会盲目追求最新与最小版本，而是从适合的角度进行考虑。
- ◎ 对 Django 的源码进行了合理归类与总结，并按照归类结果依次解读相关源码，循序渐进。
- ◎ 有对 Django 源码细节的丰富解读，直接使用实战的方式帮助读者理解相关类与方法。
- ◎ 对 Django 源码的解读均采用先上手并提问的方式，然后会带着问题去跟踪源码，最后阅读完相关模块源码后会统一对前面提出的问题进行解答。这很符合初学者的阅读习惯，使得阅读本书十分顺畅。

## 本书内容

全书共分为 8 章，分别是 Django 源码的前置知识、 Django 命令原理解析、Django 内置的 ORM 框架、Django 内置的模板系统、解读 Django 核心模块的源码、Django 的视图层、Django 的中间件原理和解读 Django 中的辅助代码。本书展示了剖析 Django 框架的完整过程，这种剖析方式同样适用于其他框架，如 Flask、Ansible 和 Scrapy 等。

本书对 Django 源码进行了全面解读，针对很多源码细节使用了大量的演示实例，以便读者更好地理解 Django 框架的核心源码。本书给读者提供了一个解读完整开源项目的案例，读者在学完本书后，可以将本书剖析 Django 源码的方式应用于其他 Python 开源项目，收获良好的学习效果。

本书的思维导图如下所示。

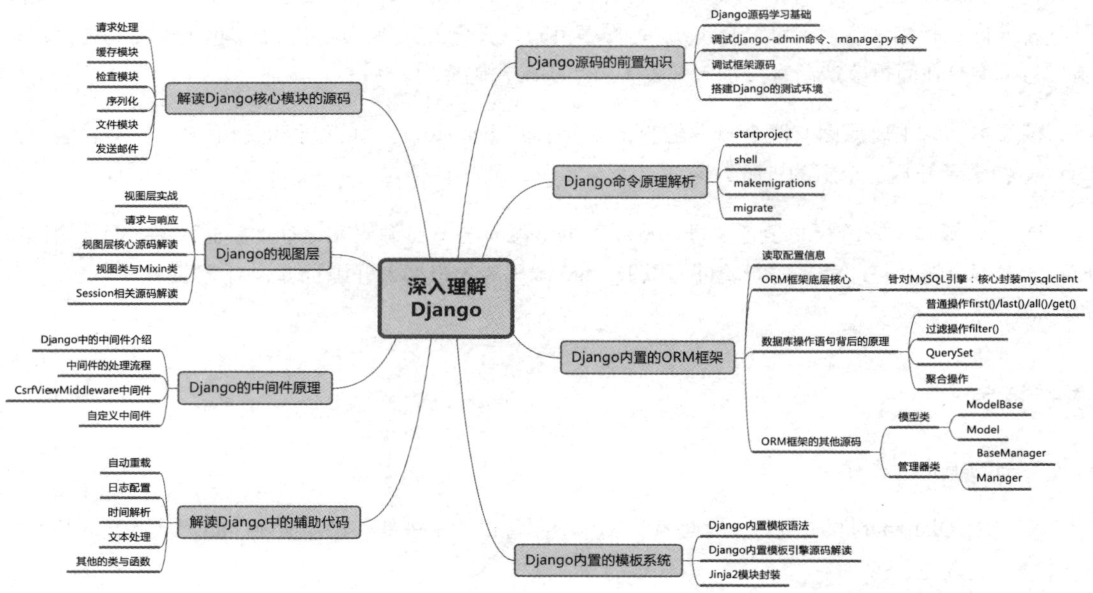


## 读者对象

想要学习本书的内容，领会其中的分析思路，读者需要具备一定的 Python 基础，同时要有坚持学习的信念，本书适合以下读者阅读：

◎ 具有一定 Python 基础的开发人员；

◎ 希望熟练使用 Django 框架进行 Web 开发，而且想深入了解 Django 核心源码的后端开发人员；
◎ 渴望进入开源世界，探索一个顶级开源项目全貌的初、中级程序员；
◎ 高等院校的老师和学生；
◎ 相关培训机构的学员。

## 配书资源获取方式

本书涉及的源码需要读者自行下载。在博文视点网站（http://www.broadview.com.cn/）搜索本书，单击进入本书页面，即可在本书页面上找到下载链接。此外，可以通过微信搜索“源码探索之旅”公众号并关注，笔者会在上面对读者的一些疑惑进行文字或视频解答。

## 致谢一

在本书的写作过程中，我充满着感激之情。

感谢我的家人！感谢父母和姐姐给我的鼓励与支持，没有他们就没有我的今天，我也不会有机会接触计算机行业。

感谢我的女朋友黄宝宝，她花费了休息时间帮我修改书稿中的不当语句和错别字，给予我充分的时间专注于核心内容。

感谢天翼云科技有限公司的领导黄润怀和我的直接上级领导林洁琬组长，他们给我们组创造了一个难得的机会，让我们能全身心地投入分布式存储领域中。正是这次机会，让我在自动化运维领域自由探索。由于工作的需要，我接触了许多基于 Python 开发的开源项目，也逐渐爱上探索开源工具内部的源码。

感谢同一届进入公司的同事：商金辉、李浩、全树强和张一飞，每天吃完晚饭在路上胡吹海侃的日子，真是令人怀恋啊!

感谢媒体存储小分队的黄鹄、陈涛、毛廷鸿、谭伟杰、吴文峰、杨佑，他们丰富的互联网经验给了我很多帮助，也让本书的写作比预期顺利很多。

感谢本书的编辑安娜，她的耐心和细心促成了本书的快速交付；还要感谢其他为本书的出版提供过帮助的编辑和朋友！没有他们的大力支持，本书也很难与读者见面。

沈　聪

## 致谢二

我们能顺利完成本书既离不开自己的付出和努力，也离不开家人和同事们的大力支持。他们的支持让我在编写本书的道路上走得更加顺利，我要向他们表示最真挚的谢意。

首先，感谢我的妻子对我的关心和支持。在技术探索和创作的历程中，她的关心让我充满了前进的动力，这也让我能够走得更远。

其次，感谢本书的合著者沈聪先生。他喜好钻研，技术能力强，我们在工作中多次合作交流，让我受益匪浅。本书的创作也凝聚了他的心血和精力。

最后，感谢我的领导林洁琬组长和其他同事。在林洁琬组长的带领下，我们项目组的业务和技术都突飞猛进，也让我有了更多的机会对 Django 框架进行实践和探索。项目组的资深运维工程师陈涛、毛廷鸿、谭伟杰、杨佑等对我们基于 Django 框架的项目开发提供了宝贵的建议和技术支持。

全树强

## 技术支持

Django 项目如今依旧频繁迭代着，在本书正式出版时，书中介绍的 Django 版本距离最新版本可能会有较大差距。尽管版本可能差距较大，但笔者发现 Django 的核心源码及相关逻辑一直未有大的改变，读者可以放心阅读。笔者完成本书所需的主要素材只有 Django 源码，许多 Django 术语（比如核心模块、中间件等）都是直接翻译英文注释而来，可能会有不当之处，加之笔者水平和精力所限，书中可能存在一些疏漏与错误，敬请各位前辈、同行、读者不吝指正。如果在阅读本书时有疑问，可以发送电子邮件到 2894577759@qq.com，笔者当尽力解答相关疑问，与读者共同成长与进步。

# 目录

# 第 1 章
# Django源码的前置知识

本章介绍 Django 源码的一些前置知识，比如搭建 Django 的调试环境、使用 Django 搭建 Web 服务等。在掌握这些前置知识后，就可以正式开始 Django 的源码学习之旅了。

## 1.1 在学习 Django 源码前，该做什么

本节介绍在学习 Django 源码前需要做的一些基本操作，主要包括两部分内容：Django 的版本选择及 Django 源码学习基础。

### 1.1.1 Django 的版本选择

截止到本书的写作日期，GitHub 上 Django 项目的最新版本是 3.1.2。相比于统治地位的 Django 2，Django 3 尚处于功能完善的初期，代码中的隐藏 bug 较多，并不适合直接在生产环境中使用。此外，笔者完整对比了 Django 2.2 和 Django 3 的最新源码，二者相比，目录结构以及功能模块并没有太大变化，只是 Django 3 在某些操作中加入了异步调用并调整了一些代码细节，代码的整体结构及一些核心思想仍旧兼容 Django 2.2。因此读者在掌握 Django 2.2 的核心源码后，即可迅速看懂 Django 3 的源码内容。目前，Django 2 仍是 Python Web 开发的主流选择，其最新版本已更新至 2.2.16。因此，基于最新的 Django 2.2 版本讲解 Django 源码较为合适。GitHub 上的 Django 版本如图 1-1 所示。

此外，Django 2 已全面拥抱 Python 3，这也是笔者选择剖析 Django 2 源码的原因之一。从 Django 2 的源码中能学到许多 Python 3 的高级用法和一些经典的设计模式，而不必考虑兼容 Python 2。可以说，Django 2 源码是深入学习 Python 3 的宝藏。

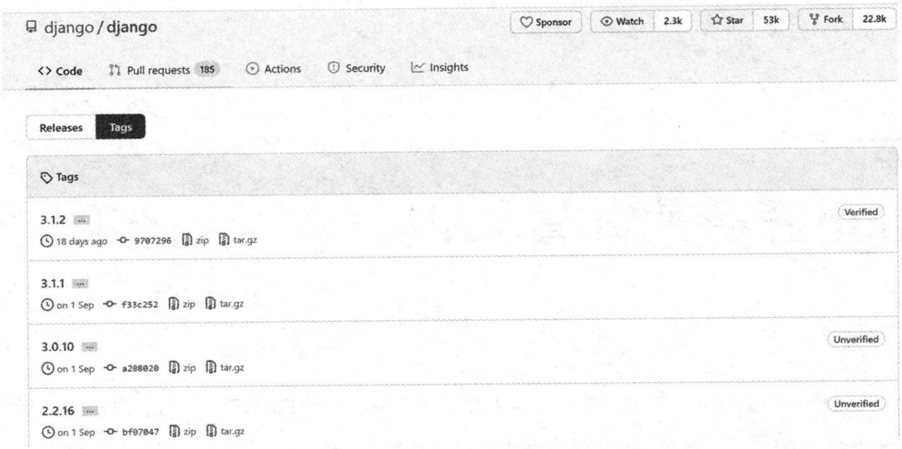

图 1-1

## 1.1.2 Django 源码学习基础

在学习 Django 源码之前，必须要做到以下两点（学习其他框架的源码也是如此）：

◎ 熟练使用 Django 框架（或者其他框架）进行日常的使用或开发。

◎ 掌握实现 Django 框架的语言基础——Python。

**注意**，Django、Flask 和 Ansible 等的语言基础是 Python，Nginx 和 Redis 等的语言基础是 C，MongoDB 和 Ceph 等的语言基础是 C++。

只有熟悉 Django 框架的用法，才能从一些基本现象入手，切入源码进行学习。在研究 Django 源码的过程中，可以发现一些更为高级的用法，从而进一步掌握 Django 框架。在使用 Django 框架的过程中，你有没有思考过以下几个问题：

◎ 创建 Django 项目的命令是 django-admin startproject project_name，它的具体执行过程是怎样的？

◎ Django 提供的 shell 交互模式（python manager.py shell）和普通的 Python 交互模式有何区别？

◎ 是否曾被 python manager.py migrate 命令所困扰？经常在不经意的改动后，使得迁移指令执行失败，最后只能完全重来一遍，费时又费力。

◎ Django 内置了简单优雅的 ORM 框架，使得 Django 操作数据库变得非常的简单，它是如何实现的？

带着问题去阅读源码是一种非常优秀的学习方式，在本书中，笔者会带领读者结合问题去源码中寻找答案，希望在学完本书后，读者能对 Django 框架有非常深刻的理解。

# 1.2　搭建 Django 的调试环境和测试环境

本节介绍如何搭建 Django 的调试环境，笔者的系统环境为 Win10。为了能学习和调试 Django 源码，你需要做两件事情：

◎ 在本机安装 Python 3 环境，Python 版本不低于 3.5。

◎ 在本机安装 Visual Studio Code（简称 VSCode）软件及 Python 插件，配置 Python 插件中的 python.pythonPath 参数。

接下来搭建 Django 的调试环境。对 Django 源码的调试分为两种：

◎ 对 Django 基本命令的调试。比如直接调用源码的命令 django-admin startproject first_django 等。

◎ 对基于 Django 框架创建的 Web 项目的调试，比如调试 python manage.py [shell | makemigrations | migrate]命令等。

此外，还可以通过创建的 Web 项目追踪 Django 框架的整个生命周期，以及它对 HTTP 请求、视图映射、数据查询、页面渲染等所做的处理。

## 1.2.1　调试 django-admin 命令

首先，从 GitHub 的 Django 项目中下载 django-2.2.16.tar.gz 包，解压缩后使用 VSCode 直接导入即可，如图 1-2 所示。

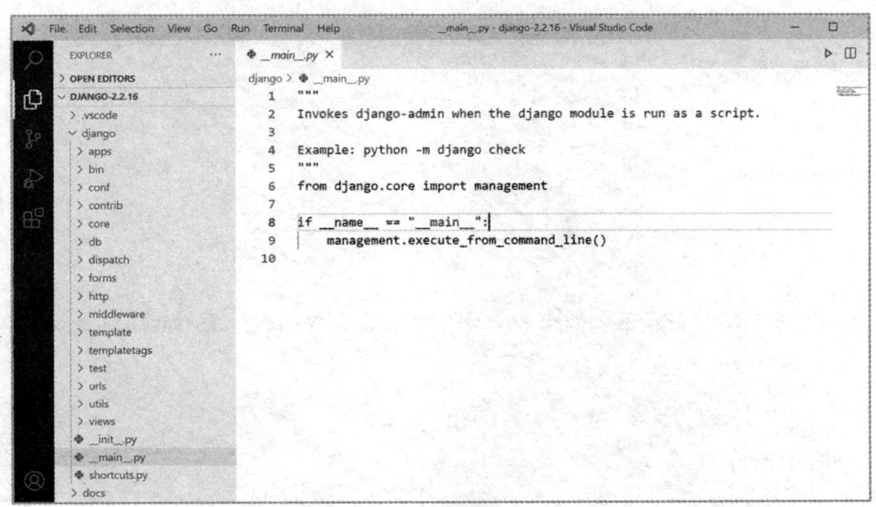


图 1-2

本次调试的命令为 django-admin startproject first_django。新建一个调试文件，单击侧边栏的 Run 图标，如果没有配置过调试信息，则会出现如图 1-3 所示界面。如果之前配置过调试信息，则直接在文件区修改.vscode 目录下的 launch.json 文件即可。

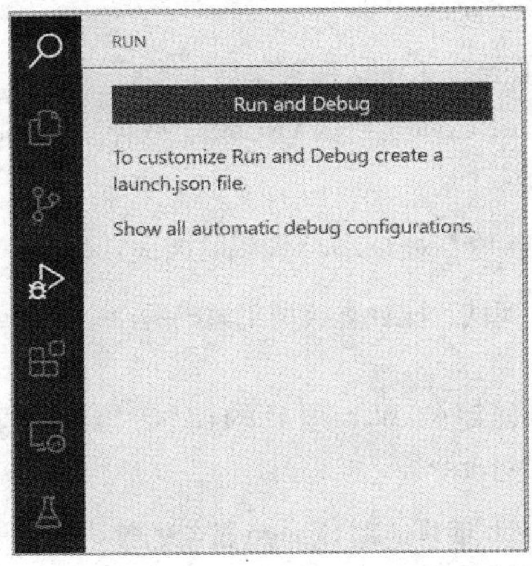


图 1-3

单击调试工作区的 create a launch.json file 链接，在弹出的窗口中单击 Python 文件调试打开的 Python 文件选项，即可创建 Django 源码调试文件，如图 1-4 所示。

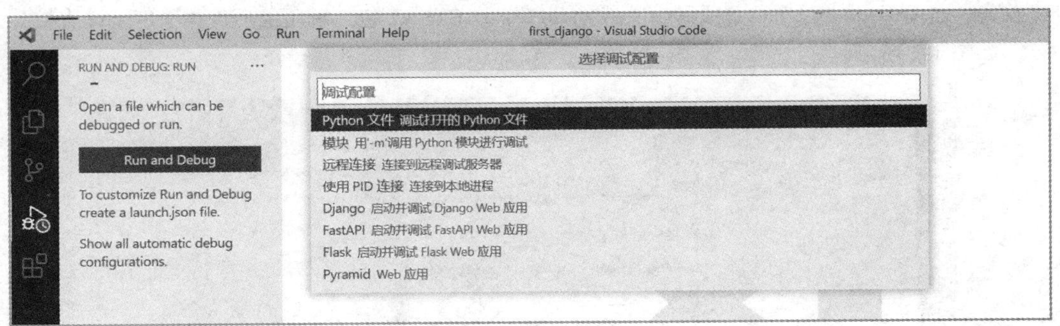


图 1-4

此时，VSCode 在项目目录下会创建一个.vscode 目录，里面有一个 launch.json 文件，其内容如下：

```
{
   "version": "0.2.0",
   "configurations": [
      {
         "name": "Python: 当前文件",
         "type": "python",
```

```
            "request": "launch",
            "program": "${file}",
            "console": "integratedTerminal"
        }
    ]
}
```

修改后的内容如下：

```
{
    "version": "0.2.0",
    "configurations": [
        {
            "name": "Python: 当前文件",
            "type": "python",
            "request": "launch",
            "program": "${file}",
            // 执行固定位置的脚本
            // "program": "${workspaceRoot}/django/bin/django-admin.py",
            "cwd": "D:/learning-notes/book-code/django-code-analyze/chap01",
            "env": {"PYTHONPATH": "${workspaceRoot}"},
            "args": [
                "startproject",
                "first_django"
            ],
            "console": "integratedTerminal"
        }
    ]
}
```

**说明**：在配置中可以写多个参数控制调试程序。下面简要说明部分参数及其值的作用，更多的参数配置可以参考 VSCode 的官方文档。

- program：提供 Python 程序入口模块的完全限定路径。推荐值为${file}。当在某个 Python 文件下按 F5 键或者是单击 Run 菜单下的 Start Debugging 命令时，${file}就是该 Python 文件的全路径。
- cwd：先进入某个目录，再执行相应的调试命令。
- env：它非常重要，没有它就无法搭建源码的调试环境。在 Django 源码中出现的导入语句，如 from django 等，希望能将其定位至下载的源码文件所在目录（django）下。因此，当调试 Python 项目的源码时，会使用该字段设定 PYTHONPATH 值。在 django-2.2.16 项目中，django 目录就是源码的根目录，而 django 目录正好在项目目录下，因此只需设置 PYTHONPATH=${workspaceRoot}即可。
- args：指定传递给 Python 程序的参数。

上面的调试配置主要是希望执行命令 cd D:/learning-notes/book-code/django-code-analyze/

chap01;django-admin startproject first_django，该命令使得新建的 Django 项目位于 chap01 目录中。为了让${file}等于 Django 源码下的 bin/django-admin.py 文件，在开始调试时就需要定位在该文件中。如果直接指定该命令文件的位置，如上面的注释语句，则无论在哪个文件上启动调试都可以。为了验证该调试配置得是否正确，可以在源码中打上断点，具体位置如图 1-5 和图 1-6 所示。

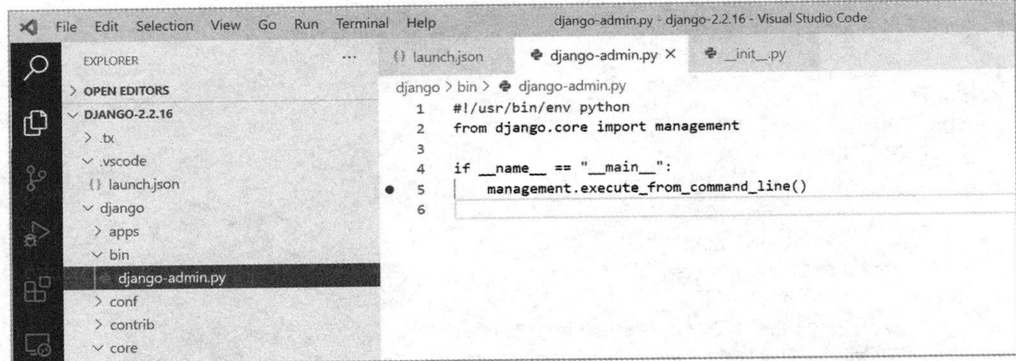


图 1-5

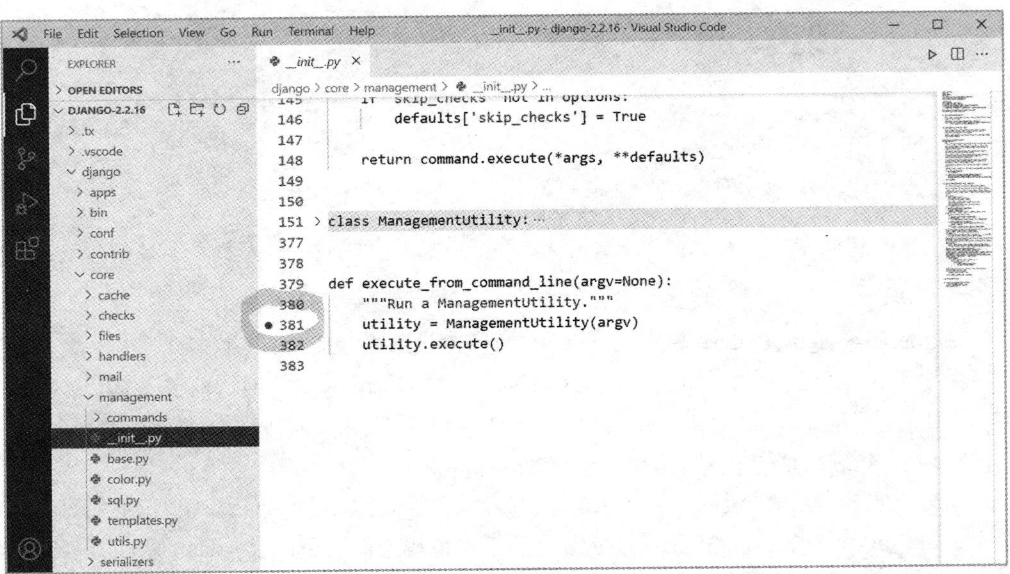


图 1-6

选中 django/bin/django-admin.py 文件，按 F5 键启动调试，此时程序停在断点一处，如图 1-7 所示。

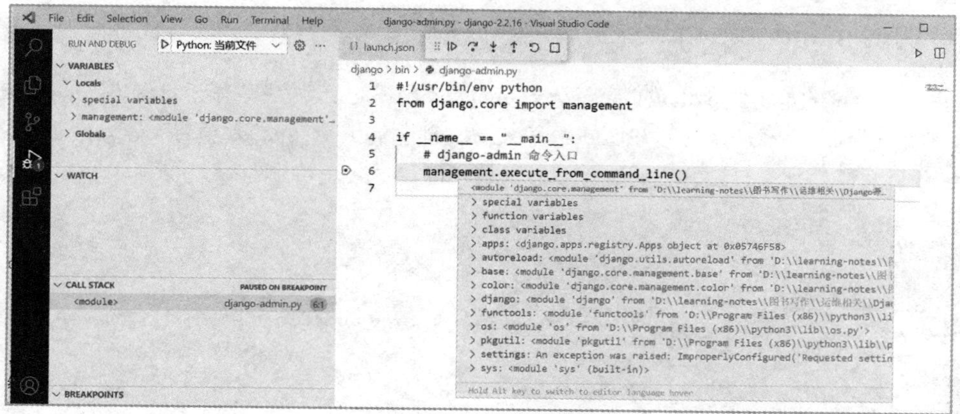

图 1-7

单击继续按钮后，程序停在断点二处，此时程序已进入 Django 内部的源码文件中，如图 1-8 所示。释放断点继续运行程序，可以看到在目录 chap01 中已生成了相应的 Django 项目文件。至此，Django 的调试环境搭建成功，如图 1-9 所示。

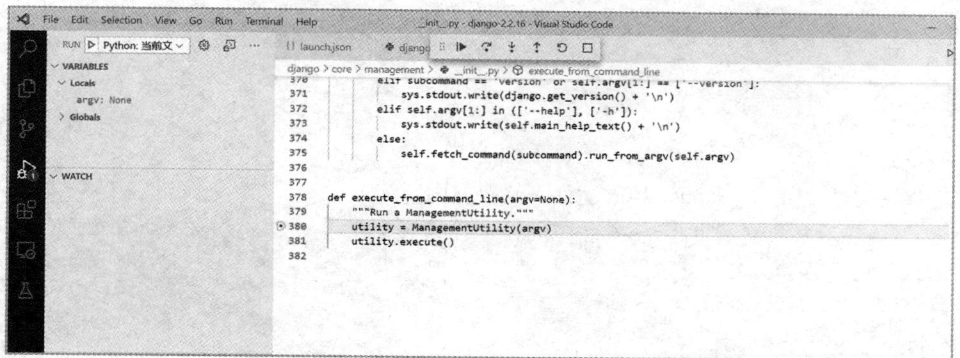

图 1-8

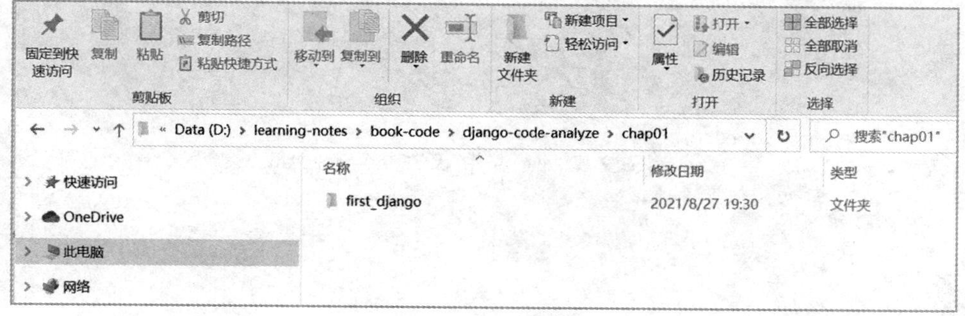

图 1-9

## 1.2.2 调试 manage.py 命令

这里主要调试的是类似 python manage.py command 这样的命令，它依赖于新创建的 Django 项目。下面调试 manage.py 命令，打开 VSCode，导入前文创建的 first_django 项目，导入结果如图 1-10 所示。

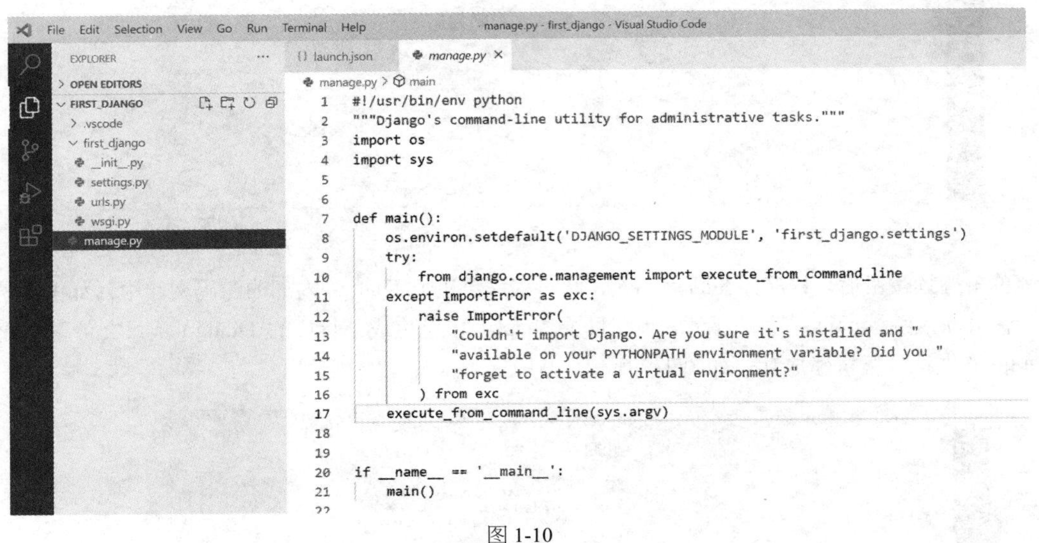

图 1-10

创建调试参数文件——lanuch.json 文件，修改其内容如下：

```
{
   "version": "0.2.0",
   "configurations": [
      {
         "name": "Python: Django",
         "type": "python",
         "request": "launch",
         "program": "${workspaceFolder}\\manage.py",
         "args": [
            "shell"
         ],
         "django": true,
         // 默认为 true，表示不进入模块源码调试，因此一定要设置为 false
         "justMyCode": false
      }
   ]
}
```

**注意**，最后的 justMyCode 参数非常重要，在 VSCode 中其默认为 true，即不进入模块源码调试。为了调试 Python 模块中的源码，这里一定要设置 justMyCode 参数为 false。此外，在 python manage.py shell 命令背后有初始化数据库信息的相关操作，因此必须在 settings.py 文件中设置可用的数据库信息：

```
# first_django/settings.py
# ……

DATABASES = {
    # 删掉默认配置
    'default': {
        'ENGINE': 'django.db.backends.mysql',
        'NAME': 'django_book',
        'HOST': '192.168.88.206',
        'PORT': 3306,
        'USER': 'store',
        'PASSWORD': 'store.1234!',
    }
}

# ……
```

**说明**：笔者在 VMWare 中创建了一个额外的虚拟机，并在其中搭建了 MySQL 服务，主机地址、端口、账号和密码如上述代码所示。搭建 MySQL 服务的方法可以参考网上的教程，此处不再赘述。此外，这里调试的是 Python 内部安装的 Django 模块的源码，和调试 django-admin 命令所使用的 Django 源码不同。首先在 Python 中安装 Django 模块：

```
pip install django==2.2.16
```

此外，必须安装 mysqlclient 模块，因为在 Django 内部是依赖第三方模块去操作 MySQL 数据库的。在 Windows 系统上安装 mysqlclient 模块略微麻烦，通常情况下，直接使用 pip install mysqlclient 命令安装会报错，此时可以从 pip 的第三方模块网站上下载对应 Python 版本及平台架构的 whl 文件直接进行安装。

接下来就可以在代码的任意位置打上断点进行调试了。比如在 manage.py 的源码文件中给 main() 函数中的 execute_from_command_line()方法打上断点，然后用鼠标定位到该方法，单击右键快捷菜单中 Go to Definition 选项，如图 1-11 所示。

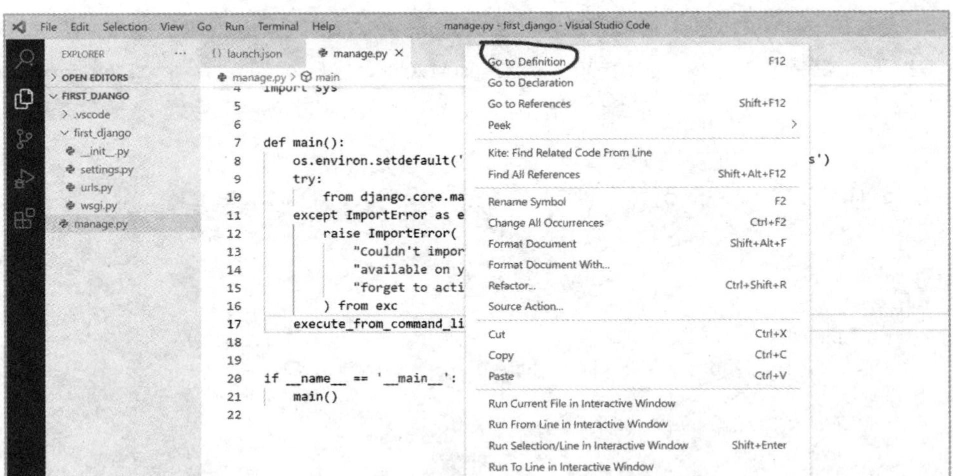

图 1-11

此时即可进入 Python 内部安装的 Django 模块的源码中，在对应调用的语句上打上断点，比如 ManagementUtility 类的实例化语句，如图 1-12 所示。

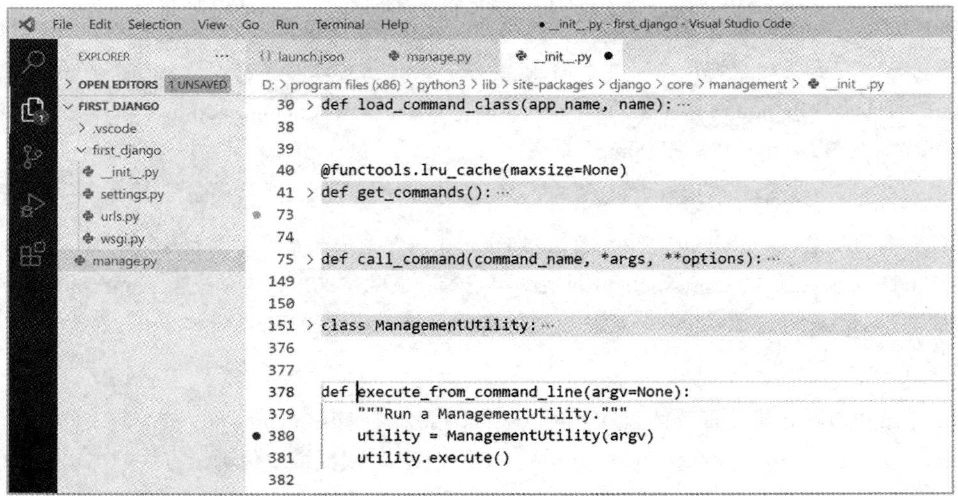

图 1-12

在所有的准备工作都完成后，单击 Run 菜单下的 Start Debugging 命令启动调试。第一次断点停在 execute_from_command_line()方法上，如图 1-13 所示。

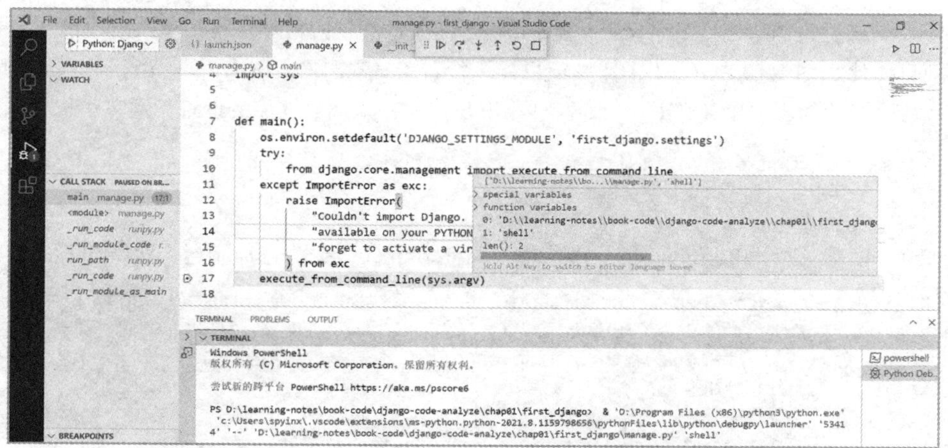


图 1-13

单击运行按钮，第二次断点停在 ManagementUtility 类的实例化语句上，如图 1-14 所示。

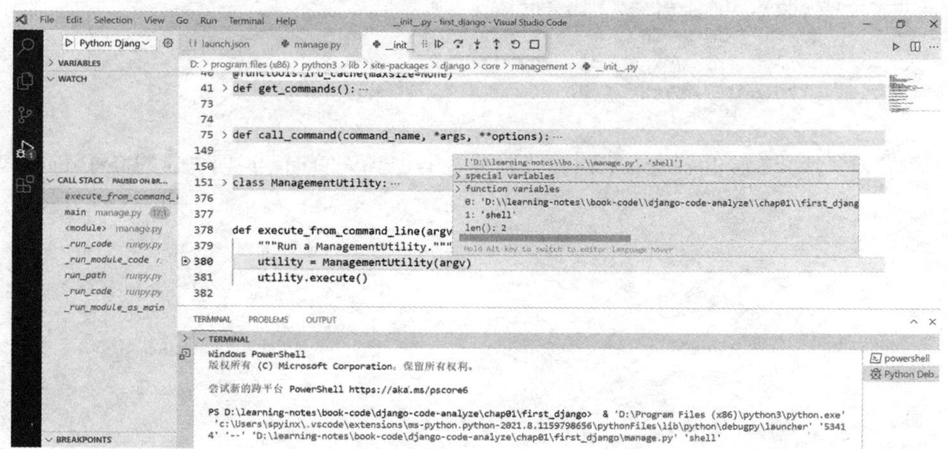


图 1-14

在断点成功进入 Django 模块的源码后，就可以继续搜索相关的语句进行断点调试了。

## 1.2.3　调试框架源码

对于 Django 框架的核心模块的源码调试，比如视图层、模板层等，同样需要基于 Django 框架创建一个简单的 Web 项目。下面通过一个简单的示例，简单介绍如何调试 Django 源码，对于其中使用的一些 Django 基本操作则不再赘述。与 1.2.2 节一样，需要先安装 Django 模块，再安装 mysqlclient 模块。

为了简单起见，继续使用 first_django 项目作为调试对象。使用 VSCode 导入该项目，由 Django 生成的项目文件如图 1-10 所示。为了让该项目能运行起来，需要修改 first_django 目录下的 settings.py 文件，将默认的数据库引擎（SQLite3）调整成 MySQL，并添加数据库的地址、端口、账号和密码。

**注意**，在启动 first_django 项目之前，一定要先创建该项目使用的数据库（在这里配置的是 django_book），否则会报错。

在 VSCode 中，单击 Terminal 菜单栏下的 New Terminal 命令，在控制台上执行如下命令：

```
PS D:\learning-notes\book-code\django-code-analyze\chap01\first_django> python
manage.py runserver 0.0.0.0:8888
Watching for file changes with StatReloader
Performing system checks...

System check identified no issues (0 silenced).

You have 17 unapplied migration(s). Your project may not work properly until you apply
the migrations for app(s): admin, auth, contenttypes, sessions.
Run 'python manage.py migrate' to apply them.
October 23, 2020 - 15:30:18
Django version 2.2.16, using settings 'first_django.settings'
Starting development server at http://0.0.0.0:8888/
Quit the server with CTRL-BREAK.
```

打开浏览器，输入 localhost:8888，可以看到如图 1-15 所示页面，说明创建的 first_django 项目启动成功。

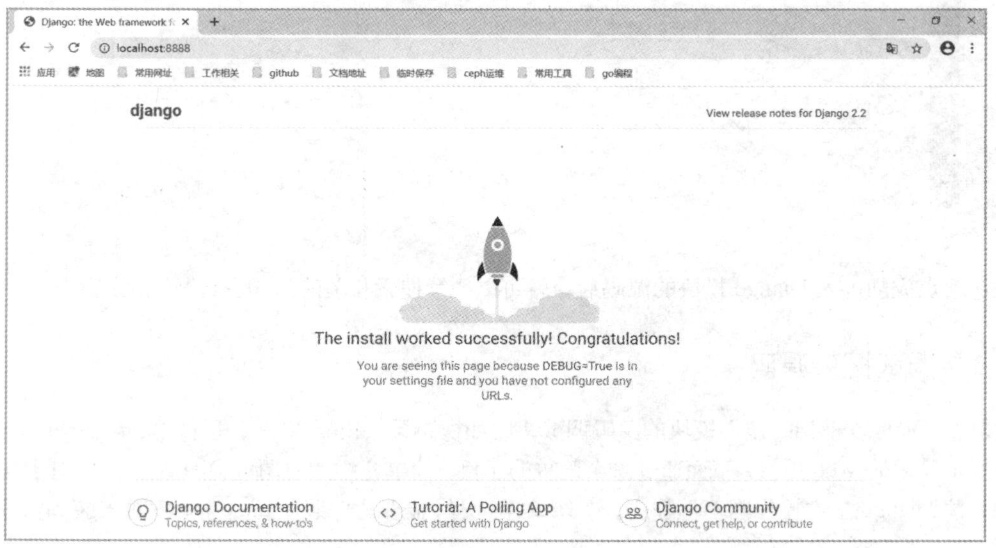


图 1-15

**说明：**在 Django 2.2 的早些版本中，在访问 Django 项目首页时会出现 DisallowedHost 异常页面，此时修改 settings.py 文件中的 ALLOWED_HOSTS 值即可，表示允许列表中的 IP 地址访问该 Web 服务。如果完全对外开放，则设置 ALLOWED_HOSTS = ['*']。

接下来，给该项目添加一个/echo/接口，这个接口可以从请求中获取 value 参数的值，然后回显在页面上。开发/echo/接口的方法非常简单，只需两个步骤：

（1）添加一个视图函数 echo()用来处理/echo/接口。

（2）添加一行 URLConf 配置，将/echo/接口映射到视图函数 echo()上进行处理。

在 first_django 目录下添加一个 views.py 文件，内容如下：

```
def echo(request, *args, **kwargs):
    data = request.GET
    return_value = data.get("value", "hello, world")
    return HttpResponse(return_value, content_type="text/plain")
```

在 first_django 目录下的 urls.py 文件中添加相应的 URLConf 配置：

```
from django.contrib import admin
from django.urls import path
#新加导入视图模块
from first_django import views

urlpatterns = [
    path('admin/', admin.site.urls),
    # 新添加的 URL 映射
    path('echo/', views.echo),
]
```

重启 first_django 项目，打开浏览器，输入 localhost:8888/echo/?value=test_echo，结果如图 1-16 所示。

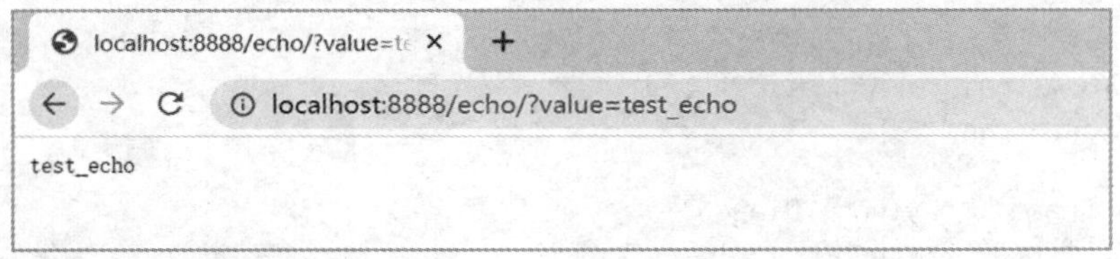


图 1-16

为了能调试该项目中的代码，需要先定位到 VSCode 的侧边栏，单击 Run 图标下的 create a launch.json file 选项，在弹出的窗口中单击 Django launch and debug a Django web application 选项，如图 1-17 所示。如果没有该选项，则需要在 VSCode 中安装 Django 插件。

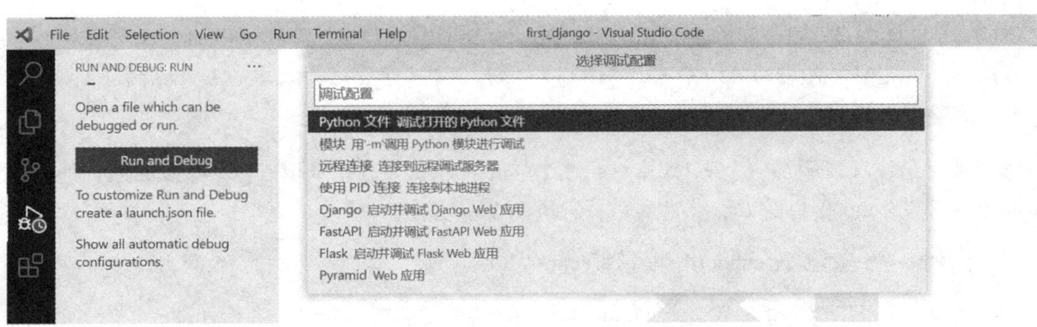


图 1-17

此时 VSCode 会生成 launch.json 文件。通过前文对 launch.json 文件中相关参数的介绍，不难理解下面的配置内容：

```
{
   "version": "0.2.0",
   "configurations": [
      {
         "name": "Python: Django",
         "type": "python",
         "request": "launch",
         "program": "${workspaceFolder}\\manage.py",
         "args": [
            "runserver"
         ],
         "django": true,
         "justMyCode": false
      }
   ]
}
```

通常为了修改启动端口，或者允许任意的 IP 地址访问，可以在 args 字段中加上相关参数：

```
{
   "version": "0.2.0",
   "configurations": [
      {
         "name": "Python: Django",
         "type": "python",
         "request": "launch",
         "program": "${workspaceFolder}\\manage.py",
         "args": [
            "runserver",
            "0.0.0.0:8888"
         ],
         "django": true,
```

```
            "justMyCode": false
        }
    ]
}
```

这样就和前文在控制台中启动服务的设置相同了。在完成这个简单的例子后，思考以下问题：

◎ 请求是如何找到对应的视图函数或者视图类去处理的？

◎ 视图函数中的第 1 个参数 request 是从哪里来的？为什么一定要放在第 1 个参数位置上？

有了问题就可以去源码中打上断点并进行调试了。这一步骤会在后续的 Django 源码分析中经常用到，即从现象入手去追踪源码，并在追踪过程中不断加深对 Django 源码的理解。

### 1.2.4 搭建 Django 的测试环境

本节基于 Python 虚拟环境搭建 Django 的测试环境，本次的实验环境为 CentOS 7 系统，后续许多测试 Django 源码中的类和函数的语句都在该虚拟环境的 Python 交互模式下进行。这里选用 pyenv 和 pyenv-virtualenv 工具构建 Python 的虚拟环境。

（1）安装这两个工具，直接从 GitHub 上下载源码即可：

```
[root@master ~]# yum install git -y
[root@master ~]# git clone https://github.com/pyenv/pyenv.git ~/.pyenv
[root@master ~]# git clone https://github.com/pyenv/pyenv-virtualenv.git
~/.pyenv/plugins/pyenv-virtualenv
```

（2）将相应的命令文件的所在路径加到系统环境变量 PATH 中：

```
[root@master ~]# cat ~/.bashrc
# ……

# 在~/.bashrc 的最后加上如下 3 行内容
export PATH="~/.pyenv/bin:$PATH"
eval "$(pyenv init -)"
eval "$(pyenv virtualenv-init -)"

[root@master ~]# source ~/.bashrc
[root@master ~]# pyenv --version
pyenv 1.2.21
```

在有了 pyenv 和它的虚拟环境管理插件后，就可以安装相应版本的 Python 虚拟环境了。由于 Django 2 只支持 Python 3，所以这里选择 Python 3.8.6 版本。由于直接使用 pyenv install 3.8.6 命令默认是从国外网站下载 Python 压缩包，速度较慢，所以建议从国内的 Python 镜像源上下载 Python-3.8.6.tar.xz 压缩包并放到 pyenv 的缓存目录下，再执行 pyenv install 3.8.6 命令。此时，pyenv 会直接从缓存目录中获取该版本的 Python 压缩包并安装，这样能极大地提高安装速度。此外，在安装 Python 之前需要安装一些依赖包，具体的操作命令如下：

```
[root@master ~]# yum -y install gcc gcc-c++ zlib zlib-devel bzip2-devel openssl-devel
ncurses-devel sqlite-devel readline-devel tk-devel gdbm-devel db4-devel libpcap-devel
libffi-devel xz-devel
[root@master ~]# v=3.8.6;wget http://mirrors.sohu.com/python/$v/Python-$v.tar.xz -P
~/.pyenv/cache/;pyenv install $v
```

在安装 Python 环境之后，即可基于 Python 3.8.6 创建虚拟环境，并命名为 django2-core-test：

```
[root@master ~]# pyenv virtualenv 3.8.6 django2-core-test
Looking in links: /tmp/tmp507vez9f
Requirement already satisfied: setuptools in
/root/.pyenv/versions/3.8.6/envs/django2-core-test/lib/python3.8/site-packages
(49.2.1)
Requirement already satisfied: pip in
/root/.pyenv/versions/3.8.6/envs/django2-core-test/lib/python3.8/site-packages
(20.2.1)
```

（3）使用 pyenv activate django2-core-test 命令激活并进入该虚拟环境中：

```
[root@master ~]# pyenv activate django2-core-test
pyenv-virtualenv: prompt changing will be removed from future release. configure `export
PYENV_VIRTUALENV_DISABLE_PROMPT=1' to simulate the behavior.
(django2-core-test) [root@master ~]# pip list
Package    Version
---------- -------
pip        20.2.1
setuptools 49.2.1
WARNING: You are using pip version 20.2.1; however, version 20.2.4 is available.
You should consider upgrading via the
'/root/.pyenv/versions/3.8.6/envs/django2-core-test/bin/python3.8 -m pip install
--upgrade pip' command.
(django2-core-test) [root@master ~]#
```

（4）在该虚拟环境中安装 Django 2 的最新版本（2.2.16）：

```
(django2-core-test) [root@master ~]# pip install django==2.2.16 -i
https://pypi.tuna.tsinghua.edu.cn/simple
# 忽略下载输出
# ……

(django2-core-test) [root@master ~]# pip list
Package    Version
---------- -------
Django     2.2.16
pip        20.2.1
pytz       2020.1
setuptools 49.2.1
sqlparse   0.4.1
```

为了能操作 MySQL 数据库，这里还需要安装 mysqlclient 模块。注意，在安装 mysqlclient 模块之前，需要安装一些必要的依赖库。在 CentOS 7 系统上，只需执行 yum install mysql-devel -y 命令即可安装 mysqlclient 模块：

```
(django2-core-test) [root@master ~]# yum install mysql-devel -y
(django2-core-test) [root@master ~]# pip install mysqlclient -i
https://pypi.tuna.tsinghua.edu.cn/simple
Looking in indexes: https://pypi.tuna.tsinghua.edu.cn/simple
Collecting mysqlclient
 Using cached
https://pypi.tuna.tsinghua.edu.cn/packages/a5/e1/e5f2b231c05dc51d9d87fa5066f90d14053
45c54b14b0b11a1c859020f21/mysqlclient-2.0.1.tar.gz (87 kB)
Using legacy 'setup.py install' for mysqlclient, since package 'wheel' is not installed.
Installing collected packages: mysqlclient
   Running setup.py install for mysqlclient ... done
Successfully installed mysqlclient-2.0.1
(django2-core-test) [root@master ~]# pip list
Package    Version
----------- -------
Django     2.2.16
mysqlclient 2.0.1
pip       20.2.1
pytz      2020.1
setuptools  49.2.1
sqlparse   0.4.1
```

进入 Python 交互模式，导入 Django 模块，之后即可调用 Django 模块下的各种类和方法进行测试了。比如，通过 django.__version__语句查看 Django 的版本信息：

```
(django2-core-test) [root@master ~]# python
Python 3.8.6 (default, Oct 18 2020, 15:33:08)
[GCC 4.8.5 20150623 (Red Hat 4.8.5-39)] on linux
Type "help", "copyright", "credits" or "license" for more information.
>>> import django
>>> django.__version__
'2.2.16'
```

## 1.3 学习 Django 源码的建议

在学习 Django 源码之前，必须要能够熟练使用 Django 框架进行 Web 项目开发，掌握 Django 的一些基本用法。而这一过程，可以通过阅读官方文档实现。最新的官方 Django 2.2 官方文档接近 2000 页，详细描述了 Django 框架的方方面面，可以说是最权威的学习 Django 框架的材料。

Django 源码并不复杂，不过略显庞大，如果漫无目的地翻看源码，很容易产生畏惧心理，因此需要合理、有序地分析 Django 源码。Django 源码的目录层次分明，每个目录下的源码都代表着其功能模块。比如，在 views 目录下是视图层的代码，这里定义了常用的 View 类、TemplateView 类等。再比如，在 db/backends 目录下定义了 Django 封装的各种数据库（MySQL、Oracle 等）操作的第三方模块（MySQL 对应着 mysqlclient 模块，Oracle 对应着 cx_Oracle 模块等），以兼容上层的调用。

在完整学习本书源码后，你就基本掌握了 Django 框架和源码。当使用 Django 框架出现报错时，你可以根据报错的位置查看对应的代码逻辑，找出报错的原因并解决，这也是提升自身解决问题能力的一个绝佳途径。

## 1.4 小结

本章先主要介绍了 Django 源码的前置知识，包括如何选择 Django 版本、在学习 Django 源码之前需要具备哪些基础，以及如何搭建 Django 的调试环境和测试环境等，然后介绍了学习 Django 源码的建议。

# 第 2 章
# Django命令原理解析

本章详细讲解 Django 2 中部分命令的执行逻辑，通过追踪这些命令的执行过程，读者可以掌握 Django 源码中定义的一些有用的类和方法。本书在 Python 交互模式下直接调用这些类和方法，以便读者理解它们的作用。此外，本章还会给出结合断点的变量结果图片，帮助读者理解部分复杂函数的输出。

## 2.1 基础方法介绍

本节介绍在追踪 Django 命令执行过程中用到的类和函数，当后面遇到这些类和函数时，不再详细分析其实现，而是直接给出其作用。

### 2.1.1 find_commands()函数

find_commands()函数会返回 Django 支持的所有命令列表，代码如下：

```
# 源码位置： django/core/management/__init__.py
# ……

def find_commands(management_dir):
    """
    给定一个管理目录的路径，返回所有可用的命令名称列表
    """
    command_dir = os.path.join(management_dir, 'commands')
    return [name for _, name, is_pkg in pkgutil.iter_modules([command_dir])
            if not is_pkg and not name.startswith('_')]

# ……
```

上面的核心语句其实是调用 pkgutil.iter_modules()方法获取参数目录下的所有模块，同时排除类似于__init__.py 这样的模块，最后返回所有命令模块的名称。这此命令模块就是 Django 支持的全部命令集合，我们可以在 Python 交互模式下手动调用 pkgutil.iter_modules()方法看看实际效果。这里针

对的是 django.core.management 包所在的目录，在该目录下有一个 commands 目录：

```
(django2-core-test) [root@master ~]# python
Python 3.8.6 (default, Oct 18 2020, 15:33:08)
[GCC 4.8.5 20150623 (Red Hat 4.8.5-39)] on linux
Type "help", "copyright", "credits" or "license" for more information.
>>> from django.core import management
>>> management.__path__
['/root/.pyenv/versions/django2-core-test/lib/python3.8/site-packages/django/core/ma
nagement']
>>> management.find_commands(management.__path__[0])
['check', 'compilemessages', 'createcachetable', 'dbshell', 'diffsettings', 'dumpdata',
'flush', 'inspectdb', 'loaddata', 'makemessages', 'makemigrations', 'migrate',
'runserver', 'sendtestemail', 'shell', 'showmigrations', 'sqlflush', 'sqlmigrate',
'sqlsequencereset', 'squashmigrations', 'startapp', 'startproject', 'test',
'testserver']
```

## 2.1.2 load_command_class()函数

在 load_command_class()函数中输入一个包路径和命令名，该函数会返回在具体命令文件中定义的 Command 对象，代码如下：

```
# 源码位置： django/core/management/__init__.py
# ……

def load_command_class(app_name, name):
    """
    给定一个命令名和应用名，返回指定路径下的 Command 类实例。
    允许向上传递导入过程中的所有异常，如 ImportError、AttributeError 等
    """
    module = import_module('%s.management.commands.%s' % (app_name, name))
    return module.Command()
```

假设输入的两个参数分别为 django.core 和 startproject，则 import_module()函数将导入 django.core.management.commands.startproject 模块，而 load_command_class()函数返回的是导入模块中的 Command 对象。下面手动测试一下该方法：

```
>>> from django.core import management
>>> management.load_command_class('django.core', 'startproject')
<django.core.management.commands.startproject.Command object at 0x7f9eec34c970>
```

## 2.1.3 get_commands()函数

这个函数比较简单，与 find_commands()函数相比，仅在结果上做了微小改动：

```
# 源码位置： django/core/management/__init__.py
# ……
```

```
@functools.lru_cache(maxsize=None)
def get_commands():
    commands = {name: 'django.core' for name in find_commands(__path__[0])}

    if not settings.configured:
        return commands

    for app_config in reversed(list(apps.get_app_configs())):
        path = os.path.join(app_config.path, 'management')
        commands.update({name: app_config.name for name in find_commands(path)})

    return commands
```

继续在 Python 交互模式下调用该函数，输出结果如下：

```
>>> from django.conf import settings
>>> settings.configured
False
>>> from django.core import management
>>> management.get_commands()
{'check': 'django.core', 'compilemessages': 'django.core', 'createcachetable':
'django.core', 'dbshell': 'django.core', 'diffsettings': 'django.core', 'dumpdata':
'django.core', 'flush': 'django.core', 'inspectdb': 'django.core', 'loaddata':
'django.core', 'makemessages': 'django.core', 'makemigrations': 'django.core',
'migrate': 'django.core', 'runserver': 'django.core', 'sendtestemail': 'django.core',
'shell': 'django.core', 'showmigrations': 'django.core', 'sqlflush': 'django.core',
'sqlmigrate': 'django.core', 'sqlsequencereset': 'django.core', 'squashmigrations':
'django.core', 'startapp': 'django.core', 'startproject': 'django.core', 'test':
'django.core', 'testserver': 'django.core'}
```

对于 startproject 命令而言，在调用 load_command_class()函数得到 Command 对象时，第 1 个参数（django.core）就是从这里获得的。

### 2.1.4　call_command()函数

从函数名即可看出该函数是用于执行命令的，即 Django 支持的那些命令，如 check、shell、startproject、migrate 等。该函数可接受一个或者多个参数，第 1 个参数为命令名称（command_name）或者一个 Command 对象。如果是后者，则可根据 Command 对象反推得到命令名称。该函数最后调用的是对应命令的 Command 对象中的 execute()方法，这个方法的核心逻辑将在后续分析命令源码时介绍，它是所有 Django 命令行的入口：

```
# 源码位置：django/core/management/__init__.py
# ……

def call_command(command_name, *args, **options):
```

```
    if isinstance(command_name, BaseCommand):
        # Command object passed in.
        command = command_name
        command_name = command.__class__.__module__.split('.')[-1]
    else:
        # Load the command object by name.
        try:
            # 对于大部分 Django 命令而言，会得到"django.core"字符串
            app_name = get_commands()[command_name]
        except KeyError:
            raise CommandError("Unknown command: %r" % command_name)

        if isinstance(app_name, BaseCommand):
            # If the command is already loaded, use it directly.
            command = app_name
        else:
            # 通过传入"django.core"字符串及相应的命令可以得到对应命令文件的 Command 对象
            command = load_command_class(app_name, command_name)

    # 比较复杂，省略
    # ……

    return command.execute(*args, **defaults)
```

在 Django 源码中，Command 类的祖先类为 BaseCommand 类，该类定义在源码的 django/core/management/base.py 文件中。下面是该类的核心实现，这里省略了部分方法和注释说明：

```
# 源码位置：django/core/management/base.py
# ……

class BaseCommand:

    # ……

    def run_from_argv(self, argv):
        self._called_from_command_line = True
        # 定义固定的命令行选项，只添加一些命令信息加以区分
        parser = self.create_parser(argv[0], argv[1])
        # 解析输入的命令行
        options = parser.parse_args(argv[2:])
        cmd_options = vars(options)
        # Move positional args out of options to mimic legacy optparse
        args = cmd_options.pop('args', ())
        # 处理默认选项
        handle_default_options(options)
        try:
            # 调用 self.execute()方法执行命令
            self.execute(*args, **cmd_options)
```

```
        except Exception as e:
            if options.traceback or not isinstance(e, CommandError):
                raise

            if isinstance(e, SystemCheckError):
                self.stderr.write(str(e), lambda x: x)
            else:
                self.stderr.write('%s: %s' % (e.__class__.__name__, e))
            sys.exit(1)
        finally:
            try:
                connections.close_all()
            except ImproperlyConfigured:
                pass

    def execute(self, *args, **options):
        # 处理颜色，以及输入和输出相关的选项
        # ......

        # 是否进行系统检查
        if self.requires_system_checks and not options.get('skip_checks'):
            self.check()
        # 是否进行迁移检查
        if self.requires_migrations_checks:
            self.check_migrations()
        # 调用 self.handler()方法
        output = self.handle(*args, **options)
        if output:
            # 是否输出执行的 SQL 语句
            if self.output_transaction:
                connection=connections[options.get('database', DEFAULT_DB_ALIAS)]
                output = '%s\n%s\n%s' % (
                    self.style.SQL_KEYWORD(connection.ops.start_transaction_sql()),
                    output,
                    self.style.SQL_KEYWORD(connection.ops.end_transaction_sql()),
                )
            self.stdout.write(output)
        return output

    # ......

    def handle(self, *args, **options):
        raise NotImplementedError('subclasses of BaseCommand must provide a handle()
method')
```

在后面分析命令时会看到，所有的命令调用的都是 Command 对象中的 run_from_argv()方法。run_from_argv()方法对命令行中的相关参数进行处理后会直接调用 self.execute()方法执行命令。在

execute()方法中，会根据类属性值及相关选项决定是否进行系统检查、迁移检查，以及是否输出执行的 SQL 语句。在 self.execute()方法中，最核心的为 self.handle()方法。handle()方法并未提供实现代码，而是要求后续继承该类的子类自行实现相应的处理逻辑。

## 2.2 startproject 命令的实现原理

本节追踪 startproject 命令的执行过程，看看第 1 个创建 Django 项目的命令是如何实现的。创建 Django 项目的命令如下：

```
(django2-core-test) [root@master django-core-test]# django-admin startproject
first_django
(django2-core-test) [root@master django-core-test]# ls
first_django
```

django-admin 命令对应的代码如下：

```
# 源码位置： django/bin/django-admin.py

#!/usr/bin/env python
from django.core import management

if __name__ == "__main__":
   # django-admin 命令入口
   management.execute_from_command_line()
```

代码转向 management 目录下__init__.py 文件中的 execute_from_command_line()方法：

```
# 源码位置： django/core/management/__init__.py
# ……

def execute_from_command_line(argv=None):
   """Run a ManagementUtility."""
   utility = ManagementUtility(argv)
   utility.execute()
```

从上面的代码可以看到，execute_from_command_line()方法先实例化 ManagementUtility 类，再调用 ManagementUtility 类中的 execute()方法。继续追踪 ManagementUtility 类及其相关方法：

```
# 源码位置： django/core/management/__init__.py
# ……

class ManagementUtility:

   def __init__(self, argv=None):
      self.argv = argv or sys.argv[:]
      self.prog_name = os.path.basename(self.argv[0])
      if self.prog_name == '__main__.py':
```

```
        self.prog_name = 'python -m django'
    self.settings_exception = None

# ……

def execute(self):
    """
    给定命令行参数，找出正在运行的相关子命令，创建适合该命令的解析器，然后运行它。
    """
    try:
        subcommand = self.argv[1]
    except IndexError:
        subcommand = 'help'  # 如果没有提供参数，转而显示 help 子命令结果

    parser = CommandParser(usage='%(prog)s subcommand [options] [args]',
                           add_help=False, allow_abbrev=False)
    parser.add_argument('--settings')
    parser.add_argument('--pythonpath')
    parser.add_argument('args', nargs='*')  # catch-all
    try:
        options, args = parser.parse_known_args(self.argv[2:])
        handle_default_options(options)
    except CommandError:
        pass  # 此时忽略任何错误选项

    try:
        settings.INSTALLED_APPS
    except ImproperlyConfigured as exc:
        self.settings_exception = exc
    except ImportError as exc:
        self.settings_exception = exc

    # ……

    # 提供自动补全功能，如果没有设置 DJANGO_AUTO_COMPLET 环境变量，该方法内部会直接 return
    self.autocomplete()

    if subcommand == 'help':
        if '--commands' in args:
            sys.stdout.write(self.main_help_text(commands_only=True) + '\n')
        elif not options.args:
            sys.stdout.write(self.main_help_text() + '\n')
        else:
            self.fetch_command(options.args[0]).print_help(self.prog_name,
options.args[0])
    elif subcommand == 'version' or self.argv[1:] == ['--version']:
        sys.stdout.write(django.get_version() + '\n')
    elif self.argv[1:] in (['--help'], ['-h']):
```

```
        sys.stdout.write(self.main_help_text() + '\n')
    else:
        # 其余命令会走到这里
        self.fetch_command(subcommand).run_from_argv(self.argv)
```

首先在初始化时设置 self.argv 值，该值保存了命令行传入的相关参数。由于前面调用 execute_from_command_line()方法时没有传入参数，所以最终是通过 sys.argv[:]得到命令行输入的。当命令行输入为 django-admin startproject first_django 时，sys.argv[:]的值为['django-admin', 'startproject', 'first_django']。在初始化后会调用 ManagementUtility 类的 execute()方法，该方法的执行逻辑如下：

◎ 得到具体执行的 subcommand 命令，在这里是 startproject 命令。

◎ 调用 parser.parse_known_args()方法，解析命令行选项得到 options，同时调用 handle_default_options()函数进一步处理选项值。

◎ 自动补全。由于这里没有设置相应的环境变量，所以 self.autocomplete()没有发挥作用。

◎ 根据 subcommand 命令进行相应处理，主要是单独处理 help（包含--help|-h 选项）和 version（包含--version 选项）。对于其他 subcommand 命令，统一调用如下语句：

```
self.fetch_command(subcommand).run_from_argv(self.argv)
```

追踪目标到 fetch_command()方法中：

```
# 源码位置： django/core/management/__init__.py
# ……

class ManagementUtility:
    # ……

    def fetch_command(self, subcommand):
        # 在 try 块之外获取命令集合，避免异常被捕获后无法向上传递
        commands = get_commands()
        try:
            app_name = commands[subcommand]
        except KeyError:
            # ……
        if isinstance(app_name, BaseCommand):
            # 如果该命令对象已经加载，则直接使用
            klass = app_name
        else:
            klass = load_command_class(app_name, subcommand)
        return klass

    # ……
```

在这段代码中，有前文讲解过的 get_commands()函数和 load_command_class()函数，因此 fetch_command()方法就很容易理解了。下面在 Python 交互模式下演示上述操作，得到针对 startproject

命令的结果，即 klass 值，操作如下：

```
(django2-core-test) [root@master django-core-test]# python
Python 3.8.6 (default, Oct 18 2020, 15:33:08)
[GCC 4.8.5 20150623 (Red Hat 4.8.5-39)] on linux
Type "help", "copyright", "credits" or "license" for more information.
>>> from django.core import management
>>> from django.core.management import load_command_class, get_commands
>>> subcommand = 'startproject'
>>> commands = get_commands()
>>> app_name = commands[subcommand]
>>> app_name
'django.core'
>>> klass = load_command_class(app_name, subcommand)
>>> klass
<django.core.management.commands.startproject.Command object at 0x7f8c9f7d59a0>
>>>
```

看到这里，klass 的值就很清楚了，它对应命令文件中定义的 Command 对象。对于本次要追踪的 startproject 命令来说，就需要到 startproject.py 文件中追踪其 Command 类的实现代码：

```
# 源码位置：django/core/management/commands/startproject.py
# ……

class Command(TemplateCommand):
    help = (
        "Creates a Django project directory structure for the given project "
        "name in the current directory or optionally in the given directory."
    )
    missing_args_message = "You must provide a project name."

    def handle(self, **options):
        project_name = options.pop('name')
        target = options.pop('directory')

        # 创建一个随机的 SECRET_KEY 并放到主配置中
        options['secret_key'] = get_random_secret_key()

        super().handle('project', project_name, target, **options)
```

**注意**，该 Command 类继承自 TemplateCommand 类，且只实现了一个 handle()方法。该 handle()方法调用了父类的 handle()方法，并传递了一些参数。前文的最后一步调用了 klass 的 run_from_argv()方法，但这里并没有，只能继续在父类中继续查找：

```
# 源码位置：django/core/management/templates.py
# ……

class TemplateCommand(BaseCommand):
```

```
    # ……
```

在 TemplateCommand 类中并没有 run_from_argv() 方法，但 TemplateCommand 类继承自 BaseCommand 类，而后者刚好定义了 run_from_argv()方法。在 2.1 节中曾介绍过 BaseCommand 类，在 BaseCommand 类中 run_from_argv()方法的执行逻辑如图 2-1 所示。

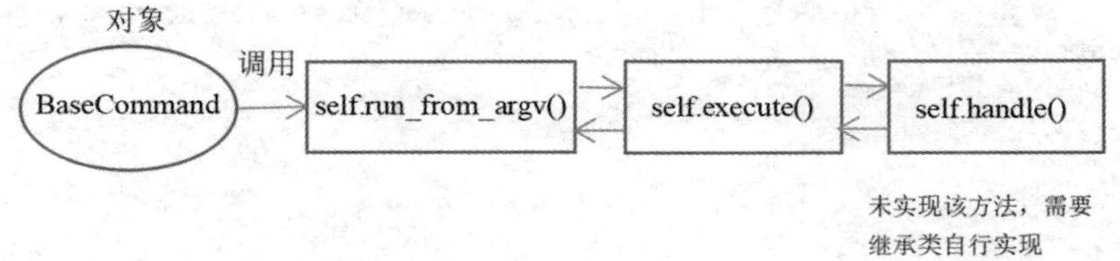


图 2-1

startproject 命令中的 Command 对象调用 run_from_argv()方法的运行流程如图 2-2 所示。

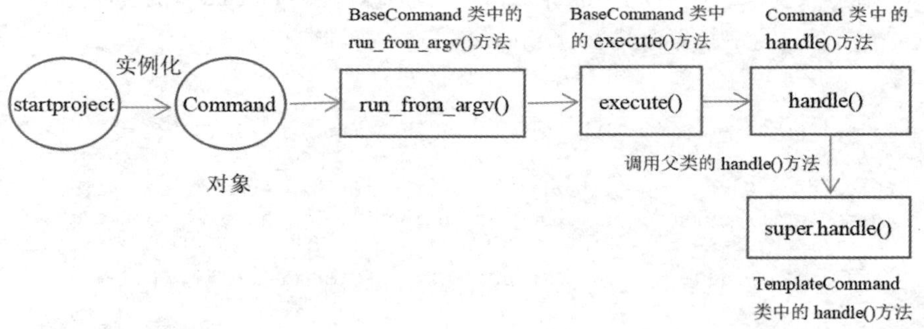


图 2-2

从图 2-2 中可知，startproject 命令的核心处理方法是 TemplateCommand 类中的 handler()方法。下面继续追踪 handler()方法的实现代码：

```
# 源码位置：django/core/management/templates.py
# ……

class TemplateCommand(BaseCommand):
    # ……

    rewrite_template_suffixes = (
        # Allow shipping invalid .py files without byte-compilation.
        ('.py-tpl', '.py'),
    )
```

```
    # ……

    def add_arguments(self, parser):
        # ……

        # 注意这里的默认值
        parser.add_argument(
            '--extension', '-e', dest='extensions',
            action='append', default=['py'],
            help='The file extension(s) to render (default: "py"). '
                 'Separate multiple extensions with commas, or use '
                 '-e multiple times.'
        )

        # ……

    def handle(self, app_or_project, name, target=None, **options):
        # 'app'或者'project'
        self.app_or_project = app_or_project
        self.paths_to_remove = []
        # 控制日志打印
        self.verbosity = options['verbosity']

        # 校验 app 或者 project 名称，核心逻辑是调用 Python 3 中字符串的 isidentifier()方法
        self.validate_name(name, app_or_project)

        # target 参数，用于在指定目录下创建 Django 项目或应用
        if target is None:
            # 如果没有指定目录，就在当前目录下创建，name 为项目名称或应用名称
            top_dir = path.join(os.getcwd(), name)
            try:
                # 创建目录
                os.makedirs(top_dir)
            except FileExistsError:
                raise CommandError("'%s' already exists" % top_dir)
            except OSError as e:
                raise CommandError(e)
        else:
            # 指定创建项目或者应用的目标目录
            top_dir = os.path.abspath(path.expanduser(target))
            # 当目录不存在时抛出异常
            if not os.path.exists(top_dir):
                raise CommandError("Destination directory '%s' does not "
                                   "exist, please create it first." % top_dir)

        # 根据 add_arguments()方法，可知 options['extensions']的默认值为['py']
        extensions = tuple(handle_extensions(options['extensions']))
        extra_files = []
```

```
for file in options['files']:
    extra_files.extend(map(lambda x: x.strip(), file.split(',')))

# 忽略一些打印信息
# ……

base_name = '%s_name' % app_or_project
base_subdir = '%s_template' % app_or_project
base_directory = '%s_directory' % app_or_project
camel_case_name = 'camel_case_%s_name' % app_or_project
camel_case_value = ''.join(x for x in name.title() if x != '_')

# 上下文信息
context = Context({
    **options,
    base_name: name,
    base_directory: top_dir,
    camel_case_name: camel_case_value,
    'docs_version': get_docs_version(),
    'django_version': django.__version__,
}, autoescape=False)

# ……

# 得到生成项目或者应用的模板文件所在目录
template_dir = self.handle_template(options['template'],
                                    base_subdir)
prefix_length = len(template_dir) + 1

for root, dirs, files in os.walk(template_dir):
    # 遍历该目录下的所有模板文件
    path_rest = root[prefix_length:]
    relative_dir = path_rest.replace(base_name, name)
    if relative_dir:
        target_dir = path.join(top_dir, relative_dir)
        if not path.exists(target_dir):
            os.mkdir(target_dir)

    for dirname in dirs[:]:
        if dirname.startswith('.') or dirname == '__pycache__':
            dirs.remove(dirname)

    for filename in files:
        if filename.endswith(('.pyo', '.pyc', '.py.class')):
            continue
        # 模板文件在 Django 中的完整路径
        old_path = path.join(root, filename)
        # 创建 Django 项目下该模板文件对应的完整路径
```

```
            new_path = path.join(top_dir, relative_dir,
                                 filename.replace(base_name, name))

            # old_suffix='.py-tpl', new_suffix='.py'
            for old_suffix, new_suffix in self.rewrite_template_suffixes:
                if new_path.endswith(old_suffix):
                    # 对于文件的新路径，需要去掉原来的模板后缀，换成新的'.py'后缀
                    new_path = new_path[:-len(old_suffix)] + new_suffix
                    break  # Only rewrite once

            if path.exists(new_path):
                # 如果需要写入的文件已存在，直接抛出异常
                # ……

            if new_path.endswith(extensions) or filename in extra_files:
                # 读取模板文件内容
                with open(old_path, 'r', encoding='utf-8') as template_file:
                    content = template_file.read()
                # 渲染模板文件，得到渲染后的内容
                template = Engine().from_string(content)
                content = template.render(context)
                # 将渲染后的内容写到相应的文件中
                with open(new_path, 'w', encoding='utf-8') as new_file:
                    new_file.write(content)
            else:
                # 非模板文件，直接复制文件即可，不需要渲染
                shutil.copyfile(old_path, new_path)

            if self.verbosity >= 2:
                # 如果 self.verbosity>=2，则可以看到下面的打印信息
                self.stdout.write("Creating %s\n" % new_path)
            try:
                # 拷贝文件权限
                shutil.copymode(old_path, new_path)
                # 使目标文件可写
                self.make_writeable(new_path)
            except OSError:
                # 打印异常信息

    # ……
```

上面的代码是 startproject 命令和 startapp 命令的核心，笔者已对大部分语句做了相关注释，以帮助读者理解整个函数的执行逻辑。下面简单说明 django-admin startproject first_django 命令在此处的执行过程：

（1）handle()方法校验项目的名称（first_django）是否合法。它的核心是 isidentifier()方法。

（2）handle()方法根据 target 参数判断是否创建项目。如果 target 参数为 None，则取当前目录加上要创建的项目名称（这里是 first_django）创建项目目录。

（3）extensions 为元组，使用其默认值即可，即('.py', )。

（4）通过 self.handle_template()函数得到生成 Django 项目的模板文件所在目录。该函数的实现源码如下：

```
# 源码位置：django/core/management/templates.py
# ……

class TemplateCommand(BaseCommand):
    # ……

    def handle_template(self, template, subdir):
        if template is None:
            # 当没有指定模板目录时，程序会直接运行到这里，使用 Django 内置的模板目录
            return path.join(django.__path__[0], 'conf', subdir)
        else:
            # ……

        raise CommandError("couldn't handle %s template %s." %
                           (self.app_or_project, template)
```

从上面的代码可以看到，当没有指定 template 时，会返回 path.join(django.__path__[0], 'conf', subdir)这个目录。下面先在虚拟环境中看看这个目录的具体路径值：

```
>>> from os import path
>>> import django
>>> path.join(django.__path__[0], 'conf', 'project_template')
'/root/.pyenv/versions/django2-core-test/lib/python3.8/site-packages/django/conf/pro
ject_template'
```

再次查看该路径下的文件树情况，直接使用 tree 命令即可。

```
(django2-core-test) [root@master ~]# tree
/root/.pyenv/versions/django2-core-test/lib/python3.8/site-packages/django/conf/proj
ect_template
/root/.pyenv/versions/django2-core-test/lib/python3.8/site-packages/django/conf/proj
ect_template
├── manage.py-tpl
└── project_name
    ├── __init__.py-tpl
    ├── settings.py-tpl
    ├── urls.py-tpl
    └── wsgi.py-tpl

1 directory, 5 files
```

接下来查看 django-admin startproject first_django 命令生成的 Django 项目的目录结构：

```
(django2-core-test) [root@master django-core-test]# tree first_django/
first_django/
├── first_django
│   ├── __init__.py
│   ├── settings.py
│   ├── urls.py
│   └── wsgi.py
└── manage.py

1 directory, 5 files
```

至此，可以想象后续的代码逻辑：先将这里的文件全部遍历出来，然后渲染，得到最终的内容，并把最终的内容写入创建 Django 项目的相应文件中。

（5）遍历得到该目录下的所有模板文件，进行渲染并写到最终的位置。渲染的关键代码如下：

```
template = Engine().from_string(content)
content = template.render(context)
```

这两句代码会调用 Django 中的模板层引擎执行渲染动作。为了更好地理解这两句代码，我们进行如下测试：

```
>>> from django.template import Engine
>>> from django.template.context import Context
>>> context = Context({'name': '奇才先生', 'planet_name': '地球'}, autoescape=False)
>>> template = Engine().from_string('hello, {{ name }}, 欢迎来到{{ planet_name }}!')
>>> template.render(context)
'hello, 奇才先生, 欢迎来到地球!'
>>> context = Context({'name': '黑洞阁下', 'planet_name': '火星'}, autoescape=False)
>>> template.render(context)
'hello, 黑洞阁下, 欢迎来到火星!'
```

至此，关于 startproject 命令的执行过程就介绍完毕了。整个过程并不复杂，而且逻辑十分清晰。接下来学习在 Django 中使用 manage.py 执行的命令，比如常用的 python manage.py shell、python manage.py migrate 等命令，了解这些命令的运行原理。

## 2.3　shell 命令的实现原理

在测试模型（Model）中进行增删改查时经常会用到 Django 中的 shell 命令。下面基于前文创建的 first_django 项目给出一个简单的 shell 操作示例，之后根据相应的现象提出问题，并通过源码追踪的方式解答这些问题。在 first_django 项目下运行 python manage.py shell 命令会报错：

```
(django2-core-test) [root@master first_django]# ls
first_django  manage.py
```

```
(django2-core-test) [root@master first_django]# python manage.py shell
Traceback (most recent call last):
# ……
   raise ImproperlyConfigured('SQLite 3.8.3 or later is required (found %s).' %
Database.sqlite_version)
django.core.exceptions.ImproperlyConfigured: SQLite 3.8.3 or later is required (found
3.7.17).
```

错误提示非常清楚，要求安装相应数据库的客户端模块。由于默认使用 SQLite3 作为数据库，而在该 Linux 系统上并没有安装相应的 SQLite3 客户端模块，所以抛出异常。与第 1 章一样，笔者使用内部搭建好的一个 MySQL 数据库，并在本机上安装 mysqlclient 模块。修改 settings.py 文件中数据库相关的配置如下：

```
# 源码位置：first_django/settings.py
# ……

DATABASES = {
    'default': {
        'ENGINE': 'django.db.backends.mysql',
        'NAME': 'django_book',
        'HOST': '192.168.88.206',
        'PORT': 3306,
        'USER': 'store',
        'PASSWORD': 'store.1234!',
    }
}

# ……
```

使用 startapp 命令创建一个 shell_test 应用（测试用）：

```
(django2-core-test) [root@master first_django]# django-admin startapp shell_test
```

在创建 shell_test 应用后，会生成若干文件，如下：

```
(django2-core-test) [root@master first_django]# cd shell_test/
(django2-core-test) [root@master shell_test]# ls
admin.py  apps.py  __init__.py  migrations  models.py  tests.py  views.py
```

其中，在 models.py 文件中保存的是模型类，这里简单创建一个 Django 图书的模型类：

```
# 源码位置： first_django/shell_test/models.py

from django.db import models

class DjangoBooks(models.Model):
    sex_choices = (
        (0, '男'),
        (1, '女'),
```

```
    )
    book_name = models.CharField('图书名', max_length=30)
    author = models.CharField('作者', max_length=30)
    sex = models.SmallIntegerField('性别', choices=sex_choices, default=0)
    price = models.FloatField('图书价格')
    isbn = models.CharField('isbn', max_length=10)
    publish_date = models.DateTimeField('出版时间', auto_now=True)

    def __str__(self):
        return "<%s, %s>" % (self.book_name, self.author)

    class Meta:
        # 通过 db_table 自定义数据表名
        db_table = 'django_books'
```

为了使 Django 能管理 shell_test 应用，需要在 settings.py 文件的 INSTALLED_APPS 列表中添加该应用：

```
# 源码位置： first_django/first_django/settings.py
# ……

INSTALLED_APPS = [
    'django.contrib.admin',
    'django.contrib.auth',
    'django.contrib.contenttypes',
    'django.contrib.sessions',
    'django.contrib.messages',
    'django.contrib.staticfiles',
    # 添加创建的应用
    'shell_test',
]

# ……
```

然后针对 shell_test 应用对数据库进行迁移操作，将 DjangoBooks 类映射到具体的数据库表中：

```
(django2-core-test) [root@master first_django]# python manage.py makemigrations
shell_test
Migrations for 'shell_test':
  shell_test/migrations/0001_initial.py
    - Create model DjangoBooks
(django2-core-test) [root@master first_django]# python manage.py migrate shell_test
System check identified some issues:

# 忽略一些告警信息
# ……
Operations to perform:
  Apply all migrations: shell_test
Running migrations:
```

```
 Applying shell_test.0001_initial... OK
(django2-core-test) [root@master first_django]#
```

这时就可以在数据库中看到和 DjangoBooks 类对应的 django_books 表了。使用 shell 命令进入 Python 交互模式，对这个表进行增删改查，操作如下：

```
(django2-core-test) [root@master first_django]# python manage.py shell
Python 3.8.6 (default, Oct 18 2020, 15:33:08)
[GCC 4.8.5 20150623 (Red Hat 4.8.5-39)] on linux
Type "help", "copyright", "credits" or "license" for more information.
(InteractiveConsole)
>>> from shell_test.models import DjangoBooks
>>> DjangoBooks.objects.all()
<QuerySet []>
>>> d1 = DjangoBooks(book_name='Django2 框架内幕', author='沈奇才', isbn='111111',
price='119.00')
>>> d1.save()
>>> DjangoBooks.objects.all()
<QuerySet [<DjangoBooks: <Django2 框架内幕, 沈奇才>>]>
```

是不是非常简单？接下来读者可以思考以下几个问题，带着这些问题去追踪源码并尝试解答：

◎ 如何通过 Python 代码实现上述交互模式？

◎ 这样的交互模式和普通的 Python 交互模式有何区别，为何前者能实现对模型层的增删改查操作，而后者在交互模式下导入 DjangoBooks 类会报错？报错的原因是什么，应如何解决？以下是直接在 Python 交互模式下导入 Django 模型类，报错如下：

```
(django2-core-test) [root@master first_django]# python
Python 3.8.6 (default, Oct 18 2020, 15:33:08)
[GCC 4.8.5 20150623 (Red Hat 4.8.5-39)] on linux
Type "help", "copyright", "credits" or "license" for more information.
>>> from shell_test.models import DjangoBooks
Traceback (most recent call last):
  File "<stdin>", line 1, in <module>
  File "/root/django-core-test/first_django/shell_test/models.py", line 4, in <module>
    class DjangoBooks(models.Model):
  File
"/root/.pyenv/versions/django2-core-test/lib/python3.8/site-packages/django/db/model
s/base.py", line 103, in __new__
    app_config = apps.get_containing_app_config(module)
  File
"/root/.pyenv/versions/django2-core-test/lib/python3.8/site-packages/django/apps/reg
istry.py", line 252, in get_containing_app_config
    self.check_apps_ready()
  File
"/root/.pyenv/versions/django2-core-test/lib/python3.8/site-packages/django/apps/reg
istry.py", line 134, in check_apps_ready
```

```
    settings.INSTALLED_APPS
  File
"/root/.pyenv/versions/django2-core-test/lib/python3.8/site-packages/django/conf/__i
nit__.py", line 79, in __getattr__
    self._setup(name)
  File
"/root/.pyenv/versions/django2-core-test/lib/python3.8/site-packages/django/conf/__i
nit__.py", line 60, in _setup
    raise ImproperlyConfigured(
django.core.exceptions.ImproperlyConfigured: Requested setting INSTALLED_APPS, but
settings are not configured. You must either define the environment variable
DJANGO_SETTINGS_MODULE or call settings.configure() before accessing settings.
>>>
```

下面带着前文提出的问题追踪 shell 命令的执行过程。我们在前面曾分析过 startproject 命令的执行过程，根据分析经验，首先在命令目录下查找 shell.py 文件：

```
# 源码位置：django/core/management/commands/shell.py
# ……

class Command(BaseCommand):

    requires_system_checks = False
    # 依次查找 Python 交互模式，找到后直接执行即可
    shells = ['ipython', 'bpython', 'python']

    # 支持额外参数的方法
    # ……


    def handle(self, **options):

        if options['command']:
            exec(options['command'])
            return

        # 当在 Linux 系统终端中执行时，sys.stdin.isatty()为 True
        if sys.platform != 'win32' and not sys.stdin.isatty() and
                select.select([sys.stdin], [], [], 0)[0]:
            exec(sys.stdin.read())
            return

        # 可以通过选项指定 shells 列表
        available_shells = [options['interface']] if options['interface'] else self.shells

        for shell in available_shells:
            try:
```

```
            # 调用对应的shell()方法，这里指的是ipython()、bpython()、python()方法中的一种
            return getattr(self, shell)(options)
        except ImportError:
            pass
    # 抛出异常
    raise CommandError("Couldn't import {} interface.".format(shell))
```

仅看这里的代码就能解决前面提出的第 1 个问题了。这里定义的 Command 类只继承了 BaseCommand 类，所以执行过程与 startproject 命令相比会简单一些。其执行流程如图 2-1 所示，只不过最后调用的 handle()方法会变成 Command 类中实现的 handle()方法。handle()方法的执行逻辑非常简单：

（1）如果有通过-c 选项输入的命令，直接执行命令后返回。

（2）对于非 Windows 平台、非终端且有 select 模块的，会通过 sys.stdin.read()读取输入数据，并在执行后返回。

（3）通过内置及外部输入得到可用的 shell()方法，在遍历后直接调用相应的 shell()方法形成交互的样式。

Django 内置了三种 Python 交互模式，分别为 ipython、bpython 和 python。通过代码可以看到，在 handle()方法返回后会依次遍历这三种模式并导入相应的模块。如果导入模块出现异常，则继续下一个模式的操作。通常情况下会使用 python 模式，因此 getattr(self, shell)会得到该 Command 类中的 python()方法。

形成交互模式的代码就在 Command 类的 python()方法中，下面看看 python()方法的具体实现：

```
# 源码位置：django/core/management/commands/shell.py
# ……

class Command(BaseCommand):

    # ……

    def python(self, options):
        import code
        imported_objects = {}
        try:
            # 导入模块，以便后续在交互模式下提供代码补全功能
            import readline
        except ImportError:
            pass
        else:
            # 添加语句的补全功能
            import rlcompleter
            readline.set_completer(rlcompleter.Completer(imported_objects).complete)
```

```
        readline_doc = getattr(readline, '__doc__', '')
        if readline_doc is not None and 'libedit' in readline_doc:
            readline.parse_and_bind("bind ^I rl_complete")
        else:
            readline.parse_and_bind("tab:complete")

    # 处理一些不重要的情况
    # ……

    # 交互模式
    code.interact(local=imported_objects)

# ……
```

python()方法是一个非常通用的方法，它主要用来导入 code 模块，实现 python 交互模式。此外，它会检查系统中是否有 readline 模块。readline 模块用于给交互模式提供代码补全功能。下面使用 python()方法完成一个简单的示例：

```
(django2-core-test) [root@master first_django]# cat test_code.py
import code

def python():
    import code
    imported_objects = {}
    try:
        import readline
    except ImportError:
        pass
    else:
        # 添加语句补全功能
        import rlcompleter
        readline.set_completer(rlcompleter.Completer(imported_objects).complete)
        readline_doc = getattr(readline, '__doc__', '')
        if readline_doc is not None and 'libedit' in readline_doc:
            readline.parse_and_bind("bind ^I rl_complete")
        else:
            readline.parse_and_bind("tab:complete")

    # 交互模式
    code.interact(
        local={'name': '沈奇才', 'age': '30'},
        banner='欢迎来到 Django 源码世界'
    )

if __name__ == '__main__':
    python()
```

在虚拟环境中运行上述 Python 脚本：

```
(django2-core-test) [root@master first_django]# python test_code.py
欢迎来到Django源码世界
>>> name
'沈奇才'
>>> age
'30'
>>> hello
Traceback (most recent call last):
  File "<console>", line 1, in <module>
NameError: name 'hello' is not defined
>>>
```

从代码中可以看出，我们成功得到了类似 Python 命令那样的交互模式。此外，code.interact()方法中的 banner 参数会被打印到交互模式之前，local 参数会作为交互模式下的本地变量被默认导入。因此，前文提出的第 1 个问题就得到了解答：Django 通过 code 模块实现了类似 Python 的交互模式，具体代码见 shell.py 文件中 Command 类的 python()方法。

第 2 个问题也比较容易解决，首先通过错误输出来定位问题，在抛出异常前，最后一行执行代码如下：

```
File
"/root/.pyenv/versions/django2-core-test/lib/python3.8/site-packages/django/conf/__i
nit__.py", line 60, in _setup
```

打开 Django 的源码工程，找到这部分代码：

```
# 源码位置：django/conf/__init__.py
# ……

ENVIRONMENT_VARIABLE = "DJANGO_SETTINGS_MODULE"

# ……

class LazySettings(LazyObject):

    def _setup(self, name=None):
        settings_module = os.environ.get(ENVIRONMENT_VARIABLE)
        if not settings_module:
            desc = ("setting %s" % name) if name else "settings"
            raise ImproperlyConfigured(
                "Requested %s, but settings are not configured. "
                "You must either define the environment variable %s "
                "or call settings.configure() before accessing settings."
                % (desc, ENVIRONMENT_VARIABLE))

        self._wrapped = Settings(settings_module)

    # ……
```

可以看到，最后抛出的异常正是这里的 ImproperlyConfigured 异常。抛出异常的原因是判断条件 not settings_module 为 True，即 settings_module 的值为 False。该 if 语句的上一句 os.environ.get (ENVIRONMENT_VARIABLE)结果为空。不妨在 python manage.py shell 和当前的 shell 交互模式下都执行这个获取环境变量的语句，看看有何不同：

```
(django2-core-test) [root@master first_django]# python
Python 3.8.6 (default, Oct 18 2020, 15:33:08)
[GCC 4.8.5 20150623 (Red Hat 4.8.5-39)] on linux
Type "help", "copyright", "credits" or "license" for more information.
>>> import os
>>> os.environ.get('DJANGO_SETTINGS_MODULE')
>>> exit()
(django2-core-test) [root@master first_django]# python manage.py shell
Python 3.8.6 (default, Oct 18 2020, 15:33:08)
[GCC 4.8.5 20150623 (Red Hat 4.8.5-39)] on linux
Type "help", "copyright", "credits" or "license" for more information.
(InteractiveConsole)
>>> import os
>>> os.environ.get('DJANGO_SETTINGS_MODULE')
'first_django.settings'
>>>
```

很明显，这里需要在环境变量中指定 DJANGO_SETTINGS_MODULE 的值，该值指定了 Django 项目的配置模块路径。这个信息非常重要，因为前面设置的数据库相关信息就保存在该模块中。是不是在这里设置 DJANGO_SETTINGS_MODULE 的值之后，就能实现和 python manage.py shell 一样的效果呢？测试结果如下：

```
(django2-core-test) [root@master first_django]# python
Python 3.8.6 (default, Oct 18 2020, 15:33:08)
[GCC 4.8.5 20150623 (Red Hat 4.8.5-39)] on linux
Type "help", "copyright", "credits" or "license" for more information.
>>> import os
>>> os.environ['DJANGO_SETTINGS_MODULE'] = 'first_django.settings'
>>> from shell_test.models import DjangoBooks
Traceback (most recent call last):
  File "<stdin>", line 1, in <module>
  File "/root/django-core-test/first_django/shell_test/models.py", line 4, in <module>
    class DjangoBooks(models.Model):
  File
"/root/.pyenv/versions/django2-core-test/lib/python3.8/site-packages/django/db/model
s/base.py", line 103, in __new__
    app_config = apps.get_containing_app_config(module)
  File
"/root/.pyenv/versions/django2-core-test/lib/python3.8/site-packages/django/apps/reg
istry.py", line 252, in get_containing_app_config
    self.check_apps_ready()
```

```
 File
"/root/.pyenv/versions/django2-core-test/lib/python3.8/site-packages/django/apps/reg
istry.py", line 135, in check_apps_ready
   raise AppRegistryNotReady("Apps aren't loaded yet.")
django.core.exceptions.AppRegistryNotReady: Apps aren't loaded yet.
>>>
```

此时又出现了一个新的报错，出错的原因是没有加载 Django 项目中的应用信息，导致在调用 Apps 对象的 check_apps_ready()方法时抛出异常：

```
# 源码位置：django/apps/registry.py
# ……

class Apps:
    # ……

    def check_apps_ready(self):
        """如果所有的应用还没有被导入，则抛出一个异常"""
        if not self.apps_ready:
            from django.conf import settings
            # 如果没有加载应用，则在获取配置中的 INSTALLED_APPS 值时，
            # 将抛出 ImproperlyConfigured 异常
            settings.INSTALLED_APPS
            raise AppRegistryNotReady("Apps aren't loaded yet.")

    # ……

apps = Apps(installed_apps=None)
```

从上面的注释可以看到，check_apps_ready()方法主要用来检查所有的应用是否被导入。导入应用这一步已经在 python manage.py shell 中执行过了，所以在其命令中导入 DjangoBooks 类时才不会报错。那么究竟是在哪一步完成的呢？其实只需反复追踪 shell 命令调用的代码，找出可能与应用导入相关的语句即可。

```
# 源码位置： django/core/management/__init__.py
# ……

class ManagementUtility:
    # ……

    def execute(self):
        # ……

        try:
            settings.INSTALLED_APPS
        except ImproperlyConfigured as exc:
            self.settings_exception = exc
```

```
        except ImportError as exc:
            self.settings_exception = exc

        if settings.configured:
            if subcommand == 'runserver' and '--noreload' not in self.argv:
                # ……

            # 在其余情况下，调用 django.setup()语句是后续成功的关键
            else:
                # 非常重要，该语句正是导致前面报错的关键语句
                django.setup()

    # ……
```

首先看 ManagementUtility 类中的 execute()方法，前文在介绍 startproject 命令时忽略了 django.setup()语句，而该语句在这里非常重要。下面是 django.setup()语句内部所做的操作：

```
# 源码位置：django/__init__.py
# ……

def setup(set_prefix=True):
    from django.apps import apps
    from django.conf import settings
    from django.urls import set_script_prefix
    from django.utils.log import configure_logging

    configure_logging(settings.LOGGING_CONFIG, settings.LOGGING)
    if set_prefix:
        set_script_prefix(
            '/' if settings.FORCE_SCRIPT_NAME is None else settings.FORCE_SCRIPT_NAME
        )
    apps.populate(settings.INSTALLED_APPS)
```

上面代码的最后一行是不是刚好和应用有关？继续看 Apps 对象的 populate()方法：

```
# 源码位置：django/apps/registry.py
# ……

class Apps:
    # ……

    def populate(self, installed_apps=None):
        if self.ready:
            return

        with self._lock:
            if self.ready:
                return
```

```
if self.loading:
    raise RuntimeError("populate() isn't reentrant")
self.loading = True

# 阶段 1：初始化应用配置并导入应用模块
for entry in installed_apps:
    if isinstance(entry, AppConfig):
        app_config = entry
    else:
        app_config = AppConfig.create(entry)
    if app_config.label in self.app_configs:
        raise ImproperlyConfigured(
            "Application labels aren't unique, "
            "duplicates: %s" % app_config.label)

    self.app_configs[app_config.label] = app_config
    app_config.apps = self

# 检查应用名重复情况
counts = Counter(
    app_config.name for app_config in self.app_configs.values())
duplicates = [
    name for name, count in counts.most_common() if count > 1]
if duplicates:
    raise ImproperlyConfigured(
        "Application names aren't unique, "
        "duplicates: %s" % ", ".join(duplicates))

self.apps_ready = True

# 阶段 2：导入模型层
for app_config in self.app_configs.values():
    app_config.import_models()

self.clear_cache()

self.models_ready = True

# 阶段 3：调用每个应用配置的 ready()方法
for app_config in self.get_app_configs():
    app_config.ready()

self.ready = True
self.ready_event.set()
```

上面的代码只需大致浏览一遍即可，无须太追究细节。只需看到，当调用 populate()方法处理应用时，每个应用都会得到一个 AppConfig 对象，除设置 self.apps_ready = True 外，还会调用每个 AppConfig 对象的 ready()方法。因此，在调用 django.setup()后，apps 的 apps_ready 属性值为 True，对于 check_apps_ready()方法自然不会再抛出异常。再次尝试在普通的 Python 交互模式下导入 DjangoBooks 类：

```
(django2-core-test) [root@master first_django]# python
Python 3.8.6 (default, Oct 18 2020, 15:33:08)
[GCC 4.8.5 20150623 (Red Hat 4.8.5-39)] on linux
Type "help", "copyright", "credits" or "license" for more information.
>>> import os
>>> os.environ['DJANGO_SETTINGS_MODULE'] = 'first_django.settings'
>>> import django
>>> django.setup()
>>> from shell_test.models import DjangoBooks
>>> DjangoBooks.objects.all()
<QuerySet [<DjangoBooks: <Django2 框架内幕，沈奇才>>]>
>>>
```

从上面的代码可以看到，在普通的 Python 命令行中也能成功导入 Django 中的模型类，只不过需要一些额外的操作，而这些操作在 shell 命令中已经提前做好了，所以可以直接导入并使用相应的模型类。

## 2.4　makemigrations 命令的实现原理

本节探索 makemigrations 命令的实现原理，掌握生成迁移文件的相关代码。这部分内容主要涉及 django/db/migrations 目录下的源码。

### 2.4.1　makemigrations 命令的基本操作示例

前面在测试 shell 命令时曾使用过 makemigrations 命令，为了能看到相应的现象，我们先删除 first_django/shell_test/migrations 目录下的所有文件，同时删除在数据库中生成的相应的表，保持干净的环境，再来操作一次：

```
(django2-core-test) [root@master first_django]# rm -rf shell_test/migrations/*
(django2-core-test) [root@master first_django]# ls shell_test/migrations/
```

重复前面的数据库迁移操作：

```
(django2-core-test) [root@master first_django]# python manage.py makemigrations
shell_test
Migrations for 'shell_test':
  shell_test/migrations/0001_initial.py
```

```
   - Create model DjangoBooks
(django2-core-test) [root@master first_django]# ls shell_test/migrations/
0001_initial.py  __init__.py
(django2-core-test) [root@master first_django]# cat
shell_test/migrations/0001_initial.py
# Generated by Django 2.2.16 on 2020-12-10 00:38

from django.db import migrations, models


class Migration(migrations.Migration):

    initial = True

    dependencies = [
    ]

    operations = [
        migrations.CreateModel(
            name='DjangoBooks',
            fields=[
                ('id', models.AutoField(auto_created=True, primary_key=True,
                                      serialize=False, verbose_name='ID')),
                ('book_name', models.CharField(max_length=30, verbose_name='图书名')),
                ('author', models.CharField(max_length=30, verbose_name='作者')),
                ('sex', models.SmallIntegerField(choices=[(0, '男'), (1, '女')],
                                              default=0, verbose_name='性别')),
                ('price', models.FloatField(verbose_name='图书价格')),
                ('isbn', models.CharField(max_length=10, verbose_name='isbn')),
                ('publish_date', models.DateTimeField(auto_now=True,
                                                    verbose_name='出版时间')),
            ],
            options={
                'db_table': 'django_books',
            },
```

注意，makemigrations 操作会在 migrations 目录下生成两个文件。其中，__init__.py 为空文件，0001_initial.py 文件中的内容如上所示，该部分内容用于后续在数据库中创建 django_books 表。

上面使用 makemigrations 命令完成了一个简单的示例，然而在正式追踪该该命令前，还需要解读 django/db/migrations 目录下的一些核心类与方法，以便快速理解相关的源码。

## 2.4.2 迁移相关的基础类与方法

### MigrationRecorder 类

在该类中定义了迁移表（django_migrations）的模型类及若干操作该表的方法，如检查该表在数据库中是否存在（has_table()）、查询迁移记录（applied_migrations()）等。该类的完整实现如下：

```
# 源码位置：django/db/migrations/recorder.py
# ……

class MigrationRecorder:

    _migration_class = None

    @classproperty
    def Migration(cls):
        """
        如果已安装的应用导入 MigrationRecorder 类，懒加载方式将避免 AppRegistryNotReady 异常
        """
        if cls._migration_class is None:
            class Migration(models.Model):
                # 迁移表字段
                app = models.CharField(max_length=255)
                name = models.CharField(max_length=255)
                applied = models.DateTimeField(default=now)

                class Meta:
                    apps = Apps()
                    app_label = 'migrations'
                    # 指定迁移表名
                    db_table = 'django_migrations'

                def __str__(self):
                    return 'Migration %s for %s' % (self.name, self.app)

            # 指定迁移类
            cls._migration_class = Migration
        return cls._migration_class

    def __init__(self, connection):
        # 数据库连接
        self.connection = connection

    @property
    def migration_qs(self):
        return self.Migration.objects.using(self.connection.alias)
```

```
    def has_table(self):
        # 检查 django_migration 表是否存在，如果存在，返回 True，否则返回 False
        return self.Migration._meta.db_table in
self.connection.introspection.table_names(self.connection.cursor())

    def ensure_schema(self):
        # 如果 django_migration 表不存在，直接返回
        if self.has_table():
            return
        try:
            with self.connection.schema_editor() as editor:
                editor.create_model(self.Migration)
        except DatabaseError as exc:
            raise MigrationSchemaMissing("Unable to create the django_migrations table
(%s)" % exc)

    def applied_migrations(self):
        if self.has_table():
            # 其实就是 SQL 语句：select app, name from django_migrations
            return {tuple(x) for x in self.migration_qs.values_list('app', 'name')}
        else:
            return set()

    def record_applied(self, app, name):
        self.ensure_schema()
        # 创建记录
        self.migration_qs.create(app=app, name=name)

    def record_unapplied(self, app, name):
        # 删除记录
        self.ensure_schema()
        self.migration_qs.filter(app=app, name=name).delete()

    def flush(self):
        # 清空 django_migrations 表
        self.migration_qs.all().delete()
```

上面的代码比较简单，主要涉及 Django 中模型类的简单操作，在第 3 章中我们将完整解读 Django 内置的 ORM 框架代码，厘清这些模型类操作语句背后的逻辑。这里先记住对模型的增删改查操作即可。

在上述源码中，在 MigrationRecorder 类的内部定义了一个模型类，映射的表名为 django_migrations。在初始化方法中必须传入对应数据库的连接信息（connection），才能知道操作的迁移表位于哪个数据库中。接下来定义对该表进行增删改查操作的方法，具体如下：

```
(django2-core-test) [root@master first_django]# python manage.py shell
Python 3.8.6 (default, Oct 18 2020, 15:33:08)
```

```
[GCC 4.8.5 20150623 (Red Hat 4.8.5-39)] on linux
Type "help", "copyright", "credits" or "license" for more information.
(InteractiveConsole)
>>> from django.db import connections
>>> connections['default']
<django.db.backends.mysql.base.DatabaseWrapper object at 0x7f115a37a880>
# 这里会获取在 settings.py 文件的数据库配置信息中根据 default 字段得到的数据库连接信息
>>> connection = connections['default']
>>> from django.db.migrations.recorder import MigrationRecorder
>>> recorder = MigrationRecorder(connection)
>>> recorder.has_table()
True
# 查询 django_migrations 表中的所有记录
>>> recorder.applied_migrations()
{('shell_test', '0001_initial')}
# 新增一条记录
>>> recorder.record_applied('test_app', '0002_xxxxx')
# 再次查询所有记录
>>> recorder.applied_migrations()
{('test_app', '0002_xxxxx'), ('shell_test', '0001_initial')}
# 删除刚刚创建的 0002_xxxxx 记录
>>> recorder.record_unapplied('test_app', '0002_xxxxx')
>>> recorder.applied_migrations()
{('shell_test', '0001_initial')}
# 清空迁移表中的所有记录
>>> recorder.flush()
>>> recorder.applied_migrations()
set()
```

### Migration 类

该类代表着一次迁移，即对应着上面迁移表中的一个记录。它有一个非常重要的属性：operations。它是一个元素为操作实例的列表，这些操作实例在 django/db/migrations/operations 目录下的代码文件中可以找到。对表字段的操作有加字段操作（AddField）、移除字段操作（RemoveField）、修改字段操作（AlterField）和重命名字段操作（RenameField）。而对表的操作有模型操作（CreateModel、DeleteModel、RenameModel、AlterModelTable）、模型选项操作和索引操作（AddIndex、RemoveIndex、AddConstraint、RemoveConstraint）。

该类的其他重要属性包括：

- ◎ dependencies：元素为（app_path, migration_name）的列表，表示该迁移类的依赖项。
- ◎ run_before：元素为（app_path, migration_name）的列表。
- ◎ replaces：包含迁移名的列表。

这些属性及其使用将在后续的代码解读中进行说明，此处不再赘述。

### MigrationGraph 类

该类用于表示迁移记录之间的相互依赖关系。在 group.py 文件中关于节点（Node）的定义如下：

```
# 源码位置：django/db/migrations/graph.py
# ……

@total_ordering
class Node:

    def __init__(self, key):
        self.key = key
        self.children = set()
        self.parents = set()

    def __eq__(self, other):
        return self.key == other

    def __lt__(self, other):
        return self.key < other

    def __hash__(self):
        return hash(self.key)

    def __getitem__(self, item):
        # n = Node(key) -> n[0]其实就是 n.key[0]
        return self.key[item]

    def __str__(self):
        return str(self.key)

    def __repr__(self):
        return '<%s: (%r, %r)>' % (self.__class__.__name__, self.key[0], self.key[1])

    def add_child(self, child):
        self.children.add(child)

    def add_parent(self, parent):
        self.parents.add(parent)


class DummyNode(Node):

    def __init__(self, key, origin, error_message):
        super().__init__(key)
        self.origin = origin
        self.error_message = error_message
```

```
    def raise_error(self):
        raise NodeNotFoundError(self.error_message, self.key, origin=self.origin)
```

上面关于节点的定义非常简单，值（key）、父辈（parents）和子孙（children）三个参数就代表了一个节点。MigrationGraph 类的实现如下：

```
# 源码位置：django/db/migrations/graph.py
# ……

class MigrationGraph:

    def __init__(self):
        # 节点图
        self.node_map = {}
        self.nodes = {}

    def add_node(self, key, migration):
        # 先确保 key 不在 node_map 中
        assert key not in self.node_map
        # 新建一个节点
        node = Node(key)
        # 添加节点到 node_map 中
        self.node_map[key] = node
        # 添加迁移对象到 nodes 中
        self.nodes[key] = migration

    def add_dummy_node(self, key, origin, error_message):
        node = DummyNode(key, origin, error_message)
        # 同样会添加到 node_map 和 nodes 属性值中
        self.node_map[key] = node
        self.nodes[key] = None

    def add_dependency(self, migration, child, parent, skip_validation=False):

        if child not in self.nodes:
            # 当 child 节点不在 self.nodes 中时，创建 dummy_node
            error_message = (
                "Migration %s dependencies reference nonexistent"
                " child node %r" % (migration, child)
            )
            self.add_dummy_node(child, migration, error_message)
        if parent not in self.nodes:
            # 当 parent 节点不在 self.nodes 中时，创建 dummy_node
            error_message = (
                "Migration %s dependencies reference nonexistent"
                " parent node %r" % (migration, parent)
            )
            self.add_dummy_node(parent, migration, error_message)
```

```
        # 添加父子关系，子节点调用 add_parent()方法添加父节点
        # 父节点调用 add_child()方法添加子节点
        self.node_map[child].add_parent(self.node_map[parent])
        self.node_map[parent].add_child(self.node_map[child])
        if not skip_validation:
            # 由于未设置跳过校验，因此需要进行校验；如果有 dummy 节点，就抛出异常
            self.validate_consistency()

    def remove_replaced_nodes(self, replacement, replaced):

        # 将要替换的 keys 转成集合以加快后续查找
        replaced = set(replaced)
        try:
            # 找到替换的节点
            replacement_node = self.node_map[replacement]
        except KeyError as err:
            # 抛出异常
            # ……

        for replaced_key in replaced:
            # 弹出对应 replaced_key 的节点
            self.nodes.pop(replaced_key, None)
            # 对应已替换节点
            replaced_node = self.node_map.pop(replaced_key, None)
            if replaced_node:
                for child in replaced_node.children:
                    # replaced_node 有子节点，需要在子节点中去掉其父节点信息
                    child.parents.remove(replaced_node)
                    if child.key not in replaced:
                        # 将 children 信息添加到 replacement_node 节点中
                        replacement_node.add_child(child)
                        # 在子节点中重新添加父节点信息
                        child.add_parent(replacement_node)

                for parent in replaced_node.parents:
                    # 处理父节点信息，移除 replaced_node 的父节点中的子节点信息
                    parent.children.remove(replaced_node)
                    if parent.key not in replaced:
                        # 如果父节点不在 replaced 列表中，则给 replacement_node 节点添加信息
                        replacement_node.add_parent(parent)
                        parent.add_child(replacement_node)

    def remove_replacement_node(self, replacement, replaced):

        # 从 nodes 属性中弹出 replacement 节点
        self.nodes.pop(replacement, None)
        try:
```

```
            # 从 node_map 中弹出 replacement 节点，相当于移除了 replacement 节点
            replacement_node = self.node_map.pop(replacement)
        except KeyError as err:
            # 抛出异常
            # ……

        #获取替换节点的集合及所有替换节点的父节点的集合
        replaced_nodes = set()
        replaced_nodes_parents = set()
        for key in replaced:
            replaced_node = self.node_map.get(key)
            if replaced_node:
                replaced_nodes.add(replaced_node)
                replaced_nodes_parents |= replaced_node.parents

        #在所有节点中将移除节点的信息全都换成替换节点的信息
        replaced_nodes -= replaced_nodes_parents
        for child in replacement_node.children:
            # 移除父节点为 replacement_node 节点的信息
            child.parents.remove(replacement_node)
            for replaced_node in replaced_nodes:
                # 给 replaced_node 节点添加子节点
                replaced_node.add_child(child)
                child.add_parent(replaced_node)

        for parent in replacement_node.parents:
            # 移除原节点中所有子节点为 replacement_node 节点的信息
            parent.children.remove(replacement_node)

    def validate_consistency(self):
        # 确保在该迁移图中没有 dummy 节点，如果有，则调用该节点的 raise_error()方法抛出异常
        [n.raise_error() for n in self.node_map.values() if isinstance(n, DummyNode)]

    # ……

    def root_nodes(self, app=None):

        # 返回所有的根节点，即在该 app 中，既无子孙节点，也无父节点
        roots = set()
        for node in self.nodes:
            # 根节点的判断条件
            if all(key[0] != node[0] for key in self.node_map[node].parents) and
(not app or app == node[0]):
                roots.add(node)
        return sorted(roots)

    def leaf_nodes(self, app=None):
```

```
        # 返回所有的叶子节点
        leaves = set()
        for node in self.nodes:
            # 叶子节点的判断条件
            if all(key[0] != node[0] for key in self.node_map[node].children) and
(not app or app == node[0]):
                leaves.add(node)
        return sorted(leaves)

    # ……
```

**注意**，在 MigrationGraph 类中，大部分方法均是操作 node_map 和 nodes 这两个属性值。接下来我们通过手工构建数据来操作该类。

（1）创建 4 个 Migration 对象，它们同属于 shell_test 应用：

```
(django2-core-test) [root@master first_django]# python manage.py shell
Python 3.8.6 (default, Oct 18 2020, 15:33:08)
[GCC 4.8.5 20150623 (Red Hat 4.8.5-39)] on linux
Type "help", "copyright", "credits" or "license" for more information.
(InteractiveConsole)
>>> from django.db.migrations.migration import Migration
>>> m1 = Migration('00001_xxxxx', 'shell_test')
>>> m2 = Migration('00002_xxxxx', 'shell_test')
>>> m3 = Migration('00003_xxxxx', 'shell_test')
>>> m4 = Migration('00004_xxxxx', 'shell_test')
```

（2）创建 MigrationGraph 对象，并将上面创建的 Migration 对象添加到 MigrationGraph 对象中：

```
>>> from django.db.migrations.graph import MigrationGraph
>>> graph = MigrationGraph()
>>> graph.add_node(('k1', 'v1'), m1)
>>> graph.add_node(('k2', 'v2'), m2)
>>> graph.add_node(('k3', 'v3'), m3)
>>> graph.add_node(('k4', 'v4'), m4)
>>> graph.nodes
{('k1', 'v1'): <Migration shell_test.00001_xxxxx>, ('k2', 'v2'): <Migration
shell_test.00002_xxxxx>, ('k3', 'v3'): <Migration shell_test.00003_xxxxx>, ('k4', 'v4'):
<Migration shell_test.00004_xxxxx>}
>>> graph.node_map
{('k1', 'v1'): <Node: ('k1', 'v1')>, ('k2', 'v2'): <Node: ('k2', 'v2')>, ('k3', 'v3'):
<Node: ('k3', 'v3')>, ('k4', 'v4'): <Node: ('k4', 'v4')>}
>>>
```

**注意**，添加节点方法（add_node()）的第 1 个参数使用了一个二元组，这其实是由在 Node 类中定义的魔法函数__repr__()决定的。add_node()方法会将第一个参数实例化 Node 类，而该参数值会赋给实例化后的 Node 对象的 key 属性。从 Node 类的__repr__()方法中可以看到，在输出 Node 对象时，会用到 key 属性值的第一个和第二个元素，因此 key 属性值必须是包含两个元素以上的数组或者元组。Node 类中__repr__()方法的源码如下：

```
# 源码位置：django/db/migrations/graph.py
# ……

@total_ordering
class Node:

    # ……

    def __repr__(self):
        # 使用 key 属性的第 0 和 1 个元素，如果 key 为二元组，显示比较方便
        return '<%s: (%r, %r)>' % (self.__class__.__name__, self.key[0], self.key[1])

    # ……
```

（3）使用 add_dependency()方法构建依赖关系：

```
>>> graph.add_dependency(None, ('k1', 'v1'), ('k4', 'v4'))
>>> graph.add_dependency(None, ('k2', 'v2'), ('k4', 'v4'))
>>> graph.add_dependency(None, ('k2', 'v2'), ('k3', 'v3'))
>>> graph.node_map[('k4', 'v4')].children
{<Node: ('k2', 'v2')>, <Node: ('k1', 'v1')>}
>>> graph.node_map[('k2', 'v2')].parents
{<Node: ('k3', 'v3')>, <Node: ('k4', 'v4')>}
```

（4）调用 MigrationGraph 对象的 root_nodes()方法和 leaf_nodes()方法：

```
>>> graph.root_nodes()
[('k1', 'v1'), ('k2', 'v2'), ('k3', 'v3'), ('k4', 'v4')]
>>> graph.leaf_nodes()
[('k1', 'v1'), ('k2', 'v2'), ('k3', 'v3'), ('k4', 'v4')]
>>>
```

从结果来看，好像没有删掉任何节点。下面根据其源码来解释，以 root_nodes()为例：

```
# 源码位置：django/db/migrations/graph.py
# ……

class MigrationGraph:

    # ……

    def root_nodes(self, app=None):

        roots = set()
        for node in self.nodes:
            if all(key[0] != node[0] for key in self.node_map[node].parents) and
(not app or app == node[0]):
```

```
            roots.add(node)
        return sorted(roots)

    # ……
```

可以看到，root_nodes()方法是遍历所有的 node 并对其进行判断，把符合根节点条件的加入 roots 列表中，最后返回排序后的 roots 列表。因此，判断是否为 root 节点的核心就在上面代码的 if 判断中：

```
条件 1：all(key[0] != node[0] for key in self.node_map[node].parents)
条件 2：not app or app == node[0]
```

if 判断可以拆成 2 个条件组合。条件 2 需要输入 app 参数，确保查找的 root 节点是本应用内的节点。由于这里没有传入 app 参数，所以条件 2 直接为 True。对于条件 1，node 表示当前搜索节点，而 key 表示该节点中的一个父节点，都是 Node 对象。而 node[0]和 key[0]的含义由 Node 类中的魔法函数__getitem__()决定：

```
# 源码位置：django/db/migrations/graph.py
# ……

@total_ordering
class Node:

    # ……

    def __getitem__(self, item):
        # 其实 n = Node(key) -> n[0]就是 n.key[0]
        return self.key[item]

    # ……
```

由代码可知，node[0]的值最终为 Node 对象中 key 属性的第 1 个元素，请看下面的操作示例：

```
>>> graph.node_map[('k1', 'v1')]
<Node: ('k1', 'v1')>
>>> graph.node_map[('k1', 'v1')].key
('k1', 'v1')
>>> graph.node_map[('k1', 'v1')][0]
'k1'
>>> graph.node_map[('k1', 'v1')][1]
'v1'
```

接着分析条件 1，对于('k1', 'v1')节点，它的父节点为[('k4', 'v4')]，node[0]='k1'。而遍历父节点后得到的 key[0]依次为'k4'，满足 all(key[0] != node[0]for key in parents)，所以判断('k1', 'v1')为一个 root 节点。再来看('k2', 'v2')节点，它的父节点为[('k3', 'v3'), ('k4', 'v4')]，因此 node[0]='k2'。而遍历父节点得到的 key[0]依次为'k3'、'k4'，同样满足条件 1，因而也被认为是 root 节点。后面的两个节点没有父节点，也满足条件 1，所以最终所有的节点都被认为是 root 节点。这并不是 Django 源码本身的问题，而是笔者在测试中随机选择的 key 参数的问题。在 Django 源码中调用 MigrationGraph 对象的 add_node()方法时传入的 key 参数如下：

```
# 源码位置：django/db/migrations/loader.py
# ……

class MigrationLoader:

    # ……

    def load_disk(self):

        self.disk_migrations = {}
        # ……

        for app_config in apps.get_app_configs():

            # ……

            for migration_name in migration_names:

                # ……

                # 更新 self.disk_migrations 值，其 key 为一个二元组，第 1 个元素表示应用名
                self.disk_migrations[app_config.label, migration_name] =
migration_module.Migration(
                    migration_name,
                    app_config.label,
                )
    # ……

    def build_graph(self):

        # ……

        for key, migration in self.disk_migrations.items():
            #从上面的 load_disk()方法中可以看出，这里的 key 是应用的 label，即 app_config.label
            self.graph.add_node(key, migration)
            # ……

        # ……
```

从上面的代码可以看到，给 MigrationGraph 对象添加节点的 key 其实是应用名。因此，上面的 root_nodes()方法和 leaf_nodes()方法获取的是同一个应用中没有依赖的节点。在清楚了上面现象的起因后，再换另一个 MigrationGraph 对象进行测试：

```
# 前面创建节点及导入相应类的步骤省略，不重复写入语句
>>> graph = MigrationGraph()
>>> graph.add_node(('k1', 'v1'), m1)
>>> graph.add_node(('k1', 'v2'), m2)
>>> graph.add_node(('k1', 'v3'), m3)
>>> graph.add_node(('k1', 'v4'), m4)
>>> graph.add_dependency(None, ('k1', 'v1'), ('k1', 'v4'))
>>> graph.add_dependency(None, ('k1', 'v2'), ('k1', 'v4'))
>>> graph.add_dependency(None, ('k1', 'v2'), ('k1', 'v3'))
>>> graph.root_nodes()
[('k1', 'v3'), ('k1', 'v4')]
>>> graph.leaf_nodes()
[('k1', 'v1'), ('k1', 'v2')]
>>>
```

这时再调用 root_nodes()方法和 leaf_nodes()方法，能否得到想要的结果？接下来介绍两个稍微复杂的方法，即 remove_replaced_nodes()方法和 remove_replacement_node()方法：

```
>>> graph.remove_replaced_nodes(('k1', 'v1'), [('k1', 'v4')])
# 原先 v1 的父节点为 v4，在调用上述方法后，将从 v1 的父节点列表中移除 v4
# 由于 v4 的子节点有 v1 和 v2，因此会将 v1 和 v2 作为 v1 的子节点
>>> graph.node_map[('k1', 'v1')].parents
{<Node: ('k1', 'v1')>}
>>> graph.node_map[('k1', 'v2')].parents
{<Node: ('k1', 'v3')>, <Node: ('k1', 'v1')>}
>>> graph.node_map[('k1', 'v1')].children
{<Node: ('k1', 'v1')>, <Node: ('k1', 'v2')>}
# 在调用 remove_replaced_nodes()方法后，replaced 中的节点全部被移除
# 在调用 remove_replaced_nodes()方法移除('k1', 'v4')节点后，
# 在迁移图的 node_map 属性中就不会有该节点的信息了
>>> graph.node_map[('k1', 'v4')]
Traceback (most recent call last):
  File "<stdin>", line 1, in <module>
KeyError: ('k1', 'v4')
```

可以看到，在调用 MigrationGraph 对象的 remove_replaced_nodes(self, replacement, replaced)方法后，replaced 中的节点将全部被移除，而其包含的父节点及子孙节点都将被转移到 replacement 节点上，该逻辑可以直接从源码中分析得到。而 remove_replacement_node(self, replacement, replaced)方法则是上一个方法的反过程，它会移除所有节点中与 replacement 节点有关的信息，然后将其子节点（注意，看源码没有处理 replacement 节点的父节点信息）重新添加到 replaced 节点集合中。为了更好地演示这个方法，下面新建一个 MigrationGraph 对象并添加节点及其依赖：

```
>>> graph = MigrationGraph()
>>> graph.add_node(('k1', 'v1'), m1)
>>> graph.add_node(('k1', 'v2'), m2)
>>> graph.add_node(('k1', 'v3'), m3)
>>> graph.add_node(('k1', 'v4'), m4)
```

```
>>> graph.add_dependency(None, ('k1', 'v1'), ('k1', 'v4'))
>>> graph.add_dependency(None, ('k1', 'v2'), ('k1', 'v1'))
>>> graph.remove_replacement_node(('k1', 'v1'), [('k1', 'v2'), ('k1', 'v3')])
# v1 节点被移除
>>> graph.node_map[(('k1', 'v1'))]
Traceback (most recent call last):
 File "<stdin>", line 1, in <module>
KeyError: ('k1', 'v1')
# v2 节点多了一个子节点信息，该信息来自原来的 v1 节点
>>> graph.node_map[(('k1', 'v2'))].children
{<Node: ('k1', 'v2')>}
# v3 节点多了一个子节点信息，该信息同样来自原来的 v1 节点
>>> graph.node_map[(('k1', 'v3'))].children
{<Node: ('k1', 'v2')>}
```

结合示例及源码分析可知，remove_replacement_node(self, replacement, replaced) 方法的执行逻辑是：移除 MigrationGraph 类中所有与 replacement 节点有关的信息，同时将所有涉及 replacement 节点的地方全部重新设置为 replaced 节点。

### MigrationLoader 类

MigrationLoader 类的源码实现如下：

```
# 源码位置：django/db/migrations/loader.py
# ……

class MigrationLoader:

    def __init__(self, connection, load=True, ignore_no_migrations=False):
        self.connection = connection
        self.disk_migrations = None
        self.applied_migrations = None
        self.ignore_no_migrations = ignore_no_migrations
        if load:
            # 该方法非常重要，用于构建迁移的依赖关系图
            self.build_graph()

    @classmethod
    def migrations_module(cls, app_label):

        if app_label in settings.MIGRATION_MODULES:
            return settings.MIGRATION_MODULES[app_label], True
        else:
            app_package_name = apps.get_app_config(app_label).name
            return '%s.%s' % (app_package_name, MIGRATIONS_MODULE_NAME), False

    def load_disk(self):
```

```
        """从磁盘上的所有已安装的应用中载入迁移数据"""

        # ……

    def get_migration(self, app_label, name_prefix):
        """返回指定的迁移对象或者抛出 NodeNotFoundError 异常"""
        return self.graph.nodes[app_label, name_prefix]

    def get_migration_by_prefix(self, app_label, name_prefix):

        # 根据应用和名称前缀搜索迁移对象
        results = []
        for migration_app_label, migration_name in self.disk_migrations:
            # 根据迁移文件的前缀判断，简单使用了 Python 字符串中的 startswith()方法
            if migration_app_label == app_label and
migration_name.startswith(name_prefix):
                results.append((migration_app_label, migration_name))
        if len(results) > 1:
            # 抛出异常
            # ……
        elif not results:
            # 抛出异常
            # ……
        else:
            #唯一的结果
            return self.disk_migrations[results[0]]

    def build_graph(self):
        # 加载磁盘上保存的数据
        self.load_disk()

        # ……

    def check_consistent_history(self, connection):

        # 得到 MigrationRecorder 对象
        recorder = MigrationRecorder(connection)
        # 获取 django_migrations 表中的迁移记录
        applied = recorder.applied_migrations()
        for migration in applied:
            # 不在本地的记录文件中，直接跳过
            if migration not in self.graph.nodes:
                continue
            for parent in self.graph.node_map[migration].parents:
                if parent not in applied:
                    if parent in self.replacements:
                        if all(m in applied for m in self.replacements[parent].replaces):
                            continue
```

```
            # 抛出异常
            raise InconsistentMigrationHistory(
                "Migration {}.{} is applied before its dependency "
                "{}.{} on database '{}'.".format(
                    migration[0], migration[1], parent[0], parent[1],
                    connection.alias,
                )
            )

    def detect_conflicts(self):

        # 检测冲突
        seen_apps = {}
        conflicting_apps = set()
        for app_label, migration_name in self.graph.leaf_nodes():
            if app_label in seen_apps:
                conflicting_apps.add(app_label)
            seen_apps.setdefault(app_label, set()).add(migration_name)
        return {app_label: seen_apps[app_label] for app_label in conflicting_apps}

    # ……
```

在 MigrationLoader 类的初始化方法中会调用 build_graph()方法（load=True）去构造所有迁移文件的关联图，这一步非常重要。在 build_graph()方法中，会在一开始就调用 load_disk()方法来加载本地的迁移文件并更新到属性 disk_migrations 中。下面通过测试来看看这些方法的输出结果：

```
(django2-core-test) [root@master first_django]# python manage.py shell
Python 3.8.6 (default, Oct 18 2020, 15:33:08)
[GCC 4.8.5 20150623 (Red Hat 4.8.5-39)] on linux
Type "help", "copyright", "credits" or "license" for more information.
(InteractiveConsole)
>>> from django.db.migrations.loader import MigrationLoader
>>> loader = MigrationLoader(None, ignore_no_migrations=True)
>>> loader.disk_migrations
{('admin', '0003_logentry_add_action_flag_choices'): <Migration
admin.0003_logentry_add_action_flag_choices>, ('admin', '0001_initial'): <Migration
admin.0001_initial>, ('admin', '0002_logentry_remove_auto_add'): <Migration
admin.0002_logentry_remove_auto_add>, ('auth', '0001_initial'): <Migration
auth.0001_initial>, ('auth', '0008_alter_user_username_max_length'): <Migration
auth.0008_alter_user_username_max_length>, ('auth',
'0003_alter_user_email_max_length'): <Migration
auth.0003_alter_user_email_max_length>, ('auth',
'0009_alter_user_last_name_max_length'): <Migration
auth.0009_alter_user_last_name_max_length>, ('auth',
'0010_alter_group_name_max_length'): <Migration
auth.0010_alter_group_name_max_length>, ('auth', '0006_require_contenttypes_0002'):
<Migration auth.0006_require_contenttypes_0002>, ('auth',
```

```
'0011_update_proxy_permissions'): <Migration auth.0011_update_proxy_permissions>,
('auth', '0002_alter_permission_name_max_length'): <Migration
auth.0002_alter_permission_name_max_length>, ('auth',
'0004_alter_user_username_opts'): <Migration auth.0004_alter_user_username_opts>,
('auth', '0005_alter_user_last_login_null'): <Migration
auth.0005_alter_user_last_login_null>, ('auth',
'0007_alter_validators_add_error_messages'): <Migration
auth.0007_alter_validators_add_error_messages>, ('contenttypes', '0001_initial'):
<Migration contenttypes.0001_initial>, ('contenttypes',
'0002_remove_content_type_name'): <Migration
contenttypes.0002_remove_content_type_name>, ('sessions', '0001_initial'): <Migration
sessions.0001_initial>, ('shell_test', '0001_initial'): <Migration
shell_test.0001_initial>}
```

load_disk()方法的源码实现如下：

```
# 源码位置：django/db/migrations/loader.py
# ......

class MigrationLoader:

    # ......

    def load_disk(self):

        self.disk_migrations = {}
        self.unmigrated_apps = set()
        self.migrated_apps = set()
        for app_config in apps.get_app_configs():
            # module_name 对应迁移模块路径，下面有相应的操作示例显示这里的结果
            module_name, explicit = self.migrations_module(app_config.label)
            if module_name is None:
                self.unmigrated_apps.add(app_config.label)
                continue
            was_loaded = module_name in sys.modules
            try:
                # 导入迁移模块
                module = import_module(module_name)
            except ImportError as e:
                # 导入异常，直接抛错
                # ......
            else:
                # ......

            self.migrated_apps.add(app_config.label)
            migration_names = {
                name for _, name, is_pkg in pkgutil.iter_modules(module.__path__)
                # 导入 module 下的模块
```

```
            if not is_pkg and name[0] not in '_~'
        }
        # 导入迁移文件中定义的Migration类
        for migration_name in migration_names:
            migration_path = '%s.%s' % (module_name, migration_name)
            try:
                # 再次导入迁移文件模块
                migration_module = import_module(migration_path)
            except ImportError as e:
                # 抛出异常
                # ……
            if not hasattr(migration_module, "Migration"):
                # 在迁移文件中没有定义Migration类，直接抛出异常
                # ……

            self.disk_migrations[app_config.label, migration_name] =
migration_module.Migration(
                migration_name,
                app_config.label,
            )
```

下面对 load_disk()方法进行拆解，先厘清第 1 个 for 循环语句的含义，操作示例如下：

```
>>> from django.apps import apps
>>> apps.get_app_configs()
odict_values([<AdminConfig: admin>, <AuthConfig: auth>, <ContentTypesConfig:
contenttypes>, <SessionsConfig: sessions>, <MessagesConfig: messages>,
<StaticFilesConfig: staticfiles>, <AppConfig: shell_test>])
>>> from django.db.migrations.loader import MigrationLoader
>>> loader = MigrationLoader(None, ignore_no_migrations=True, load=False)
>>> for app_config in apps.get_app_configs():
...     print(loader.migrations_module(app_config.label))
...
('django.contrib.admin.migrations', False)
('django.contrib.auth.migrations', False)
('django.contrib.contenttypes.migrations', False)
('django.contrib.sessions.migrations', False)
('django.contrib.messages.migrations', False)
('django.contrib.staticfiles.migrations', False)
('shell_test.migrations', False)
>>>
```

从上面的代码中不难看出，for 循环中的 module_name 其实就是应用的迁移模块路径。从这里也可以知道 Django 框架中各应用的默认的迁移文件位置。以 auth 应用为例，其默认的迁移文件位置如图 2-3 所示。

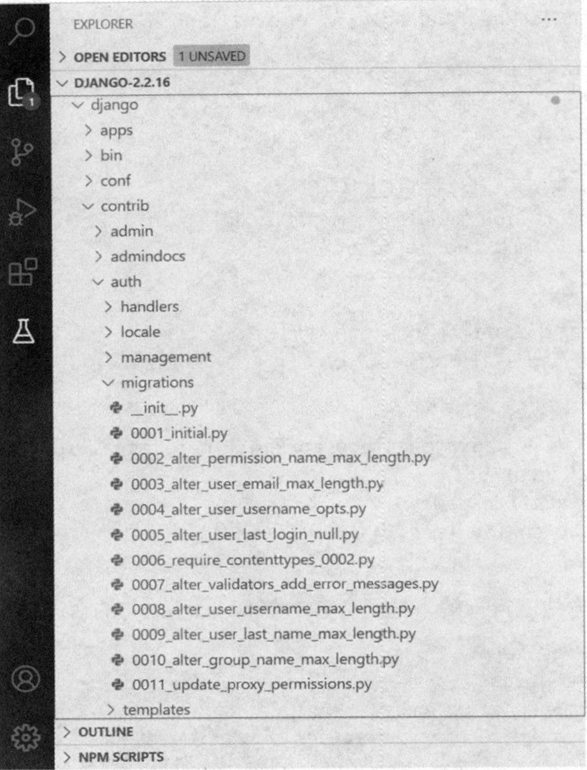


图 2-3

继续执行 load_disk()方法中 for 循环语句的后半段，以 django.contrib.auth.migrations 为例：

```
>>> from importlib import import_module
>>> module = import_module('django.contrib.auth.migrations')
>>> module
<module 'django.contrib.auth.migrations' from
'/root/.pyenv/versions/django2-core-test/lib/python3.8/site-packages/django/contrib/
auth/migrations/__init__.py'>
>>> import pkgutil
# 模块名不能以_或者~开头，因而排除了__init__.py 文件
>>> migration_names = {name for _, name, is_pkg in pkgutil.iter_modules(module.__path__)
if not is_pkg and name[0] not in '_~'}
>>> migration_names
{'0002_alter_permission_name_max_length', '0001_initial',
'0005_alter_user_last_login_null', '0006_require_contenttypes_0002',
'0007_alter_validators_add_error_messages', '0008_alter_user_username_max_length',
'0010_alter_group_name_max_length', '0009_alter_user_last_name_max_length',
'0011_update_proxy_permissions', '0003_alter_user_email_max_length',
'0004_alter_user_username_opts'}
```

这里再一次用到了 pkgutil 模块。通过 pkgutil.iter_modules()方法可以找到 migration_names 路径下的所有迁移文件（过滤掉以~或者_开头的文件）。

load_disk()方法的最后一部分就是遍历找到迁移文件并导入该迁移文件，同时得到该迁移文件中定义的迁移对象，并将该迁移对象记录到对象的 disk_migrations 属性中：

```
# 再次导入迁移模块
migration_module = import_module(migration_path)

# 实例化迁移模块中的Migration 类并保存
self.disk_migrations[app_config.label, migration_name] = migration_module.Migration(
                                                        migration_name,
                                                        app_config.label,
                                                     )
```

接着看 MigrationLoader 对象加载得到的 MigrationGraph 对象，它是通过调用 build_graph()方法得到的。先来看手工测试结果：

```
>>> from django.db.migrations.loader import MigrationLoader
>>> loader = MigrationLoader(None, ignore_no_migrations=True)
>>> loader.graph
<MigrationGraph: nodes=18, edges=18>
>>> loader.graph.node_map[('auth', '0001_initial')]
<Node: ('auth', '0001_initial')>
>>> loader.graph.node_map[('auth', '0001_initial')].parents
{<Node: ('contenttypes', '0001_initial')>}
>>> loader.graph.node_map[('auth', '0001_initial')].children
{<Node: ('auth', '0002_alter_permission_name_max_length')>, <Node: ('admin',
'0001_initial')>}
>>>
```

这些依赖结果都可以从具体迁移文件的 Migration 类中得到。下面分别查看上面代码中涉及的三个迁移文件：

```
# 源码位置：django/contrib/auth/migrations/0001_initial.py
# ……

class Migration(migrations.Migration):

   dependencies = [
      ('contenttypes', '__first__'),
   ]

   # ……

# 源码位置：django/contrib/auth/migrations/0002_alter_permission_name_max_length.py
# ……
```

```
class Migration(migrations.Migration):

    dependencies = [
        ('auth', '0001_initial'),
    ]

    # ……

# 源码位置：django/contrib/admin/migrations/0001_initial.py
# ……

class Migration(migrations.Migration):

    dependencies = [
        migrations.swappable_dependency(settings.AUTH_USER_MODEL),
        ('contenttypes', '__first__'),
    ]
```

前面两个比较好理解：('auth', '0001_initial')依赖('contenttypes', '0001_initial') ，__first__表示的正是第 1 个迁移文件；('auth', '0002_alter_permission_name_max_length')依赖('auth', '0001_initial')。最后，('admin', '0001_initial') 除依赖 ('contenttypes', '0001_initial') 外，还依赖 migrations.swappable_dependency(settings.AUTH_USER_MODEL)语句的结果。通过全局搜索可知，Django 中默认的 settings.AUTH_USER_MODEL 值如下：

```
# 源码位置：django/conf/global_settings.py
# ……

AUTH_USER_MODEL = 'auth.User'

# ……
```

直接在 shell 命令行中执行如下语句：

```
>>> from django.db.migrations.loader import MigrationLoader
>>> loader = MigrationLoader(None, ignore_no_migrations=True)
>>> from django.db import migrations
>>> migrations.swappable_dependency('auth.User')
('auth', '__first__')
```

从结果可知，('admin', '0001_initial')还依赖('auth', '0001_initial')，于是就有了前面迁移节点的 parents 和 children 属性值。

```
# 源码位置：django/db/migrations/loader.py
# ……

class MigrationLoader:

    # ……
```

```
def add_internal_dependencies(self, key, migration):

    for parent in migration.dependencies:
        # 忽略__first__对同一应用的引用.
        if parent[0] == key[0] and parent[1] != '__first__':
            self.graph.add_dependency(migration, key, parent, skip_validation=True)

def add_external_dependencies(self, key, migration):
    # 处理迁移文件中 Migration 类的 dependencies 属性值，依赖节点为其父节点
    for parent in migration.dependencies:
        # 忽略内部依赖
        if key[0] == parent[0]:
            continue
        parent = self.check_key(parent, key[0])
        if parent is not None:
            # 在 MigrationGrap 对象中添加迁移节点的依赖关系
            self.graph.add_dependency(migration, key, parent, skip_validation=True)
    for child in migration.run_before:
        child = self.check_key(child, key[0])
        if child is not None:
            self.graph.add_dependency(migration, child, key, skip_validation=True)

def build_graph(self):

    self.load_disk()

    # 载入数据库中的迁移记录数据
    if self.connection is None:
        self.applied_migrations = set()
    else:
        # 在实例化时传入 connection，才会去查询迁移表中的记录
        recorder = MigrationRecorder(self.connection)
        self.applied_migrations = recorder.applied_migrations()

    self.graph = MigrationGraph()
    self.replacements = {}
    for key, migration in self.disk_migrations.items():
        # 添加节点
        self.graph.add_node(key, migration)
        # 记录迁移类中的 replaces 属性值
        if migration.replaces:
            self.replacements[key] = migration

    for key, migration in self.disk_migrations.items():
        # 处理相同应用内的依赖
        self.add_internal_dependencies(key, migration)
```

```
        for key, migration in self.disk_migrations.items():
            # 处理不同应用之间的依赖
            self.add_external_dependencies(key, migration)

        # 处理迁移类中出现的 replacements 属性
        # ……

        try:
            # 在 MigrationGraph 类中介绍过，检查是否有 dummy 节点
            self.graph.validate_consistency()
        except NodeNotFoundError as exc:
            # 打印异常信息，最终抛出异常
            # ……

        # 确保没有
        self.graph.ensure_not_cyclic()

    # ……
```

在上面的 build_graph()方法中省略了对迁移类（Migration）中 replacements 属性的处理。在默认的迁移文件及 shell_test 应用的迁移类中，并不涉及 replacements 属性值：

```
>>> from django.db.migrations.loader import MigrationLoader
>>> loader = MigrationLoader(None, ignore_no_migrations=True)
>>> loader.replacements
{}
>>>
```

此外，在 build_graph()方法的最后调用了 MigrationGraph 对象中的两个方法：validate_consistency()和 ensure_not_cyclic()。前一个方法的实现比较简单，就是检查是否有 dummy 节点，有则直接抛错；后一个方法的含义是确保迁移图的节点之间不存在循环依赖关系。下面给出一个简单循环关系的示例，最后调用 ensure_not_cyclic()方法抛出异常：

```
>>> from django.db.migrations.migration import Migration
>>> m1 = Migration('00001_xxxxx', 'shell_test')
>>> m2 = Migration('00002_xxxxx', 'shell_test')
>>> m3 = Migration('00003_xxxxx', 'shell_test')
>>> from django.db.migrations.graph import MigrationGraph
>>> graph = MigrationGraph()
>>> graph.add_node(('k1', 'v1'), m1)
>>> graph.add_node(('k1', 'v2'), m2)
>>> graph.add_node(('k1', 'v3'), m3)
# v1 的父节点为 v3
>>> graph.add_dependency(None, ('k1', 'v1'), ('k1', 'v3'))
# v3 的父节点为 v2
>>> graph.add_dependency(None, ('k1', 'v3'), ('k1', 'v2'))
# v2 的父节点为 v1
```

```
>>> graph.add_dependency(None, ('k1', 'v2'), ('k1', 'v1'))
# 此时，节点之间存在循环关系，在调用 ensure_not_cyclic()方法后会抛出异常
>>> graph.ensure_not_cyclic()
Traceback (most recent call last):
  File "<console>", line 1, in <module>
  File
"/root/.pyenv/versions/django2-core-test/lib/python3.8/site-packages/django/db/migra
tions/graph.py", line 274, in ensure_not_cyclic
    raise CircularDependencyError(", ".join("%s.%s" % n for n in cycle))
django.db.migrations.exceptions.CircularDependencyError: k1.v1, k1.v2, k1.v3
```

在掌握了前面这些基础知识后，就可以正式追踪 makemigrations 命令和 migrate 命令了。

### 2.4.3　追踪 makemigrations 命令

makemigrations 命令对应的代码如下：

```
# 源码位置： django/core/management/commands/makemigrations.py
# ……

class Command(BaseCommand):
    help = "Creates new migration(s) for apps."

    # ……

    @no_translations
    def handle(self, *app_labels, **options):
        # ……

    def write_migration_files(self, changes):
        # ……

    def handle_merge(self, loader, conflicts):
        # ……
```

从上面的代码可以看到，在 makemigrations 命令中，Command 类直接继承自 BaseCommand 类，所以它的执行流程相对比较简单，重点是 Command 类中的 handle()方法。handle()方法的实现如下：

```
# 源码位置： django/core/management/commands/makemigrations.py
# ……

class Command(BaseCommand):

    # ……

    @no_translations
    def handle(self, *app_labels, **options):
        # 获取相关参数
```

```
    # ……

    # 确保要求执行迁移的应用存在
    app_labels = set(app_labels)
    has_bad_labels = False
    for app_label in app_labels:
        try:
            # 这里的 apps 就是前面遇到的 Apps 对象
            apps.get_app_config(app_label)
        except LookupError as err:
            self.stderr.write(str(err))
            has_bad_labels = True
    if has_bad_labels:
        # 没有不正常的应用，直接退出
        sys.exit(2)

    # 1. 得到 MigrationLoader 对象
    loader = MigrationLoader(None, ignore_no_migrations=True)

    # 2. 检测迁移文件的一致性
    consistency_check_labels = {config.label for config in apps.get_app_configs()}
    # 获取数据库配置信息，没有设置 settings.DATABASE_ROUTERS，得到的值为['default']
    aliases_to_check = connections if settings.DATABASE_ROUTERS
else [DEFAULT_DB_ALIAS]
    for alias in sorted(aliases_to_check):
        # connections 将在第 3 章介绍，这里可以简单理解为根据设置的数据库信息得到一个连接对象，
        # 后续 Django 将根据该连接对象对数据库进行操作

        connection = connections[alias]
        if (connection.settings_dict['ENGINE'] != 'django.db.backends.dummy' and any(
                router.allow_migrate(connection.alias, app_label,
                                     model_name=model._meta.object_name)
                for app_label in consistency_check_labels
                for model in apps.get_app_config(app_label).get_models()
        )):
            # 检测数据表中的一致性
            loader.check_consistent_history(connection)

    # 3. 检测冲突
    conflicts = loader.detect_conflicts()

    if app_labels:
        conflicts = {
            app_label: conflict for app_label, conflict in conflicts.items()
            if app_label in app_labels
        }

    # 4. 当存在冲突且设置不合并时直接抛出异常
```

```
if conflicts and not self.merge:
    name_str = "; ".join(
        "%s in %s" % (", ".join(names), app)
        for app, names in conflicts.items()
    )
    raise CommandError(
        "Conflicting migrations detected; multiple leaf nodes in the "
        "migration graph: (%s).\nTo fix them run "
        "'python manage.py makemigrations --merge'" % name_str
    )

# 设置合并且没有冲突
if self.merge and not conflicts:
    self.stdout.write("No conflicts detected to merge.")
    return

# 5. 如果设置合并，而且存在冲突，
# 则直接调用 self.handle_merge()方法处理并返回
if self.merge and conflicts:
    return self.handle_merge(loader, conflicts)

# ……

# 如果想执行一次空迁移，则对每个应用都执行一次
# ……

# 6. 检查变化，代码比较复杂
changes = autodetector.changes(
    graph=loader.graph,
    trim_to_apps=app_labels or None,
    convert_apps=app_labels or None,
    migration_name=self.migration_name,
)

# 7. 处理变化，当没有变化时提示
if not changes:
    # 没有变化，需要提示
    if self.verbosity >= 1:
        # 看 makemigrations 后面有没有应用，是一个还是多个
        if app_labels:
            if len(app_labels) == 1:
                self.stdout.write("No changes detected in app '%s'" %
                                  app_labels.pop())
            else:
                self.stdout.write("No changes detected in apps '%s'" %
                                 ("', '".join(app_labels)))
        else:
            self.stdout.write("No changes detected")
```

```
    else:
        # 有变化，写入数据文件中
        self.write_migration_files(changes)
        if check_changes:
            sys.exit(1)
```

针对 handle()方法，笔者整理了 7 处代码进行讲解：

第 1 处代码非常简单，得到 MigrationLoader 对象，在前面已多次使用过。

第 2 处代码是检测迁移文件之间的一致性。该代码段中的 aliases_to_check 默认为['default']。这个列表中的元素实际上对应着 settings.py 文件中 DATABASES 的 key，而该 key 对应的 value 值正是以字典形式保存的数据库连接信息。在 DATABASES 中配置的数据库连接信息会在一开始全部处理后保存在 django.db.connections 中，需要通过 key 来获取相应的数据库连接信息，最后通过 MigrationLoader 对象的 check_consistent_history(connection)方法检测迁移文件之间的一致性。该方法会根据 connection 搜索该数据库迁移表中的数据进行判断，具体的判断逻辑可以通过直接阅读方法的源码得到。第一次执行 makemigrations 命令时，数据库中的迁移表还未生成，没有任何迁移数据，所以检查直接通过。

第 3 处代码是检测迁移文件之间的冲突。detect_conflicts()方法的源码如下：

```
# 源码位置：django/db/migrations/loader.py
# ……

class MigrationLoader:

    # ……

    def detect_conflicts(self):

        seen_apps = {}
        conflicting_apps = set()
        # 遍历迁移图中的叶子节点
        for app_label, migration_name in self.graph.leaf_nodes():
            # 判断是否有相同应用的迁移文件出现
            if app_label in seen_apps:
                conflicting_apps.add(app_label)
            seen_apps.setdefault(app_label, set()).add(migration_name)
        return {app_label: seen_apps[app_label] for app_label in conflicting_apps}

    # ……
```

从上面的代码可以看到，当在同一应用下存在多个属于叶子节点的迁移文件时，会被检测为冲突。下面看一个操作示例：

```
>>> from django.db.migrations.migration import Migration
>>> m1 = Migration('00001_xxxxx', 'shell_test')
>>> m2 = Migration('00002_xxxxx', 'shell_test')
>>> m3 = Migration('00001_xxxxx', 'auth')
>>> from django.db.migrations.graph import MigrationGraph
>>> graph = MigrationGraph()
>>> graph.add_node(('shell_test', '00001_xxxxx'), m1)
>>> graph.add_node(('shell_test', '00002_xxxxx'), m2)
>>> graph.add_node(('auth', '00001_xxxxx'), m3)
>>> graph.leaf_nodes()
[('auth', '00001_xxxxx'), ('shell_test', '00001_xxxxx'), ('shell_test', '00002_xxxxx')]
>>> from django.db.migrations.loader import MigrationLoader
# 使 load=False，不加载本地的迁移文件和生成迁移图
>>> loader = MigrationLoader(None, ignore_no_migrations=True, load=False)
# 手工加入迁移图
>>> loader.graph = graph
>>> loader.detect_conflicts()
{'shell_test': {'00001_xxxxx', '00002_xxxxx'}}
# 添加外部应用依赖，迁移图的叶子节点不变，
# 具体可以参考 2.4.2 节对该方法的介绍
>>> graph.add_dependency(None, ('shell_test', '00001_xxxxx'), ('auth', '00001_xxxxx'))
>>> graph.leaf_nodes()
[('auth', '00001_xxxxx'), ('shell_test', '00001_xxxxx'), ('shell_test', '00002_xxxxx')]
# 添加内部应用依赖，('shell_test', '00002_xxxxx')不再是叶子节点
>>> graph.add_dependency(None, ('shell_test', '00001_xxxxx'), ('shell_test',
'00002_xxxxx'))
# 再次检查迁移图的叶子节点
>>> graph.leaf_nodes()
[('auth', '00001_xxxxx'), ('shell_test', '00001_xxxxx')]
>>> loader.graph = graph
>>> loader.detect_conflicts()
{}
```

在上面的示例中，给 shell_test 应用添加了 2 个迁移文件，并都加到 MigrationLoader 对象中。由于没有添加依赖，所以在 shell_test 应用下有 2 个叶子节点，会被检测为冲突。在给 ('shell_test', '00001_xxxxx)' 添加内部应用依赖后，在每个应用下只有 1 个叶子节点，不会被检测为冲突。下面看本次命令追踪的检测冲突情况：

```
>>> loader = MigrationLoader(None, ignore_no_migrations=True)
>>> loader.graph.leaf_nodes()
[('admin', '0003_logentry_add_action_flag_choices'), ('auth',
'0011_update_proxy_permissions'), ('contenttypes', '0002_remove_content_type_name'),
('sessions', '0001_initial'), ('shell_test', '0001_initial')]
>>> loader.detect_conflicts()
{}
```

从上面的结果可以看出，在 Django 中，默认应用的迁移文件都不存在冲突。在笔者手工创建的

shell_test 应用下只有一个迁移文件，也不存在冲突。

第 4 处代码比较简单，就是当检测到有冲突且设置为不合并（merge=False）时，直接抛出异常。

第 5 处代码也比较简单，如果有冲突且设置了合并参数，就对这些迁移文件进行合并处理。

第 6 处代码是检查应用中的模型类与项目已有的迁移文件中保存的模型信息是否一致。这里是通过一个 changes 检测机制得到相应变化的。该检测机制比较复杂，代码量较大，本书后续不会直接分析该自动检测类，而是观察检测结果，为后面分析迁移文件的生成代码做准备。

第 7 处代码是根据上一步得到的变化信息进行处理。如果没有变化，提示 No changes detected。如果有变化，调用 write_migration_files()方法生成一个新的迁移文件。

通过上面的分析可以清楚地看到，迁移文件的生成需要 2 个条件：得到 changes 和调用 write_migration_files()方法。为了演示 changes 的输出，这里先删除 shell_test 应用下的所有迁移数据：

```
(django2-core-test) [root@master first_django]# rm -f shell_test/migrations/*.py
```

再手工生成 MigrationAutodetector 对象，并调用该对象的 changes()方法获取 shell_test 应用下的迁移变化数据：

```
>>> from django.apps import apps
>>> from django.db.migrations.autodetector import MigrationAutodetector
>>> from django.db.migrations.questioner import InteractiveMigrationQuestioner
>>> from django.db.migrations.loader import MigrationLoader
>>> from django.db.migrations.state import ProjectState
# 按照 handle()方法的核心步骤依次执行
>>> app_labels = set({'shell_test'})
>>> loader = MigrationLoader(None, ignore_no_migrations=True)
>>> questioner = InteractiveMigrationQuestioner(specified_apps=app_labels,
dry_run=False)
>>> autodetector =
MigrationAutodetector(loader.project_state(),ProjectState.from_apps(apps),questioner
)
>>> changes =
autodetector.changes(graph=loader.graph,trim_to_apps=app_labels,convert_apps=app_lab
els,migration_name=None)
>>> changes
{'shell_test': [<Migration shell_test.0001_initial>]}
```

从上面的代码可以看到，当使用 makemigrations 命令给 shell_test 应用生成迁移文件时，得到的 changes 为一个字典形式，每个 key 代表着对应的应用标签，其 value 值为一个列表，列表的元素为 Migration 对象。这些 Migration 对象将被写入对应的迁移文件中。文件名既可以由外部指定，也可以由 Django 默认指定。最后看 write_migration_files()方法，其源码如下：

```
# 源码位置： django/core/management/commands/makemigrations.py
# ……
```

```
class Command(BaseCommand):

    # ……

    def write_migration_files(self, changes):
        """
        Take a changes dict and write them out as migration files.
        """
        directory_created = {}
        for app_label, app_migrations in changes.items():
            # 打印信息
            # ……
            for migration in app_migrations:
                # 生成 MigrationWriter 对象
                writer = MigrationWriter(migration, self.include_header)
                # 打印信息
                # ……
                if not self.dry_run:
                    # 得到迁移文件目录
                    migrations_directory = os.path.dirname(writer.path)
                    if not directory_created.get(app_label):
                        # 如果目录没有被创建过，创建迁移文件目录
                        if not os.path.isdir(migrations_directory):
                            os.mkdir(migrations_directory)
                        # 在 migrations 目录下创建空的__init__.py 文件
                        init_path=os.path.join(migrations_directory, "__init__.py")
                        if not os.path.isfile(init_path):
                            open(init_path, "w").close()
                        # 每个应用只创建一次目录即可，在执行创建动作后需要在字典中记录，避免重复创建
                        directory_created[app_label] = True
                    # 生成迁移文件的内容
                    migration_string = writer.as_string()
                    # 将生成的迁移文件写入迁移文件中
                    with open(writer.path, "w", encoding='utf-8') as fh:
                        fh.write(migration_string)
                elif self.verbosity == 3:
                    # 如果 makemigrations 命令参数为--dry-run --verbosity 3
                    # 则不会生成迁移文件，而是仅仅打印迁移文件的内容
                    self.stdout.write(self.style.MIGRATE_HEADING(
                        "Full migrations file '%s':" % writer.filename) + "\n"
                    )
                    self.stdout.write("%s\n" % writer.as_string())
    # ……
```

write_migration_files()方法通过循环遍历 changes 来生成迁移文件，而迁移文件的内容正是其中保存的 Migration 类。其中，MigrationWriter 对象的 as_string()方法可以将传入的 Migration 对象转成

字符串形式。下面使用 as_string()方法输出上面得到的 Migration 对象：

```
# 接着上面的操作
>>> migration = changes['shell_test'][0]
>>> migration
<Migration shell_test.0001_initial>
>>> from django.db.migrations.writer import MigrationWriter
>>> writer = MigrationWriter(migration)
>>> print(writer.as_string())
# Generated by Django 2.2.16 on 2021-01-14 04:21

from django.db import migrations, models


class Migration(migrations.Migration):

    initial = True

    dependencies = [
    ]

    operations = [
        migrations.CreateModel(
            name='DjangoBooks',
            fields=[
                ('id', models.AutoField(auto_created=True, primary_key=True,
serialize=False, verbose_name='ID')),
                ('book_name', models.CharField(max_length=30, verbose_name='图书名')),
                ('author', models.CharField(max_length=30, verbose_name='作者')),
                ('sex', models.SmallIntegerField(choices=[(0, '男'), (1, '女')], default=0,
verbose_name='性别')),
                ('price', models.FloatField(verbose_name='图书价格')),
                ('isbn', models.CharField(max_length=10, verbose_name='isbn')),
                ('publish_date', models.DateTimeField(auto_now=True, verbose_name='出版时
间')),
            ],
            options={
                'db_table': 'django_books',
            },
        ),
    ]
```

上面输出的结果正是第一次执行 makemigrations 命令时得到的迁移文件内容。下面给出一些简单的操作示例，以帮助读者更好地理解 MigrationWriter 类的功能：

```
(django2-core-test) [root@master first_django]# python manage.py shell
Python 3.8.6 (default, Oct 18 2020, 15:33:08)
```

```
[GCC 4.8.5 20150623 (Red Hat 4.8.5-39)] on linux
Type "help", "copyright", "credits" or "license" for more information.
(InteractiveConsole)
>>> from django.db.migrations.migration import Migration
>>> m1 = Migration('00001_xxxx', 'shell_test')
>>> from django.db.migrations.writer import MigrationWriter
>>> writer = MigrationWriter(m1)
>>> m1.dependencies = [('auth', '0001_initial'), ['xxxx', '11111']]
>>> writer = MigrationWriter(m1)
>>> print(writer.as_string())
# Generated by Django 2.2.16 on 2021-01-14 04:33

from django.db import migrations


class Migration(migrations.Migration):

    dependencies = [
        ('auth', '0001_initial'),
        ['xxxx', '11111'],
    ]

    operations = [
    ]

>>> writer = MigrationWriter(m1, include_header=False)
>>> print(writer.as_string())
from django.db import migrations


class Migration(migrations.Migration):

    dependencies = [
        ('auth', '0001_initial'),
        ['xxxx', '11111'],
    ]

    operations = [
    ]
```

在追踪 makemigrations 命令的执行过程中，最难的部分在于检测当前模型类与已存在迁移文件中的模型类的变化部分（即得到 changes 值），而这一工作最终是由 MigrationAutodetector 对象的 changes()方法来承担的。至此，我们已基本掌握了 makemigrations 命令。

## 2.5 migrate 命令的实现原理

在 2.4 节中，生成 shell_test 应用下的迁移文件后，执行 migrate 命令，即可将迁移文件中的表信息及其数据映射到项目配置的数据库中：

```
(django2-core-test) [root@master first_django]# python manage.py migrate shell_test
System check identified some issues:

# 打印告警信息
# ……
Operations to perform:
  Apply all migrations: shell_test
Running migrations:
  Applying shell_test.0001_initial... OK
```

在执行上述操作后，可以看到数据库中的 django_books 表已经被创建，一同被创建的还有 django_migrations 表。在 django_migrations 表中只有一条记录，如图 2-4 所示。

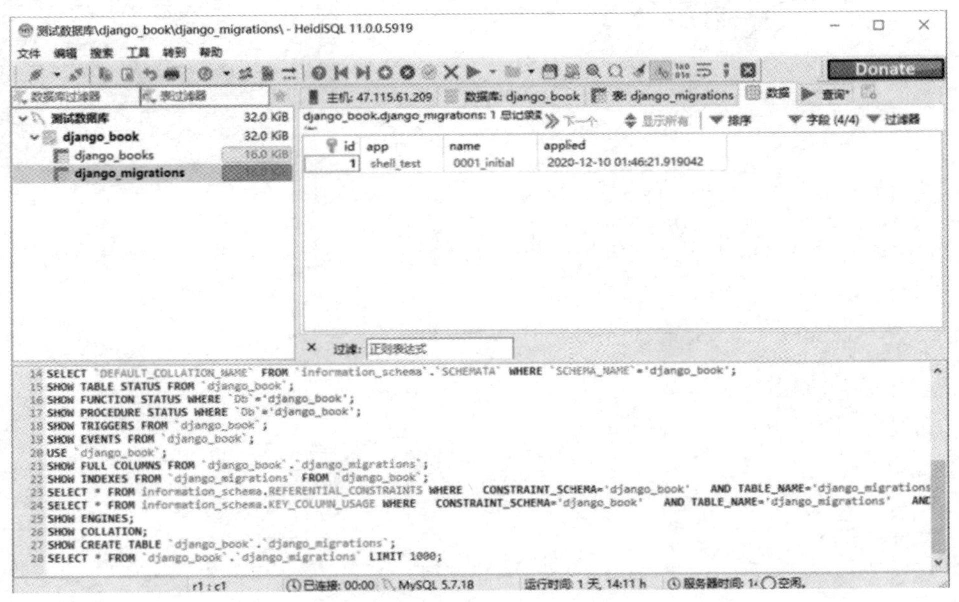

图 2-4

在对该命令有了初步印象后，思考以下两个问题：

◎ migrate 命令是如何将迁移文件中的信息映射到数据库中的？

◎ 在执行 migrate 命令时，可能会出现哪些异常？出现这些异常的原因是什么，又该如何复现呢？

下面我们就带着这些问题开启 migrate 命令的源码追踪之旅。migrate 命令对应的源码如下：

```
# 源码位置：django/core/management/commands/migrate.py
# ……

class Command(BaseCommand):

    # ……

    @no_translations
    def handle(self, *args, **options):

        # ……

        # 获取要操作的数据库信息
        db = options['database']
        connection = connections[db]

        # 对 MySQL 而言，没有任何操作
        connection.prepare_database()
        # 1.实例化 MigrationExecutor 类
        executor = MigrationExecutor(connection, self.migration_progress_callback)

        # 调用 MigrationLoader 对象的 check_consistent_history()方法检测一致性
        executor.loader.check_consistent_history(connection)

        # 调用 MigrationLoader 对象的 detect_conflicts()方法检测冲突，具体细节可以参考 2.4.3 节
        conflicts = executor.loader.detect_conflicts()
        if conflicts:
            # 如果存在冲突，抛出异常信息
            # ……

        # run_syncdb 选项，默认为 False。该选项的含义在后面可以看到，即只创建表并不迁移
        run_syncdb = options['run_syncdb']
        target_app_labels_only = True
        if options['app_label']:
            # Validate app_label.
            app_label = options['app_label']
            try:
                apps.get_app_config(app_label)
            except LookupError as err:
                raise CommandError(str(err))

            # 处理一些选项异常情况
            # ……

        # 2.得到迁移文件
        if options['app_label'] and options['migration_name']:
            # 指定应用及对应的迁移文件
            migration_name = options['migration_name']
            if migration_name == "zero":
                targets = [(app_label, None)]
```

```
            else:
                try:
                    # 根据应用获取对应的迁移对象
                    migration = executor.loader.get_migration_by_prefix(app_label,
                                                                         migration_name)
                except AmbiguityError:
                    # 如果一个应用下有多个迁移匹配，抛出异常
                    # ……
                except KeyError:
                    # 如果没有匹配应用中的迁移文件，抛出异常
                    # ……

                targets = [(app_label, migration.name)]
            target_app_labels_only = False
        elif options['app_label']:
            # 找出当前应用的叶子节点的迁移文件
            targets = [key for key in executor.loader.graph.leaf_nodes()
                            if key[0] == app_label]
        else:
            # 找出所有处于叶子节点的迁移文件
            targets = executor.loader.graph.leaf_nodes()

        # 3. targets 保存的就是待迁移的列表，查询迁移表 django_migrations
        plan = executor.migration_plan(targets)

        if options['plan']:
            # 对于 plan 选项，只显示将要执行的迁移文件，然后返回
            self.stdout.write('Planned operations:', self.style.MIGRATE_LABEL)
            if not plan:
                self.stdout.write('  No planned migration operations.')
            for migration, backwards in plan:
                self.stdout.write(str(migration), self.style.MIGRATE_HEADING)
                # 4. 循环打印要执行的操作信息，比如创建表、修改表字段等
                for operation in migration.operations:
                    message, is_error = self.describe_operation(operation, backwards)
                    style = self.style.WARNING if is_error else None
                    self.stdout.write('    ' + message, style)
            return

        run_syncdb = options['run_syncdb'] and executor.loader.unmigrated_apps

        # 打印一些有用的信息
        # ……
        # 获取一些迁移前的状态信息
        pre_migrate_state =
executor._create_project_state(with_applied_migrations=True)
        pre_migrate_apps = pre_migrate_state.apps
        emit_pre_migrate_signal(
            self.verbosity, self.interactive, connection.alias, apps=pre_migrate_apps,
plan=plan,
```

```
)

# Run the syncdb phase.
if run_syncdb:
    # ……
    if options['app_label']:
        self.sync_apps(connection, [app_label])
    else:
        self.sync_apps(connection, executor.loader.unmigrated_apps)

# 开始迁移
if self.verbosity >= 1:
    # 打印开始迁移
    self.stdout.write(self.style.MIGRATE_HEADING("Running migrations:"))

if not plan:
    # 没有待迁移的表，打印相关信息
    if self.verbosity >= 1:
        self.stdout.write("  No migrations to apply.")

        # 得到自动检测迁移过程的对象
        autodetector = MigrationAutodetector(
            executor.loader.project_state(),
            ProjectState.from_apps(apps),
        )
        # 检测当前模型和当前生成迁移文件之间是否有差异
        changes = autodetector.changes(graph=executor.loader.graph)
        if changes:
            # 如果有差异，则提示
            self.stdout.write(self.style.NOTICE(
                "  Your models have changes that are not yet reflected "
                "in a migration, and so won't be applied."
            ))
            self.stdout.write(self.style.NOTICE(
                "  Run 'manage.py makemigrations' to make new "
                "migrations, and then re-run 'manage.py migrate' to "
                "apply them."
            ))
    # 设置相关参数
    fake = False
    fake_initial = False
else:
    fake = options['fake']
    fake_initial = options['fake_initial']

# 5.执行迁移动作，创建表及相关的迁移文件
post_migrate_state = executor.migrate(
    targets, plan=plan, state=pre_migrate_state.clone(), fake=fake,
    fake_initial=fake_initial,
)
```

```
    # ……
```

上面是整个 migrate 命令的执行过程，省略了部分细节代码，同时对重要的代码段进行了相应注释。为了帮助读者理解 migrate 命令的执行流程，笔者整理出 5 处代码进行讲解：

第 1 处代码是实例化 MigrationExecutor 类，这一步非常重要，因为后续迁移文件的一致性检测、冲突检测，以及最终的解析迁移文件等动作都依赖于得到的 MigrationExecutor 类。MigrationExecutor 类的源码如下：

```
# 源码位置： django/db/migrations/executor.py
# ……

class MigrationExecutor:

    def __init__(self, connection, progress_callback=None):
        self.connection = connection
        # 根据 connection 实例化 MigrationLoader 类
        self.loader = MigrationLoader(self.connection)
        self.recorder = MigrationRecorder(self.connection)
        self.progress_callback = progress_callback

    # ……
```

可以看到，MigrationExecutor 对象的 loader 属性即前面介绍过的 MigrationLoader 对象。在 migrate 命令的后续执行过程中会调用该对象的 check_consistent_history()方法和 detect_conflicts()方法，分别检测迁移文件之间的一致性和冲突。这两个方法的细节已在 2.4.3 节中介绍过，不再赘述。

第 2 处代码是通过输入的应用（可能没有）及相应的迁移文件（可能没有）得到最终的迁移目标文件列表的。以 python manage.py shell_test 命令为例，由于第一次使用 makemigrations 命令生成了迁移文件 0001_initial.py，因此代码将进入 elif options['app_label']分支中执行，最终得到 targets=[('shell_test', '0001_initial')]。

第 3 处代码是调用 MigrationExecutor 对象的 migration_plan()方法得到相应的迁移计划。该方法的代码如下：

```
# 源码位置： django/db/migrations/executor.py
# ……

class MigrationExecutor:

    # ……

    def migration_plan(self, targets, clean_start=False):
        """
        给一个目标集合，返回迁移对象的列表
```

```
"""
plan = []
if clean_start:
    applied = set()
else:
    # 默认 clean_start 为 False，进入这里
    applied = set(self.loader.applied_migrations)
for target in targets:
    # 如果 target 是(app_label, None)，则表示不进行迁移操作
    if target[1] is None:
        for root in self.loader.graph.root_nodes():
            if root[0] == target[0]:
                # 在同一个应用内
                for migration in self.loader.graph.backwards_plan(root):
                    if migration in applied:
                        # 给 plan 添加对应的迁移对象
                        plan.append((self.loader.graph.nodes[migration], True))
                        # 从 applied 中移除该迁移对象
                        applied.remove(migration)
    elif target in applied:
        # 对于已经存在的记录，这里的逻辑会略微复杂，迁移的冲突往往出现在这里
        next_in_app = sorted(
            n for n in
            self.loader.graph.node_map[target].children
            # 同一个应用，该迁移节点的子节点
            if n[0] == target[0]
        )
        for node in next_in_app:
            # 遍历 target 的所有子孙节点，如果已经出现在迁移表中，
            # 则加入 plan 列表，同时把元素的第 2 个字段标记为 True
            for migration in self.loader.graph.backwards_plan(node):
                if migration in applied:
                    plan.append((self.loader.graph.nodes[migration], True))
                    applied.remove(migration)
    else:
        # 对于未应用到数据库中的迁移对象
        for migration in self.loader.graph.forwards_plan(target):
            if migration not in applied:
                plan.append((self.loader.graph.nodes[migration], False))
                applied.add(migration)
return plan
```

对于首次进行迁移或者是未应用到数据库中的迁移对象时，程序会在最后的 else 分支进行处理。以迁移命令 python manage.py shell_test 为例，由于是第一次执行迁移命令，相关的表都还未创建，此时 applied 为空集合。通过前面的分析已知 target 的值为[('shell_test', '0001_initial')]，于是在进入 for 循环遍历后，target[1]有值且不在 applied 中，因此会进入 else 分支执行。在 else 分支的代码段中，要理解 self.loader.graph.forwards_plan(target)语句的含义，该语句的含义是根据当前的完整迁移图找

出 target 节点所依赖的迁移节点，而这些迁移节点所代表的迁移文件必须要先应用到数据库中，然后才能迁移。为了方便理解，这里先删除前面创建的 django_books 表和 django_migrations 表，然后创建一个新的应用 book_sales：

```
(django2-core-test) [root@master first_django]# django-admin startapp book_sales
(django2-core-test) [root@master first_django]# ls
book_sales  first_django  manage.py  shell_test
```

（1）将应用 book_sales 添加到 settings.py 文件的 INSTALLED_APPS 变量中：

```
# 源码位置： first_django/first_django/settings.py
# ……

INSTALLED_APPS = [
    'django.contrib.admin',
    'django.contrib.auth',
    'django.contrib.contenttypes',
    'django.contrib.sessions',
    'django.contrib.messages',
    'django.contrib.staticfiles',
    'shell_test',
    'book_sales',
]

# ……
```

（2）在应用 book_sales 中创建一个模型类 BookSales。该模型类表示前面 Django 图书的整体销售情况，模型定义如下：

```
# 源码位置：first_django/book_sales/models.py
from shell_test.models import DjangoBooks
from django.db import models

# Create your models here.

class BookSales(models.Model):
    '''图书销售情况'''

    classify = models.CharField('图书分类', max_length=100)
    isbn = models.OneToOneField(to=DjangoBooks, on_delete=models.CASCADE)
    sales_volume = models.IntegerField('总销量', default=0)

    def __str__(self):
        return "<%s, %s>" % (self.isbn, self.sales_volume)

    class Meta:
        # 通过 db_table 自定义数据表名
        db_table = 'book_sales'
```

这里的 isbn 字段关联了 DjangoBook 模型类中的 isbn 字段，因此模型类 BookSales 依赖模型类 DjangoBook。

(3) 对该应用执行迁移命令：

```
(django2-core-test) [root@master first_django]# python manage.py makemigrations shell_test
Migrations for 'shell_test':
  shell_test/migrations/0001_initial.py
    - Create model DjangoBooks
(django2-core-test) [root@master first_django]# python manage.py makemigrations book_sales
Migrations for 'book_sales':
  book_sales/migrations/0001_initial.py
- Create model BookSales
(django2-core-test) [root@master first_django]# python manage.py migrate book_sales
# 一些告警信息
# ……
Operations to perform:
  Apply all migrations: book_sales
Running migrations:
  Applying shell_test.0001_initial... OK
  Applying book_sales.0001_initial... OK
```

通过上面的输出可以看出，当迁移应用 book_sales 中的模型时，会自动发现其依赖应用的模型文件。此时先迁移依赖模型，再迁移自身。这段逻辑刚好对应 migration_plan()方法中最后的 else 分支，最终生成的迁移记录如图 2-5 所示。

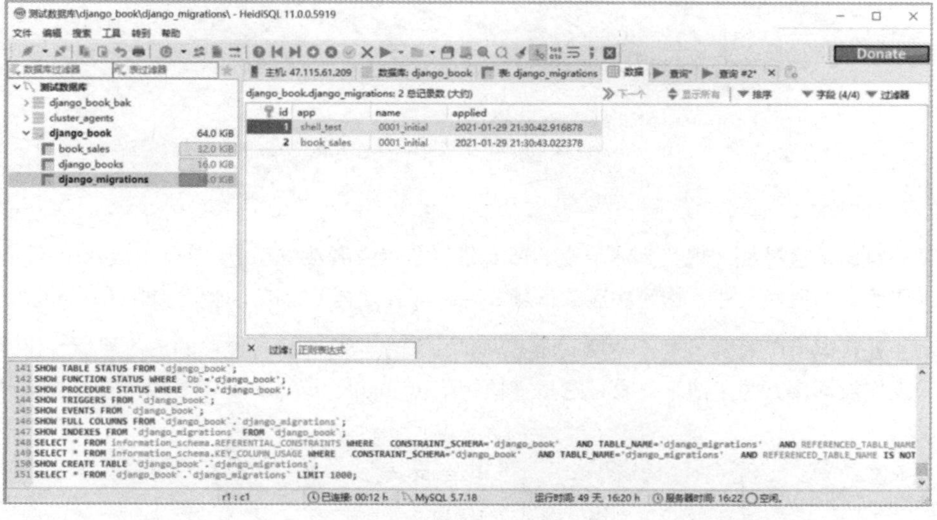

图 2-5

Django 是如何找到迁移节点的所有依赖节点的？直接看 MigrationGraph 类中的 forwards_plan() 方法即可：

```
# 源码位置: django/db/migrations/graph.py
# ……

class MigrationGraph:
    # ……

    def forwards_plan(self, target):

        if target not in self.nodes:
            # 如果不是磁盘上的迁移文件，直接抛错
            raise NodeNotFoundError("Node %r not a valid node" % (target,), target)
        # 返回所有该节点依赖的迁移节点
        return self.iterative_dfs(self.node_map[target])

    def iterative_dfs(self, start, forwards=True):
        # 使用一个简单的算法，找出该节点的所有依赖节点
        # 如果 forwards=True，找它的所有依赖节点，即父节点和祖先节点
        # 如果 forwards=False，找所有依赖它的节点，即子孙节点
        visited = []
        visited_set = set()
        stack = [(start, False)]
        while stack:
            node, processed = stack.pop()
            if node in visited_set:
                pass
            elif processed:
                visited_set.add(node)
                visited.append(node.key)
            else:
                stack.append((node, True))
                stack += [(n, False) for n in sorted(node.parents if forwards else
node.children)]
        return visited
```

上面的算法是通过栈实现找到该节点的所有依赖节点或者被依赖节点的。而查找的核心是，从磁盘上读取所有的迁移文件，构建相应的依赖关系，即各迁移节点的 parents 属性和 children 属性。

从第 4 处代码可以看到，对于使用 plan 选项的，只显示将要执行迁移的动作然后直接返回，并不会真的操作数据库。为了演示效果，这里直接删除 django_migrations 表中的两条记录，再次执行 migrate 命令：

```
(django2-core-test) [root@master first_django]# python manage.py migrate book_sales
--plan
# 打印告警信息，可忽略
```

```
# ……
Planned operations:
shell_test.0001_initial
    Create model DjangoBooks
book_sales.0001_initial
    Create model BookSales
```

查看数据库，可以看到并没有生成迁移记录，说明程序在打印了迁移操作后就直接返回了。

第 5 处代码是迁移命令的核心，这里将生成相关模型类对应的表及迁移记录。它直接调用 MigrationExecutor 对象的 migrate()方法，因此继续追踪 migrate()方法的实现源码：

```
# 源码位置： django/db/migrations/executor.py
# ……

class MigrationExecutor:

    # ……

    def migrate(self, targets, plan=None, state=None, fake=False, fake_initial=False):
        """
        将计划迁移的文件映射到数据库中
        """
        # 如果没有创建 django_migrations 表，则创建它
        self.recorder.ensure_schema()

        # 如果 plan 输入为 None，则根据 targets 参数调用一次 migration_plan()方法获取 plan
        if plan is None:
            plan = self.migration_plan(targets)
        # 获取完整的迁移记录表，包括所有应用，这里 clean_start 为 True，即迁移记录设为空
        full_plan=self.migration_plan(self.loader.graph.leaf_nodes(), clean_start=True)

        # 这里非常重要，得到的 plan 其实是一组（迁移对象，True|False）这样的列表
        # 下面判断 plan 中所有元素的第 2 个字段是否全为 False 或者全为 True
        all_forwards = all(not backwards for mig, backwards in plan)
        all_backwards = all(backwards for mig, backwards in plan)

        if not plan:
            if state is None:
                state = self._create_project_state(with_applied_migrations=True)
        elif all_forwards == all_backwards:
            # 当 plan 中的元素的第 2 个字段既有 True，又有 False 时，表示有冲突，直接抛出错误
            raise InvalidMigrationPlan(
                "Migration plans with both forwards and backwards migrations "
                "are not supported. Please split your migration process into "
                "separate plans of only forwards OR backwards migrations.",
                plan
            )
```

```
        elif all_forwards:
            # 一般第一次执行迁移时，程序会到这里执行
            if state is None:
                state = self._create_project_state(with_applied_migrations=True)
            state = self._migrate_all_forwards(state, plan, full_plan, fake=fake,
fake_initial=fake_initial)
        else:
            state = self._migrate_all_backwards(plan, full_plan, fake=fake)

        self.check_replacements()

        return state
```

这里的逻辑代码从总体上看并不复杂，但需要理解 plan 的输出。下面简单介绍一下在第 3 处代码中得到的 plan 结果。前面分析过，plan 是一串迁移计划的组合，该迁移计划是一个二元组，即(迁移对象，True|False)。其中，当第 2 个元素为 True 时，表示该迁移文件（target）在迁移表中，且其子孙迁移节点(同一应用)也已经出现在记录表中(对于 target 第 2 个元素为 None 的情况先不讨论)。而当第 2 个元素为 False 时，表示该迁移文件并未在迁移表中出现过。当第一次执行迁移指令时，迁移文件（包括其依赖的迁移文件）得到的 plan=[(迁移对象 1，False), (迁移对象 2, False), ..., (迁移对象 n, False)]。此时，all_forwards 的值为 True，而 all_backwards 的值为 False。最终 Django 会调用 else all_forwars 分支下的 state = self._migrate_all_forwards()语句，该语句是迁移操作的核心。继续追踪 _migrate_all_forwards()方法的实现源码：

```
# 源码位置：django/db/migrations/executor.py
# ……

class MigrationExecutor:

    # ……

    def _migrate_all_forwards(self, state, plan, full_plan, fake, fake_initial):

        migrations_to_run = {m[0] for m in plan}
        for migration, _ in full_plan:
            if not migrations_to_run:
                # 没有迁移计划，直接 break
                break
            if migration in migrations_to_run:
                # 如果 migration 在待迁移表中，则代码继续向下执行
                if 'apps' not in state.__dict__:
                    if self.progress_callback:
                        self.progress_callback("render_start")
                    state.apps # Render all -- performance critical
                    if self.progress_callback:
                        self.progress_callback("render_success")
```

```
            # 最终针对该迁移文件执行迁移操作
            state = self.apply_migration(state, migration, fake=fake,
fake_initial=fake_initial)
            migrations_to_run.remove(migration)

      return state

   # ……
```

在上面的代码中，会遍历所有的迁移对象。如果迁移对象在迁移表中，则调用 apply_migration() 方法执行迁移操作。继续追踪 apply_migration()方法的实现源码：

```
# 源码位置： django/db/migrations/executor.py
# ……

class MigrationExecutor:

   # ……

   def apply_migration(self, state, migration, fake=False, fake_initial=False):
      """执行一次迁移操作"""

      # 先标记迁移记录为 False
      migration_recorded = False
      if self.progress_callback:
         self.progress_callback("apply_start", migration, fake)
      if not fake:
         if fake_initial:
            applied, state = self.detect_soft_applied(state, migration)
            if applied:
               fake = True
         if not fake:
            with self.connection.schema_editor(atomic=migration.atomic) as
schema_editor:
               # 将迁移文件中的内容按照约定规则更新到数据库中
               state = migration.apply(state, schema_editor)
               # 添加迁移记录到 django_migrations 表中
               self.record_migration(migration)
               # 标记迁移记录为 True
               migration_recorded = True
      if not migration_recorded:
         self.record_migration(migration)
      # 报告进度
      if self.progress_callback:
         # 回调，通知迁移完成
         self.progress_callback("apply_success", migration, fake)
      return state
```

从上面的代码可以看到真正操作数据库的动作：self.record_migration()方法用于更新迁移记录，它的实现比较简单，正是借助 2.4 节中介绍的 MigrationRecorder 对象完成的；apply()方法位于 Migration 类中，该方法通过遍历 Migration 对象的 operations 属性实现对数据库的映射，其实现的具体细节不再介绍，有兴趣的读者可以自行深入研究。

## 2.6 小结

本章选择了 4 个典型命令进行深入分析，先从基础的类与函数开始介绍，到后续完整追踪一个 Django 命令，如 makemigrations 命令，这其中涉及不少其他模块的类与函数，这部分内容将在第 3 章中讲解。

# 第 3 章 Django内置的ORM框架

本章将完整解析 Django 内置的 ORM 框架源码，从解析数据库配置信息开始到模型类的各种链式调用结束，帮助读者理解 Django 框架中模型层背后的实现原理。

## 3.1 读取 Django 项目的配置信息

数据库的配置信息位于 Django 项目的 settings.py 文件中，项目执行命令或者启动时会读取这里的配置信息。除此之外，在 Django 内部还有一个默认的全局配置文件：django/conf/global_settings.py。如果在项目的 settings.py 文件中没有找到数据库的配置信息，则使用 Django 内部默认的全局配置文件中的信息。这种配置模式在框架中十分常见，例如，著名的爬虫框架 Scrapy 便是如此。下面思考两个问题：

◎ Django 是如何识别项目的 settings.py 文件的？

◎ Django 内部是如何实现先读取 settings.py 文件中的配置信息，如果没有找到，再读取 global_settings.py 文件中的配置信息的？

接下来带着这些问题进入 Django 源码的 conf 目录下进行学习。该目录下的源码文件如下：

◎ conf/app_template 和 conf/project_template 目录：该目录下的文件分别为创建 Django 项目和 Django 应用的模板文件，在 2.2 节中曾介绍过。

◎ conf/local 目录：该目录下的文件为 Django 支持的语言集。

◎ conf/urls/__init__.py：对一些视图层的方法取别名。

◎ conf/global_settings.py：默认的全局配置文件。

◎ conf/__init__.py：读取配置的核心代码。

在对上面的源码文件进行梳理后，读者可以将学习重心放到 conf/__init__.py 文件中。在 conf/__init__.py 文件中一共定义了 4 个类，其中比较重要的是 LazySettings 类和 Settings 类。

## 3.1.1 Settings 类

Settings 类的定义如下：

```
# 源码位置：django/conf/__init__.py
# ……
from django.conf import global_settings
# ……

class Settings:
    def __init__(self, settings_module):
        # 设置对象在配置中出现的属性，注意关键点：dir()函数和setattr()函数
        for setting in dir(global_settings):
            if setting.isupper():
                setattr(self, setting, getattr(global_settings, setting))

        # 保存配置文件的模块路径
        self.SETTINGS_MODULE = settings_module

        # 导入该模块
        mod = importlib.import_module(self.SETTINGS_MODULE)

        tuple_settings = (
            "INSTALLED_APPS",
            "TEMPLATE_DIRS",
            "LOCALE_PATHS",
        )
        self._explicit_settings = set()

        # 读取载入配置的属性
        for setting in dir(mod):
            if setting.isupper():
                # 获取该配置的值
                setting_value = getattr(mod, setting)

                # 如果项目配置了tuple_settings中的配置，其值必须是元组或者列表形式，否则直接抛出异常
                if (setting in tuple_settings and
                        not isinstance(setting_value, (list, tuple))):
                    raise ImproperlyConfigured"The %s setting must be a list or a tuple. " %
setting)
                # 设置对象在配置中出现的属性，这会覆盖global_settings中的原属性值
                setattr(self, setting, setting_value)
                # 在集合中添加该属性
                self._explicit_settings.add(setting)

        # 如果对象中得到的SECRET_KEY属性为空字符串，则抛出异常
        if not self.SECRET_KEY:
            raise ImproperlyConfigured("The SECRET_KEY setting must not be empty.")
```

```
        # 打印一些版本的告警信息
        # ……

        # 设置时区
        if hasattr(time, 'tzset') and self.TIME_ZONE:
            zoneinfo_root = Path('/usr/share/zoneinfo')
            zone_info_file = zoneinfo_root.joinpath(*self.TIME_ZONE.split('/'))
            if zoneinfo_root.exists() and not zone_info_file.exists():
                raise ValueError("Incorrect timezone setting: %s" % self.TIME_ZONE)
            os.environ['TZ'] = self.TIME_ZONE
            time.tzset()

    # ……

    def __repr__(self):
        return '<%(cls)s "%(settings_module)s">' % {
            'cls': self.__class__.__name__,
            'settings_module': self.SETTINGS_MODULE,
        }
```

从 Settings 类中可以直接得到第 2 个问题的答案。在该类的初始化方法中，先在全局的配置模块（global_settings 模块）中遍历大写的属性及其值并添加到该 Settings 对象中。这里用到了 dir()和 setattr()这两个非常常见且十分重要的函数，前者表示获取模块的属性，后者表示给对象设置属性及属性值。这样 Settings 对象就具备了 global_settings 模块中的属性，并且其值和 global_settings 模块中的相同。之后再导入传入的 settings_module 模块，按同样的方式设置该 Settings 对象的属性。如果 settings_module 模块和 global_settings 模块中的属性有交叉，则以 settings_module 模块的为准（因为是后设置的）。接着是一些必须要设置的属性值，比如 SECRET_KEY 值等。如果不在 settings_module 模块中设置，即默认为空字符串，则会直接抛出异常。下面看一下该类的一个使用示例，代码如下：

```
(django2-core-test) [root@master first_django]# cat first_django/settings.py | egrep
"DEBUG|SECRET_KEY|FILE_UPLOAD_MAX_MEMORY_SIZE"
SECRET_KEY = 'd(ld-9zgy=cbhqb_^4&ityb8eoif1#p5@zw8ig5-ni8gxe-)f7'
DEBUG = True
(django2-core-test) [root@master first_django]# cat
~/.pyenv/versions/django2-core-test/lib/python3.8/site-packages/django/conf/global_s
ettings.py | egrep "DEBUG|SECRET|FILE_UPLOAD_MAX_MEMORY_SIZE"
DEBUG = False
DEBUG_PROPAGATE_EXCEPTIONS = False
#   * See debug comments, when DEBUG is true
SECRET_KEY = ''
FILE_UPLOAD_MAX_MEMORY_SIZE = 2621440  # i.e. 2.5 MB
(django2-core-test) [root@master first_django]# python
Python 3.8.6 (default, Oct 18 2020, 15:33:08)
[GCC 4.8.5 20150623 (Red Hat 4.8.5-39)] on linux
Type "help", "copyright", "credits" or "license" for more information.
```

```
>>> from django.conf import Settings
>>> settings = Settings("first_django.settings")
>>> settings.DEBUG
True
>>> settings.SECRET_KEY
'd(ld-9zgy=cbhqb_^4&ityb8eoif1#p5@zw8ig5-ni8gxe-)f7'
>>> settings.FILE_UPLOAD_MAX_MEMORY_SIZE
2621440
>>>
```

上面的示例演示了在配置文件中定义的三个变量值：DEBUG、SECRET_KEY 和 FILE_UPLOAD_MAX_MEMORY_SIZE。其中，前两个变量在 first_django.settings 模块中已定义，而第三个变量没有在 first_django.settings 模块中定义。因此，最终得到的 Settings 对象的前两个属性值为 first_django.settings 模块中定义的变量值，而第三个属性值为 django.conf.global_settings 模块中定义的变量值。

### 3.1.2 LazySettings 类

LazySettings 类的定义如下：

```
# 源码位置：django/conf/__init__.py
# ……
from django.conf import global_settings
from django.utils.functional import LazyObject, empty
# ……

class LazySettings(LazyObject):

    def _setup(self, name=None):
        # 核心代码，先从环境变量中取 settings_module 的值
        settings_module = os.environ.get(ENVIRONMENT_VARIABLE)
        if not settings_module:
            #如果在环境变量中未设置该值，直接抛错
            desc = ("setting %s" % name) if name else "settings"
            raise ImproperlyConfigured(
                "Requested %s, but settings are not configured. "
                "You must either define the environment variable %s "
                "or call settings.configure() before accessing settings."
                % (desc, ENVIRONMENT_VARIABLE))
        # 从这里可以看到，该对象的_wrapped 属性值为 Settings 对象
        self._wrapped = Settings(settings_module)

    def __repr__(self):

        if self._wrapped is empty:
            return '<LazySettings [Unevaluated]>'
```

```
        return '<LazySettings "%(settings_module)s">' % {
            'settings_module': self._wrapped.SETTINGS_MODULE,
        }

    def __getattr__(self, name):

        # 懒加载模式的体现，只有当调用某个属性时，才获取并设置其属性值
        if self._wrapped is empty:
            self._setup(name)
        val = getattr(self._wrapped, name)
        self.__dict__[name] = val
        return val

    def __setattr__(self, name, value):

        if name == '_wrapped':
            self.__dict__.clear()
        else:
            self.__dict__.pop(name, None)
        super().__setattr__(name, value)

    def __delattr__(self, name):

        super().__delattr__(name)
        self.__dict__.pop(name, None)

    def configure(self, default_settings=global_settings, **options):

        if self._wrapped is not empty:
            raise RuntimeError('Settings already configured.')
        holder = UserSettingsHolder(default_settings)
        for name, value in options.items():
            setattr(holder, name, value)
        self._wrapped = holder

    @property
    def configured(self):
        return self._wrapped is not empty

    # ......
```

上面的代码非常有意思，它展示了一种懒加载的代码模式。Settings 类在实例化时会设置很多属性及属性值，LazySettings 类则是在第一次访问该对象的某个属性时才去设置相应的属性及属性值。这种模式是如何做到的呢？实际上，在调用 LazySettings 类的某个属性时，会进入类的魔法函数__setattr__()中。该魔法函数先校验该对象的_wrapped 属性，在第一次进入该魔法函数时，_wrapped 属性值为空对象，即 self._wrapped is empty 语句为 True,，于是调用_setup()方法。而_setup()方法会先

从环境变量中读取 settings_module 的值，接着根据 settings_module 的值实例化 Settings 类，并赋给 LazySettings 对象的_wrapped 属性。在这个赋值语句中，LazySettings 类加载了 settings_module 和 global_settings 中的所有配置信息并保存在了_wrapped 属性中。接下来在获取_wrapped 属性值时，就不会加载所有的配置了，而是直接从_wrapped 属性中获取。以下是对该类的一些操作示例：

```
(django2-core-test) [root@master first_django]# python
Python 3.8.6 (default, Oct 18 2020, 15:33:08)
[GCC 4.8.5 20150623 (Red Hat 4.8.5-39)] on linux
Type "help", "copyright", "credits" or "license" for more information.
>>> from django.conf import LazySettings
>>> from django.utils.functional import empty
>>> import os
>>> os.environ.setdefault('DJANGO_SETTINGS_MODULE', 'first_django.settings')
'first_django.settings'
>>> settings = LazySettings()
>>> settings._wrapped is empty
True
>>> settings.DEBUG
True
>>> settings._wrapped is empty
False
>>> settings._wrapped
<Settings "first_django.settings">
>>> settings._wrapped.DEBUG
True
>>> settings._wrapped.SECRET_KEY
'd(ld-9zgy=cbhqb_^4&ityb8eoif1#p5@zw8ig5-ni8gxe-)f7'
>>> settings.FILE_UPLOAD_MAX_MEMORY_SIZE
2621440
```

上面的代码演示了 LazySettings 对象在第一次和后续调用属性时_wrapped 属性值的区别。

在 2.3 节介绍 shell 命令的执行过程时曾遇到这个设置环境变量的语句，它出现在 Django 项目的 manage.py 文件中。下面是 first_django 项目的 manage.py 文件中的内容：

```
# 源码位置：first_django/manage.py
# ……

def main():
    os.environ.setdefault('DJANGO_SETTINGS_MODULE', 'first_django.settings')
    try:
        from django.core.management import execute_from_command_line
    except ImportError as exc:
        raise ImportError(
            "Couldn't import Django. Are you sure it's installed and "
            "available on your PYTHONPATH environment variable? Did you "
            "forget to activate a virtual environment?"
```

```
        ) from exc
    execute_from_command_line(sys.argv)


if __name__ == '__main__':
    main()
```

实际上，在 manage.py 文件中设置的环境变量正是用在配置变量读取上的，即这里的 LazySettings 类中。在__init__.py 文件中还有一条非常关键的语句：

```
# 源码位置：django/conf/__init__.py
# ……

settings = LazySettings()
```

有了这条语句，LazySettings 类的演示操作就非常简单了。这里将演示导入该对象并读取 first_django 项目中关于数据库的配置信息，这和后面要讲的内容息息相关：

```
(django2-core-test) [root@master first_django]# python
Python 3.8.6 (default, Oct 18 2020, 15:33:08)
[GCC 4.8.5 20150623 (Red Hat 4.8.5-39)] on linux
Type "help", "copyright", "credits" or "license" for more information.
>>> import os
>>> os.environ.setdefault('DJANGO_SETTINGS_MODULE', 'first_django.settings')
'first_django.settings'
>>> from django.conf import settings
>>> settings.DATABASES['default']
{'ENGINE': 'django.db.backends.mysql', 'NAME': 'django_book', 'HOST': '192.168.88.206',
'PORT': 3306, 'USER': 'store', 'PASSWORD': 'store.1234!'}
```

这样即可读取 settings.py 文件中的配置信息了。至此，前文提出的两个问题就迎刃而解了。

## 3.2 ORM 框架的底层核心

本节探索 Django 中 ORM 框架的底层核心，这里涉及的源码主要位于 django/db 目录中。前文曾介绍过，想要在 settings.py 文件中使用 MySQL 数据库，就必须安装 mysqlclient 模块，这是否意味着 Django 就是单纯封装 mysqlclient 模块来进行数据库的交互的？

### 3.2.1 mysqlclient 模块中的常用方法

为了更好地理解 ORM 框架的源码，建议读者必须掌握 mysqlclient 模块的用法。下面看一个简单的示例：

```
(django2-core-test) [root@master first_django]# python
Python 3.8.6 (default, Oct 18 2020, 15:33:08)
```

```
[GCC 4.8.5 20150623 (Red Hat 4.8.5-39)] on linux
Type "help", "copyright", "credits" or "license" for more information.
>>> import MySQLdb
# 如果要插入中文，一定要带上 use_unicode 和 charset 参数
>>> conn = MySQLdb.connect(host='192.168.88.206', port=3306, user='store',
passwd='store.1234!', db='django_book', use_unicode=1, charset='utf8')
>>> sql = "insert into django_books(book_name, author, sex, price, isbn,
publish_date)values('Django 源码笔记', 'spyinx', 1, 119, 223454566, '2021-09-10')"
>>> cur = conn.cursor()
>>> cur.execute(sql)
1
>>> conn.commit()
>>> sql = 'select * from django_books'
>>> cur.execute(sql)
1
>>> cur.fetchone()
(4, DDjango 源码笔记', 'spyinx', 1, 119.0, '223454566', datetime.datetime(2021, 9, 10, 0,
0))
>>> cur.fetchone()
>>>
```

上面的代码演示了基于 mysqlclient 模块对 MySQL 数据库进行的新增和查询操作。在 mysqlclient 模块中，最常用的几个方法如下：

◎ MySQLdb.connect()方法：连接 MySQL 数据库，在这里输入 MySQL 数据库的地址、端口、账号和密码，以及要使用的数据库。

◎ conn.cursor()：创建游标，固定做法。

◎ cursor.execute()：通过游标的 execute()方法可以执行 SQL 语句，其返回值表示的是操作的记录数。这里的 cursor 表示的是上一步创建的游标。

◎ conn.commit()：提交数据库操作的动作，比如新增数据、修改数据和删除数据等。一定要使用 commit()方法提交，否则前面的操作将不生效。如果想在每次调用 execute()方法后都自动提交，则可以使用 conn.autocommit(True)语句。

◎ cursor.fetchone()：只取一条记录，游标后移一位。

◎ cursor.fetchmany()：取多条记录，参数为取的记录数，在执行后游标移到相应位置。

◎ cursor.fetchall()：取出 SQL 执行的所有记录，游标移至末尾。

### 3.2.2 ORM 框架的源码解析

在了解了 mysqlclient 模块的常用方法后，就可以开始全面学习 ORM 框架了，先来看 django/db/__init__.py 文件中的源码内容。在 2.4 节介绍 makemigrations 命令时曾出现导入 connections 变量的语句，当时并未详细说明，只知道它可以用来保存各种数据库的连接信息，下面详细分析这

个变量：

```
# 源码位置：django/db/__init__.py
# ……
# 非常重要
connections = ConnectionHandler()
# ……
```

从上面的语句可知，connections 变量是一个 ConnectionHandler 对象。继续追踪 ConnectionHandler 对象的源码，它位于 django/db/utils.py 文件中，内容如下：

```
# 源码位置：django/db/utils.py
# ……
from django.conf import settings
# ……
DEFAULT_DB_ALIAS = 'default'
DJANGO_VERSION_PICKLE_KEY = '_django_version'
# ……

class ConnectionHandler:
   def __init__(self, databases=None):

      self._databases = databases
      self._connections = local()

   @cached_property
   def databases(self):
      if self._databases is None:
         # 获取 settings 模块中的 DATABASES 属性值，前文介绍过
         self._databases = settings.DATABASES
      # 如果在配置中没有设置 DATABASES 属性值，就添加一个默认值
      if self._databases == {}:
         self._databases = {
            DEFAULT_DB_ALIAS: {
               'ENGINE': 'django.db.backends.dummy',
            },
         }
      # 如果在配置中没有'default'字段，直接抛出异常
      if DEFAULT_DB_ALIAS not in self._databases:
         raise ImproperlyConfigured("You must define a '%s' database." %
DEFAULT_DB_ALIAS)
      # 如果 DATABASES['default']为空，赋给默认值
      if self._databases[DEFAULT_DB_ALIAS] == {}:
         self._databases[DEFAULT_DB_ALIAS]['ENGINE'] = 'django.db.backends.dummy'
      return self._databases

   def ensure_defaults(self, alias):
```

```
        # 给 alias 对应的数据库信息添加一些额外的默认信息
        try:
            conn = self.databases[alias]
        except KeyError:
            raise ConnectionDoesNotExist("The connection %s doesn't exist" % alias)

        # 由于 conn 为字典，所以这些更新会被反映到 self.databases[alias]中
        conn.setdefault('ATOMIC_REQUESTS', False)
        conn.setdefault('AUTOCOMMIT', True)
        conn.setdefault('ENGINE', 'django.db.backends.dummy')
        if conn['ENGINE'] == 'django.db.backends.' or not conn['ENGINE']:
            conn['ENGINE'] = 'django.db.backends.dummy'
        conn.setdefault('CONN_MAX_AGE', 0)
        conn.setdefault('OPTIONS', {})
        conn.setdefault('TIME_ZONE', None)
        for setting in ['NAME', 'USER', 'PASSWORD', 'HOST', 'PORT']:
            conn.setdefault(setting, '')

    def prepare_test_settings(self, alias):

        try:
            conn = self.databases[alias]
        except KeyError:
            raise ConnectionDoesNotExist("The connection %s doesn't exist" % alias)

        test_settings = conn.setdefault('TEST', {})
        for key in ['CHARSET', 'COLLATION', 'NAME', 'MIRROR']:
            test_settings.setdefault(key, None)

    def __getitem__(self, alias):
        if hasattr(self._connections, alias):
            return getattr(self._connections, alias)

        self.ensure_defaults(alias)
        self.prepare_test_settings(alias)
        db = self.databases[alias]
        # 导入数据库配置的引擎模块
        backend = load_backend(db['ENGINE'])
        conn = backend.DatabaseWrapper(db, alias)
        setattr(self._connections, alias, conn)
        return conn

    def __setitem__(self, key, value):
        setattr(self._connections, key, value)

    def __delitem__(self, key):
        delattr(self._connections, key)
```

```
    def __iter__(self):
        return iter(self.databases)

    def all(self):
        return [self[alias] for alias in self]

    def close_all(self):
        for alias in self:
            try:
                connection = getattr(self._connections, alias)
            except AttributeError:
                continue
            connection.close()
```

ConnectionHandler 类的源码并不复杂，其中定义了许多魔法函数。只有理解这些魔法函数的功能，才能更好地理解 ConnectionHandler 类。下面先来看看 databases()方法的实现代码，该方法的一个核心就是调用 LazySetting 对象，即源码中的 settings 模块，借助该对象可以获取项目中配置的 DATABASES 信息，具体操作如下：

```
(django2-core-test) [root@master first_django]# python
Python 3.8.6 (default, Oct 18 2020, 15:33:08)
[GCC 4.8.5 20150623 (Red Hat 4.8.5-39)] on linux
Type "help", "copyright", "credits" or "license" for more information.
>>> import os
>>> os.environ.setdefault('DJANGO_SETTINGS_MODULE', 'first_django.settings')
'first_django.settings'
>>> from django.db.utils import ConnectionHandler
>>> connections = ConnectionHandler()
>>> connections._databases
>>> connections.databases
{'default': {'ENGINE': 'django.db.backends.mysql', 'NAME': 'django_book', 'HOST':
'192.168.88.206', 'PORT': 3306, 'USER': 'store', 'PASSWORD': 'store.1234!'}}
>>> connections._databases
{'default': {'ENGINE': 'django.db.backends.mysql', 'NAME': 'django_book', 'HOST':
'192.168.88.206', 'PORT': 3306, 'USER': 'store', 'PASSWORD': 'store.1234!'}}
```

提示：这里有两个地方需要注意。

（1）@cached_property 装饰器可以将只有 self 参数的方法转换成缓存在实例中的属性。所以这里不能使用 connections.databases()这样的形式访问 databases。

（2）这里使用的是 Python 直接进入命令行而不是 python manage.py shell，所以要一开始就配置文件的环境变量，否则无法导入 settings 模块。

此外，ensure_defaults()函数和 prepare_test_settings()函数只是单纯地给 databases[alias]添加一些额外信息。接着上面的交互模式继续执行如下操作：

```
>>> connections.ensure_defaults('default')
>>> connections._databases
{'default': {'ENGINE': 'django.db.backends.mysql', 'NAME': 'django_book', 'HOST':
'192.168.88.206', 'PORT': 3306, 'USER': 'store', 'PASSWORD': 'store.1234!',
'ATOMIC_REQUESTS': False, 'AUTOCOMMIT': True, 'CONN_MAX_AGE': 0, 'OPTIONS': {},
'TIME_ZONE': None}}
>>> connections.prepare_test_settings('default')
>>> connections._databases
{'default': {'ENGINE': 'django.db.backends.mysql', 'NAME': 'django_book', 'HOST':
'192.168.88.206', 'PORT': 3306, 'USER': 'store', 'PASSWORD': 'store.1234!',
'ATOMIC_REQUESTS': False, 'AUTOCOMMIT': True, 'CONN_MAX_AGE': 0, 'OPTIONS': {},
'TIME_ZONE': None, 'TEST': {'CHARSET': None, 'COLLATION': None, 'NAME': None, 'MIRROR':
None}}}
```

接下来解析魔法函数，这里先给出一些基础代码，在理解了这些基础代码后就可以继续学习上面的魔法函数了：

```
(django2-core-test) [root@master first_django]# cat test_item.py
class Test:

   def __init__(self, data):
      self.data = data

   def __getitem__(self, alias):
      print('调用__getitem__()函数')
      return self.data[alias]

   def __setitem__(self, key, value):
      print('调用__setitem__()函数')
      self.data[key] = value

data = {}
test = Test(data)

test['x'] = 'test'
print(test.data)

print(test['x'])
```

直接使用 Python 解释器执行该脚本，结果如下：

```
(django2-core-test) [root@master first_django]# python test_item.py
调用__setitem__()函数
{'x': 'test'}
调用__getitem__()函数
test
```

从上面的代码可以看到，魔法函数__getitem__()在类对象被索引时调用，而__setitem__()函数在给类对象的索引赋值时调用。接下来看 ConnectionHandler 类中的__getitem__()函数，它的执行逻辑如下：

◎ 如果_connections 属性有 alias 索引的值，则直接获取其索引值并返回，否则执行下面的操作。
◎ 调用 ensure_defaults()函数和 prepare_test_settings()函数，给对应的数据库设置一些默认信息，正如上面演示的那样。
◎ 导入数据库配置的引擎模块，得到 backend 模块。后面在分析 load_backend()函数时可以得到 backend 模块的默认具体路径。
◎ 根据数据库信息及导入的 backend 模块，得到一个该数据库的连接信息 conn。
◎ 将得到的数据库连接信息 conn 设置到_connections 属性的 alias 索引中。

为了更好地理解__getitem__()函数中的语句，先来看看 load_backend()函数的源码：

```
# 源码位置：django/db/utils.py
# ……

def load_backend(backend_name):
    # 这个 backend 在 Django 1.9 中进行了重命名
    if backend_name == 'django.db.backends.postgresql_psycopg2':
        backend_name = 'django.db.backends.postgresql'

    try:
        return import_module('%s.base' % backend_name)
    except ImportError as e_user:
        # 导入异常处理
        # ……
```

在默认情况下加载的数据库引擎模块路径为 django.db.backends.mysql.base，最终得到的 conn 就是在该模块下定义的 DatabaseWrapper 对象。进入该模块的源码，就可以看到 ORM 框架的底层核心了：

```
# 源码位置： django/db/mysql/base.py
# ……

try:
    # 给导入的 MySQLdb 模块取一个别名
    import MySQLdb as Database
except ImportError as err:
    raise ImproperlyConfigured(
        'Error loading MySQLdb module.\n'
        'Did you install mysqlclient?'
    ) from err
```

```
from MySQLdb.constants import CLIENT, FIELD_TYPE             # isort:skip
from MySQLdb.converters import conversions

# ......

class CursorWrapper:

    # ......

    def __init__(self, cursor):
        self.cursor = cursor

    def execute(self, query, args=None):
        try:
            return self.cursor.execute(query, args)
        except Database.OperationalError as e:
            # 继续抛出异常
            # ......

    def executemany(self, query, args):
        try:
            return self.cursor.executemany(query, args)
        except Database.OperationalError as e:
            # 继续抛出异常
            # ......

    def __getattr__(self, attr):
        return getattr(self.cursor, attr)

    def __iter__(self):
        return iter(self.cursor)

class DatabaseWrapper(BaseDatabaseWrapper):

    # ......
    Database = Database
    # ......

    def get_connection_params(self):
        #以下获取配置中的数据库信息，将其作为 MySQLdb.connect()方法的参数
        kwargs = {
            'conv': django_conversions,
            'charset': 'utf8',
        }
        settings_dict = self.settings_dict
        if settings_dict['USER']:
            kwargs['user'] = settings_dict['USER']
        if settings_dict['NAME']:
```

```
            kwargs['db'] = settings_dict['NAME']
        if settings_dict['PASSWORD']:
            kwargs['passwd'] = settings_dict['PASSWORD']
        if settings_dict['HOST'].startswith('/'):
            kwargs['unix_socket'] = settings_dict['HOST']
        elif settings_dict['HOST']:
            kwargs['host'] = settings_dict['HOST']
        if settings_dict['PORT']:
            kwargs['port'] = int(settings_dict['PORT'])
        kwargs['client_flag'] = CLIENT.FOUND_ROWS
        options = settings_dict['OPTIONS'].copy()
        isolation_level = options.pop('isolation_level', 'read committed')
        if isolation_level:
            isolation_level = isolation_level.lower()
            if isolation_level not in self.isolation_levels:
                # 抛出异常
                # ……
        self.isolation_level = isolation_level
        kwargs.update(options)
        return kwargs

    def get_new_connection(self, conn_params):
        # 其实就是MySQLdb.connect()方法
        return Database.connect(**conn_params)

    def init_connection_state(self):
        assignments = []
        if self.features.is_sql_auto_is_null_enabled:
            assignments.append('SET SQL_AUTO_IS_NULL = 0')

        if self.isolation_level:
            assignments.append('SET SESSION TRANSACTION ISOLATION LEVEL %s' %
                               self.isolation_level.upper())

        if assignments:
            with self.cursor() as cursor:
                cursor.execute('; '.join(assignments))

    def create_cursor(self, name=None):
        # 创建游标，与仅使用mysqlclient模块的方法一致
        cursor = self.connection.cursor()
        # 对mysqlclient模块中的游标对象进行进一步封装
        return CursorWrapper(cursor)

    def _rollback(self):
        try:
            BaseDatabaseWrapper._rollback(self)
        except Database.NotSupportedError:
```

```
        pass

    def _set_autocommit(self, autocommit):
        with self.wrap_database_errors:
            # 设置自动 commit
            self.connection.autocommit(autocommit)

    def disable_constraint_checking(self):

        self.cursor().execute('SET foreign_key_checks=0')
        return True

    def enable_constraint_checking(self):

        self.needs_rollback, needs_rollback = False, self.needs_rollback
        try:
            self.cursor().execute('SET foreign_key_checks=1')
        finally:
            self.needs_rollback = needs_rollback

    # ……

    def is_usable(self):
        try:
            # 简单调用 ping()方法，探测数据库是否可用
            self.connection.ping()
        except Database.Error:
            return False
        else:
            return True

    @cached_property
    def mysql_server_info(self):
        with self.temporary_connection() as cursor:
            cursor.execute('SELECT VERSION()')
            # 返回第一条结果
            return cursor.fetchone()[0]

    @cached_property
    def mysql_version(self):
        match = server_version_re.match(self.mysql_server_info)
        if not match:
            # 抛出异常
            # ……
        return tuple(int(x) for x in match.groups())

    @cached_property
```

```
    def mysql_is_mariadb(self):
        return 'mariadb' in self.mysql_server_info.lower()
```

在看完上面的代码后，可以发现这里的 Database 就是 MySQLdb 模块。再来看 DatabaseWrapper 类中的 create_cursor()函数，该函数会调用类对象的 connection 属性的 cursor()函数。而 connection 属性在当前类中并没有定义，因此需要继续查找其父类 BaseDatabaseWrapper 的实现源码：

```
# 源码位置：django/db/backends/base.py
# ……

class BaseDatabaseWrapper:
    # ……

    def __init__(self, settings_dict, alias=DEFAULT_DB_ALIAS):

        self.connection = None

        # ……

    def connect(self):
        """连接数据库"""

        # 检查配置的有效性
        self.check_settings()

        # ……

        # 建立连接
        conn_params = self.get_connection_params()
        self.connection = self.get_new_connection(conn_params)
        self.set_autocommit(self.settings_dict['AUTOCOMMIT'])
        self.init_connection_state()
        connection_created.send(sender=self.__class__, connection=self)
        # ……

    def check_settings(self):
        if self.settings_dict['TIME_ZONE'] is not None:
            # 抛出时钟异常情况
            # ……

    def ensure_connection(self):
        """Guarantee that a connection to the database is established."""
        if self.connection is None:
            with self.wrap_database_errors:
                self.connect()

    # ……
```

从父类 BaseDatabaseWrapper 的源码可以看到，首先初始化__init__()函数中的设置 self.connection=None；其次在 connect()函数中调用了 self.get_connection_params()函数，以获取数据库的连接参数（如 MySQL 服务地址、端口、账号及密码等）；然后调用 get_new_connection()函数获取连接对象并赋给 self.connection；最后回到 django/db/mysql/base.py 中继续学习 DatabaseWrapper 类。由于在 DatabaseWrapper 类中并没有 connect()函数，因此只有调用 connect()函数（在父类中定义的该方法），才能给实例的 connection 属性赋值，而该值正是 MySQLdb.connect()方法返回的数据库连接对象。

现在再来看在 DatabaseWrapper 类中定义的 create_cursor()函数，在该函数中得到的 cursor 对象正是前面得到的数据库连接对象调用 cursor()方法得到的结果，只不过其返回的结果对该游标对象进行了封装，得到 CursorWrapper 对象。而 CursorWrapper 对象的核心正是这个 cursor 对象。通过该类编写的魔法函数，可知这个 CursorWrapper 对象和 mysqlclient 中的 cursor 对象的功能几乎一致，只不过增加了对 execute()函数和 executemany()函数的异常处理。

### 3.2.3 DatabaseWrapper 类的实战案例

接下来介绍 Python 交互模式下 DatabaseWrapper 类的实战案例，这里采用先分析结果后执行验证的顺序进行讲解。此次演示只是牛刀小试，后面的章节将在本节的基础上详细分析 ORM 框架的操作原理。

#### 创建 DatabaseWrapper 对象并调用 connect()函数

前文在演示 ConnectionHandler 类的使用时，得到了一个字典结果：

```
>>> connections._databases
{'default': {'ENGINE': 'django.db.backends.mysql', 'NAME': 'django_book', 'HOST':
'192.168.88.206', 'PORT': 3306, 'USER': 'store', 'PASSWORD': 'store.1234!',
'ATOMIC_REQUESTS': False, 'AUTOCOMMIT': True, 'CONN_MAX_AGE': 0, 'OPTIONS': {},
'TIME_ZONE': None, 'TEST': {'CHARSET': None, 'COLLATION': None, 'NAME': None, 'MIRROR':
None}}}
```

接下来，使用保存了数据库信息的字典数据初始化 DatabaseWrapper 类，并比较该类中通过 connections 属性和 MySQLdb.connect()方法得到的连接对象。从前面的源码分析中可知，它们都属于同一个类：

```
(django2-core-test) [root@master first_django]# python
Python 3.8.6 (default, Oct 18 2020, 15:33:08)
[GCC 4.8.5 20150623 (Red Hat 4.8.5-39)] on linux
Type "help", "copyright", "credits" or "license" for more information.
>>> databases = {'default': {'ENGINE': 'django.db.backends.mysql', 'NAME': 'django_book',
'HOST': '192.168.88.206', 'PORT': 3306, 'USER': 'store', 'PASSWORD': 'store.1234!',
'ATOMIC_REQUESTS': False, 'AUTOCOMMIT': True, 'CONN_MAX_AGE': 0, 'OPTIONS': {},
```

```
'TIME_ZONE': None, 'TEST': {'CHARSET': None, 'COLLATION': None, 'NAME': None, 'MIRROR':
None}}}
>>> from django.db.backends.mysql.base import DatabaseWrapper
>>> alias = 'default'
>>> db = databases[alias]
>>> wrapper = DatabaseWrapper(db, alias)
# 一定要导入环境变量才能保证不出错，前面已多次强调
>>> import os
>>> os.environ.setdefault('DJANGO_SETTINGS_MODULE', 'first_django.settings')
'first_django.settings'
# 连接
>>> wrapper.connect()
# 得到 connection 属性
>>> wrapper.connection
<_mysql.connection open to '192.168.88.206' at 0x1618f10>
# 获取连接数据库的相关参数
>>> params = wrapper.get_connection_params()
>>> import MySQLdb
>>> conn = MySQLdb.connect(**params)
>>> conn
<_mysql.connection open to '192.168.88.206' at 0x2472820>
>>> type(conn)
<class 'MySQLdb.connections.Connection'>
```

继续调用 mysql_version()方法。注意，由于该方法前面加了@cached_property 装饰器，所以只需用访问属性的方式即可调用。那么，wrapper.mysql_version 的结果应该是什么呢？从字面的含义很容易猜出应该是该数据库的版本信息。mysql_version()方法最终解析的是 mysql_server_info()方法的结果。而 mysql_server_info()方法的逻辑非常简单：先获取游标，再执行 SELECT VERSION()方法，最后通过游标的 fetchone()方法获取第 1 条结果，同时取得结果的第 1 个元素。下面使用 MySQLdb 模块进行测试：

```
>>> import MySQLdb
>>> conn = MySQLdb.connect(host='192.168.88.206', port=3006, user='store',
password='store.1234!')
>>> cursor = conn.cursor()
>>> cursor.execute('SELECT VERSION()')
1
>>> data = cursor.fetchone()
>>> data
('5.7.18-1-log',)
>>> data[0]
'5.7.18-1-log'
```

上面演示的是 mysql_server_info()方法的模拟结果。接下来看看在 mysql_version()方法中对该结果的加工操作。接着上面的交互模式，继续执行如下操作：

```
>>> import re
>>> server_version_re = re.compile(r'(\d{1,2})\.(\d{1,2})\.(\d{1,2})')
>>> match = server_version_re.match(data[0])
>>> match
<re.Match object; span=(0, 6), match='5.7.18'>
>>> tuple(int(x) for x in match.groups())
(5, 7, 18)
```

从上面的代码可以看到，wrapper.mysql_version 的结果应该为(5,7,18)，是由 MySQL 的三个版本号组成的元组：

```
>>> wrapper.mysql_version
(5, 7, 18)
```

接下来继续分析 cursor 的来龙去脉，它是 mysqlclient 模块中的游标对象吗？分析一下源码就清楚了：

```
# 源码位置： django/db/backends/base/base.py
# ……

class BaseDatabaseWrapper:

   # ……

   def _prepare_cursor(self, cursor):
      """
      验证连接是否可用，然后获取封装后的数据库游标对象
      """
      self.validate_thread_sharing()
      if self.queries_logged:
         wrapped_cursor = self.make_debug_cursor(cursor)
      else:
         wrapped_cursor = self.make_cursor(cursor)
      # 4. 这里得到的 cursor 是 mysqlclient 模块中的游标对象
      # 不过最终返回的是经过 Django 封装的游标，它与 mysqlclient 模块中的游标功能类似
      return wrapped_cursor

   def _cursor(self, name=None):
      # 确保数据库连接成功
      self.ensure_connection()
      with self.wrap_database_errors:
         # 3. 注意这里调用的 create_cursor()方法在继承类中已经实现
         return self._prepare_cursor(self.create_cursor(name))

   # ……

   def cursor(self):
      """Create a cursor, opening a connection if necessary."""
```

```
        # 2. 调用 self._cursor()得到 cursor
        return self._cursor()

    def make_cursor(self, cursor):
        """Create a cursor without debug logging."""
        # 5. 最终返回的是在 utils 模块中定义的 CursorWrapper 对象，
        # 通过前面的分析可知，该对象和 mysqlclient 模块中的游标对象并无太大区别
        return utils.CursorWrapper(cursor, self)

    @contextmanager
    def temporary_connection(self):

        must_close = self.connection is None
        try:
            # 1. 调用 self.cursor()方法得到 cursor
            with self.cursor() as cursor:
                # 使用 yield 返回该 cursor
                yield cursor
        finally:
            if must_close:
                self.close()

    # ……
```

**注意**，在上面的代码中，使用数字 1～5 简单标记了从 temporary_connection()方法开始的代码执行顺序。

上面的逻辑关系比较清楚，最终在 mysql_server_info()方法中得到的 cursor 并不是 mysqlclient 模块中的游标对象，而是由 Django 封装的 CursorWrapper 对象（和前面提到的 CursorWrapper 对象不同，前者是在 django/db/utils.py 中定义的），该对象是在 django/db/backends/utils.py 中定义的：

```
# 源码位置： django/db/backends/utils.py
# ……

class CursorWrapper:
    def __init__(self, cursor, db):
        # 这个 cursor 就是 mysqlclient 模块中的游标对象
        self.cursor = cursor
        # BaseDatabaseWrapper 对象或其子类对象
        self.db = db

    WRAP_ERROR_ATTRS = frozenset(['fetchone', 'fetchmany', 'fetchall', 'nextset'])

    def __getattr__(self, attr):
        # 关键代码
        cursor_attr = getattr(self.cursor, attr)
        if attr in CursorWrapper.WRAP_ERROR_ATTRS:
```

```
            return self.db.wrap_database_errors(cursor_attr)
        else:
            return cursor_attr

    def __iter__(self):
        with self.db.wrap_database_errors:
            yield from self.cursor

    def __enter__(self):
        return self

    def __exit__(self, type, value, traceback):
        try:
            self.close()
        except self.db.Database.Error:
            pass

    def callproc(self, procname, params=None, kparams=None):
        if kparams is not None and not self.db.features.supports_callproc_kwargs:
            raise NotSupportedError(
                'Keyword parameters for callproc are not supported on this '
                'database backend.'
            )
        self.db.validate_no_broken_transaction()
        with self.db.wrap_database_errors:
            if params is None and kparams is None:
                return self.cursor.callproc(procname)
            elif kparams is None:
                return self.cursor.callproc(procname, params)
            else:
                params = params or ()
                return self.cursor.callproc(procname, params, kparams)

    def execute(self, sql, params=None):
        return self._execute_with_wrappers(sql, params, many=False,
executor=self._execute)

    def executemany(self, sql, param_list):
        return self._execute_with_wrappers(sql, param_list, many=True,
executor=self._executemany)

    def _execute_with_wrappers(self, sql, params, many, executor):
        # 游标对象的核心调用方法
        context = {'connection': self.db, 'cursor': self}
        for wrapper in reversed(self.db.execute_wrappers):
            executor = functools.partial(wrapper, executor)
        # 最终调用传过来的 executor 参数指定的方法
        return executor(sql, params, many, context)
```

```
    def _execute(self, sql, params, *ignored_wrapper_args):
        self.db.validate_no_broken_transaction()
        with self.db.wrap_database_errors:
            # 看到这里即可明白，该 CursorWrapper 对象的 execute()方法
            # 就是调用 mysqlclient 游标对象的 execute()方法
            if params is None:
                return self.cursor.execute(sql)
            else:
                return self.cursor.execute(sql, params)

    def _executemany(self, sql, param_list, *ignored_wrapper_args):
        self.db.validate_no_broken_transaction()
        with self.db.wrap_database_errors:
            # 该 CursorWrapper 对象的 executemany()方法最终调用
            # mysqlclient 游标对象的 executemany()方法
            return self.cursor.executemany(sql, param_list)

# ……
```

这里封装的 CursorWrapper 类在实例化时需要传入两个参数：cursor（mysqlclient 模块中的游标对象）和 db（BaseDatabaseWrapper 对象及其子类对象）。该类完全兼容 mysqlclient 模块中游标类的所有属性与方法，其中，execute()方法和 executemany()方法会分别调用 self.cursor 的 execute()方法和 executemany()方法并返回结果。而对于其他属性和方法，如 fetchone()方法等，则是通过魔法函数 __getattr__()获取的。通过阅读该魔法函数的源码可知，对于非 WRAP_ERROR_ATTRS 集合中的属性值，直接通过 getattr()方法获取 self.cursor 中对应的属性值即可。如果想要获取['fetchone', 'fetchmany', 'fetchall', 'nextset']这些属性，就需要调用 self.db.wrap_database_errors(cursor_attr)。注意，此时 cursor_attr 已经通过 getattr() 方法从 self.cursor 中获取对应的属性值了。继续追踪 self.db 中的 wrap_database_errors()方法，它的定义在 BaseDatabaseWrapper 类中：

```
# 源码位置: django/db/backends/base/base.py
# ……

class BaseDatabaseWrapper:

    # ……

    @cached_property
    def wrap_database_errors(self):

        return DatabaseErrorWrapper(self)
```

继续追踪 DatabaseErrorWrapper 类的实现，它位于 django/db/utils.py 文件中：

```
# 源码位置: django/db/utils.py
```

```
class DatabaseErrorWrapper:

    def __init__(self, wrapper):

        self.wrapper = wrapper

    def __enter__(self):
        pass

    def __exit__(self, exc_type, exc_value, traceback):
        # 处理异常，这是该类的核心
        if exc_type is None:
            return
        for dj_exc_type in (
                DataError,
                OperationalError,
                IntegrityError,
                InternalError,
                ProgrammingError,
                NotSupportedError,
                DatabaseError,
                InterfaceError,
                Error,
        ):
            db_exc_type = getattr(self.wrapper.Database, dj_exc_type.__name__)
            if issubclass(exc_type, db_exc_type):
                dj_exc_value = dj_exc_type(*exc_value.args)
                # Only set the 'errors_occurred' flag for errors that may make
                # the connection unusable.
                if dj_exc_type not in (DataError, IntegrityError):
                    self.wrapper.errors_occurred = True
                raise dj_exc_value.with_traceback(traceback) from exc_value

    def __call__(self, func):
        # 注意，考虑到性能，我们故意在这里不使用@warps，理由见#21109
        def inner(*args, **kwargs):
            # 处理 func()方法的最终目的是为了加上 with self 语句，以用于处理一些自定义的异常
            with self:
                return func(*args, **kwargs)
        return inner
```

由上面两处源码可知，wrap_database_errors()方法返回的是 DatabaseErrorWrapper 对象，因此 self.db.wrap_database_errors(cursor_attr)实际上调用的是 DatabaseErrorWrapper 类中的魔法函数 __call__()，而该魔法函数其实就是返回 func 方法的一个封装形式，使得原方法在 with self 语句下执行并返回相应的结果。以获取 fetchone 属性值为例，在魔法函数__call__()中，传入的 func 参数正是 mysqlclient 模块中游标类的 fetchone 方法，这里得到的是 inner 方法。inner 方法只是对传入的 func

方法进行了封装，最终调用执行的仍然是 mysqlclient 模块中的 fetchone 方法。

因此，整个在 DatabaseWrapper 类中得到的 cursor 与 mysqlclient 模块中的游标对象相比有两处升级（调用 cursor()方法得到的 cursor 值）：

◎ 用 django/db/mysql/base.py 中的 CursorWrapper 对象封装 mysqlclient 模块中的游标对象。这个对应封装的方法为 DatabaseWrapper 对象中的 create_cursor()方法。

◎ 假设上一步得到的是 x_cursor，调用父类中的_prepare_cursor()方法继续处理 x_cursor，在该方法中继续调用 make_cursor()方法，最终使用在 django/db/backends/utils.py 中定义的 CursorWrapper 类进一步封装 x_cursor。

接下来在 Python 命令行中进行相关类的操作，以验证上面的分析结果：

```
(django2-core-test) [root@master first_django]# python manage.py shell
Python 3.8.6 (default, Oct 18 2020, 15:33:08)
[GCC 4.8.5 20150623 (Red Hat 4.8.5-39)] on linux
Type "help", "copyright", "credits" or "license" for more information.
(InteractiveConsole)
>>> from django.db import connections
# 对于该语句，下面有说明
>>> wrapper = connections['default']
>>> wrapper
<django.db.backends.mysql.base.DatabaseWrapper object at 0x7f25576267f0>
# 通过 cursor()方法得到的是两次封装后的 cursor
>>> wrapper.cursor()
<django.db.backends.utils.CursorWrapper object at 0x7f2555c587f0>
# wrapper.cursor()对象中传入的 cursor 是上面分析的 x_cursor（别名）
>>> wrapper.cursor().cursor
<django.db.backends.mysql.base.CursorWrapper object at 0x7f2555cc5e20>
>>>
```

注意，在上面的演示代码中有两个地方需要注意。

（1）使用 python manage.py shell 方式进入交互模式，可以直接通过导入 connections 模块得到 DatabaseWrapper 对象，不用像前文那样麻烦，需要导入环境变量等。

（2）有些人在执行 wrapper.cursor()语句后得到的可能是 CursorDebugWrapper 对象，这是因为 first_django 项目的 settings.py 文件中的 DEBUG 被设置为 True。如果将其设置为 False 后再次执行上面的语句，就可以得到 CursorWrapper 对象了。通过源码可知，CursorDebugWrapper 类继承了 CursorWrapper 类，并重写了 execute()方法和 executemany()方法。这两个方法主要是记录并打印方法的执行时间，以便调试。

上面介绍的是针对 MySQL 的底层原理。除 MySQL 外，还可以选择 Oracle、PostgreSQL、SQLite3 等数据库。通过在项目的 settings.py 文件中选择不同的数据库 ENGINE，就可以和前文分析的一样，借助对应的 Python 模块封装一层，提供统一对外的 DatabaseWrapper 类及方法了。这样就形成了 Django 的一大特色：支持多种数据库。这样的编程模式在 Python 中十分常见，Ansible 源码和 Scrapy 源码均是如此。

现在回到最开始追踪源码的部分，即 ConnectionHandler 类的魔法函数__getitem__()中。在上一个案例中，由于在 Django 项目中设置的数据库是 MySQL，因此魔法函数__getitem__()返回的 conn 其实是 django.db.backends.mysql.base.DatabaseWrapper 对象，借助这个对象可以完成很多操作，和使用 mysqlclient 模块一样：

```
(django2-core-test) [root@master first_django]# python manage.py shell
Python 3.8.6 (default, Oct 18 2020, 15:33:08)
[GCC 4.8.5 20150623 (Red Hat 4.8.5-39)] on linux
Type "help", "copyright", "credits" or "license" for more information.
(InteractiveConsole)
>>> from django.db import connections
>>> conn = connections['default']
# 调用 connect()方法连接数据库
>>> conn.connect()
# 创建游标，确保和数据库已连接
>>> cursor = conn.cursor()
# 执行 SQL 语句
>>> cursor.execute('select * from django_books')
1
# 获取结果
>>> cursor.fetchone()
(4, 'Django 源码笔记', 'spyinx', 1, 119.0, '223454566', datetime.datetime(2021, 9, 10, 0, 0))
```

至此，关于 ORM 框架的核心部分就分析完了。Django 为 MySQL 数据库封装了 mysqlclient 模块，升级了相应的游标类，并提供了统一的对外操作接口，而对数据库的操作最后都通过调用 mysqlclient 模块中的相关类与方法完成。有兴趣的读者可以根据各自熟悉的数据库分析 Django 底层的封装代码，此处不再赘述。

## 3.3 Django 中数据库操作背后的原理

在 Django 中操作数据库对于不熟悉 SQL 编写的程序员而言，真是一大享受。Django 为开发者提供了足够简单的规则，以及漂亮的链式写法，可以轻而易举地实现对数据库的各种增删改查操作。然而，美中不足的是，Django 对于两个或者多个表的关联查询支持得并不好，且 Django 提供的数据库操作规则在熟练 SQL 编写的人员看来十分鸡肋。幸运的是，Django 提供了直接执行 SQL 语句的原始方式，开发者可以选择更为合适的数据库操作方式。

本节所有的操作都基于 MySQL 数据库，熟悉其他数据库的读者只需修改 settings.py 文件中的数据库引擎字段即可，这些前台操作屏蔽了底层数据库之间的差异。

### 3.3.1　在 Django 中执行原生 SQL 语句

在 Django 中执行原生 SQL 语句的操作非常简单，在 3.2 节的最后演示了如何导入 connections 模块来执行 SQL 语句，这就是一种简单的执行原生 SQL 语句的方式。另外，在 django/db/__init__.py 中还提供了 connection 变量，它的定义如下：

```
# 源码位置: django/db/__init__.py
# ……

class DefaultConnectionProxy:

    def __getattr__(self, item):
        return getattr(connections[DEFAULT_DB_ALIAS], item)

    def __setattr__(self, name, value):
        return setattr(connections[DEFAULT_DB_ALIAS], name, value)

    def __delattr__(self, name):
        return delattr(connections[DEFAULT_DB_ALIAS], name)

    def __eq__(self, other):
        return connections[DEFAULT_DB_ALIAS] == other


# For backwards compatibility. Prefer connections['default'] instead.
connection = DefaultConnectionProxy()

# ……
```

其中，DEFAULT_DB_ALIAS 变量在前文中已有介绍，它的值正是'default'字符串。根据在 DefaultConnectionProxy 类中定义的魔法函数可以很明确地知道：connection 从代码角度来看正是 connections['default']。下面继续在 python manage.py shell 命令行中演示对 connection 的操作，代码如下：

```
>>> from django.db import connection
>>> connection
<django.db.DefaultConnectionProxy object at 0x7f3db6db53a0>
>>> cursor = connection.cursor()
>>> cursor.execute('select * from django_books')
1
>>> cursor.fetchall()
((4, 'Django 源码笔记', 'spyinx', 1, 119.0, '223454566', datetime.datetime(2021, 9, 10,
0, 0)),)
# 执行插入 SQL 语句
```

```
>>> cursor.execute("insert into django_books(book_name, author, sex, price,
isbn)values('Django 源码笔记(第 2 版)', 'spyinx', 1, 129.0, 'xxxxx')")
# 一定要提交，或者设置自动提交
>>> connection.commit()
# 再次执行查询 SQL 语句，返回两条结果
>>> cursor.execute('select * from django_books')
2
# 查看返回的两条结果
>>> cursor.fetchall()
((4, 'Django 源码笔记', 'spyinx', 1, 119.0, '223454566', datetime.datetime(2021, 9, 10,
0, 0)), (5, 'Django 源码笔记(第 2 版)', 'spyinx', 1, 129.0, 'xxxxx', None))
# 执行删除的 SQL 语句
>>> cursor.execute("delete from django_books where id=5")
1
>>> connection.commit()
>>> cursor.execute('select * from django_books')
1
>>> cursor.fetchall()
((4, 'Django 源码笔记', 'spyinx', 1, 119.0, '223454566', datetime.datetime(2021, 9, 10,
0, 0)),)
```

上面演示了如何使用 Django 内置变量 connection 执行原生 SQL 语句来操作数据库。这些操作背后的原理在 3.2 节中已经分析得很清楚了，不再赘述。

### 3.3.2 ORM 框架的基本操作

继续使用 first_django 项目中的 DjangoBooks 类在本节进行演示。为了简单起见，本书后面会一直使用通过 python manage.py shell 命令行得到的交互模式来演示相关数据库操作。

新增操作：

```
(django2-core-test) [root@master first_django]# python manage.py shell
Python 3.8.6 (default, Oct 18 2020, 15:33:08)
[GCC 4.8.5 20150623 (Red Hat 4.8.5-39)] on linux
Type "help", "copyright", "credits" or "license" for more information.
(InteractiveConsole)
>>> from shell_test.models import DjangoBooks
>>> DjangoBooks.objects.all()
<QuerySet [<DjangoBooks: <Django 源码笔记, spyinx>>]>
>>> d1 = DjangoBooks()
>>> d1.book_name = '测试名'
>>> d1.author = 'test1'
>>> d1.sex = 0
>>> d1.price = 59
>>> d1.isbn =166777881
>>> d1.publish_date = '2021-01-22 19:22:10'
>>> d1.save()
```

```
>>> DjangoBooks.objects.all()
<QuerySet [<DjangoBooks: <Django 源码笔记, spyinx>>, <DjangoBooks: <测试名, test1>>]>
>>> DjangoBooks(book_name='test2', author='xyz', sex=1, price=110, isbn=22888,
publish_date='2021-07-11 12:23:22').save()
>>> DjangoBooks.objects.all()
<QuerySet [<DjangoBooks: <Django 源码笔记, spyinx>>, <DjangoBooks: <测试名, test1>>,
<DjangoBooks: <test2, xyz>>]>
```

修改操作：

```
>>> d = DjangoBooks.objects.get(author='test1')
>>> d
<DjangoBooks: <测试名, test1>>
>>> d.book_name = '修改图书名'
>>> d.save()
>>> DjangoBooks.objects.all()
<QuerySet [<DjangoBooks: <Django 源码笔记, spyinx>>, <DjangoBooks: <修改图书名, test1>>,
<DjangoBooks: <test2, xyz>>]>
>>> DjangoBooks.objects.get(author='test1').update(book_name= '再次修改图书名')
<QuerySet [<DjangoBooks: <Django 源码笔记, spyinx>>, <DjangoBooks: <再次修改图书名, test1>>,
<DjangoBooks: <test2, xyz>>]>
```

删除操作：

```
>>> DjangoBooks.objects.get(author='test1')
<DjangoBooks: <修改图书名, test1>>
>>> DjangoBooks.objects.get(author='test1').delete()
(1, {'shell_test.DjangoBooks': 1})
>>> DjangoBooks.objects.all()
<QuerySet [<DjangoBooks: <Django 源码笔记, spyinx>>, <DjangoBooks: <test2, xyz>>]>
```

多样化的查询操作：

```
# 在数据库中重新准备几条记录，以便于操作演示
(django2-core-test) [root@master first_django]# python manage.py shell
Python 3.8.6 (default, Oct 18 2020, 15:33:08)
[GCC 4.8.5 20150623 (Red Hat 4.8.5-39)] on linux
Type "help", "copyright", "credits" or "license" for more information.
(InteractiveConsole)
>>> from shell_test.models import DjangoBooks
>>> DjangoBooks.objects.all().last()
<DjangoBooks: <中国电信云公司媒体存储产品研发之路, kkkk>>
>>> DjangoBooks.objects.all()[1:4]
<QuerySet [<DjangoBooks: <测试之道, sister>>, <DjangoBooks: <Django 源码笔记, spyinx>>,
<DjangoBooks: <test2, xyz>>]>
>>> DjangoBooks.objects.filter(sex=1)
<QuerySet [<DjangoBooks: <ansible 源码完全剖析, spyinx>>, <DjangoBooks: <Django 源码笔记,
spyinx>>, <DjangoBooks: <test2, xyz>>]>
>>> DjangoBooks.objects.filter(sex=0).filter(book_name__startswith="南")
<QuerySet [<DjangoBooks: <南昌大学, xxxx>>]>
```

```
>>> DjangoBooks.objects.filter(sex=0).filter(price__gt=80)
<QuerySet [<DjangoBooks: <中国电信云公司媒体存储产品研发之路, kkkk>>]>
>>> DjangoBooks.objects.filter(sex=0).order_by('price')
<QuerySet [<DjangoBooks: <南昌大学, xxxx>>, <DjangoBooks: <测试之道, sister>>, <DjangoBooks:
<中国电信云公司媒体存储产品研发之路, kkkk>>]>
>>> DjangoBooks.objects.filter(sex=0).order_by('-price')
<QuerySet [<DjangoBooks: <中国电信云公司媒体存储产品研发之路, kkkk>>, <DjangoBooks: <测试之
道, sister>>, <DjangoBooks: <南昌大学, xxxx>>]>
>>> DjangoBooks.objects.filter(sex=0).filter(price__gt=80).values_list('book_name',
'price')
<QuerySet [('中国电信云公司媒体存储产品研发之路', 89.0)]>
>>> DjangoBooks.objects.filter(sex=0).values_list('book_name', 'price')
<QuerySet [('测试之道', 79.0), ('南昌大学', 49.0), ('中国电信云公司媒体存储产品研发之路', 89.0)]>
>>> DjangoBooks.objects.filter(sex=0).filter(price__gt=60).values_list('book_name',
'price')
<QuerySet [('测试之道', 79.0), ('中国电信云公司媒体存储产品研发之路', 89.0)]>
```

上面对单表的增删改查操作是不是很简单，也很神奇？有没有思考过如下问题：

- ◎ QuerySet 是什么？
- ◎ Django 中的模型类为什么支持这样的链式操作？
- ◎ 链式表达式是如何被翻译成 SQL 语句的，又该如何查看？
- ◎ 为什么能对查询结果使用切片？当使用 Django 操作数据库时，懒查询指的是什么？
- ◎ DjangoBooks 为什么要通过 objects 属性才能调用 all()、filter()等方法，为什么要通过 objects 属性才能进行链式操作？

带着上面的问题，我们从 ORM 框架的几个基础类的源码开始介绍。

### 基础类 RawSQL

第 1 个要介绍的是基础类 RawSQL，它位于 django/db/models/sql/query.py 文件中。

```
# 源码位置：django/db/models/sql/query.py
# ……

class RawQuery:
    """原始 SQL."""

    def __init__(self, sql, using, params=None):

        self.params = params or ()
        # 原始 SQL 语句
        self.sql = sql
        # 选择使用的数据库，默认的为"default"
        self.using = using
        self.cursor = None
```

```
        self.low_mark, self.high_mark = 0, None  # Used for offset/limit
        self.extra_select = {}
        self.annotation_select = {}

    def chain(self, using):
        # 链式，起始就是再次返回该类的实例化对象
        return self.clone(using)

    def clone(self, using):
        # 再次实例化该类
        return RawQuery(self.sql, using, params=self.params)

    def get_columns(self):
        # 获取表结构，根据 SQL 语句获取
        if self.cursor is None:
            self._execute_query()
        converter = connections[self.using].introspection.identifier_converter
        return [converter(column_meta[0])
                for column_meta in self.cursor.description]

    def __iter__(self):
        self._execute_query()
        if not connections[self.using].features.can_use_chunked_reads:
            result = list(self.cursor)
        else:
            result = self.cursor
        return iter(result)

    def __repr__(self):
        return "<%s: %s>" % (self.__class__.__name__, self)

    @property
    def params_type(self):
        # params 属性类型
        return dict if isinstance(self.params, Mapping) else tuple

    def __str__(self):
        return self.sql % self.params_type(self.params)

    def _execute_query(self):
        # 获取对应数据库的连接信息
        connection = connections[self.using]

        params_type = self.params_type
        adapter = connection.ops.adapt_unknown_value
        # 整理查询参数
        if params_type is tuple:
```

```
        params = tuple(adapter(val) for val in self.params)
    elif params_type is dict:
        params = {key: adapter(val) for key, val in self.params.items()}
    else:
        raise RuntimeError("Unexpected params type: %s" % params_type)

    # 获取游标对象
    self.cursor = connection.cursor()
    # 调用游标对象的execute()方法执行 SQL 语句
    self.cursor.execute(self.sql, params)
```

这里核心的方法是_execute_query()方法和 get_columns()方法，其中，后者是调用前一个方法来获取查询语句中表的结构的。先看_execute_query()方法，重点看该方法的最后两句：从 connection 中获取游标对象和调用游标对象的 execute()方法执行 SQL 语句。这里需要介绍在 mysqlclient 模块中定义的游标类的 description 属性值，可以通过源码来理解：

```
# 源码位置：MySQLdb/cursors.py
# ……

class BaseCursor(object):

    def __init__(self, connection):
        self.connection = connection
        self.description = None
        self.description_flags = None
        self.rowcount = -1
        self.arraysize = 1
        self._executed = None

        # ……
```

从源码中的注释可以看到：MySQLdb 模块中游标对象的 description 属性值指的是最后一个查询语句对应的表结构信息。不妨做一个简单的测试，如下：

```
(django2-core-test) [root@master first_django]# python manage.py shell
Python 3.8.6 (default, Oct 18 2020, 15:33:08)
[GCC 4.8.5 20150623 (Red Hat 4.8.5-39)] on linux
Type "help", "copyright", "credits" or "license" for more information.
(InteractiveConsole)
>>> from django.db import connection
>>> cursor = connection.cursor()
>>> cursor.execute('select * from django_books')
0
>>> cursor.description
(('id', 3, 0, 11, 11, 0, 0), ('book_name', 253, 0, 90, 90, 0, 0), ('author', 253, 0, 90,
90, 0, 0), ('sex', 2, 0, 6, 6, 0, 0), ('price', 5, 0, 22, 22, 31, 0), ('isbn', 253, 0,
30, 30, 0, 0), ('publish_date', 12, 0, 26, 26, 6, 0))
```

当明白了 description 属性值后，就可以测试 get_columns()中的代码段输出了：

```
>>> [converter(column_meta[0]) for column_meta in cursor.description]
['id', 'book_name', 'author', 'sex', 'price', 'isbn', 'publish_date']
```

**注意**，这里导入的 connection 其实就是 connections['default']。

最后看 RawQuery 类的一个完整调用案例，如下：

```
(django2-core-test) [root@master first_django]# python manage.py shell
Python 3.8.6 (default, Oct 18 2020, 15:33:08)
[GCC 4.8.5 20150623 (Red Hat 4.8.5-39)] on linux
Type "help", "copyright", "credits" or "license" for more information.
(InteractiveConsole)
>>> from django.db.models.sql.query import RawQuery
>>> raw_query = RawQuery("select * from django_books where author = %s ", 'default',
('spyinx',))
>>> raw_query.get_columns()
['id', 'book_name', 'author', 'sex', 'price', 'isbn', 'publish_date']
>>> raw_query
<RawQuery: select * from django_books where author = spyinx >
>>> raw_query._execute_query()
>>> raw_query.cursor.fetchall()
((1, 'Django 源码笔记', 'spyinx', 0, 109.0, '11111', datetime.datetime(2021, 2, 4, 21, 42,
43)), (2, 'Ansible 核心源码剖析与项目实战', 'spyinx', 0, 119.0, '22222',
datetime.datetime(2021, 2, 4, 21, 43, 14)))
```

**注意**，为了演示效果，在 django_books 表中插入了三条数据，其中，两条数据的作者为 spyinx。

### 基础类 Query

第 2 个要介绍的是基础类 Query，同样位于 django/db/models/sql/query.py 文件中。对于该类定义的方法非常多，不适合一一学习。这里将通过代码实践来了解该类。先看初始化方法：

```
# 源码位置： django/db/models/sql/query.py
# ……

class Query:
    """单一的 SQL 查询"""

    alias_prefix = 'T'
    subq_aliases = frozenset([alias_prefix])

    compiler = 'SQLCompiler'  # SQL 翻译器

    def __init__(self, model, where=WhereNode):
        # 模型
        self.model = model
        self.alias_refcount = {}
```

```
    self.alias_map = OrderedDict()
    self.external_aliases = set()
    self.table_map = {}
    self.default_cols = True
    self.default_ordering = True
    self.standard_ordering = True
    self.used_aliases = set()
    self.filter_is_sticky = False
    self.subquery = False

    #与 SQL 查询相关的属性
    self.select = ()
    self.where = where()
    self.where_class = where
    #用于表征该查询的分组、结果排序、唯一性等属性
    self.group_by = None
    self.order_by = ()
    self.low_mark, self.high_mark = 0, None  # Used for offset/limit
    self.distinct = False
    self.distinct_fields = ()
    self.select_for_update = False
    self.select_for_update_nowait = False
    self.select_for_update_skip_locked = False
    self.select_for_update_of = ()

    self.select_related = False
    # 防止无限递归
    self.max_depth = 5

    self.values_select = ()

    # 与 SQL 聚合查询相关的属性
    self._annotations = None
    self.annotation_select_mask = None
    self._annotation_select_cache = None

    # 用于联合查询的属性
    self.combinator = None
    self.combinator_all = False
    self.combined_queries = ()

    # 用于扩展的属性
    self._extra = None  # Maps col_alias -> (col_sql, params).
    self.extra_select_mask = None
    self._extra_select_cache = None

    self.extra_tables = ()
    self.extra_order_by = ()
```

```
        self.deferred_loading = (frozenset(), True)

        self._filtered_relations = {}

        self.explain_query = False
        self.explain_format = None
        self.explain_options = {}

# ……
```

Query 类定义了非常多的属性，这些属性分别关联着 SQL 查询中的一些子语句，如 where 属性关联着 SQL 查询的 WHERE 语句、group_by 属性关联着 SQL 查询的 GROUP BY 语句等。在该查询类的初始化方法中传入的 model 参数正是在 Django 项目中定义的模型类。下面以在 first_django 项目中创建的 DjangoBooks 模型类为例进行说明：

```
(django2-core-test) [root@master first_django]# python manage.py shell
Python 3.8.6 (default, Oct 18 2020, 15:33:08)
[GCC 4.8.5 20150623 (Red Hat 4.8.5-39)] on linux
Type "help", "copyright", "credits" or "license" for more information.
(InteractiveConsole)
>>> from django.db.models.sql.query import Query
>>> from shell_test.models import DjangoBooks
>>> query = Query(model=DjangoBooks)
>>> query.get_meta().db_table
'django_books'
```

这里可以通过 Query 对象的 get_meta()方法获取对应的 Django 模型类(model)中定义的元信息，比如前面在元类中定义了 db_table 属性，这里通过 query.get_meta().db_table 就能获取其属性值，也就是对应的表名。如果在前面的模型中没有定义这个元类，还能获取 db_table 的属性值吗？手动将 DjangoBooks 类中的 Meta 信息注释后，再次执行上述操作，结果如下：

```
# 和前面一样的操作
# ……
>>> query.get_meta().db_table
'shell_test_djangobooks'
```

可以看到，这里同样获取了 db_table 的属性值，其值由应用名和模型类（小写）组成。这是当没有设置 db_table 属性时 Django 默认设置的值。

接下来看 Query 类中一个非常重要的方法：

```
class Query:
    """单一的 SQL 查询"""
    # ……
    compiler = 'SQLCompiler'
    # ……
```

```
    def get_compiler(self, using=None, connection=None):
        if using is None and connection is None:
            raise ValueError("Need either using or connection")
        if using:
            # 一般就是"defaults"，这里的 connection 即默认数据库的连接对象
            connection = connections[using]
        return connection.ops.compiler(self.compiler)(self, connection, using)

# ……
```

get_compiler()方法获取一个编译对象（Compiler），专门用于将这里的 Query 对象翻译成对应数据库中的 SQL 语句。connection 指未指定 using 参数时，默认数据库配置的连接对象。那么 connection 的 ops 属性是什么呢？先追踪 connection 所属的类：

```
# 源码位置：django/db/backends/mysql/base.py
# ……

class DatabaseWrapper(BaseDatabaseWrapper):

    # ……

    ops_class = DatabaseOperations

    # ……
```

在 DatabaseWrapper 类中并没有直接定义 ops 属性，只有一个 ops_class 属性。继续追踪其父类：

```
# 源码位置：django/db/backends/base/base.py
# ……

class BaseDatabaseWrapper:

    # ……

    def __init__(self, settings_dict, alias=DEFAULT_DB_ALIAS):

        # ……
        self.ops = self.ops_class(self)
        self.validation = self.validation_class(self)

    # ……
```

到这里就已经很清楚了，在 DatabaseWrapper 类中指定的 ops_class 属性就是用来初始化 ops 属性的。也就是说，connection.ops 即 DatabaseOperations 对象，该类位于 django/db/backends/mysql 目录下的 operations.py 文件中。下面看在 DatabaseOperations 类中定义的 compiler()方法源码，仍然找不到，继续去父类中找，结果如下：

```
# 源码位置：django/db/backends/base/operations
# ……

class BaseDatabaseOperations:

    compiler_module = "django.db.models.sql.compiler"

    # ……

    def compiler(self, compiler_name):

        if self._cache is None:
            self._cache = import_module(self.compiler_module)
        return getattr(self._cache, compiler_name)

    # ……
```

从上述代码可知，调用 query.get_compiler()方法的结果应该为在 django/db/models/sql/compiler.py 文件中定义的 SQLCompiler 类，下面验证一下：

```
# 导入 Query 类及得到 query 的语句忽略，和前面一样
# ……
>>> query.get_compiler(using='default')
<django.db.backends.mysql.compiler.SQLCompiler object at 0x7fc47130e490>
```

这里需要说明的是，由于代码之间的关系错综复杂，所以不能一味地只看源码，一定要上手测试，否则很容易陷入一些错误的假象中。例如，很容易认为 compiler_module 就是这里定义的值，但实际上调用的是 DatabaseOperations 对象中的 compiler()方法，而在该类中已经定义过 compiler_module 的属性值：

```
# 源码位置：django/db/backends/mysql/operations.py
# ……
class DatabaseOperations(BaseDatabaseOperations):
compiler_module = "django.db.backends.mysql.compiler"

# ……
```

看到这里后便不难解释 query.get_compiler(using='default')的输出结果了。继续深究这个编译对象，可以说它是整个 Django 底层 ORM 框架的核心：

```
# 源码位置：django/db/backends/mysql/compiler.py
from django.db.models.sql import compiler


class SQLCompiler(compiler.SQLCompiler):
    def as_subquery_condition(self, alias, columns, compiler):
        qn = compiler.quote_name_unless_alias
        qn2 = self.connection.ops.quote_name
```

```
        sql, params = self.as_sql()
        return '(%s) IN (%s)' % (', '.join('%s.%s' % (qn(alias), qn2(column)) for column
in columns), sql), params


class SQLInsertCompiler(compiler.SQLInsertCompiler, SQLCompiler):
    pass


class SQLDeleteCompiler(compiler.SQLDeleteCompiler, SQLCompiler):
    pass


class SQLUpdateCompiler(compiler.SQLUpdateCompiler, SQLCompiler):
    pass


class SQLAggregateCompiler(compiler.SQLAggregateCompiler, SQLCompiler):
    pass
```

从这里可以看到，SQLCompiler 类只是简单继承了 compiler.SQLCompiler，几乎没做修改。此时，再次将关注点转向 compiler.SQLCompiler 这个核心的 SQL 编译类：

```
# 源码位置：django/db/models/sql/compiler.py
# ……

class SQLCompiler:

    def __init__(self, query, connection, using):

        self.query = query          # 关联的 Query 对象
        self.connection = connection  # 数据库连接对象
        self.using = using
        self.quote_cache = {'*': '*'}
        self.select = None
        self.annotation_col_map = None
        self.klass_info = None
        self.ordering_parts = re.compile(r'(.*)\s(ASC|DESC)(.*)')
        self._meta_ordering = None

    # ……
```

从 SQLCompiler 的初始化方法中可以看到 query 属性，它是前面的 Query 对象吗？只需看一下 Query 类中 get_compiler()方法的最后一句即可：

```
return connection.ops.compiler(self.compiler)(self, connection, using)
```

通过前面的分析可知，connection.ops.compiler(self.compiler)得到的正是这里的 SQLCompiler 类，

在实例化该类时传入的第 1 个参数为 self，该值正是 SQLCompiler 对象中的 query 属性。这很容易理解，SQLCompiler 是编译器，它需要编译的是 Query 对象，然后根据 Query 对象调用 as_sql()方法，最终得到 Query 对象对应的完整 SQL 语句。关于 SQLCompiler 类的源码学习先告一段落，后续在使用到该类中的方法时再进行介绍。

接下来学习 QuerySet 类，同样先介绍 RawQuerySet 这个类，其初始化方法如下：

```
# 源码位置：django/db/models/query.py
# ……

class RawQuerySet:

    def __init__(self, raw_query, model=None, query=None, params=None,
                 translations=None, using=None, hints=None):
        # 原始的 SQL 语句
        self.raw_query = raw_query
        self.model = model
        self._db = using
        self._hints = hints or {}
        # 会根据 SQL 语句生成 RawQuery 对象
        self.query = query or sql.RawQuery(sql=raw_query, using=self.db, params=params)
        self.params = params or ()
        self.translations = translations or {}
        self._result_cache = None
        self._prefetch_related_lookups = ()
        self._prefetch_done = False

    # ……
```

在 RawQuerySet 类的初始化方法中，raw_query 是原生的 SQL 语句，model 是关联的模型对象，query 可有可无，当没有输入 query 时，会根据 SQL 语句生成 RawQuery 对象并赋给 query 属性。首先，根据上面的参数初始化 RawQuerySet 类：

```
(django2-core-test) [root@master first_django]# python manage.py shell
Python 3.8.6 (default, Oct 18 2020, 15:33:08)
[GCC 4.8.5 20150623 (Red Hat 4.8.5-39)] on linux
Type "help", "copyright", "credits" or "license" for more information.
(InteractiveConsole)
>>> from django.db.models.query import RawQuerySet
>>> from shell_test.models import DjangoBooks
# 这里设置的 raw_query 和 params 参数可以参考 RawQuery 初始化部分的代码
>>> raw_queryset = RawQuerySet(raw_query="select id, book_name, author, price from
django_books where author='spyinx'",model=DjangoBooks)
>>> raw_queryset.query
<RawQuery: select id, book_name, author, price from django_books where author='spyinx'>
```

在 RawQuerySet 类中有两个属性，用于获取表相关的信息，它们的逻辑比较简单：

```
# 源码位置：django/db/models/query.py
# ……

class RawQuerySet:

    # ……

    # 前面介绍过该装饰器，此处不再赘述
    @cached_property
    def columns(self):
        # 通过 RawQuery 对象的 get_columns()方法查询关联表的全部字段
        columns = self.query.get_columns()
        # 调整与字段名不匹配的列名
        for (query_name, model_name) in self.translations.items():
            # 忽略对不存在的列名的翻译
            try:
                index = columns.index(query_name)
            except ValueError:
                pass
            else:
                columns[index] = model_name
        return columns

    @cached_property
    def model_fields(self):
        converter = connections[self.db].introspection.identifier_converter
        model_fields = {}
        # 通过模型类返回字段信息
        for field in self.model._meta.fields:
            name, column = field.get_attname_column()
            model_fields[converter(column)] = field
        return model_fields
```

直接在前面的交互模式下继续调用这两个方法：

```
# 匹配输入 SQL 语句的字段
>>> raw_queryset.columns
['id', 'book_name', 'author', 'price']
# 获取关联模型类中定义的全部字段
>>> raw_queryset.model_fields
{'id': <django.db.models.fields.AutoField: id>, 'book_name':
<django.db.models.fields.CharField: book_name>, 'author':
<django.db.models.fields.CharField: author>, 'sex':
<django.db.models.fields.SmallIntegerField: sex>, 'price':
<django.db.models.fields.FloatField: price>, 'isbn':
<django.db.models.fields.CharField: isbn>, 'publish_date':
<django.db.models.fields.DateTimeField: publish_date>}
```

如何能得到 raw_query 语句的结果呢？在笔者的数据库上单独执行 raw_query 语句，可以查询到

2 条记录（总共 3 条），如图 3-1 所示。

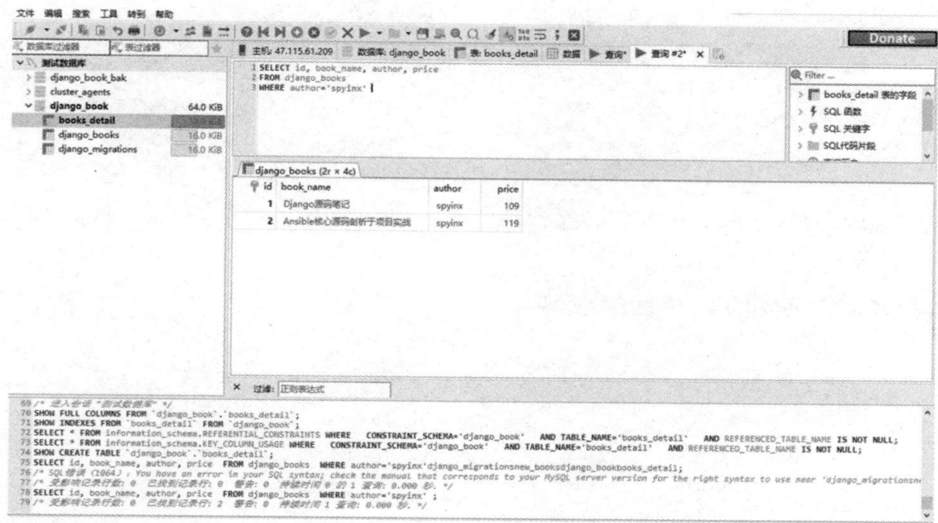

图 3-1

在 RawQuerySet 类中有一个_fetch_all()方法，和前面的游标获取结果类似。大胆猜测通过该方法就能执行 raw_query 语句并获取其查询结果。问题是，在调用该方法后，从哪里获取结果呢？这些疑问都可以在源码中找到答案：

```
# 源码位置：django/db/models/query.py
# ……

class RawQuerySet:

   # ……

   def _prefetch_related_objects(self):
      prefetch_related_objects(self._result_cache, *self._prefetch_related_lookups)
      self._prefetch_done = True

   # ……

   def _fetch_all(self):
      if self._result_cache is None:
         # 从这里的结果可以猜测一下，结果保存在 self._result_cache 中
         self._result_cache = list(self.iterator())
      if self._prefetch_related_lookups and not self._prefetch_done:
         self._prefetch_related_objects()

   # ……
```

从上面的_fetch_all()方法的实现源码中大致可以猜到，对象的_result_cache 是用来保存结果的。这时就需要在 Django 的 Shell 模式下验证一下：

```
>>> raw_queryset._fetch_all()
>>> raw_queryset._result_cache
[<DjangoBooks: <Django 源码笔记, spyinx>>, <DjangoBooks: <Ansible 核心源码剖析与项目实战,
spyinx>>]
>>>
```

在看到现象后再结合_fetch_all()方法的源码，可知获取结果的语句为 list(self.iterator())。继续追踪这里的 iterator()方法，其源码较为复杂：

```
# 源码位置：django/db/models/query.py
# ……

class RawQuerySet:

    # ……

    def iterator(self):
        # 在外层缓存数据，提高性能
        db = self.db
        # 获取 SQLCompiler 对象
        compiler = connections[db].ops.compiler('SQLCompiler')(
            self.query, connections[db], db
        )

        # 生成迭代器，可以用 for 循环迭代 query
        query = iter(self.query)

        try:
            model_init_names, model_init_pos, annotation_fields =
self.resolve_model_init_order()
            # 注意，如果查询的字段没有主键，可能会报错
            if self.model._meta.pk.attname not in model_init_names:
                raise InvalidQuery('Raw query must include the primary key')
            # 模型类
            model_cls = self.model
            fields = [self.model_fields.get(c) for c in self.columns]

            # 编译器的一些转换功能
            # ……

            # 其实迭代 query 就是迭代得到的数据
            for values in query:
                model_init_values = [values[pos] for pos in model_init_pos]
                instance = model_cls.from_db(db, model_init_names, model_init_values)
                if annotation_fields:
```

```
                for column, pos in annotation_fields:
                    setattr(instance, column, values[pos])
                yield instance
        finally:
            # query 迭代结束，准备关闭数据库的查询连接
            if hasattr(self.query, 'cursor') and self.query.cursor:
                self.query.cursor.close()

    # ……
```

在上述代码中，最关键的一条语句是 query=iter(self.query)。self.query 其实就是 RawQuery 对象，iter()方法的作用是生成迭代对象，对应会调用 RawQuery 类中的_iter_()方法得到一个可迭代的对象。继续追踪 RawQuery 类中的_iter_()方法的源码，如下：

```
# 源码位置：django/db/models/sql/query.py
# ……

class RawQuery:

    # ……

    def __iter__(self):
        # 这里会调用 self.cursor.execute()方法去查询数据库，获取结果
        self._execute_query()
        # 对 MySQL 数据库来说，connections[self.using].features.can_use_chunked_reads 为 True
        if not connections[self.using].features.can_use_chunked_reads:
            result = list(self.cursor)
        else:
            # 一般会执行这个语句
            result = self.cursor

        # 其实返回的是 iter(self.result)
        return iter(result)

    # ……
```

从上面的代码可知，iter(self.query)返回的迭代对象为 iter(self.cursor)。而从 3.2 节分析的结果可知，针对 MySQL 数据库而言，self.cursor 是 django/db/backends/utils.py 中的 CursorWrapper 对象，其中定义的_iter_()方法如下：

```
# 源码位置：django/db/backends/utils.py
# ……

class CursorWrapper:
    def __init__(self, cursor, db):
        self.cursor = cursor
        self.db = db
```

```
    # ……

    def __iter__(self):
        with self.db.wrap_database_errors:
            yield from self.cursor
```

这里传入的 cursor 参数正是在 django/db/backends/mysql/base.py 中定义的 CursorWrapper 对象。继续追踪在该 CursorWrapper 对象中定义的_iter_()方法，如下：

```
# 源码位置： django/db/backends/mysql/base.py
# ……

class CursorWrapper:
    # ……

    def __init__(self, cursor):
        # 这里的 self.cursor 便是 mysqlclient 模块中的游标对象
        self.cursor = cursor

    # ……

    def __iter__(self):
        return iter(self.cursor)
```

最终，在兜兜转转一圈之后，发现从 query=iter(self.query)语句中得到的迭代对象其实就是执行了 SQL 语句（调用了游标对象的 execute()方法）后的游标对象，而且是 mysqlclient 模块中的游标对象。下面进入 mysqlclient 模块的源码，查看这个游标对象定义的_iter_()方法，如下：

```
# 源码位置：MYSQLdb/cursors.py
# ……

class BaseCursor(object):

    # ……

    def __iter__(self):
        return iter(self.fetchone, None)

    # ……
```

看到这里是不是恍然大悟？在得到游标对象 self.cursor 后，再次使用 for 循环对该游标对象迭代时，就等价于每次在调用 cursor.fetchone()方法后都获取一条结果数据，直到数据全部取完。有了这些知识后，再看如下测试操作：

```
(django2-core-test) [root@master first_django]# python manage.py shell
Python 3.8.6 (default, Oct 18 2020, 15:33:08)
[GCC 4.8.5 20150623 (Red Hat 4.8.5-39)] on linux
```

```
Type "help", "copyright", "credits" or "license" for more information.
(InteractiveConsole)
>>> from django.db.models.query import RawQuerySet
>>> from shell_test.models import DjangoBooks
>>> raw_queryset = RawQuerySet(raw_query="select id, book_name, author, price from django_books where author='spyinx'",model=DjangoBooks)
>>> raw_queryset.query.cursor
<django.db.backends.utils.CursorWrapper object at 0x7f388a474130>
# 上面的 cursor 对象是加强封装的 cursor，它实例化输入的 cursor 参数为如下位置的对象
>>> raw_queryset.query.cursor.cursor
<django.db.backends.mysql.base.CursorWrapper object at 0x7f388a467be0>
>>> query = iter(raw_queryset.query)
>>> for values in query:
...     print(values)
...
(1, 'Django 源码笔记', 'spyinx', 109.0)
(2, 'Ansible 核心源码剖析与项目实战', 'spyinx', 119.0)
>>>
```

在通过 iter(raw_queryset.query)得到游标对象后，只需通过 for 循环遍历该游标对象，就可以直接获取 RawQuerySet 对象中 raw_query 的执行结果了。现在再看 iterator()方法是不是就非常简单了？另外，需要说明的是：如果出现了 InvalidQuery 异常且提示的错误信息为'Raw query must include the primary key'，不用担心，只需查看一下源码就能明白原因。我们先使用下面的语句复现该异常：

```
(django2-core-test) [root@master first_django]# python manage.py shell
Python 3.8.6 (default, Oct 18 2020, 15:33:08)
[GCC 4.8.5 20150623 (Red Hat 4.8.5-39)] on linux
Type "help", "copyright", "credits" or "license" for more information.
(InteractiveConsole)
>>> from django.db.models.query import RawQuerySet
>>> from shell_test.models import DjangoBooks
>>> raw_queryset = RawQuerySet(raw_query="select book_name, author from django_books where author='spyinx'",model=DjangoBooks)
>>> raw_queryset._fetch_all()
Traceback (most recent call last):
  File "<console>", line 1, in <module>
  File "/root/.pyenv/versions/django2-core-test/lib/python3.8/site-packages/django/db/models/query.py", line 1382, in _fetch_all
    self._result_cache = list(self.iterator())
  File "/root/.pyenv/versions/django2-core-test/lib/python3.8/site-packages/django/db/models/query.py", line 1410, in iterator
    raise InvalidQuery('Raw query must include the primary key')
django.db.models.query_utils.InvalidQuery: Raw query must include the primary key
```

从上面的异常可知，抛出异常的位置是 RawQuerySet 类中的 iterator()方法：

```
# 源码位置：django/db/models/query.py
# ……

class RawQuerySet:

    # ……

    def iterator(self):

        # ……

        try:
            model_init_names, model_init_pos, annotation_fields =
self.resolve_model_init_order()
            if self.model._meta.pk.attname not in model_init_names:
                # 抛出异常的位置
                raise InvalidQuery('Raw query must include the primary key')
            # ……
        finally:
            # ……

    # ……
```

为了理解上述异常，先调用异常 RawQuerySet 对象的 resolve_model_init_order()方法，查看其输出结果：

```
# 接上面的交互模式，不再重复输入 Python 语句
>>> raw_queryset.resolve_model_init_order()
(['book_name', 'author'], [0, 1], [])
>>> raw_queryset.model._meta.pk.attname
'id'
```

在前面代码的 if 判断中，self.model._meta.pk.attname 的值为主键名'id'，而通过 resolve_model_init_order()方法得到的 model_init_names 为(['book_name', 'author'], [0, 1], [])。if 判断成立，程序抛出异常。这里的逻辑表明，在输入的 raw_query 中一定要带上主键，否则会抛出异常。当然，如果写'SELECT *'的语句也是可以的，resolve_model_init_order()方法会默认指定全部表字段，如下：

```
>>> raw_queryset = RawQuerySet(raw_query="select * from django_books where
author='spyinx'",model=DjangoBooks)
>>> raw_queryset.resolve_model_init_order()
(['id', 'book_name', 'author', 'sex', 'price', 'isbn', 'publish_date'], [0, 1, 2, 3, 4,
5, 6], [])
```

接下来就是剖析 QuerySet 类的源码了，这里的内容较多，请耐心阅读：

```
# 源码位置：django/db/models/query.py
# ……
```

```
class QuerySet:
    """Represent a lazy database lookup for a set of objects."""

    def __init__(self, model=None, query=None, using=None, hints=None):
        self.model = model       # 关联的模型，必须要有
        self._db = using         # 数据库选择，默认是'default'
        self._hints = hints or {}
        self.query = query or sql.Query(self.model) # 可以指定，也可以不指定 Query 对象
        self._result_cache = None   # 该 QuerySet 对象代表的查询结果
        self._sticky_filter = False
        self._for_write = False
        self._prefetch_related_lookups = ()
        self._prefetch_done = False
        self._known_related_objects = {}    # {rel_field: {pk: rel_obj}}
        self._iterable_class = ModelIterable  # 后续会用到
        self._fields = None       # 关联模型表的字段

    # ……
```

以上是 QuerySet 类的初始化方法，比较简单。它和 RawQuery 类的初始化方法类似，都必须输入关联的模型类，对于 Query 对象则可有可无。当没有输入 Query 对象时，会根据模型类去实例化 Query 类。

接着在 QuerySet 类中定义了非常多的魔法函数，比如__getstate__()、__setstate__ ()、__repr__()和__len__()等，这些魔法函数的作用在后续用到时会分析，这里先忽略。继续看在 QuerySet 类中定义的一些针对数据库查询的方法，比如 all()、first()、count()和 filter()等。

### 1. all()方法：获取该查询对象对应的所有结果

如果想要理解 Django 中 DjangoBooks.objects.all()语句的核心逻辑，就需要查看 QuerySet 类中 all()方法的源码，其内容如下：

```
# 源码位置：django/db/models/query.py
# ……

class QuerySet:

    # ……

    def all(self):
        """
        返回一个新的 QuerySet 对象，是当前 QuerySet 对象的一个副本
        """
        return self._chain()
```

```
    # ……

    def _chain(self, **kwargs):
        """
        返回当前 QuerySet 的一个副本，用于做其他操作
        """
        obj = self._clone()
        if obj._sticky_filter:
            obj.query.filter_is_sticky = True
            obj._sticky_filter = False
        obj.__dict__.update(kwargs)
        return obj

    def _clone(self):
        """
        返回当前 QuerySet 对象的一个副本，这是除 deepcopy()方法外的另一种轻量级选择
        """
        c = self.__class__(model=self.model, query=self.query.chain(), using=self._db,
hints=self._hints)
        c._sticky_filter = self._sticky_filter
        c._for_write = self._for_write
        c._prefetch_related_lookups = self._prefetch_related_lookups[:]
        c._known_related_objects = self._known_related_objects
        c._iterable_class = self._iterable_class
        c._fields = self._fields
        return c

    # ……
```

从上面的代码可以看到，QuerySet 类中的_clone()方法类似于 copy 模块中的 deepcopy()函数，但是_clone()方法更为轻量，其复制的数据并没有 deepcopy()函数多。_chain()方法会基于_clone()方法返回当前 QuerySet 对象的副本，同时可以更新一些新的数据。最终，QuerySet 对象中的 all()方法只是简单调用_chain()方法并返回其结果，因此，all()方法只是单纯返回当前 QuerySet 对象的一个副本。下面再来看本节开头部分的操作示例，如下：

```
(django2-core-test) [root@master first_django]# python manage.py shell
Python 3.8.6 (default, Oct 18 2020, 15:33:08)
[GCC 4.8.5 20150623 (Red Hat 4.8.5-39)] on linux
Type "help", "copyright", "credits" or "license" for more information.
(InteractiveConsole)
>>> from shell_test.models import DjangoBooks
>>> d = DjangoBooks.objects.all()
>>> type(d)
<class 'django.db.models.query.QuerySet'>
>>> d
<QuerySet [<DjangoBooks: <Django 源码笔记, spyinx>>, <DjangoBooks: <Ansible 核心源码剖析与
项目实战, spyinx>>, <DjangoBooks: <Python 自动化运维, test>>]>
```

在学习了 QuerySet 类中 all()方法的源码后，再看上面的结果，是不是会有些疑惑？d 得到的应该是当前 QuerySet 对象的一个副本，而且没有任何查询数据库的动作，应该没有查询结果啊？其实，上面的操作给读者造成了一个假象，认为打印 d 有结果输出，那么结果一定是在上一条语句中得到的。但经过仔细观察可以发现，当在交互模式下输入 d 时，是需要等一会儿才有输出的（可能零点几秒）。我们可以测试一下，首先，停止 MySQL 服务；然后，输入上面的语句进行测试。最终可以发现，在交互模式下输入 d 时会因为连接 MySQL 数据库失败而抛出异常。

这种只有在用到时才真正去查询或者执行的操作模式在 Django 源码中十分常见，在 3.1 节中读取配置文件的方式也是如此。我们姑且称之为懒执行模式，该模式在 QuerySet 类的定义中已有注释：

```
Represent a lazy database lookup for a set of objects.
```

那 Django 是如何实现这种懒查询模式的呢？答案就在 QuerySet 类定义的那些魔法函数中。先看打印 d 时的调用过程，Python 解释器会调用 QuerySet 对象中定义的魔法函数__repr__()，其内容如下：

```
# 源码位置：django/db/models/query.py
# ……

REPR_OUTPUT_SIZE = 20

# ……

class QuerySet:

    # ……

    def __repr__(self):
        # 下面的一行语句，正是去数据库中获取结果并存入 data 中
        data = list(self[:REPR_OUTPUT_SIZE + 1])
        # 整理结果，如果超出了设定的最多显示数，多出的部分用特定字符代替
        if len(data) > REPR_OUTPUT_SIZE:
            data[-1] = "...(remaining elements truncated)..."
        return '<%s %r>' % (self.__class__.__name__, data)

    # ……
```

上述代码又继续调用了 QuerySet 对象的魔法函数__getitem__()，通过该函数查询数据库，最终获取相应的查询结果。这里还设置了最大显示数（REPR_OUTPUT_SIZE），当超过该值时，将使用特定的字符串代替。魔法函数__getitem__()的源码如下：

```
# 源码位置：django/db/models/query.py
# ……

class QuerySet:
```

```
# ……

def __getitem__(self, k):

    # 索引 k 必须是 int 或者是 slice 类型的
    if not isinstance(k, (int, slice)):
        raise TypeError

    # assert 判断
    assert ((not isinstance(k, slice) and (k >= 0)) or
            (isinstance(k, slice) and (k.start is None or k.start >= 0) and
             (k.stop is None or k.stop >= 0))), \
        "Negative indexing is not supported."

    # 如果缓存中有数据，直接返回
    if self._result_cache is not None:
        return self._result_cache[k]

    #如果 k 是切片类型，需要得到 start、stop 参数
    if isinstance(k, slice):
        # 得到当前 QuerySet 对象的副本
        qs = self._chain()
        # 设置索引的开始值
        if k.start is not None:
            start = int(k.start)
        else:
            start = None
        # 设置索引的结束值
        if k.stop is not None:
            stop = int(k.stop)
        else:
            stop = None
        # 调用 Query 对象的 set_limits()方法
        qs.query.set_limits(start, stop)
        # 获取结果的核心语句
        return list(qs)[::k.step] if k.step else qs

    # 直接索引
    qs = self._chain()
    qs.query.set_limits(k, k + 1)
    # 最终在这里调用_fetch_all()方法查询数据库获取结果
    qs._fetch_all()
    return qs._result_cache[0]

# ……
```

对于前面调用的 all()方法来说，调用魔法函数__getitem__()传入的 k 参数为 slice 类型

（self[:REPR_OUTPUT_SIZE + 1]），即 slice(None, 21, None)。如果在_result_cache 属性中保存了请求结果，则直接返回，否则进入 if 判断。在该 if 判断中得到 start 和 stop 值，并调用 Query 对象的 set_limits()方法，最终通过语句 list(qs)[::k.step] if k.step else qs 得到结果并返回。在最后的 return 语句中又出现了 list()方法，该方法要求输入一个可迭代的对象，这里会进入 QuerySet 对象中的魔法函数__iter__()中。下面通过一个简单的示例理解这里的调用过程：

```
(django2-core-test) [root@master first_django]# cat test_iter.py
class TestIter:

    def __init__(self, start, stop, step=1):
        self._result_cache = [i ** 2 for i in range(start - 1, stop, step)]

    def __iter__(self):
        print('进入迭代')
        return iter(self._result_cache)

test_iter = TestIter(1, 5)
print(list(test_iter)[1:4])
```

上述代码定义了一个简单的 TestIter 类，然后实现了魔法函数__iter__()。当对 TestIter 对象应用 list()函数时，会通过魔法函数__iter__()得到可迭代值，并将其转成 list 类型，最后对该列表切片。该脚本的运行结果如下：

```
(django2-core-test) [root@master first_django]# python test_iter.py
进入迭代
[1, 4, 9]
```

有了上面的基础之后，可知最终获取数据库结果的为魔法函数__iter__()，其源码如下：

```
# 源码位置：django/db/models/query.py
# ……

class QuerySet:

    # ……

    def __iter__(self):
        # 调用_fetch_all()方法，获取结果
        self._fetch_all()
        return iter(self._result_cache)

    # ……

    def _fetch_all(self):
        if self._result_cache is None:
            # 最关键的一句，获取结果
            self._result_cache = list(self._iterable_class(self))
```

```
        if self._prefetch_related_lookups and not self._prefetch_done:
            self._prefetch_related_objects()

    # ……
```

看到上面的代码，是不是有似曾相识的感觉？前文介绍的 RawQuerySet 类中也有_fetch_all()方法，且内容也十分接近。其中，获取数据库中的数据最关键的语句为 list(self._iterable_class(self))。这里的_iterable_class 在类的初始化方法__init__()中定义过，其值为 ModelIterable，因此 self._iterable_class(self)语句的含义就是实例化 ModelIterable 类。接着从对该得到的 ModelIterable 对象使用 list()方法，继续把目光转向 ModelIterable 类中的魔法函数__iter__()，其源码内容如下：

```
# 源码位置：django/db/models/query.py
# ……

class BaseIterable:
    def __init__(self, queryset, chunked_fetch=False,
chunk_size=GET_ITERATOR_CHUNK_SIZE):
        # 初始化的第 1 个参数就是 QuerySet 对象
        self.queryset = queryset
        self.chunked_fetch = chunked_fetch   # 一般默认值 False
        self.chunk_size = chunk_size         # 默认为 100


class ModelIterable(BaseIterable):
"""迭代每一条记录产生模型的实例"""

    def __iter__(self):
        # QuerySet 对象，是在初始化时得到的
        queryset = self.queryset
        # 选择的数据库
        db = queryset.db
        # 从 queryset 中得到 SQLCompiler 对象，前面分析过该类的具体位置
        compiler = queryset.query.get_compiler(using=db)
        # 执行 query，得到查询结果，保存到 results 中
        results = compiler.execute_sql(chunked_fetch=self.chunked_fetch,
chunk_size=self.chunk_size)
        # 得到选择的字段信息、关联模型表信息及相关聚合查询信息
        select, klass_info, annotation_col_map = (compiler.select, compiler.klass_info,
                                                  compiler.annotation_col_map)
        model_cls = klass_info['model']
        select_fields = klass_info['select_fields']
        model_fields_start, model_fields_end = select_fields[0], select_fields[-1] + 1
        # 其实就是打印表字段列表
        # ['id', 'book_name', 'author', 'sex', 'price', 'isbn', 'publish_date']
        init_list = [f[0].target.attname
                     for f in select[model_fields_start:model_fields_end]]
        # 对于 DjangoBooks.objects.all()语句来说，以下两个变量值均为[]
```

```
        related_populators = get_related_populators(klass_info, select, db)
        known_related_objects = [
            (field, related_objs, operator.attrgetter(*[
                field.attname
                if from_field == 'self' else
                queryset.model._meta.get_field(from_field).attname
                for from_field in field.from_fields
            ])) for field, related_objs in queryset._known_related_objects.items()
        ]
        for row in compiler.results_iter(results):
            # 实例化一个模型对象，row[model_fields_start:model_fields_end]就对应一条记录
            obj = model_cls.from_db(db, init_list,
row[model_fields_start:model_fields_end])

            # 一些与 related_populators 和 known_related_objects 相关的处理语句
            # ……

            yield obj
```

在上面的代码中有一些比较复杂的部分，比如 related_populators 和 known_related_objects，这里涉及关联表的查询。对 DjangoBooks.objects.all()语句而言（在模型表中没有关联其他模型表），这两个值都是[]，因此在对结果（results）进行处理时去掉了和这两个参数相关的代码段。此时，该方法的逻辑已经非常清晰了：通过 SQLCompiler 对象的 execute_sql()方法获取数据库的查询结果，然后在将每条结果都转成对应的模型对象后使用 yield 方式返回。

再次进入 SQLCompiler 类中定义的 execute_sql()方法，其源码如下：

```
# 源码位置：django/db/models/sql/compiler.py
# ……

class SQLCompiler:

    # ……

    def execute_sql(self, result_type=MULTI, chunked_fetch=False,
chunk_size=GET_ITERATOR_CHUNK_SIZE):

        result_type = result_type or NO_RESULTS
        try:
            # 通过这里的 as_sql()方法可以得到对应的查询的 SQL 语句
            sql, params = self.as_sql()
            if not sql:
                raise EmptyResultSet
        except EmptyResultSet:
            if result_type == MULTI:
                return iter([])
            else:
```

```
        Return
    # 在默认情况下 chunked_fetch 为 False
    if chunked_fetch:
        cursor = self.connection.chunked_cursor()
    else:
        # 获取封装过的游标对象
        cursor = self.connection.cursor()
    try:
        # 核心语句，调用游标对象的 execute()方法执行 SQL 语句
        cursor.execute(sql, params)
    except Exception:
        # 异常情况处理
        cursor.close()
        raise

    if result_type == CURSOR:
        return cursor
    if result_type == SINGLE:
        try:
            val = cursor.fetchone()
            if val:
                return val[0:self.col_count]
            return val
        finally:
            # 关闭连接
            cursor.close()
    if result_type == NO_RESULTS:
        cursor.close()
        return

    # 在默认情况下 result_type=MULTI
    result = cursor_iter(
        cursor, self.connection.features.empty_fetchmany_value,
        self.col_count if self.has_extra_select else None,
        chunk_size,  # 默认为 100
    )

    #在默认情况下 chunked_fetch=False,
    self.connection.features.can_use_chunked_reads=True
    #if not chunked_fetch or not self.connection.features.can_use_chunked_reads:
        try:
            # 返回结果，再次使用 list()函数
            return list(result)
        finally:
            # done with the cursor
            cursor.close()
    return result
```

```
    # ……

# ……

def cursor_iter(cursor, sentinel, col_count, itersize):

    try:
        # 每次都取 itersize 条数据
        for rows in iter((lambda: cursor.fetchmany(itersize)), sentinel):
            yield rows if col_count is None else [r[:col_count] for r in rows]
    finally:
        cursor.close()
```

在上面的代码中，从 execute_sql()方法中可以得到两条重要信息：

◎ 对 SQLCompiler 类中的 as_sql()方法不难猜出其含义，该方法正是根据 Query 对象结合参数 params 生成同样含义的 SQL 语句。

◎ 最终通过游标对象的 execute()方法去执行编译得到的参数 sql 和 params，得到数据库查询结果。

### 实战案例

下面通过案例演示上面的代码段，以便更好地理解源码。首先测试 as_sql()方法：

```
(django2-core-test) [root@master first_django]# python manage.py shell
Python 3.8.6 (default, Oct 18 2020, 15:33:08)
[GCC 4.8.5 20150623 (Red Hat 4.8.5-39)] on linux
Type "help", "copyright", "credits" or "license" for more information.
(InteractiveConsole)
>>> from shell_test.models import DjangoBooks
>>> d = DjangoBooks.objects.all()
#从 ModelIterable 类的魔法函数__iter__()中获取 SQLCompiler 对象的语句
>>> compiler = d.query.get_compiler(using='default')
>>> compiler
<django.db.backends.mysql.compiler.SQLCompiler object at 0x7f7b953c9ac0>
>>> compiler.as_sql()
('SELECT `django_books`.`id`, `django_books`.`book_name`, `django_books`.`author`,
`django_books`.`sex`, `django_books`.`price`, `django_books`.`isbn`,
`django_books`.`publish_date` FROM `django_books`', ())
>>>
```

可以看到，DjangoBooks.objects.all()翻译过来就是很简单的 SQL 语句，连 where 条件都没有。下面使用 execute_sql()方法执行数据库查询：

```
# 获取 sql 和 params
>>> sql, params = compiler.as_sql()
>>> cursor = compiler.connection.cursor()
```

```
>>> cursor
<django.db.backends.utils.CursorWrapper object at 0x7f7b953f18b0>
# 执行 SQL 语句
>>> cursor.execute(sql, params)
3
>>> from django.db.models.sql.compiler import cursor_iter
# 迭代查询结果，内部使用 cursor.fetchmany()方法
>>> result = cursor_iter(cursor, compiler.connection.features.empty_fetchmany_value,
None, 100)
>>> result
<generator object cursor_iter at 0x7f7b953c6c80>
# 迭代结果，由于设置每次最多获取 100 条记录，所以迭代一次后就停止迭代了
>>> next(result)
((1, 'Django 源码笔记', 'spyinx', 0, 109.0, '11111', datetime.datetime(2021, 2, 4, 21, 42,
43)), (2, 'Ansible 核心源码剖析与项目实战', 'spyinx', 0, 119.0, '22222',
datetime.datetime(2021, 2, 4, 21, 43, 14)), (3, 'Python 自动化运维', 'test', 1, 79.0,
'333333', datetime.datetime(2021, 2, 4, 21, 43, 37)))
>>> next(result)
Traceback (most recent call last):
  File "<console>", line 1, in <module>
StopIteration
>>>
```

在上述交互模式下，笔者简单模拟了 SQLCompiler 对象中的 execute_sql()方法。由于默认的 chunk_size 值为 100，所以在迭代一次后，结果已经取完。再次调用 next()函数时，将抛出 StopIteration 异常。此外，可以直接对得到的结果使用 list()函数，如下所示：

```
# 接着前面的操作，一些重复的操作不再执行
>>> cursor = compiler.connection.cursor()
>>> cursor.execute(sql, params)
3
>>> from django.db.models.sql.compiler import cursor_iter
>>> result = cursor_iter(cursor, compiler.connection.features.empty_fetchmany_value,
None, 100)
# 直接使用 list()函数
>>> list(result)
[((1, 'Django 源码笔记', 'spyinx', 0, 109.0, '11111', datetime.datetime(2021, 2, 4, 21,
42, 43)), (2, 'Ansible 核心源码剖析与项目实战', 'spyinx', 0, 119.0, '22222',
datetime.datetime(2021, 2, 4, 21, 43, 14)), (3, 'Python 自动化运维', 'test', 1, 79.0,
'333333', datetime.datetime(2021, 2, 4, 21, 43, 37)))]
>>>
```

此外，可以尝试设置 chunk_size 为 2，再次运行上面的代码，结果如下：

```
>>> cursor = compiler.connection.cursor()
>>> cursor.execute(sql, params)
3
```

```
>>> result = cursor_iter(cursor, compiler.connection.features.empty_fetchmany_value,
None, 2)
>>> list(result)
[((1, 'Django源码笔记', 'spyinx', 0, 109.0, '11111', datetime.datetime(2021, 2, 4, 21,
42, 43)), (2, 'Ansible核心源码剖析与项目实战', 'spyinx', 0, 119.0, '22222',
datetime.datetime(2021, 2, 4, 21, 43, 14))), ((3, 'Python自动化运维', 'test', 1, 79.0,
'333333', datetime.datetime(2021, 2, 4, 21, 43, 37)),)]
```

由于把 chunk_size 设置为 2，所以第一次迭代只有 2 条记录，第二次迭代有 1 条记录。最终 list(result)的结果应该有 2 个元素，分别为两次迭代的结果。因此在 SQLCompiler 对象中 execute_sql()方法的结果已经非常明确了。继续看 ModelIterable 类的魔法函数__iter__()对得到的结果的后续处理，即将得到的每条记录都转成对应的模型对象。下面完整演示一遍这个过程：

```
(django2-core-test) [root@master first_django]# python manage.py shell
Python 3.8.6 (default, Oct 18 2020, 15:33:08)
[GCC 4.8.5 20150623 (Red Hat 4.8.5-39)] on linux
Type "help", "copyright", "credits" or "license" for more information.
(InteractiveConsole)
>>> from shell_test.models import DjangoBooks
>>> queryset = DjangoBooks.objects.all()
>>> compiler = queryset.query.get_compiler(using='default')
>>> results = compiler.execute_sql(chunked_fetch=False, chunk_size=100)
>>> results
[((1, 'Django源码笔记', 'spyinx', 0, 109.0, '11111', datetime.datetime(2021, 2, 4, 21,
42, 43)), (2, 'Ansible核心源码剖析与项目实战', 'spyinx', 0, 119.0, '22222',
datetime.datetime(2021, 2, 4, 21, 43, 14)), (3, 'Python自动化运维', 'test', 1, 79.0,
'333333', datetime.datetime(2021, 2, 4, 21, 43, 37)))]
>>> select, klass_info, annotation_col_map = (compiler.select,
compiler.klass_info,compiler.annotation_col_map)
>>> select
[(Col(django_books, shell_test.DjangoBooks.id), ('`django_books`.`id`', []), None),
(Col(django_books, shell_test.DjangoBooks.book_name), ('`django_books`.`book_name`',
[]), None), (Col(django_books, shell_test.DjangoBooks.author),
('`django_books`.`author`', []), None), (Col(django_books, shell_test.DjangoBooks.sex),
('`django_books`.`sex`', []), None), (Col(django_books, shell_test.DjangoBooks.price),
('`django_books`.`price`', []), None), (Col(django_books, shell_test.DjangoBooks.isbn),
('`django_books`.`isbn`', []), None), (Col(django_books,
shell_test.DjangoBooks.publish_date), ('`django_books`.`publish_date`', []), None)]
>>> klass_info
{'model': <class 'shell_test.models.DjangoBooks'>, 'select_fields': [0, 1, 2, 3, 4, 5,
6]}
>>> annotation_col_map
{}
>>> model_cls = klass_info['model']
>>> select_fields = klass_info['select_fields']
>>> model_fields_start, model_fields_end = select_fields[0], select_fields[-1] + 1
```

```
>>> init_list = [f[0].target.attname for f in
select[model_fields_start:model_fields_end]]
>>> init_list
['id', 'book_name', 'author', 'sex', 'price', 'isbn', 'publish_date']
>>> for row in compiler.results_iter(results):
...     obj = model_cls.from_db('default', init_list,
row[model_fields_start:model_fields_end])
...     print(obj, type(obj))
...
<Django 源码笔记, spyinx> <class 'shell_test.models.DjangoBooks'>
<Ansible 核心源码剖析与项目实战, spyinx> <class 'shell_test.models.DjangoBooks'>
<Python 自动化运维, test> <class 'shell_test.models.DjangoBooks'>
```

通过上面的操作，简单实现了魔法函数__iter__()的整个过程（忽略了对 related_populators 和 known_related_objects 两个参数的处理）。在掌握了上面这些基础知识后，再学习 QuerySet 类中的其他方法就非常容易了。接下来介绍其他常见的 QuerySet 方法，之后便可以解决本节一开始提出的那些问题了。

### 2. first()/last()方法：获取查询结果的第一条记录或最后一条记录

这里要分析的查询语句为 DjangoBooks.objects.all().first()（first()也可以换成 last()，它们的分析过程是一样的）。首先看 first()的实现源码，如下：

```
# 源码位置：django/db/models/sql/compiler.py
# ……

class SQLCompiler:

    # ……

    def first(self):
        """返回查询结果的第 1 个对象，如果没有则为 None"""
        for obj in (self if self.ordered else self.order_by('pk'))[:1]:
            return obj

    def last(self):
        """返回查询结果的最后一个对象，如果没有则为 None"""
        for obj in (self.reverse() if self.ordered else self.order_by('-pk'))[:1]:
            return obj

    # ……
```

语句 self if self.ordered else self.order_by('pk')表示根据是否设置排序字段来获取 QuerySet 对象。如果没有设置排序字段，则会得到当前 QuerySet 对象的一个副本并给其 query 属性添加排序字段'pk'：

```
# 源码位置：django/db/models/sql/compiler.py
# ……
```

```
class SQLCompiler:

    # ……

    def order_by(self, *field_names):
        """返回一个新的QuerySet对象，并设置其query属性的排序信息"""
        assert self.query.can_filter(), \
            "Cannot reorder a query once a slice has been taken."
        obj = self._chain()
        obj.query.clear_ordering(force_empty=False)
        obj.query.add_ordering(*field_names)
        return obj

    # ……
```

在 first()方法中，最关键的地方在于[:1]，正是因为这个索引操作，导致得到的 QuerySet 对象会调用魔法函数__getitem__()，这样就又回到了前面的分析链上。后续的追踪不再赘述，参考前面打印 QuerySet 对象的追踪过程即可。

注意，和 all()方法不同，这里在调用 first()方法后会直接去数据库中查找符合条件的第一条数据，而 all()方法只是返回新的 QuerySet 对象，并不会查询数据库。对于 first()方法，这里是通过索引操作成功获取第一条数据的。而对于 last()方法，则是对排序的字段再逆向排序后取第一条数据，即可达到取原来查询结果中最后一条数据的效果。

如果想查看在调用 first()方法后对应执行的 SQL 语句，可以在源码中加 print()语句。首先找到 QuerySet 类中的魔法函数__getitem__()，如下所示：

```
# 源码位置：django/db/models/query.py
# ……

class QuerySet:

    # ……

    def __getitem__(self, k):

        # ……

        if isinstance(k, slice):
            # ……
            qs.query.set_limits(start, stop)
            # 通过前面的分析可知，最终first()或者last()方法会调用至这里
            print('待执行的SQL语句:{}'.format(qs.query.sql_with_params()))
            return list(qs)[::k.step] if k.step else qs
```

```
        qs = self._chain()
        qs.query.set_limits(k, k + 1)
        qs._fetch_all()
        return qs._result_cache[0]

    # ……
```

在魔法函数__getitem__()中，k 为 slice 类型，且值为 slice(None, 1, None)，因此，程序会进入 if isinstance(k, slice)语句下执行。这里可以通过 Query 对象的 sql_with_params()方法得到执行的 SQL 语句，该方法最终调用的是 SQLCompiler 对象中的 as_sql()方法。这里添加的 print()语句如上面的代码所示，同时将该语句添加到虚拟环境的 Django 模块源码中。再次进入 Django 的 shell 模式下执行如下操作：

```
(django2-core-test) [root@master first_django]# python manage.py shell
Python 3.8.6 (default, Oct 18 2020, 15:33:08)
[GCC 4.8.5 20150623 (Red Hat 4.8.5-39)] on linux
Type "help", "copyright", "credits" or "license" for more information.
(InteractiveConsole)
>>> from shell_test.models import DjangoBooks
>>> DjangoBooks.objects.all().first()
待执行的 SQL 语句:('SELECT `django_books`.`id`, `django_books`.`book_name`,
`django_books`.`author`, `django_books`.`sex`, `django_books`.`price`,
`django_books`.`isbn`, `django_books`.`publish_date` FROM `django_books` ORDER BY
`django_books`.`id` ASC  LIMIT 1', ())
<DjangoBooks: <Django 源码笔记, spyinx>>
>>> DjangoBooks.objects.all().last()
待执行的 SQL 语句:('SELECT `django_books`.`id`, `django_books`.`book_name`,
`django_books`.`author`, `django_books`.`sex`, `django_books`.`price`,
`django_books`.`isbn`, `django_books`.`publish_date` FROM `django_books` ORDER BY
`django_books`.`id` DESC  LIMIT 1', ())
<DjangoBooks: <Python 自动化运维, test>>
```

至此，关于 first()和 last()方法背后的原理就已介绍完毕。接下来开始探索 filter()方法的源码。

### 3．filter()方法：查询过滤方法，非常强大，可以实现非常多的查询方式

下面解析 DjangoBooks.objects.all().filter(author__contains='st')语句的执行过程。filter()方法的源码实现如下：

```
# 源码位置：django/db/models/query.py
# ……

class QuerySet:

    # ……

    def filter(self, *args, **kwargs):
```

```
        return self._filter_or_exclude(False, *args, **kwargs)

    def exclude(self, *args, **kwargs):
        return self._filter_or_exclude(True, *args, **kwargs)

    def _filter_or_exclude(self, negate, *args, **kwargs):
        if args or kwargs:
            assert self.query.can_filter(), \
                "Cannot filter a query once a slice has been taken."
        # 创建当前QuerySet对象的副本
        clone = self._chain()
        if negate:
            clone.query.add_q(~Q(*args, **kwargs))
        else:
            # 针对filter()方法，传入的negate为False
            clone.query.add_q(Q(*args, **kwargs))
        return clone

    # ……
```

上面的代码都比较简单，filter()方法调用的是_filter_or_exclude()方法，传入的 negate 参数值为 False。因此，filter()方法的核心过程即创建当前 QuerySet 对象的副本 clone，并给该副本对象添加 q 查询（调用 query 属性的 add_q()方法）。add_q()方法传入的是一个 Q 对象，Q 类的实现源码如下：

```
# 源码位置：django/db/models/query_utils.py
# ……

class Q(tree.Node):
    """
    将过滤器封装成可以使用&或者|等逻辑组合的对象
    """
    # 连接类型
    AND = 'AND'
    OR = 'OR'
    default = AND
    conditional = True

    def __init__(self, *args, _connector=None, _negated=False, **kwargs):
        # 调用父类的初始化方法
        super().__init__(children=[*args, *sorted(kwargs.items())],
connector=_connector, negated=_negated)

    # 组合方法，Q类中最核心的方法
    def _combine(self, other, conn):
        # 必须是Q对象才能组合，否则直接抛出异常
        if not isinstance(other, Q):
            raise TypeError(other)
```

```
        if not other:
            # 如果待组合的other是空的（初始化参数为空），则返回Q对象自身的一个深拷贝
            return copy.deepcopy(self)
        elif not self:
            # 如果Q对象为空，则返回other的深拷贝
            return copy.deepcopy(other)

        obj = type(self)()          # 创建一个空的Q对象
        obj.connector = conn        # 连接器，是AND、OR等
        obj.add(self, conn)         # 添加子节点
        obj.add(other, conn)        # 添加other子节点
        return obj

    # 魔法函数，非常重要
    def __or__(self, other):
        return self._combine(other, self.OR)

    def __and__(self, other):
        return self._combine(other, self.AND)

    def __invert__(self):
        obj = type(self)()  # 得到空的Q对象
        obj.add(self, self.AND)
        obj.negate()
        return obj

    # ……

# ……
```

从上面的代码可以看到 Q 类的功能，它通过魔法函数使得 Q 对象支持&、|和~三个逻辑操作。由于 Q 类继承自 tree.Node 类，所以这里为了更好地理解 Q 类源码，继续追踪其父类（Node）的实现，如下：

```
# 源码位置：django/utils/tree.py

import copy

from django.utils.hashable import make_hashable


class Node:

    default = 'DEFAULT'

    def __init__(self, children=None, connector=None, negated=False):
        # Node对象的三个属性
        self.children = children[:] if children else []
```

```
        self.connector = connector or self.default
        self.negated = negated

    @classmethod
    def _new_instance(cls, children=None, connector=None, negated=False):
        obj = Node(children, connector, negated)
        obj.__class__ = cls
        return obj

    def __str__(self):
        template = '(NOT (%s: %s))' if self.negated else '(%s: %s)'
        return template % (self.connector, ', '.join(str(c) for c in self.children))

    def __repr__(self):
        return "<%s: %s>" % (self.__class__.__name__, self)

    def __deepcopy__(self, memodict):
        obj = Node(connector=self.connector, negated=self.negated)
        obj.__class__ = self.__class__
        obj.children = copy.deepcopy(self.children, memodict)
        return obj

    def __len__(self):
        """返回子节点数"""
        return len(self.children)

    def __bool__(self):
        """判断节点是否有子节点"""
        return bool(self.children)

    def __contains__(self, other):
        """如果 other 是该节点的子节点，返回 True"""
        return other in self.children

    def __eq__(self, other):
        """判断节点是否相等，三个属性必须完全相同"""
        return (
            self.__class__ == other.__class__ and
            (self.connector, self.negated) == (other.connector, other.negated) and
            self.children == other.children
        )

    def __hash__(self):
        return hash((self.__class__, self.connector, self.negated,
*make_hashable(self.children)))

    def add(self, data, conn_type, squash=True):
        if data in self.children:
```

```
            # 如果节点已经是该对象的子节点，直接返回
            return data
        if not squash:
            # 当 squash 为 False 时，只添加 data 到该对象的 children 属性中并返回 data
            self.children.append(data)
            return data
        if self.connector == conn_type:
            # 同一个连接类型
            if (isinstance(data, Node) and not data.negated and
                    (data.connector == conn_type or len(data) == 1)):
                self.children.extend(data.children)
                return self
            else:
                self.children.append(data)
                return data
        else:
            obj = self._new_instance(self.children, self.connector,
                                     self.negated)
            self.connector = conn_type
            self.children = [obj, data]
            return data

    def negate(self):
        """对 negated 属性取反"""
        self.negated = not self.negated
```

为了进一步熟悉和掌握 Q 类，下面使用该类完成一些简单的案例：

```
(django2-core-test) [root@master first_django]# python manage.py shell
Python 3.8.6 (default, Oct 18 2020, 15:33:08)
[GCC 4.8.5 20150623 (Red Hat 4.8.5-39)] on linux
Type "help", "copyright", "credits" or "license" for more information.
(InteractiveConsole)
>>> from django.utils.tree import Node
>>> n1 = Node(connector='AND')
>>> n2 = Node(connector='AND')
>>> n3 = Node(connector='AND')
>>> n1.add(n2, 'AND', squash=False)
<Node: (AND: )>
>>> n1.children
[<Node: (AND: )>]
>>> n2 in n1
True
# 这次在 add()方法中传入 conn_type= 'OR'
>>> n3.add(n2, 'OR')
<Node: (AND: )>
>>> n3.children
[<Node: (AND: )>, <Node: (AND: )>]
```

```
# 查看 n3 的子节点
>>> len(n3)
2
# 查看 n3 是否有子节点
>>> bool(n3)
True
>>> bool(n2)
False
```

在简单操作 Node 节点后，可以尝试操作 Q 对象，在实战中进一步理解 Q 类：

```
(django2-core-test) [root@master first_django]# python manage.py shell
Python 3.8.6 (default, Oct 18 2020, 15:33:08)
[GCC 4.8.5 20150623 (Red Hat 4.8.5-39)] on linux
Type "help", "copyright", "credits" or "license" for more information.
(InteractiveConsole)
>>> from django.db.models.query_utils import Q
>>> q1 = Q(author__contains='st')
>>> q2 = Q(price__gt=100)
>>> q1.deconstruct()
('django.db.models.Q', (), {'author__contains': 'st'})
>>> q2.deconstruct()
('django.db.models.Q', (), {'price__gt': 100})
>>> q_and = q1 & q2
>>> q_and
<Q: (AND: ('author__contains', 'st'), ('price__gt', 100))>
>>> q_and.children
[('author__contains', 'st'), ('price__gt', 100)]
>>> q_and.deconstruct()
('django.db.models.Q', (('author__contains', 'st'), ('price__gt', 100)), {})
>>> q_or = q1 | q2
>>> q_or.children
[('author__contains', 'st'), ('price__gt', 100)]
>>> q_or.deconstruct()
('django.db.models.Q', (('author__contains', 'st'), ('price__gt', 100)), {'_connector':
'OR'})
>>> q_not = ~q1
>>> q_not.deconstruct()
('django.db.models.Q', (), {'author__contains': 'st', '_negated': True})
```

上面用到了在 Q 类中定义的 deconstruct()方法，该方法主要用于打印 Q 类的一些基本信息，比如相关的参数信息、连接方式及是否取反等。它的源码实现如下：

```
# 源码位置：django/db/models/query_utils.py
# ……

class Q(tree.Node):

    # ……
```

```
def deconstruct(self):
    # path 为当前对象所在的模块全路径
    path = '%s.%s' % (self.__class__.__module__, self.__class__.__name__)
    if path.startswith('django.db.models.query_utils'):
        # 将模块路径'django.db.models.query_utils'转换成'django.db.models.Q'
        path = path.replace('django.db.models.query_utils', 'django.db.models')
    # 相关参数信息
    args, kwargs = (), {}
    if len(self.children) == 1 and not isinstance(self.children[0], Q):
        child = self.children[0]
        kwargs = {child[0]: child[1]}
    else:
        # 当有多个 Q 对象时，给 args 变量赋值
        args = tuple(self.children)
        # 有多个 Q 对象且连接器属性非默认值，此时需设置 kwargs 值
        if self.connector != self.default:
            kwargs = {'_connector': self.connector}
    if self.negated:
        # 如果是取反的属性，也需要加入 kwargs 中
        kwargs['_negated'] = True
    return path, args, kwargs
```

根据这里的源码，就能理解前面关于 Q 对象的输出了。在掌握了 Q 对象的基本用法后，继续回到 clone.query.add_q(Q(*args, **kwargs))语句。下一步需要追踪的是 Query 对象中的 add_q()方法，它的源码实现如下：

```
# 源码位置：django/db/models/sql/query.py
# ……

class Query:

    # ……

    def add_q(self, q_object):

        # 没有内连接
        existing_inner = {a for a in self.alias_map if self.alias_map[a].join_type == INNER}
        # 处理过滤条件，核心语句
        clause, _ = self._add_q(q_object, self.used_aliases)
        if clause:
            self.where.add(clause, AND)
        self.demote_joins(existing_inner)

    # ……
```

后续的代码比较难理解，因此这里采用先看现象后理解源码的方式。首先使用三个过滤参数测

试_add_q()方法的结果，这三个参数分别为作者名包含'st'字符串、图书价格大于 100 元、性别为男性（0）。测试结果如下：

```
(django2-core-test) [root@master first_django]# python manage.py shell
Python 3.8.6 (default, Oct 18 2020, 15:33:08)
[GCC 4.8.5 20150623 (Red Hat 4.8.5-39)] on linux
Type "help", "copyright", "credits" or "license" for more information.
(InteractiveConsole)
>>> from shell_test.models import DjangoBooks
>>> queryset = DjangoBooks.objects.all()
# 克隆原 QuerySet 对象，出现在_filter_or_exclude()方法源码中
>>> clone = queryset._chain()
# 从 QuerySet 中获取 Query 对象
>>> query = clone.query
# 以下两个变量来自 Qeury 类的 add_q()方法。由于模型没有关联性，所以 alias_map 为空字典
>>> query.alias_map
OrderedDict()
>>> query.used_aliases
set()
>>> from django.db.models.query_utils import Q
# filter()方法中的参数都原封不动地传入，这里将参数全部放入 Q 类中进行初始化
>>> clause, _ = query._add_q(Q(author__contains='st', price__gt=100, sex=0),
query.used_aliases)
>>> clause
<WhereNode: (AND: <django.db.models.lookups.Contains object at 0x7f005052afa0>,
<django.db.models.lookups.GreaterThan object at 0x7f005052ab50>,
<django.db.models.lookups.Exact object at 0x7f005052ab80>)>
# 对 Q 对象使用&操作，最终得到的仍是一个 Q 对象
>>> clause, _ = query._add_q(Q(author__contains='st') & Q(price__gt=100) & Q(sex=0),
query.used_aliases)
>>> clause
<WhereNode: (AND: <django.db.models.lookups.Contains object at 0x7f005052ac70>,
<django.db.models.lookups.GreaterThan object at 0x7f005052aa00>,
<django.db.models.lookups.Exact object at 0x7f005052aa90>)>
# 对 Q 对象使用&和|混合操作，最终得到的仍是一个 Q 对象
>>> clause, _ = query._add_q(Q(author__contains='st') & Q(price__gt=100) | Q(sex=0),
query.used_aliases)
>>> clause
<WhereNode: (OR: (AND: <django.db.models.lookups.Contains object at 0x7f005052a700>,
<django.db.models.lookups.GreaterThan object at 0x7f005052af70>),
<django.db.models.lookups.Exact object at 0x7f005052adc0>)>
```

从上面的结果中可以得出一个非常重要的信息：过滤器中的参数全部都由 Query 对象中的 add_q()方法来处理，最终得到 WhereNode 对象。在 WhereNode 对象中包括众多的条件对象，例如，判断是否包含字符的 Contains 对象、判断大于的 GreaterThan 对象及判断等于的 Exact 对象，这些对象都在源码的 django.db.models.lookups 路径下定义。此外，针对 Q 查询的逻辑操作，在 WhereNode 对象中

都有记录。例如，在上面测试的最后一个_add_q()语句中，笔者使用了两个 Q 对象取与后再与一个 Q 对象取或操作，得到的 WhereNode 节点刚好反映了这一关系，如代码输出中所示，Contains 对象与 GreaterThan 对象先做 AND 操作，接着和 Exact 对象做 OR 操作。关于_add_q()方法的源码及条件类（Exact、GreaterThan 等）涉及的相关源码不再进一步分析了，有兴趣的读者可以自行学习。最后，将上面演示的条件查询用 SQLCompiler 编译成 SQL 语句，操作如下：

```
(django2-core-test) [root@master first_django]# python manage.py shell
Python 3.8.6 (default, Oct 18 2020, 15:33:08)
[GCC 4.8.5 20150623 (Red Hat 4.8.5-39)] on linux
Type "help", "copyright", "credits" or "license" for more information.
(InteractiveConsole)
>>> from django.db.models.query_utils import Q
>>> from shell_test.models import DjangoBooks
>>> queryset = DjangoBooks.objects.all()
>>> clone = queryset._chain()
>>> query = clone.query
>>> clause, _ = query._add_q(Q(author__contains='st') & Q(price__gt=100) | Q(sex=0),
query.used_aliases)
>>> query.where.add(clause, 'AND')
<WhereNode: (OR: (AND: <django.db.models.lookups.Contains object at 0x7f5d0fd85ac0>,
<django.db.models.lookups.GreaterThan object at 0x7f5d0fd85b80>),
<django.db.models.lookups.Exact object at 0x7f5d0fd85580>)>
# sql_with_params()方法返回二元组(sql, params)
>>> query.sql_with_params()
('SELECT `django_books`.`id`, `django_books`.`book_name`, `django_books`.`author`,
`django_books`.`sex`, `django_books`.`price`, `django_books`.`isbn`,
`django_books`.`publish_date` FROM `django_books` WHERE ((`django_books`.`author` LIKE
BINARY %s AND `django_books`.`price` > %s) OR `django_books`.`sex` = %s)', ('%st%', 100.0,
0))
```

通过这个完整的操作示例，读者可以了解 filter()方法中的参数是如何解析的。filter()方法会被 QuerySet 类中的很多方法调用，比如下面的 get()方法。

#### 4. get()方法：查询单条数据，可附带简单的查询条件

通常，在 Django 项目的查询数据库的操作中会经常使用 get()方法，通过传入查询参数，实现简单条件查询。注意，get()方法最终只能得到一条记录，如果查询无结果或者有多个结果，就会抛出异常。这些逻辑在 get()方法的源码中已经十分明确，具体如下：

```
# 源码位置：django/db/models/query.py
# ……

class QuerySet:

    # ……
```

```
def get(self, *args, **kwargs):

    # 将查询参数写入新的 QuerySet 对象中
    clone = self.filter(*args, **kwargs)
    if self.query.can_filter() and not self.query.distinct_fields:
        clone = clone.order_by()
    # 核心部分，调用魔法函数__len__()，查询数据库中的数据
    num = len(clone)
    if num == 1:
        # 查询结果只有一条记录，注意这里的返回结果
        return clone._result_cache[0]
    if not num:
        # 没有对应的查询结果，抛出异常
        raise self.model.DoesNotExist(
            "%s matching query does not exist." %
            self.model._meta.object_name
        )
    # 查询存在多条记录，抛出异常
    raise self.model.MultipleObjectsReturned(
        "get() returned more than one %s -- it returned %s!" %
        (self.model._meta.object_name, num)
    )

# ……
```

上述代码的逻辑非常清晰，其中，get()方法的核心是先调用 filter()方法得到新的包含查询条件的 QuerySet 对象，再调用 len()方法真正获取查询结果并保存在_result_cache 属性中。这里获取数据库查询结果的语句是 len(clone)，实现原理同样基于魔法函数__len__()：

```
# 源码位置：django/db/models/query.py
# ……

class QuerySet:

    # ……

    def __len__(self):
        # 在调用_fetcha_all()方法后，对应 QuertSet 对象的查询结果就保存在了_result_cache 属性中
        self._fetch_all()
        return len(self._result_cache)

    # ……
```

在调用_fech_all()方法后，对应 QuertSet 对象的查询结果就保存在了_result_cache 属性中。在理解了这些之后，再来看一个非常简单的示例，即模拟 get()方法的调用过程：

```
(django2-core-test) [root@master first_django]# python manage.py shell
Python 3.8.6 (default, Oct 18 2020, 15:33:08)
[GCC 4.8.5 20150623 (Red Hat 4.8.5-39)] on linux
Type "help", "copyright", "credits" or "license" for more information.
(InteractiveConsole)
>>> from shell_test.models import DjangoBooks
>>> queryset = DjangoBooks.objects.all()
# 将get()方法中的查询参数对应传给filter()方法，得到新的QuerySet对象
>>> clone = queryset.filter(author='test')
# if判断
>>> queryset.query.can_filter() and not queryset.query.distinct_fields
True
>>> queryset.query.can_filter()
True
>>> queryset.query.distinct_fields
()
>>> clone = clone.order_by()
#魔法函数 __len__()中的语句
>>> clone._fetch_all()
>>> clone._result_cache[0]
<DjangoBooks: <Python自动化运维, test>>
```

### 3.3.3 答疑解惑

在学完上面的内容后，就可以对 3.3.2 节中一开始提出的 4 个问题进行解答了，但是最后一个问题的答案将在学完 3.4 节后得到。

（1）QuerySet 是一个表示查询结果的类，它定义了非常多的方法，比如 get()、filter()、first()等。通过 SQLCompiler 类可以将链式操作最终翻译成待执行的 SQL 语句。

（2）支持链式操作的原理很简单，该方法只需返回一个新的 QuerySet 对象即可。但并不是所有的方法都会返回 QuerySet 对象，比如前面介绍的 get()、first()、last()等方法，它们会直接返回最终的查询结果。

（3）将 QuerySet 对象中的 query 属性（Query 对象，代表着数据库的查询）翻译成 SQL 语句是 SQLCompiler 类的工作，这里的代码略微复杂，可以直接通过调用 SQLCompiler 类中的 as_sql()方法看到最终的编译结果。查看方式有两种：

◎ 对于返回为 QuerySet 对象的语句，可以先通过其 query 属性获取 SQLCompiler 对象，再调用该编译对象的 as_sql()方法即可。操作示例如下：

```
(django2-core-test) [root@master first_django]# python manage.py shell
Python 3.8.6 (default, Oct 18 2020, 15:33:08)
[GCC 4.8.5 20150623 (Red Hat 4.8.5-39)] on linux
Type "help", "copyright", "credits" or "license" for more information.
```

```
(InteractiveConsole)
>>> from shell_test.models import DjangoBooks
>>> queryset = DjangoBooks.objects.all().filter(author__contains='st')
>>> queryset.query.get_compiler(using='default').as_sql()
('SELECT `django_books`.`id`, `django_books`.`book_name`, `django_books`.`author`,
`django_books`.`sex`, `django_books`.`price`, `django_books`.`isbn`,
`django_books`.`publish_date` FROM `django_books` WHERE `django_books`.`author` LIKE
BINARY %s', ('%st%',))
```

◎ 对于如 DjangoBooks.objects.all()[1:2]或者是调用 first()、last()、order_by()方法等的语句，可以在 ModelIterable 类的魔法函数__iter__()中添加 print()语句，打印最终将执行的 SQL 语句，具体如下：

```
# 源码位置：django/db/models/query.py
# ……

class ModelIterable(BaseIterable):

   def __iter__(self):
      queryset = self.queryset
      db = queryset.db
      compiler = queryset.query.get_compiler(using=db)
      # 添加打印 SQL 的语句
      print("最终执行的 SQL 语句为:{}".format(compiler .as_sql()))

      # ……
```

在虚拟环境的 Django 源码中添加如下 print()语句，然后在 Python 交互模式下进行如下测试：

```
(django2-core-test) [root@master first_django]# python manage.py shell
Python 3.8.6 (default, Oct 18 2020, 15:33:08)
[GCC 4.8.5 20150623 (Red Hat 4.8.5-39)] on linux
Type "help", "copyright", "credits" or "license" for more information.
(InteractiveConsole)
>>> from shell_test.models import DjangoBooks
>>> DjangoBooks.objects.all().filter(author__contains='sp').order_by('price')
最终执行的 SQL 语句为:('SELECT `django_books`.`id`, `django_books`.`book_name`,
`django_books`.`author`, `django_books`.`sex`, `django_books`.`price`,
`django_books`.`isbn`, `django_books`.`publish_date` FROM `django_books` WHERE
`django_books`.`author` LIKE BINARY %s ORDER BY `django_books`.`price` ASC  LIMIT 21',
('%sp%',))
<QuerySet [<DjangoBooks: <Django 源码笔记, spyinx>>, <DjangoBooks: <Ansible 核心源码剖析与
项目实战, spyinx>>]>
>>> DjangoBooks.objects.all().filter(author__contains='sp').order_by('-price')
最终执行的 SQL 语句为:('SELECT `django_books`.`id`, `django_books`.`book_name`,
`django_books`.`author`, `django_books`.`sex`, `django_books`.`price`,
`django_books`.`isbn`, `django_books`.`publish_date` FROM `django_books` WHERE
```

```
`django_books`.`author` LIKE BINARY %s ORDER BY `django_books`.`price` DESC  LIMIT 21',
('%sp%',))
<QuerySet [<DjangoBooks: <Ansible 核心源码剖析与项目实战, spyinx>>, <DjangoBooks: <Django
源码笔记, spyinx>>]>
```

（4）能使用 Django.objects.all()[1:2]这样的语句的原因是在 QuerySet 类中定义了魔法函数 __getitem__()，该方法在 3.3.2 节中已经详细分析过。此外，从代码层面看，懒查询方式指的是没有真正执行查询动作，只是返回新的 QuerySet 对象。只有在使用到该查询结果或者是打印结果时才真正去数据库中执行对应的 SQL 语句。这样做的好处是避免了在链式操作中对数据库进行反复查询，而是将完整链式操作合成一条 SQL 语句，只执行一次数据库操作。

## 3.3.4 ORM 框架的聚合操作

聚合函数（aggregation function）又被称为组函数。在默认情况下聚合函数会把当前所在表当作一个组进行统计。关于聚合函数操作，下面给出几个简单的 SQL 示例：

```
# 1. 统计女作者的图书平均价格
SELECT AVG(price) FROM django_books WHERE 1=1 and sex = 1;

# 2. 按 author 字段分组，计算同一作者出版书籍的平均价格并按平均价格从高到低排序
SELECT author, FORMAT(AVG(price), 2) * 1 AS avg_price FROM django_books WHERE 1=1 GROUP
BY author ORDER BY avg_price desc

# 3. 按 author 字段分组，计算同一作者出版书籍数并按该数量大小从高到低排序
SELECT author, COUNT(*) AS count FROM django_books WHERE 1=1 GROUP BY author ORDER BY
count desc

# 4. 将图书价格大于 80 的书籍按作者分类并将作者的书籍以分号连接起来
SELECT author, GROUP_CONCAT(price SEPARATOR ';') FROM django_books WHERE 1=1 and price >
80 GROUP BY author
```

为了能更好地演示聚合操作效果，笔者手动向 django_books 表中插入 10 条数据，如图 3-2 所示。接着进入 Django 的 shell 模式下完成上述示例中的 SQL 语句：

django_book.django_books: 10 总记录数 (大约)　　》下一个

| id | book_name | author | sex | price | isbn | publish_date |
|---|---|---|---|---|---|---|
| 1 | Django源码笔记 | spyinx | 0 | 109 | 11111 | 2021-02-04 21:42:43.000000 |
| 2 | Ansible核心源码剖析与项目实战 | spyinx | 0 | 119 | 22222 | 2021-02-04 21:43:14.000000 |
| 3 | Python自动化运维 | test | 1 | 79 | 333333 | 2021-02-04 21:43:37.000000 |
| 4 | 中国电信媒体存储自主研发之路 | group | 0 | 89 | 1234 | 2021-02-11 20:22:00.000000 |
| 5 | 大国崛起 | xyz | 0 | 59 | 7761 | 2021-02-10 11:12:00.000000 |
| 6 | 小郭历险记 | mmmm | 1 | 39 | 222221 | 2021-02-14 10:23:00.000000 |
| 7 | Django项目实战 | hhhhh | 0 | 79 | 1123456 | 2019-07-18 11:17:00.000000 |
| 8 | 精通Django框架 | zzzzz | 1 | 129 | 1234321 | 2020-01-18 10:27:00.000000 |
| 9 | Django实战 | test | 1 | 89 | 1111112 | 2009-02-28 11:37:00.000000 |
| 10 | Django3.0从入门到精通 | group | 0 | 40 | 1234257 | 2021-04-18 10:27:00.000000 |

图 3-2

```
(django2-core-test) [root@master first_django]# python manage.py shell
Python 3.8.6 (default, Oct 18 2020, 15:33:08)
[GCC 4.8.5 20150623 (Red Hat 4.8.5-39)] on linux
Type "help", "copyright", "credits" or "license" for more information.
(InteractiveConsole)
>>> from django.db.models import Avg, Count
>>> from shell_test.models import DjangoBooks
# 实现示例 1
>>> DjangoBooks.objects.all().filter(sex=1).aggregate(avg_price=Avg('price'))
{'avg_price': 84.0}
# 实现示例 2
>>>
DjangoBooks.objects.values('author').annotate(avg_price=Avg('price')).order_by('-avg
_price')
<QuerySet [{'author': 'zzzzz', 'avg_price': 129.0}, {'author': 'spyinx', 'avg_price':
114.0}, {'author': 'test', 'avg_price': 84.0}, {'author': 'hhhhh', 'avg_price': 79.0},
{'author': 'xyz', 'avg_price': 59.0}, {'author': 'group', 'avg_price': 44.5}, {'author':
'mmmm', 'avg_price': 39.0}]>
# 实现示例 3
>>>
DjangoBooks.objects.values('author').annotate(count=Count('*')).order_by('-count')
<QuerySet [{'author': 'spyinx', 'count': 2}, {'author': 'test', 'count': 2}, {'author':
'group', 'count': 2}, {'author': 'hhhhh', 'count': 1}, {'author': 'xyz', 'count': 1},
{'author': 'zzzzz', 'count': 1}, {'author': 'mmmm', 'count': 1}]>
```

最后一个聚合函数在 Django 内部并没有实现，可以先忽略。下面从源码角度介绍上面两种查询语句的实现原理。待掌握了相关的源码实现后，就可以通过额外手段使得其支持 group_concat()聚合函数。

QuerySet 类中的 aggregate()方法的实现源码如下：

```
# 源码位置：django/db/models/query.py
# ……

class QuerySet:

    # ……

    def aggregate(self, *args, **kwargs):

        if self.query.distinct_fields:
            # 在前面的模型操作中不能出现 distinct()方法，否则会在这里抛出异常
            raise NotImplementedError("aggregate() + distinct(fields) not implemented.")

        # 检验传入的参数，确保传入的聚合函数符合要求，有'resolve_expression'属性值
        self._validate_values_are_expressions((*args, *kwargs.values()),
method_name='aggregate')
        for arg in args:
```

```
            try:
                # 默认名称，对于直接传入的 Avg('price')，该值为 price__avg
                arg.default_alias
            except (AttributeError, TypeError):
                raise TypeError("Complex aggregates require an alias")
            # 没有指定别名的参数的默认名称
            kwargs[arg.default_alias] = arg

        # 创建新的 Query 对象副本
        query = self.query.chain()
        for (alias, aggregate_expr) in kwargs.items():
            # 对于前面第 1 个 SQL 语句而言，alias=avg_price，aggregate_expr=Avg('price')
            query.add_annotation(aggregate_expr, alias, is_summary=True)
            if not query.annotations[alias].contains_aggregate:
                raise TypeError("%s is not an aggregate expression" % alias)
        # 核心语句
        return query.get_aggregation(self.db, kwargs)

    # ……

    @staticmethod
    def _validate_values_are_expressions(values, method_name):
        # 检验传过来的聚合对象是否有 resolve_expression 属性值，没有则属于异常参数
        invalid_args = sorted(str(arg) for arg in values if not hasattr(arg,
'resolve_expression'))
        if invalid_args:
            raise TypeError(
                'QuerySet.%s() received non-expression(s): %s.' % (
                    method_name,
                    ', '.join(invalid_args),
                )
            )
```

上述代码不难理解，先校验传入 aggregate()方法中的聚合对象是否有 resolve_expression 属性，然后遍历传入的聚合对象，并通过 Query 对象的 add_annotation()方法添加到 Query 对象的 annotations 属性中。最后调用 Query 对象的 get_aggregation()方法获得聚合查询结果。此外，已按位置传入的参数都有默认的参数名称，操作示例如下：

```
>>> from django.db.models.aggregates import Avg
>>> DjangoBooks.objects.all().aggregate(Avg('price'))
{'price__avg': 79.1}
```

下面以输入参数 avg_price=Avg('price')为例，在 Shell 模式下直接调用 aggregate()方法中最核心的语句，实现与 aggregate()方法相同的效果，具体操作如下：

```
(django2-core-test) [root@master first_django]# python manage.py shell
Python 3.8.6 (default, Oct 18 2020, 15:33:08)
```

```
[GCC 4.8.5 20150623 (Red Hat 4.8.5-39)] on linux
Type "help", "copyright", "credits" or "license" for more information.
(InteractiveConsole)
>>> from shell_test.models import DjangoBooks
>>> queryset = DjangoBooks.objects.all()
>>> query = queryset.query.chain()
>>> from django.db.models import Avg
# 调用 Query 对象中的 add_annotation()方法，添加聚合操作
>>> query.add_annotation(Avg('price'), 'avg_price', is_summary=True)
# 可以通过 annotations 属性查看输入的所有聚合操作
>>> query.annotations
OrderedDict([('avg_price', Avg(Col(django_books, shell_test.DjangoBooks.price)))])
# 通过 get_aggregation()方法执行该聚合操作，获得最终结果
>>> query.get_aggregation('default', {'avg_price': Avg('price')})
{'avg_price': 79.1}
```

关于 add_annotation()方法此处不再深入研究，重点学习 get_aggregation()方法如何用于查询数据库并返回相应的结果：

```
# 源码位置：django/db/models/query.py
# ……

class QuerySet:

   # ……

   def get_aggregation(self, using, added_aggregate_names):
      """
      返回聚合结果的字典值
      """
      if not self.annotation_select:
         return {}
      has_limit = self.low_mark != 0 or self.high_mark is not None
      # 判断是否有聚合对象
      has_existing_annotations = any(
         annotation for alias, annotation
         in self.annotations.items()
         if alias not in added_aggregate_names
      )

      if (isinstance(self.group_by, tuple) or has_limit or has_existing_annotations or
            self.distinct or self.combinator):
         # ……
      else:
         outer_query = self
         self.select = ()
         self.default_cols = False
         self._extra = {}
```

```
        outer_query.clear_ordering(True)
        outer_query.clear_limits()
        outer_query.select_for_update = False
        outer_query.select_related = False
        # 获取编译对象
        compiler = outer_query.get_compiler(using)
        # 执行 SQL 语句，获取结果
        result = compiler.execute_sql(SINGLE)
        if result is None:
            result = [None] * len(outer_query.annotation_select)

        converters = compiler.get_converters(outer_query.annotation_select.values())
        result = next(compiler.apply_converters((result,), converters))

        # 整理对应结果为字典形式
        return dict(zip(outer_query.annotation_select, result))

    # ……
```

这里以语句 DjangoBooks.objects.all().filter(sex=1).aggregate(avg_price=Avg('price'))为例解读该部分源码。对于该执行语句而言，在进入 aggregate()方法后，得到的 has_existing_annotations 为 False，没有分组等其他操作。因此 aggregate()方法中的第 1 个 if 判断为 False，程序会进入 else 分支。此时，outer_query 为当前 Query 对象，接下来获取该 Query 对象对应的编译器 compiler，然后调用编译对象的 execute_sql()方法执行相应的聚合 SQL 语句并获取结果。最后整理语句并返回，也就是前面看到的结果。下面继续在 Django 的交互模式下单步演示上面的核心语句，如下：

```
(django2-core-test) [root@master first_django]# python manage.py shell
Python 3.8.6 (default, Oct 18 2020, 15:33:08)
[GCC 4.8.5 20150623 (Red Hat 4.8.5-39)] on linux
Type "help", "copyright", "credits" or "license" for more information.
(InteractiveConsole)
>>> from django.db.models.aggregates import Avg
>>> from shell_test.models import DjangoBooks
>>> queryset = DjangoBooks.objects.all().filter(sex=1)
>>> query = queryset.query.chain()
>>> query.add_annotation(Avg('price'), 'avg_price', is_summary=True)
>>> query.annotation_select
OrderedDict([('avg_price', Avg(Col(django_books, shell_test.DjangoBooks.price)))])
# else 分支中的语句
>>> outer_query = query
>>> query.select = ()
>>> query.default_cols = False
>>> query._extra = {}
# if-else 之后给 outer_query 赋值的语句
>>> outer_query.clear_ordering(True)
```

```
>>> outer_query.clear_limits()
>>> outer_query.select_for_update = False
>>> outer_query.select_related = False
# 获取编译对象
>>> compiler = outer_query.get_compiler('default')
# 打印最终执行的 SQL 语句
>>> compiler.as_sql()
('SELECT AVG(`django_books`.`price`) AS `avg_price` FROM `django_books` WHERE
`django_books`.`sex` = %s', (1,))
>>> result = compiler.execute_sql('single')
>>> result
(84.0,)
>>> converters = compiler.get_converters(outer_query.annotation_select.values())
>>> result = next(compiler.apply_converters((result,), converters))
# 转换成数组形式
>>> result
[84.0]
# 最终和 outer_query.annotation_select 字段对应形成字典
>>> dict(zip(outer_query.annotation_select, result))
{'avg_price': 84.0}
```

以上便是示例语句在 aggregate()方法中执行的全过程，通过打印最终执行的 SQL 语句，可以帮助读者更好地理解该聚合方法。接下来分析带分组的聚合操作，具体语句如下：

```
DjangoBooks.objects.values('author').annotate(avg_price=Avg('price')).order_by('-avg
_price')
```

这里思考一个问题：上述语句如何通过 values()方法传入的参数实现 SQL 中的 GROUP BY 功能？首先分析 QuerySet 类中的 values()方法：

```
# 源码位置：django/db/models/query.py
# ……

class QuerySet:

   # ……

   def _values(self, *fields, **expressions):
      clone = self._chain()
      if expressions:
         clone = clone.annotate(**expressions)
      clone._fields = fields
      # 注意这里的 set_values()方法
      clone.query.set_values(fields)
      return clone

   def values(self, *fields, **expressions):
      fields += tuple(expressions)
```

```
        # 创建当前 QuerySet 的新副本
        clone = self._values(*fields, **expressions)
        clone._iterable_class = ValuesIterable
        return clone

    # ……
```

上面关于 values()方法的内容比较简单，它的核心就是调用 set_values()方法生成新的 QuerySet 对象副本。而在 set_values()方法中，最核心的便是 clone.query.set_values(fields)语句。当调用 DjangoBooks.objects.values('author')时，这里的 fields 为('author',)。继续追踪 Query 对象中的 set_values()方法的源码：

```
# 源码位置：django/db/models/sql/query.py
# ……

class Query:

    # ……

    def set_values(self, fields):
        self.select_related = False
        self.clear_deferred_loading()
        self.clear_select_fields()

        # ……

        if fields:
            # 从下面的逻辑可以看出，field_names 是将 fields 转成了数组形式
            field_names = []
            extra_names = []
            annotation_names = []
            if not self._extra and not self._annotations:
                # 对分析的语句而言，self._extra 和 self._annotations 均为 None，进入这里执行
                field_names = list(fields)
            else:
                # ……
            self.set_extra_mask(extra_names)
            self.set_annotation_mask(annotation_names)
        else:
            field_names = [f.attname for f in self.model._meta.concrete_fields]

        self.values_select = tuple(field_names)
        # 注意这里调用的 add_fields()方法
        self.add_fields(field_names, True)

    # ……
```

set_values()方法的主要功能是对 Query 对象的属性进行赋值操作，最后调用了 add_fields()方法。self._extra 和 self._annotations 均为 None，所以会直接进入 if 语句下执行，将 fields（在 values()方法中，参数会组成元组形式）转成列表格式，从而得到 field_names 并传入 add_fields()方法中。继续追踪 add_fields()方法的源码：

```
# 源码位置：django/db/models/sql/query.py
# ……

class Query:

    # ……

    def set_select(self, cols):
        self.default_cols = False
        # 添加到 select 属性中
        self.select = tuple(cols)

    # ……

    def add_fields(self, field_names, allow_m2m=True):
        """
        将给定的模型字段按指定顺序添加到 select 集中，后续会翻译成 select xx, xxxx from
        """
        alias = self.get_initial_alias()
        opts = self.get_meta()

        try:
            cols = []
            for name in field_names:
                join_info = self.setup_joins(name.split(LOOKUP_SEP), opts, alias,
allow_many=allow_m2m)
                targets, final_alias, joins = self.trim_joins(
                    join_info.targets,
                    join_info.joins,
                    join_info.path,
                )
                for target in targets:
                    cols.append(join_info.transform_function(target, final_alias))
            if cols:
                调用 set_select()方法，设置 Query 的 select 属性
                self.set_select(cols)
        except MultiJoin:
            raise FieldError("Invalid field name: '%s'" % name)
        except FieldError:
            # 异常处理
            # ……
```

```
    # ……
```

从前面分析的 Django 表达式可知，field_names 的值为['author']。下面根据该值模拟 add_fields()方法的执行过程，如下：

```
(django2-core-test) [root@master first_django]# python manage.py shell
Python 3.8.6 (default, Oct 18 2020, 15:33:08)
[GCC 4.8.5 20150623 (Red Hat 4.8.5-39)] on linux
Type "help", "copyright", "credits" or "license" for more information.
(InteractiveConsole)
>>> from shell_test.models import DjangoBooks
>>> queryset = DjangoBooks.objects.all()
>>> query = queryset.query
>>> alias = query.get_initial_alias()
>>> alias
'django_books'
>>> opts = query.get_meta()
>>> opts
<Options for DjangoBooks>
# LOOKUP_SEP 就是双下划线字符串，这一点可直接从 django/db/models/constants.py 文件中得到
>>> join_info = query.setup_joins('author'.split('__'), opts, alias, allow_many=True)
>>> join_info
JoinInfo(final_field=<django.db.models.fields.CharField: author>,
targets=(<django.db.models.fields.CharField: author>,), opts=<Options for DjangoBooks>,
joins=['django_books'], path=[], transform_function=<function
Query.setup_joins.<locals>.final_transformer at 0x7f7116ed91f0>)
>>> targets, final_alias, joins = query.trim_joins(join_info.targets, join_info.joins,
join_info.path)
>>> targets
(<django.db.models.fields.CharField: author>,)
>>> final_alias
'django_books'
>>> cols = []
>>> cols.append(join_info.transform_function(targets[0], final_alias))
>>> cols
[Col(django_books, shell_test.DjangoBooks.author)]
>>> query.select
()
>>> query.set_select(cols)
>>> query.select
(Col(django_books, shell_test.DjangoBooks.author),)
```

由上可知，在 QuerySet 对象中调用 values()方法传入的参数最终会反映到 Query 对象的 select 属性中，先记住这里的现象，后续会看到该属性在 annotate()方法中的作用。继续追踪 annotate()方法的源码：

```
# 源码位置：django/db/models/query.py
# ……
```

```
class QuerySet:

    # ……

    def annotate(self, *args, **kwargs):

        # 1. 校验传入的聚合对象是否正确
        self._validate_values_are_expressions(args + tuple(kwargs.values()),
                        method_name='annotate')
        annotations = OrderedDict()  # To preserve ordering of args
        # 对于调用 annotate(avg_price=Avg('annotate'))来说，args 为空元组()
        for arg in args:
            # 由于 args 为空元组，所以下面不会执行，代码并不复杂
            # ……

        # 2. 调用 annotate(avg_price=Avg('annotate'))方法，传入的参数会被保存到 kwargs 中
        annotations.update(kwargs)

        # 得到当前 QuerySet 对象的一个副本
        clone = self._chain()
        # 如果前面是 values('author')，那么 names=('author',)
        names = self._fields
        if names is None:
            # 如果没有，获取全部字段名称
            names = set(chain.from_iterable(
                (field.name, field.attname) if hasattr(field, 'attname') else (field.name,)
                for field in self.model._meta.get_fields()
            ))

        # 3. 核心是调用 Query 对象的 add_annotation()方法
        for alias, annotation in annotations.items():
            # 这个异常容易复现，就是 alias 不能在 names 中出现
            if alias in names:
                raise ValueError("The annotation '%s' conflicts with a field on "
                        "the model." % alias)
            if isinstance(annotation, FilteredRelation):
                clone.query.add_filtered_relation(annotation, alias)
            else:
                # 和 aggregate()方法一样，调用 Query 对象的 add_annotation()方法添加聚合操作
                clone.query.add_annotation(annotation, alias, is_summary=False)

        # 4. 添加分组参数
        for alias, annotation in clone.query.annotations.items():
            if alias in annotations and annotation.contains_aggregate:
                if clone._fields is None:
                    clone.query.group_by = True
                else:
```

```
                # clone._fields 为('author',)
                clone.query.set_group_by()
            break

    return clone

# ……
```

在上面的代码中，笔者整理了 4 处代码段进行说明：

- ◎ 第 1 处代码段和之前的 aggregate()方法一样，会调用_validate_values_are_expressions()方法去校验传入的聚合对象（Avg('price')）是否正确，该方法会校验从 annotate()方法中传入的所有参数是否有 resolve_expression 属性。如果有不符合要求的参数传入，直接抛出异常。
- ◎ 第 2 处代码段比较简单，就是将 annotate()方法中传入的聚合对象及别名保存到变量 annotations 中。
- ◎ 第 3 处代码段也容易理解，就是单纯遍历 annotations 中的聚合对象及其别名，最终会调用 Query 对象 add_annotation()方法，annotation 为聚合对象，alias 为其对应的别名。
- ◎ 第 4 处代码段是添加分组属性，这里非常重要。由于前面调用 values('author')得到的是 QuerySet 对象，clone._fileds 的值为('author', )，因此这一步的核心是调用 Query 对象的 set_group_by()方法。继续追踪 set_group_by()方法的源码：

```
# 源码位置：django/db/models/sql/query.py
# ……

class Query:

    # ……

    def set_group_by(self):

        # 这里正好用到了 self.select 属性
        group_by = list(self.select)
        if self.annotation_select:
            for annotation in self.annotation_select.values():
                # 注意这里的 annotation 正是位于 aggregates.py 中的聚合对象
                for col in annotation.get_group_by_cols():
                    group_by.append(col)
        # 最终得到的 group_by 属性值
        self.group_by = tuple(group_by)

    # ……
```

注意，一开始会根据 Query 对象的 select 属性初次遍历 group_by。前面提到过，通过 values()传入的参数最终会被添加到 self.select 属性上：

```
>>> queryset = DjangoBooks.objects.values('author', 'sex')
>>> queryset.query.select
(Col(django_books, shell_test.DjangoBooks.author), Col(django_books,
shell_test.DjangoBooks.sex))
```

注意，在 set_group_by()方法中，for 循环里的 annotation 变量其实是 django/db/models 目录下 aggregates.py 文件中定义的聚合对象，例如前面分析的 Avg 对象。对于 Django 内置的聚合类来说，get_group_by_cols()方法被定义在父类中且返回的为空列表：

```
# 源码位置：django/db/models/aggregates.py
# ……

class Aggregate(Func):

    # ……

    def get_group_by_cols(self):
        return []

    # ……

class Avg(FixDurationInputMixin, NumericOutputFieldMixin, Aggregate):
    function = 'AVG'
    name = 'Avg'

# ……
```

可以简单测试一下 Avg 类：

```
>>> from django.db.models.aggregates import Avg
>>> annotation = Avg('price')
>>> annotation.get_group_by_cols()
[]
```

最终 self.group_by 就是 self.select，代码如下：

```
>>> from shell_test.models import DjangoBooks
>>> queryset = DjangoBooks.objects.values('author')
>>> queryset.query.select
(Col(django_books, shell_test.DjangoBooks.author),)
# 在 Query 对象中的 group_by 为 None
>>> queryset.query.group_by is None
True
>>> from django.db.models.aggregates import Avg
# 在调用 QuerySet 对象的 annotate()方法后，新的 QuerySet 对象中的 query 就有了 group_by 值
>>> clone = queryset.annotate(avg_price=Avg('price'))
>>> clone.query.group_by
(Col(django_books, shell_test.DjangoBooks.author),)
# 最后打印 SQL 语句，可以看到出现的 GROUP BY 语句
```

```
>>> clone.query.sql_with_params()
('SELECT `django_books`.`author`, AVG(`django_books`.`price`) AS `avg_price` FROM
`django_books` GROUP BY `django_books`.`author` ORDER BY NULL', ())
```

从上面的代码可以看到，Query 对象中 group_by 值的来源和 Query 对象的 select 属性值一致。从打印的 SQL 语句中可以看出，GROUP BY 语句正是来自 Query 对象的 group_by 属性值。

接下来思考如何使 Django 的聚合操作支持 GROUP_CONCAT()函数。首先看在 Django 内部是如何支持 Avg()函数和 Count()函数的：

```
# 源码位置：django/db/models/aggregates.py
# ……

class Aggregate(Func):
    # 模板表达式
    template = '%(function)s(%(distinct)s%(expressions)s)'
    contains_aggregate = True
    name = None
    filter_template = '%s FILTER (WHERE %%(filter)s)'
    window_compatible = True
    allow_distinct = False

    # ……

    def get_group_by_cols(self):
        return []

    # ……

class Avg(FixDurationInputMixin, NumericOutputFieldMixin, Aggregate):
    function = 'AVG'
    name = 'Avg'


class Count(Aggregate):
    function = 'COUNT'
    name = 'Count'
    # 输出类型，对于 Count()函数而言，输出结果是整型
    output_field = IntegerField()
    allow_distinct = True

    def __init__(self, expression, filter=None, **extra):
        if expression == '*':
            expression = Star()
        if isinstance(expression, Star) and filter is not None:
            raise ValueError('Star cannot be used with filter. Please specify a field.')
        super().__init__(expression, filter=filter, **extra)
```

```
    def convert_value(self, value, expression, connection):
        return 0 if value is None else value

# ……
```

从上面的示例可以看到，在定义聚合类中有以下几个重要属性：

◎ function：对应的聚合函数。

◎ template：聚合语句模板。对于 GROUP_CONCAT()函数来说，它还需要支持 SEPARATOR，所以这个属性值需要调整。

◎ 参考 Count 类的实现可知，由于 Count()函数可用于计算记录数，是一个整型数据，所以这里指定了输出字段类型 output_field 为 IntegerField。而 group_concat()函数输出的是字符串格式，所以需要将 output_field 调整为 CharField。

综上所述，我们可以定义一个 Concat 类，使该类继承 Django 中的 Aggregate 类并调整上面介绍的三个属性值。相应的代码如下：

```
(django2-core-test) [root@master first_django]# cat test_concat.py
from django.db.models.aggregates import Aggregate
from django.db.models.fields import CharField

class Concat(Aggregate):
    function = 'GROUP_CONCAT'
    template = '%(function)s(%(distinct)s%(expressions)s%(separator)s)'
    output_field=CharField()

    def __init__(self, expression, distinct=False, separator=',', **extra):
        super(Concat, self).__init__(
            expression,
            distinct='DISTINCT ' if distinct else '',
            separator=' SEPARATOR "%s"' % separator,
            **extra
        )
```

接着在 Django 的 shell 模式下测试定义的聚合对象：

```
(django2-core-test) [root@master first_django]# python manage.py shell
Python 3.8.6 (default, Oct 18 2020, 15:33:08)
[GCC 4.8.5 20150623 (Red Hat 4.8.5-39)] on linux
Type "help", "copyright", "credits" or "license" for more information.
(InteractiveConsole)
>>> from shell_test.models import DjangoBooks
>>> from test_concat import Concat
>>> queryset =
DjangoBooks.objects.all().filter(price__gt=80).values('author').annotate(prices=Conc
at('price', separator=';'))
fields=('author',)
```

```
>>> queryset
<QuerySet [{'author': 'group', 'prices': '89'}, {'author': 'spyinx', 'prices': '109;119'},
{'author': 'test', 'prices': '89'}, {'author': 'zzzzz', 'prices': '129'}]>
>>> queryset.query.sql_with_params()
('SELECT `django_books`.`author`, GROUP_CONCAT(`django_books`.`price` SEPARATOR ";") AS
`prices` FROM `django_books` WHERE `django_books`.`price` > %s GROUP BY
`django_books`.`author` ORDER BY NULL', (80.0,))
```

可以看到，通过简单继承 Django 中的 Aggregate 类就能实现在 MySQL 等数据库中所支持的聚合函数，十分方便。

## 3.4 ORM 框架的部分源码解读

3.3 节重点介绍了 QuerySet 类，它依赖于关联的模型类，因此，本节将从模型类开始介绍，并解答在 3.3.2 节中提出的第 5 个问题。前面在定义 Django 中的模型类时，都会继承 django.db.models.Model 类，它的源码如下：

```
# 源码位置：django/db/models/base.py
# ……

class Model(metaclass=ModelBase):

    # ……
```

从上面的源码可以看到，在 Model 类中设置了 metaclass=ModelBase，关于 metaclass 的相关概念读者可以自行学习。由于在 Model 类定义中指定创建该类的 metaclass，因此 Model 类及其子类都会由 ModelBase 类中定义的魔法函数__new__()创建。在 ModelBase 类中最核心的魔法函数__new__()的源码如下：

```
# 源码位置：django/db/models/base.py
# ……

class ModelBase(type):
    """Metaclass for all models."""
    def __new__(cls, name, bases, attrs, **kwargs):
        super_new = super().__new__  # 即 type.__new__

        # 1. bases 是创建该类的父类，到 Model 类为止
        parents = [b for b in bases if isinstance(b, ModelBase)]
        if not parents:
            # 没有 parents，即该类为 Model 类
            return super_new(cls, name, bases, attrs)

        module = attrs.pop('__module__')  # 模块路径，字符串
        new_attrs = {'__module__': module}
```

```
classcell = attrs.pop('__classcell__', None)
if classcell is not None:
    new_attrs['__classcell__'] = classcell
# 2. 这里将模型类中定义的 Meta 类取出来赋给 attr_meta 变量，如果没有定义，为 None
attr_meta = attrs.pop('Meta', None) # 注意调用 pop()方法弹出
contributable_attrs = {}
# 3. 遍历原来类中定义的所有属性，不包括 Meta，前面已经弹出
for obj_name, obj in list(attrs.items()):
    # 判断 obj 中是否有'has_contribute_to_class'属性
    if _has_contribute_to_class(obj):
        contributable_attrs[obj_name] = obj
    else:
        new_attrs[obj_name] = obj
# 4. 创建类，传入类名、继承的父类及初始属性值
# new_attrs 排除了值为 Field 对象的属性
new_class = super_new(cls, name, bases, new_attrs, **kwargs)

abstract = getattr(attr_meta, 'abstract', False)
meta = attr_meta or getattr(new_class, 'Meta', None)
base_meta = getattr(new_class, '_meta', None)

app_label = None

# 5. 查找模型类对应的 app 配置信息
# 例如对于 DjangoBooks 模型类来说，得到的 app_config 就是'shell_test'
app_config = apps.get_containing_app_config(module)

if getattr(meta, 'app_label', None) is None:
    if app_config is None:
        if not abstract:
            raise RuntimeError(
                "Model class %s.%s doesn't declare an explicit "
                "app_label and isn't in an application in "
                "INSTALLED_APPS." % (module, name)
            )

    else:
        app_label = app_config.label

# 将 app_label 添加到模型类的_meta 属性中
new_class.add_to_class('_meta', Options(meta, app_label))
# ……

# 6. 将剩余的属性（那些包含 contribute_to_class()方法的属性）添加到 new_class 中
for obj_name, obj in contributable_attrs.items():
    new_class.add_to_class(obj_name, obj)

# 在该模型类中定义的所有 Field 字段
```

```
    new_fields = chain(
        new_class._meta.local_fields,
        new_class._meta.local_many_to_many,
        new_class._meta.private_fields
    )
    # 得到在该模型类中定义的所有 Field 字段名，类似于得到表的所有字段名
    field_names = {f.name for f in new_fields}

    # 处理代理模型类
    # ……

    # 7. 处理多表继承中关于父链接的字段
    parent_links = {}
    for base in reversed([new_class] + parents):
        # 在概念上等同于`if base is Model`.
        if not hasattr(base, '_meta'):
            continue

        if base != new_class and not base._meta.abstract:
            continue
        # 定位 OneToOneField 实例.
        for field in base._meta.local_fields:
            if isinstance(field, OneToOneField):
                related = resolve_relation(new_class, field.remote_field.model)
                parent_links[make_model_tuple(related)] = field

    # 8. 追踪从父类中继承的字段
    inherited_attributes = set()
    # ……

    new_class._meta.indexes = [copy.deepcopy(idx) for idx in new_class._meta.indexes]

    # 如果是抽象类，设置相关属性，然后返回该类
    if abstract:
        attr_meta.abstract = False
        new_class.Meta = attr_meta
        return new_class

    # 9. 调用_prepare()方法
    new_class._prepare()
    new_class._meta.apps.register_model(new_class._meta.app_label, new_class)
    return new_class
```

这里去掉了魔法函数__new__()中部分不重要的代码，同时整理出 9 处代码段，相关含义在注释中进行了说明。这里需要关注第 9 处代码段中的语句：new_class._prepare()，它调用了_prepare()方法，在该方法中，最核心的操作就是给新建的类添加 objects 属性。_prepare()方法的源码如下：

```
# 源码位置：django/db/models/base.py
# ……

class ModelBase(type):

    # ……

    def _prepare(cls):

        # 元信息，在魔法函数__new__()中设置
        opts = cls._meta
        opts._prepare(cls)

        # ……

        # 没有设置对应的managers
        if not opts.managers:
            # 不允许在字段中出现objects名称
            if any(f.name == 'objects' for f in opts.fields):
                raise ValueError(
                    "Model %s must specify a custom Manager, because it has a "
                    "field named 'objects'." % cls.__name__
                )

            # 得到manager对象
            manager = Manager()
            manager.auto_created = True
            # 添加objects属性，这里非常关键
            cls.add_to_class('objects', manager)

        for index in cls._meta.indexes:
            if not index.name:
                index.set_name_with_model(cls)

        class_prepared.send(sender=cls)

    @property
    def _base_manager(cls):
        return cls._meta.base_manager

    @property
    def _default_manager(cls):
        return cls._meta.default_manager
```

可以看到，在_prepare()方法中给新类添加了 objects 属性，该属性值为 manager 对象。下面用自定义的 DjangoBooks 类来演示 objects 属性，如下：

```
>>> from shell_test.models import DjangoBooks
>>> DjangoBooks.objects
<django.db.models.manager.Manager object at 0x7f8a8ca5ad00>
>>> DjangoBooks._base_manager
<django.db.models.manager.Manager object at 0x7f8a8ca5aeb0>
>>> DjangoBooks._default_manager
<django.db.models.manager.Manager object at 0x7f8a8ca5ad00>
```

但是在 Model 类中就不会有 objects 属性，如下：

```
>>> from django.db.models.base import Model
>>> Model.objects
Traceback (most recent call last):
  File "<console>", line 1, in <module>
AttributeError: type object 'Model' has no attribute 'objects'
```

原因很简单，查看魔法函数__new__()的第 1 处代码可知，由于在 Model 类的定义中直接带上了 metaclass=ModelBase，而从该处代码中得到的 parents 为空列表，所以程序将直接返由 type 创建的新类，而不会调用后面的_prepare()方法。

下面将目标转向 Manager 类，它位于 django/db/models/manager.py 文件中，源码如下：

```
# 源码位置：django/db/models/manager.py
# ……

class Manager(BaseManager.from_queryset(QuerySet)):
    pass

# ……
```

Manager 类直接继承了 BaseManager.from_queryset(QuerySet)的结果。这个结果是什么呢？继续看 BaseManager 类中的 from_queryset()方法，源码如下：

```
# 源码位置：django/db/models/manager.py
# ……
class BaseManager:
    # ……

    @classmethod
    def from_queryset(cls, queryset_class, class_name=None):
        if class_name is None:
            class_name = '%sFrom%s' % (cls.__name__, queryset_class.__name__)
        # 使用 type 方法创建新类
        return type(class_name, (cls,), {
            '_queryset_class': queryset_class,
            **cls._get_queryset_methods(queryset_class),
        })

    # ……
```

from_queryset()方法非常简单，就是用 type 方法创建一个新类。其中，默认的类名为 BaseManagerFromQuerySet，且该类具有 QuerySet 类中的全部方法：

```
>>> from django.db.models.manager import BaseManager
>>> from django.db.models.query import QuerySet
>>> new_class = BaseManager.from_queryset(QuerySet)
>>> new_class
<class 'django.db.models.manager.BaseManagerFromQuerySet'>
>>> new_class()._queryset_class
<class 'django.db.models.query.QuerySet'>
```

注意，关于通过 type 创建类的此处不再赘述，请读者自行学习。下面给出几个简单的示例：

```
>>> class Hello:  # 定义了一个 Hello 类
...     a = 1     # 类属性
...     def hello(self):   # 对象方法
...         print('a={}'.format(self.a))
...
# 创建 TestHello 类，该类继承 Hello 类，此外该类具有属性'b'，其值为 2
>>> new_class = type('TestHello', (Hello, ), {'b': 2})
>>> new_class
<class 'TestHello'>
>>> obj = new_class()
>>> obj.a
1
>>> obj.b
2
>>> obj.hello()
a=1
```

综合上面的结果可知，BaseManager.from_queryset(QuerySet)的结果为一个新类，该新类继承了 BaseManager 类且具有 QuerySet 类中的全部方法。这里可以测试一下在 BaseManager 类中定义的 _get_queryset_methods()方法，以便理解其含义。具体操作如下：

```
>>> from django.db.models.query import QuerySet
>>> from django.db.models.manager import BaseManager
>>> BaseManager._get_queryset_methods(QuerySet)
# QuerySet 中的全部函数，由于内容太多，故忽略
# ……
>>> BaseManager._get_queryset_methods(QuerySet).keys()  # 对应的键是函数名
dict_keys(['_insert', '_update', 'aggregate', 'annotate', 'bulk_create', 'bulk_update',
'complex_filter', 'count', 'create', 'dates', 'datetimes', 'defer', 'difference',
'distinct', 'earliest', 'exclude', 'exists', 'explain', 'extra', 'filter', 'first', 'get',
'get_or_create', 'in_bulk', 'intersection', 'iterator', 'last', 'latest', 'none', 'only',
'order_by', 'prefetch_related', 'raw', 'reverse', 'select_for_update', 'select_related',
'union', 'update', 'update_or_create', 'using', 'values', 'values_list'])
>>> BaseManager._get_queryset_methods(QuerySet).get('_insert')
```

```
<function
BaseManager._get_queryset_methods.<locals>.create_method.<locals>.manager_method at
0x7fd0051d75e0>
```

在上面的代码中，在 BaseManager._get_queryset_methods(QuerySet)结果中键对应的方法源码在哪里定义？

```
# 源码位置：django/db/models/manager.py
# ……

class BaseManager:
    # ……

    @classmethod
    def _get_queryset_methods(cls, queryset_class):
        def create_method(name, method):
            def manager_method(self, *args, **kwargs):
                # 其实就是获取 QuerySet 对象中相应名称的方法
                return getattr(self.get_queryset(), name)(*args, **kwargs)
            manager_method.__name__ = method.__name__
            manager_method.__doc__ = method.__doc__
            # 返回方法名变量
            return manager_method

        # 方法字典，将来用于创建新类
        new_methods = {}
        for name, method in inspect.getmembers(queryset_class,
predicate=inspect.isfunction):
            # 只处理不在本类中的属性
            if hasattr(cls, name):
                continue
            # 获取对应方法
            queryset_only = getattr(method, 'queryset_only', None)
            if queryset_only or (queryset_only is None and name.startswith('_')):
                # 跳过设置了 queryset_only=True 的方法
                # 同时跳过没有设置 queryset_only 且方法名以'_'开头的方法
                continue
            # 形成上面看到的字典结果
            new_methods[name] = create_method(name, method)
        return new_methods

    # ……

    def get_queryset(self):
        return self._queryset_class(model=self.model, using=self._db, hints=self._hints)

    # ……
```

注意，在创建的新类中，传入的 queryset_class 属性值为 QuerySet 类，因此调用 BaseManager 的 get_queryset()方法将得到一个 QuerySet 对象。该对象会关联到模型及相应的数据库。从上面的代码中可以很明显地看到，最终在 new_methods 中保存的相关方法其实都来自 QuerySet 类。接下来对上面的代码进行拆解，以帮助读者更好地理解_get_queryset_methods()方法的含义：

```
>>> import inspect
>>> from django.db.models.query import QuerySet
>>> from django.db.models.manager import BaseManager
>>> queryset_class = QuerySet
>>> for name, method in inspect.getmembers(queryset_class,
predicate=inspect.isfunction):
...     if hasattr(BaseManager, name):  # 模拟 if hasattr(cls, name)语句
...         print('存在{}方法，跳过'.format(name))
...         continue
...     queryset_only = getattr(method, 'queryset_only', None)
...     if queryset_only or (queryset_only is None and name.startswith('_')):
...         continue
...     print('{}:{}'.format(name, method))
...
存在魔法函数__init__（），跳过
存在魔法函数__repr__（）方法，跳过
_insert:<function QuerySet._insert at 0x7f44d1270d30>
_update:<function QuerySet._update at 0x7f44d126fa60>
aggregate:<function QuerySet.aggregate at 0x7f44d1268ee0>
存在 all()方法，跳过
annotate:<function QuerySet.annotate at 0x7f44d1270790>
bulk_create:<function QuerySet.bulk_create at 0x7f44d126f1f0>
bulk_update:<function QuerySet.bulk_update at 0x7f44d126f280>
complex_filter:<function QuerySet.complex_filter at 0x7f44d1270310>
count:<function QuerySet.count at 0x7f44d1268f70>
create:<function QuerySet.create at 0x7f44d126f0d0>
dates:<function QuerySet.dates at 0x7f44d126fee0>
datetimes:<function QuerySet.datetimes at 0x7f44d126ff70>
defer:<function QuerySet.defer at 0x7f44d1270a60>
difference:<function QuerySet.difference at 0x7f44d1270550>
distinct:<function QuerySet.distinct at 0x7f44d12708b0>
earliest:<function QuerySet.earliest at 0x7f44d126f5e0>
exclude:<function QuerySet.exclude at 0x7f44d12701f0>
exists:<function QuerySet.exists at 0x7f44d126faf0>
explain:<function QuerySet.explain at 0x7f44d126fc10>
extra:<function QuerySet.extra at 0x7f44d1270940>
filter:<function QuerySet.filter at 0x7f44d1270160>
first:<function QuerySet.first at 0x7f44d126f700>
get:<function QuerySet.get at 0x7f44d126f040>
get_or_create:<function QuerySet.get_or_create at 0x7f44d126f310>
in_bulk:<function QuerySet.in_bulk at 0x7f44d126f820>
intersection:<function QuerySet.intersection at 0x7f44d12704c0>
```

```
iterator:<function QuerySet.iterator at 0x7f44d1268e50>
last:<function QuerySet.last at 0x7f44d126f790>
latest:<function QuerySet.latest at 0x7f44d126f670>
none:<function QuerySet.none at 0x7f44d1270040>
only:<function QuerySet.only at 0x7f44d1270af0>
order_by:<function QuerySet.order_by at 0x7f44d1270820>
prefetch_related:<function QuerySet.prefetch_related at 0x7f44d1270700>
raw:<function QuerySet.raw at 0x7f44d126fca0>
reverse:<function QuerySet.reverse at 0x7f44d12709d0>
select_for_update:<function QuerySet.select_for_update at 0x7f44d12705e0>
select_related:<function QuerySet.select_related at 0x7f44d1270670>
union:<function QuerySet.union at 0x7f44d1270430>
update:<function QuerySet.update at 0x7f44d126f9d0>
update_or_create:<function QuerySet.update_or_create at 0x7f44d126f3a0>
using:<function QuerySet.using at 0x7f44d1270b80>
values:<function QuerySet.values at 0x7f44d126fdc0>
values_list:<function QuerySet.values_list at 0x7f44d126fe50>
```

由于上面的代码没有将得到的 name 和 method 变量再次放入 create_method()方法中二次处理，所以显示的结果是 QuerySet 类中的相关方法。而最终经过 create_method()方法处理后，得到的是一个方法地址，而该地址最终执行的仍然是 QuerySet 类中对应方法的源码，这从 create_method()方法中的 getattr(self.get_queryset(), name)(*args, **kwargs)语句就能看出。

至此，根据上面的分析结果可以简单总结一下 Manager 类的特点，具体如下：

（1）它继承了一个由 type 方法创建的新类。

（2）该新类继承了 BaseManager 类且具有 QuerySet 类中大部分的方法。

下面在 Django 的 shell 模式下测试 Manager 类，并展现其相关特点，具体操作如下：

```
>>> from django.db.models.manager import Manager
>>> from shell_test.models import DjangoBooks
>>> manager = Manager()
>>> manager.model = DjangoBooks
>>> manager._db = 'default'
>>> manager.all()
<QuerySet [<DjangoBooks: <Django 源码笔记, spyinx>>, <DjangoBooks: <Ansible 核心源码剖析与项目实战, spyinx>>, <DjangoBooks: <Python 自动化运维, test>>, <DjangoBooks: <中国电信媒体存储自主研发之路, group>>, <DjangoBooks: <大国崛起, xyz>>, <DjangoBooks: <小郭历险记, mmmm>>, <DjangoBooks: <Django 项目实战, hhhhh>>, <DjangoBooks: <精通 Django 框架, zzzzz>>, <DjangoBooks: <Django 实战, test>>, <DjangoBooks: <Django3.0 从入门到精通, group>>]>
>>> manager.all().filter(author='spyinx')
<QuerySet [<DjangoBooks: <Django 源码笔记, spyinx>>, <DjangoBooks: <Ansible 核心源码剖析与项目实战, spyinx>>]>
```

看到上面的结果是不是有点能理解 DjangoBooks.objects.all().filter(author='spyinx')语句了？DjangoBooks 是一个模型类（该类继承 Model 类），因此 DjangoBooks 在创建时就具有 objects 属性，

且其值为一个 Manager 对象。而该 Manager 对象的 model 属性正好关联这里的 DjangoBooks 模型类，源码如下：

```
>>> from shell_test.models import DjangoBooks
>>> DjangoBooks.objects
<django.db.models.manager.Manager object at 0x7fab6619dbb0>
>>> DjangoBooks.objects.model
<class 'shell_test.models.DjangoBooks'>
```

下面思考一个问题：DjangoBooks.objects.all().filter(author='spyinx')语句的结果是一个 Manager 对象，可以看到它的 model 属性值为 DjangoBooks 类，那么这个 Manager 对象的 model 属性值是在哪里设置的？我们回到 ModelBase 类中设置新建类的 objects 属性值的地方，如下：

```
# 源码位置：django/db/models/base.py
# ……

class ModelBase(type):

    # ……

    def _prepare(cls):

        # ……

        if not opts.managers:
            # 不允许在字段中出现 objects 名称
            # ……

            # 得到 Manager 对象
            manager = Manager()
            manager.auto_created = True
            # 添加 objects 属性
            cls.add_to_class('objects', manager)

        # ……
```

可以看到，这里是通过调用 add_to_class()方法来设置 objects 属性值的。继续看 add_to_class()方法的实现源码，如下：

```
# 源码位置：django/db/models/base.py
# ……

def _has_contribute_to_class(value):
    # value 不能是类，同时该值有 contribute_to_class 属性
    return not inspect.isclass(value) and hasattr(value, 'contribute_to_class')

class ModelBase(type):
```

```
    # ……

    def add_to_class(cls, name, value):
        if _has_contribute_to_class(value):
            value.contribute_to_class(cls, name)
        else:
            setattr(cls, name, value)

    # ……
```

假设这里要创建的是 DjangoBooks 类，注意，前面调用_prepare()方法的语句为 new_class._prepare()。因此，传入_prepare()方法中的 cls 其实是 DjangoBooks 类，对应 add_to_class()方法中的 cls 同样是 DjangoBooks 类，而此时 name 为“objects”，value 为一个 Manager 对象。注意，在 Manager 类中其实是定义了 contribute_to_class()方法的，于是 if 语句_has_contribute_to_class(value)为 True（value 为 Manager 对象且有相应属性），最终执行 value.contribute_to_class(cls, name)语句。在 Manager 类中定义的 contribute_to_class()方法位于父类 BaseManager 中，内容如下：

```
# 源码位置：django/db/models/manager.py
# ……

class BaseManager:

    # ……

    def contribute_to_class(self, model, name):
        self.name = self.name or name
        self.model = model  # 设置模型类

        setattr(model, name, ManagerDescriptor(self))  # 同时给模型类设置 objects 属性值

        model._meta.add_manager(self)

    # ……
```

看了这里的源码，一切就非常清楚了。继续前面的假设，此时调用 contribute_to_class()方法的 model 和 name 分别为 DjangoBook 模型类和字符串“objects”。因此，最后该 Manager 对象的 model 属性值被设置为了 DjangoBook 模型类。此外，该方法还会对 DjangoBook 模型类设置 objects 属性值，设置的语句正是 setattr(model, name, ManagerDescriptor(self))。下面拆解执行 contribute_to_class()方法中的语句，以便读者理解这个属性值的设置过程，具体操作如下：

```
>>> from django.db.models.manager import ManagerDescriptor
>>> from shell_test.models import DjangoBooks
>>> from django.db.models.manager import Manager
>>> del DjangoBooks.objects  # 删除 DjangoBooks 原先设置的 objects 属性
>>> DjangoBooks.objects    # 确认无法使用 DjangoBooks.objects 语句
```

```
Traceback (most recent call last):
  File "<console>", line 1, in <module>
AttributeError: type object 'DjangoBooks' has no attribute 'objects'
>>> name = 'objects'        # 以下模拟 contribute_to_class()方法中的语句
>>> manager = Manager()
>>> manager.name = manager.name or name
>>> manager.model = DjangoBooks
>>> setattr(DjangoBooks, 'objects', ManagerDescriptor(manager))  # 设置 objects 属性
>>> DjangoBooks.objects     # 设置后的结果
<django.db.models.manager.Manager object at 0x7ff5c8544400>
>>> DjangoBooks.objects.all()  # 同样可以继续查看数据库的记录
<QuerySet [<DjangoBooks: <Django 源码笔记, spyinx>>, <DjangoBooks: <Ansible 核心源码剖析与项目实战, spyinx>>, <DjangoBooks: <Python 自动化运维, test>>, <DjangoBooks: <中国电信媒体存储自主研发之路, group>>, <DjangoBooks: <大国崛起, xyz>>, <DjangoBooks: <小郭历险记, mmmm>>, <DjangoBooks: <Django 项目实战, hhhhh>>, <DjangoBooks: <精通 Django 框架, zzzzz>>, <DjangoBooks: <Django 实战, test>>, <DjangoBooks: <Django3.0 从入门到精通, group>>]>
```

看了上面的操作，读者可能会产生这样一个疑问：在源码中将模型类的 objects 属性赋值为 ManagerDescriptor 对象，为什么输出的是 Manager 对象？ManagerDescriptor 类的实现源码如下：

```
# 源码位置：django/db/models/manager.py
# ……

class ManagerDescriptor:

   def __init__(self, manager):
      self.manager = manager

   def __get__(self, instance, cls=None):

      # 异常情况
      # ……

      return cls._meta.managers_map[self.manager.name]

# ……
```

在该类中，只有初始化方法__init__()和魔法函数__get__()。一个类只要实现了魔法函数__get__()、__set__()和__delete__()中的任意一个，就可以称它为描述器（descriptor）。如果只定义了魔法函数__get__()，一般称它为非资料描述器（non-data descriptor）。这里要明确一点：拥有这个魔法函数的类，必须产生一个实例，并且这个实例是另外一个类的类属性。下面给出一个简单的例子：

```
(django2-core-test) [root@master first_django]# cat test_manager_descriptor.py
from django.db.models.manager import Manager

manager = Manager
```

```
class ManagerDescriptor:
   def __get__(self, instance, cls):
      print('调用魔法函数__get__():', instance, cls)
      return manager  # 返回Manager对象

class TestBooks:
   objects = ManagerDescriptor()

print(TestBooks.objects)
```

运行上面的 Python 脚本，结果如下：

```
(django2-core-test) [root@master first_django]# python test_manager_descriptor.py
调用魔法函数__get__(): None <class '__main__.TestBooks'>
<class 'django.db.models.manager.Manager'>
```

这里我们定义了一个只包含魔法函数__get__()的 ManagerDescriptor 类，然后又定义了一个 TestBooks 类。在 TestBooks 类中定义了一个 objects 属性，它的值为一个 ManagerDescriptor 对象。在执行 TestBooks.objects 语句时，将调用 ManagerDescriptor 类中的魔法函数__get__()并返回该方法的结果，即一个 Manager 对象。这里的 TestBooks 类类似于前面的 DjangoBooks 类，ManagerDescriptor 类也模拟了 Django 源码中的 ManagerDescriptor 类。因此在 Django 源码中，ManagerDescriptor 类的魔法函数__get__()的结果一定是前面看到的 Manager 对象。继续在前面的交互模式下测试魔法函数__get__()的最后一句，如下：

```
# 接着前面的交互模式继执行，注意这里的对象和DjangoBooks.objects是同一个
>>> DjangoBooks._meta.managers_map['objects']
<django.db.models.manager.Manager object at 0x7ff5c8544400>
```

可以看到，这里 ManagerDescriptor 类的魔法函数__get__()的输出正是前面看到的 objects 属性值。

接下来先学习 Field 及其子类源码，这些是理解 Model 类的前提，同时是经常在定义模型类中用到的类，相关源码均位于 django/db/models/fields 目录中：

```
# 源码位置：django/db/models/fields/__init__.py
# ……

@total_ordering
class Field(RegisterLookupMixin):
    """所有field类型的基类"""

    # 一些类属性定义
    # ……

    def __init__(self, verbose_name=None, name=None, primary_key=False,
                 max_length=None, unique=False, blank=False, null=False,
                 db_index=False, rel=None, default=NOT_PROVIDED, editable=True,
                 serialize=True, unique_for_date=None, unique_for_month=None,
```

```
                 unique_for_year=None, choices=None, help_text='', db_column=None,
                 db_tablespace=None, auto_created=False, validators=(),
                 error_messages=None):

        # ……

    def __str__(self):

        if not hasattr(self, 'model'):
            return super().__str__()
        model = self.model
        app = model._meta.app_label
        return '%s.%s.%s' % (app, model._meta.object_name, self.name)

    # ……
```

Field 类是前面见到的 BooleanField、CharField、DateField、DecimalField 等类的父类。从初始化方法__init__()中可以看到，在实例化 Field 类时可以传入的一些参数，如 name、primary_key、max_length、blank 等，关于它们的含义与作用将在后续的源码分析中说明。这里在魔法函数__str__()中可以看到语句 model = self.model，由于每个 Field 对象都是定义在模型类中的，因此 Field 对象一定会关联一个模型类，而这里 self.model 是在哪里赋值的呢？这就需要回到前面 ModelBase 类的源码中进行说明：

```
# 源码位置：django/db/models/base.py
# ……

class ModelBase(type):

    """所有模型类的元类"""

    def __new__(cls, name, bases, attrs, **kwargs):

        # ……

        contributable_attrs = {}
        for obj_name, obj in list(attrs.items()):
            if _has_contribute_to_class(obj): # Field 类中定义了 contribute_to_class()方法
                contributable_attrs[obj_name] = obj
            else:
                new_attrs[obj_name] = obj
        new_class = super_new(cls, name, bases, new_attrs, **kwargs)

        # ……

        for obj_name, obj in contributable_attrs.items():
            # 对所有有 contribute_to_class()方法的属性都调用模型类的 add_to_class()方法
```

```
        new_class.add_to_class(obj_name, obj)

    # ……
```

注意，在 Field 类中定义了 contribute_to_class()方法，因此 contributable_attrs 至少包括模型类中的 Field 对象（以及 Field 继承类的对象）。后面会对 contributable_attrs 中的这些属性调用 add_to_class() 方法，即调用这些属性的 contribute_to_class()方法（前面分析过），即调用在 Field 类中定义的 contribute_to_class()方法（AutoField 类和 DateField 类等会重写这个方法，因此对于这些类将调用它们自己实现的 contribute_to_class()方法）。在 Field 类中，ontribute_to_class()方法的源码如下：

```
# 源码位置：django/db/models/base.py
# ……

@total_ordering
class Field(RegisterLookupMixin):

    # ……

    def contribute_to_class(self, cls, name, private_only=False):

        self.set_attributes_from_name(name)
        self.model = cls   # 在这里设置 model 属性值
        # 在模型类的_meta 中添加该 field 信息
        if private_only:
            cls._meta.add_field(self, private=True)
        else:
            cls._meta.add_field(self)
        if self.column:
            if not getattr(cls, self.attname, None):
                # 重新在模型类中设置该属性
                setattr(cls, self.attname, DeferredAttribute(self.attname))
        if self.choices:
            # 对于由选项值的字段，将添加一个 get_xxx_name()方法，用于显示该字段值
            if not hasattr(cls, 'get_%s_display' % self.name):
                setattr(
                    cls,
                    'get_%s_display' % self.name,
                    partialmethod(cls._get_FIELD_display, field=self),
                )

    # ……
```

上面正是关联模型类与该 Field 对象的地方，下面以 DjangoBooks 类为例进行测试，操作如下：

```
>>> from shell_test.models import DjangoBooks
>>> DjangoBooks.book_name  # 该值果然是一个 DeferredAttribute 对象
<django.db.models.query_utils.DeferredAttribute object at 0x7f3f17d2ea90>
```

```
# sex 属性值是一个选项值，因此在模型对象中会有一个 get_sex_display()方法用于翻译该字段值
book = DjangoBooks.objects.all().get(book_name='Django 源码笔记')
>>> book
<DjangoBooks: <Django 源码笔记, spyinx>>
>>> book.sex
0
>>> book.get_sex_display()
'男'
```

从上面的代码可以看到，在模型类中定义的所有 Field 相关的对象均被保存在了_meta 属性中（调用_meta 属性的 add_field()方法）。而该_meta 属性值为一个 Option 对象，相关源码内容如下：

```
# 源码位置：django/db/models/options.py
# ……

calss Options:

   # ……

   def __init__(self, meta, app_label=None):

      # ……

   def add_field(self, field, private=False):
      # 保存 Field 对象
      if private:
         self.private_fields.append(field)  # 保存到 Options 对象的 private_fields 属性中
      elif field.is_relation and field.many_to_many:
         # 保存到 Options 对象的 local_many_to_many 属性中
         self.local_many_to_many.insert(bisect(self.local_many_to_many, field), 
field)
      else:
         # 保存到 Options 对象的 local_fields 属性中
         self.local_fields.insert(bisect(self.local_fields, field), field)
         self.setup_pk(field)

      if field.is_relation and hasattr(field.remote_field, 'model') and 
field.remote_field.model:
         try:
            field.remote_field.model._meta._expire_cache(forward=False)
         except AttributeError:
            pass
         self._expire_cache()
      else:
         self._expire_cache(reverse=False)
```

从 add_field()方法的源码可以看到，最终模型类中的 Field 对象将被分类保存到_meta 属性的 private_fields、.local_many_to_many 和 local_fields 属性中。以 DjangoBooks 类为例，操作如下：

```
>>> DjangoBooks._meta
<Options for DjangoBooks>
>>> DjangoBooks._meta.private_fields
[]
>>> DjangoBooks._meta.local_many_to_many
[]
>>> DjangoBooks._meta.local_fields
[<django.db.models.fields.AutoField: id>, <django.db.models.fields.CharField:
book_name>, <django.db.models.fields.CharField: author>,
<django.db.models.fields.SmallIntegerField: sex>, <django.db.models.fields.FloatField:
price>, <django.db.models.fields.CharField: isbn>,
<django.db.models.fields.DateTimeField: publish_date>]
>>> DjangoBooks._meta.local_fields[0].model  # 关联到对应的模型类
<class 'shell_test.models.DjangoBooks'>
```

接下来我们继续以实战的方式介绍在 Field 基类中定义的一些常用方法，并对部分方法进行源码解读：

```
# 接着上面的操作，不再重复导入 DjangoBooks
>>> local_fields = DjangoBooks._meta.local_fields
>>> local_fields[0].unique # 属性方法，判断是否是主键
True
>>> local_fields[1].unique
False
>>> local_fields[1]      # 调用魔法函数__repr__()
<django.db.models.fields.CharField: book_name>
>>> print(local_fields[1]) # 调用魔法函数__str__()
shell_test.DjangoBooks.book_name
>>> local_fields[0].check()
[]
```

Field 类中的 check()方法用于校验该 Field 对象，例如校验该 Field 字段名是否合法。如果有 choices 属性，会校验该 choices 值。此外，如果设置了 validators 属性值，还会调用自定义的校验器去校验对该对象，源码如下：

```
# 源码位置：django/db/models/fields/__init__.py
# ……

@total_ordering
class Field(RegisterLookupMixin):

    # ……

    def check(self, **kwargs):
        return [
            *self._check_field_name(),
            *self._check_choices(),
            *self._check_db_index(),
```

```
            *self._check_null_allowed_for_primary_keys(),
            *self._check_backend_specific_checks(**kwargs),
            *self._check_validators(),
            *self._check_deprecation_details(),
        ]

    # ……

    def _check_validators(self):
        errors = []
        for i, validator in enumerate(self.validators):
            # 对于自定义的校验器，先判断该校验器是否可调用，如果不可调用，抛出异常
            if not callable(validator):
                errors.append(
                    checks.Error(
                        "All 'validators' must be callable.",
                        hint=(
                            "validators[{i}] ({repr}) isn't a function or "
                            "instance of a validator class.".format(
                                i=i, repr=repr(validator),
                            )
                        ),
                        obj=self,
                        id='fields.E008',
                    )
                )
        return errors

    # ……
```

通过阅读源码可以看到，校验方法 self._check_field_name()、self._check_choices()等均返回列表值，并在这些值的前面加上*号后放到 check()方法返回的数组中。那加上这个*号有何含义呢？下面简单测试一下就清楚了，操作如下：

```
>>> list1 = []
>>> list2 = ['a', 2, 3]
>>> list3 = []
>>> list = [*list1, *list2, *list3]
>>> list
['a', 2, 3]
```

其他的检查方法都比较简单，读者可以自行阅读相关检查方法的源码进行学习。需要说明的是，对于输入的自定义校验器，需要是可枚举的且其中的元素必须是可调用的，否则会校验不过。例如下面的操作：

```
>>> local_fields[1].validators = 1
>>> local_fields[1].check()  # validators 属性值必须是可迭代的
```

```
Traceback (most recent call last):
 File "<console>", line 1, in <module>
 File
"/root/.pyenv/versions/django2-core-test/lib/python3.8/site-packages/django/db/model
s/fields/__init__.py", line 1048, in check
   *super().check(**kwargs),
 File
"/root/.pyenv/versions/django2-core-test/lib/python3.8/site-packages/django/db/model
s/fields/__init__.py", line 206, in check
   *self._check_validators(),
 File
"/root/.pyenv/versions/django2-core-test/lib/python3.8/site-packages/django/db/model
s/fields/__init__.py", line 331, in _check_validators
   for i, validator in enumerate(self.validators):
TypeError: 'int' object is not iterable
>>> local_fields[1].check()  # validators 属性值中的元素必须是可调用的
[<Error: level=40, msg="All 'validators' must be callable.", hint="validators[0] (1) isn't
a function or instance of a validator class.", obj=<django.db.models.fields.CharField:
book_name>, id='fields.E008'>]
```

这些校验器是在哪里调用的呢？只需在 Field 类中简单浏览一下就能找到下面的源码：

```
# 源码位置：django/db/models/fields/__init__.py
# ……

@total_ordering
class Field(RegisterLookupMixin):

   # ……

   @cached_property
   def validators(self):
      return [*self.default_validators, *self._validators] # 返回校验器

   def run_validators(self, value):
      if value in self.empty_values:
         return

      errors = []
      for v in self.validators:  # 遍历校验器
         try:
            v(value)    # 校验
         except exceptions.ValidationError as e:
            # 异常处理
            # ……

      if errors:  # 任意一个校验器校验异常，都会使得 errors 有值
         raise exceptions.ValidationError(errors)
```

继续介绍 Field 类中的其他方法，操作如下：

```
>>> local_fields[1].get_col('book_name')    # 返回 Col 对象
Col(book_name, shell_test.DjangoBooks.book_name)
>>> local_fields[1].deconstruct()          # 展示该 field 信息
('book_name', 'django.db.models.CharField', [], {'verbose_name': '图书名', 'max_length':
30})
>>> obj = DjangoBooks.objects.all().filter(author='spyinx')[0]
>>> local_fields[1].get_filter_kwargs_for_object(obj)  # 获取模型对象中对应 field 字段的值
{'book_name': 'Django 源码笔记'}
>>> local_fields[2].get_filter_kwargs_for_object(obj)
{'author': 'spyinx'}
>>> local_fields[3].get_filter_kwargs_for_object(obj)
{'sex': 0}
>>> local_fields[1].value_to_string(obj)    # 只获取模型对象中该字段的值，转成 str 类型
'Django 源码笔记'
>>> local_fields[2].value_to_string(obj)
'spyinx'
>>> local_fields[3].value_to_string(obj)
'0'
>>> local_fields[3].value_from_object(obj) # 获取模型对象中该字段的原始值
0
>>> local_fields[1].get_attname_column()  # 返回(属性名，关联表字段名)
('book_name', 'book_name')
>>> local_fields[2].get_default()         # 返回该字段设置的默认值
''
>>> local_fields[3].get_default()
0
>>> local_fields[3].get_choices()
[('', '---------'), (0, '男'), (1, '女')]
>>> local_fields[3]._get_flatchoices()
[(0, '男'), (1, '女')]
>>> local_fields[2].get_internal_type()    # 各 Field 类都有自己的实现，返回对应的类型名
'CharField'
>>> local_fields[3].get_internal_type()
'SmallIntegerField'
```

AutoField、BooleanField 和 CharField 等类都继承自 Field 类，有的甚至还继承了包含某一功能的 mixin。所有子类都对 Field 类中定义的部分方法进行了重写，有兴趣的读者可以参考相关方法的源码，理解各 Field 类的特点。

继续回到模型类 Model 上来，在该类中有非常多的方法，本节只介绍部分核心的方法源码，其余方法的分析方法与之类似，操作如下：

```
>>> from shell_test.models import DjangoBooks
>>> new_obj._meta.fields   # 获取模型类中定义的各 Field 对象
```

```
(<django.db.models.fields.AutoField: id>, <django.db.models.fields.CharField:
book_name>, <django.db.models.fields.CharField: author>,
<django.db.models.fields.SmallIntegerField: sex>, <django.db.models.fields.FloatField:
price>, <django.db.models.fields.CharField: isbn>,
<django.db.models.fields.DateTimeField: publish_date>)
>>> DjangoBooks._meta.concrete_fields
(<django.db.models.fields.AutoField: id>, <django.db.models.fields.CharField:
book_name>, <django.db.models.fields.CharField: author>,
<django.db.models.fields.SmallIntegerField: sex>, <django.db.models.fields.FloatField:
price>, <django.db.models.fields.CharField: isbn>,
<django.db.models.fields.DateTimeField: publish_date>)
>>> obj = DjangoBooks.objects.all().filter(author='spyinx')[0]
>>> obj
<DjangoBooks: <Django 源码笔记, spyinx>>
>>> obj._get_pk_val()  # 获取主键值
1
>>> obj.__dict__      # 模型对象的字典属性值，会在许多方法中用到
{'_state': <django.db.models.base.ModelState object at 0x7f875b26eaf0>, 'id': 1,
'book_name': 'Django 源码笔记', 'author': 'spyinx', 'sex': 0, 'price': 109.0, 'isbn':
'11111', 'publish_date': datetime.datetime(2021, 2, 4, 21, 42, 43)}
>>> obj._meta.get_field('book_name') # Options 对象的 get_field()方法,
                                     # 获取对应名称的 Field 对象
<django.db.models.fields.CharField: book_name>
>>> getattr(obj, 'book_name') # 出自 serializable_value()方法中的核心语句，获取对应字段的值
'Django 源码笔记'
>>> obj.serializable_value('book_name')
'Django 源码笔记'
>>> obj.serializable_value('sex')
0
```

上面的这些操作主要是帮助读者理解在 Model 类中定义的一些简单方法，如 from_db()、get_deferred_fields()、serializable_value()等方法。接下来详细介绍一个非常重要的方法，即 save()方法，该方法的效果如下：

```
>>> from shell_test.models import DjangoBooks
# 查询在 django_books 表中有没有 author 为'new_spyinx'的记录
>>> DjangoBooks.objects.filter(author='new_spyinx')
<QuerySet []>
>>> new_obj = DjangoBooks(book_name='我是奇才', author='new_spyinx', sex=1, price=99)
>>> new_obj.save()
>>> DjangoBooks.objects.filter(author='new_spyinx')  # 成功保存到数据库中
<QuerySet [<DjangoBooks: <我是奇才, new_spyinx>>]>
```

从上面的代码可以看到，save()方法居然可以直接使用模型对象保存数据，这背后的原理是什么？在 Model 类中 save()方法的源码如下：

```
# 源码位置：django/db/models/base.py
# ……
```

```
class Model(metaclass=ModelBase):

    # ……

    def save(self, force_insert=False, force_update=False, using=None,
             update_fields=None):
        # 遍历模型对象中的 Field 字段
        for field in self._meta.concrete_fields:
            if field.is_relation and field.is_cached(self): # 如果有关联，获取关联对象
                obj = getattr(self, field.name, None)
                if obj and obj.pk is None:
                    if not field.remote_field.multiple:
                        field.remote_field.delete_cached_value(obj)
                    raise ValueError(
                        "save() prohibited to prevent data loss due to "
                        "unsaved related object '%s'." % field.name
                    )
                if obj and getattr(obj, field.target_field.attname) != getattr(self,
field.attname):
                    field.delete_cached_value(self)

        #usging 用于获取相应的数据库配置信息
        using = using or router.db_for_write(self.__class__, instance=self)
        if force_insert and (force_update or update_fields):
            raise ValueError("Cannot force both insert and updating in model saving.")

        deferred_fields = self.get_deferred_fields()
        if update_fields is not None:
            # 更新字段
            # ……

        elif not force_insert and deferred_fields and using == self._state.db:
            field_names = set()
            for field in self._meta.concrete_fields:
                # 添加字段名
                if not field.primary_key and not hasattr(field, 'through'):
                    field_names.add(field.attname)
            loaded_fields = field_names.difference(deferred_fields)
            if loaded_fields:
                update_fields = frozenset(loaded_fields)
        # 核心调用方法，update_fields 就是统一要操作数据表中的字段列表
        self.save_base(using=using, force_insert=force_insert,
                       force_update=force_update, update_fields=update_fields)

    save.alters_data = True

# ……
```

为了帮助读者理解上述代码，下面新建一个有关联字段的模型类 DjangoDetail，其定义如下：

```
# 源码位置：shell_test/models.py
# ……

class DjangoDetail(models.Model):

   book_detail = models.TextField('图书详情')
   book_brief = models.CharField('图书简介', max_length=200)
   isbn = models.OneToOneField(to=DjangoBooks, on_delete=models.CASCADE) # 关联字段

   class Meta:
      db_table = 'books_detail'

# ……
```

首先看在 save()方法中出现的第 1 个 for 循环操作，我们通过以下测试语句理解在该循环内所做的工作：

```
>>> from shell_test.models import DjangoBooks
>>> from shell_test.models import DjangoDetail
>>> book_obj = DjangoBooks.objects.all().filter(author='spyinx')[0]
>>> book_detail_obj = DjangoDetail(book_detail='详情', book_brief='简介', isbn=book_obj)
>>> book_fields = book_obj._meta.concrete_fields
>>> book_detail_fields = book_detail_obj._meta.concrete_fields
>>> book_fields[1]
<django.db.models.fields.CharField: book_name>
>>> book_fields[1].is_relation
False
>>> book_detail_fields[3]   # 有关联的字段
<django.db.models.fields.related.OneToOneField: isbn>
>>> book_detail_fields[3].is_relation
True
>>> book_detail_fields[3].is_cached(book_detail_obj)
True
# 模拟第 1 个 for 循环中获取关联的模型对象
>>> obj = getattr(book_detail_obj, book_detail_fields[3].name)
>>> obj
<DjangoBooks: <Django 源码笔记, spyinx>>
>>> obj.pk
1
>>> book_detail_fields[3].target_field.attname
'id'
>>> book_detail_fields[3].attname
'isbn_id'
>>> getattr(book_detail_obj, field.target_field.attname)
>>> getattr(book_detail_obj, field.attname)
1
```

如果是操作 DjangoBooks 对象，由于在它的属性中没有关联的 Field 对象，因此对于第 1 个 for 循环而言不会做任何操作。如果是操作 DjangoDetail 对象，则会有关联的 isbn 属性，最终会执行 field.delete_cached_value(self)语句。接下来是大家比较熟悉的 using 语句，它的默认值为“default”，主要用于获取配置中设置的数据库连接信息。请看如下操作：

```
>>> from django.db import router
>>> using = None or router.db_for_write(book_detail_obj.__class__,
instance=book_detail_obj)
>>> using
'default'
```

接下来的 if 和 else 分支语句的作用是得到需要更新的字段 update_fields，如果这些字段有输入值，会做一定处理。以 DjangoDetail 对象为例，在下面的交互模式下模拟 else 分支下的语句，具体如下：

```
>>> field_names = set()
>>> for field in book_detail_obj._meta.concrete_fields:
...     if not field.primary_key and not hasattr(field, 'through'):
...         field_names.add(field.attname)
...
>>> print(field_names)
{'isbn_id', 'book_detail', 'book_brief'}
>>> deferred_fields = book_detail_obj.get_deferred_fields()
>>> deferred_fields  # 为空
set()
>>> loaded_fields = field_names.difference(deferred_fields)
>>> loaded_fields
{'isbn_id', 'book_detail', 'book_brief'}
>>> update_fields = frozenset(loaded_fields)
>>> update_fields
frozenset({'isbn_id', 'book_detail', 'book_brief'})
```

**注意**，在实际运行中可以看到，deferred_fields 为空，因此上面的 else 分支不会执行，最终传递到 save_base()方法中的 update_fields 参数为 None。

接下来调用保存在数据库中的核心方法 save_base()，继续追踪该方法的源码，如下：

```
# 源码位置：django/db/models/base.py
# ……

class Model(metaclass=ModelBase):

    # ……


    def save_base(self, raw=False, force_insert=False,
                  force_update=False, using=None, update_fields=None):
```

```
        using = using or router.db_for_write(self.__class__, instance=self)
        # 一些 assert 语句
        # ……

        cls = origin = self.__class__  # 对应的模型类
        if cls._meta.proxy:
            cls = cls._meta.concrete_model
        meta = cls._meta         # 元信息
        if not meta.auto_created:
            pre_save.send(
                sender=origin, instance=self, raw=raw, using=using,
                update_fields=update_fields,
            )
        if meta.parents:
            context_manager = transaction.atomic(using=using, savepoint=False)
        else:
            context_manager = transaction.mark_for_rollback_on_error(using=using)
        with context_manager:
            parent_inserted = False
            # raw 的含义在本方法的注释中已做说明，...，当 raw 为 False 时才会执行下面的语句
            if not raw:
                parent_inserted = self._save_parents(cls, using, update_fields)
            # 保存的核心语句，调用_save_table()方法
            updated = self._save_table(
                raw, cls, force_insert or parent_inserted,
                force_update, using, update_fields,
            )
        self._state.db = using
        self._state.adding = False

        # 发送信号，保存模型对象完成
        if not meta.auto_created:
            post_save.send(
                sender=origin, instance=self, created=(not updated),
                update_fields=update_fields, raw=raw, using=using,
            )

    # ……
```

在上述源码中，raw 的默认传入值为 False，在 save_base()方法的注释中已经对该参数进行了说明。当把该值设置为 True 时，将不会保存任何父模型，也不会在保存之前对当前值进行任何更改。从源码可以看到，保存当前模型对象的核心语句为调用 self._save_table()方法，该方法的实现源码如下：

```
# 源码位置：django/db/models/base.py
# ……
```

```
class Model(metaclass=ModelBase):
    # ……

    def _save_table(self, raw=False, cls=None, force_insert=False,
                    force_update=False, using=None, update_fields=None):
        meta = cls._meta
        # 非主键字段
        non_pks = [f for f in meta.local_concrete_fields if not f.primary_key]

        if update_fields:
            # 如果设置了 update_fields 值，将重新给 non_pks 赋值
            non_pks = [f for f in non_pks
                       if f.name in update_fields or f.attname in update_fields]

        pk_val = self._get_pk_val(meta)  # 获取 pk 值
        if pk_val is None:
            pk_val = meta.pk.get_pk_value_on_save(self)  # 获取 pk_val 值
            setattr(self, meta.pk.attname, pk_val)        # 设置 pk 字段值
        pk_set = pk_val is not None
        if not pk_set and (force_update or update_fields):
            # 如果是更新操作，必须要设置 pk 字段的值，否则无法找到更新的记录
            raise ValueError("Cannot force an update in save() with no primary key.")

        updated = False
        # 先尝试更新操作，如果没有更新成功，就执行插入操作
        if pk_set and not force_insert:
            # base_qs 其实是一个 QuerySet 对象，等价于 xxx.objects.all()的结果
            base_qs = cls._base_manager.using(using)
            # 这里的 f.pre_save()在某些 Field 中有用，比如 FileField 对象
            # FileField 对象需要将上传的文件保存到指定位置后再继续后面的操作
            values = [(f, None, (getattr(self, f.attname)
                        if raw else f.pre_save(self, False))) for f in non_pks]
            forced_update = update_fields or force_update
            # 更新操作的核心调用语句
            updated = self._do_update(base_qs, using, pk_val, values, update_fields,
                                      forced_update)
            if force_update and not updated:
                # 强制更新但更新不成功，抛出异常
                raise DatabaseError("Forced update did not affect any rows.")
            if update_fields and not updated:
                # 如果设置了更新字段且没有更新成功，抛出异常
                raise DatabaseError("Save with update_fields did not affect any rows.")

        # 如果更新不成功，就执行插入操作，调用_do_insert()方法
        if not updated:
            if meta.order_with_respect_to:  # 默认为 None
```

```
            field = meta.order_with_respect_to
            filter_args = field.get_filter_kwargs_for_object(self)
            order_value =
cls._base_manager.using(using).filter(**filter_args).count()
            self._order = order_value

        fields = meta.local_concrete_fields
        if not pk_set:
            fields = [f for f in fields if f is not meta.auto_field]

        update_pk = meta.auto_field and not pk_set
        # 最终插入方式
        result = self._do_insert(cls._base_manager, using, fields, update_pk, raw)
        if update_pk:
            setattr(self, meta.pk.attname, result)
    return updated
```

这里以前面操作 DjangoBooks 对象的 save()方法为例，继续测试_save_table()方法中的相关语句，以便读者更好地理解这些源码，具体操作如下：

```
>>> from shell_test.models import DjangoBooks
>>> book_obj = DjangoBooks(book_name='测试', author='xxxh', sex=0, price=89)
>>> meta = DjangoBooks._meta
>>> meta
<Options for DjangoBooks>
>>> non_pks = [f for f in meta.local_concrete_fields if not f.primary_key]
>>> non_pks
[<django.db.models.fields.CharField: book_name>, <django.db.models.fields.CharField:
author>, <django.db.models.fields.SmallIntegerField: sex>,
<django.db.models.fields.FloatField: price>, <django.db.models.fields.CharField: isbn>,
<django.db.models.fields.DateTimeField: publish_date>]
>>> pk_val = book_obj._get_pk_val(meta)
>>> pk_val  # 该值为 None
>>> pk_val = meta.pk.get_pk_value_on_save(book_obj)
>>> pk_val  # 同样为 None
>>> meta.pk.attname  # 该 pk 字段的名称
'id'
>>> setattr(book_obj, meta.pk.attname, pk_val)  # 设置 pk 字段的属性值
>>> book_obj.id   # 同样为 None
>>> pk_set = pk_val is not None   # 由于 pk_val 为 None，所以 pk_set 为 False
>>> pk_set
False
>>> DjangoBooks._base_manager  # 模型类的_base_manager 属性值默认是一个 Manager 对象
<django.db.models.manager.Manager object at 0x7f64dd6e9e50>
#这里的 using()方法来自 QuerySet 类，该方法返回一个 QuerySet 对象
# 等价于执行 DjangoBooks.objects.all()语句
>>> DjangoBooks._base_manager.using('default')
```

```
<QuerySet [<DjangoBooks: <Django 源码笔记, spyinx>>, <DjangoBooks: <Ansible 核心源码剖析与项目实战, spyinx>>, <DjangoBooks: <Python 自动化运维, test>>, <DjangoBooks: <中国电信媒体存储自主研发之路, group>>, <DjangoBooks: <大国崛起, xyz>>, <DjangoBooks: <小郭历险记, mmmm>>, <DjangoBooks: <Django 项目实战, hhhhh>>, <DjangoBooks: <精通 Django 框架, zzzzz>>, <DjangoBooks: <Django 实战, test>>, <DjangoBooks: <Django3.0 从入门到精通, group>>, <DjangoBooks: <我是奇才, new_spyinx>>]>
>>> raw = False
>>> values = [(f, None, (getattr(book_obj, f.attname) if raw else f.pre_save(book_obj, False))) for f in non_pks]
>>> values
[(<django.db.models.fields.CharField: book_name>, None, '测试'),
(<django.db.models.fields.CharField: author>, None, 'xxxh'),
(<django.db.models.fields.SmallIntegerField: sex>, None, 0),
(<django.db.models.fields.FloatField: price>, None, 89),
(<django.db.models.fields.CharField: isbn>, None, ''),
(<django.db.models.fields.DateTimeField: publish_date>, None, datetime.datetime(2021, 5, 15, 21, 37, 17, 257485))]
>>> meta.order_with_respect_to  # 在默认情况下为 None
>>> fields = meta.local_concrete_fields
>>> fields
(<django.db.models.fields.AutoField: id>, <django.db.models.fields.CharField: book_name>, <django.db.models.fields.CharField: author>,
<django.db.models.fields.SmallIntegerField: sex>, <django.db.models.fields.FloatField: price>, <django.db.models.fields.CharField: isbn>,
<django.db.models.fields.DateTimeField: publish_date>)
>>> fields = [f for f in fields if f is not meta.auto_field] # pk_set 为 False 时执行
>>> fields
[<django.db.models.fields.CharField: book_name>, <django.db.models.fields.CharField: author>, <django.db.models.fields.SmallIntegerField: sex>,
<django.db.models.fields.FloatField: price>, <django.db.models.fields.CharField: isbn>,
<django.db.models.fields.DateTimeField: publish_date>]
>>> update_pk = meta.auto_field and not pk_set
>>> update_pk
True
```

从前面的源码可以看到，在_save_table()方法中，最核心的语句就是调用 self._do_update()方法和 self._do_insert()方法。继续追踪这两个方法的源码，如下：

```
# 源码位置：django/db/models/base.py
# ……

class Model(metaclass=ModelBase):

    # ……

    def _do_update(self, base_qs, using, pk_val, values, update_fields, forced_update):
        # 调用 QuerySet 对象的 filter()方法，根据 pk 参数找到待更新的记录
        filtered = base_qs.filter(pk=pk_val)
```

```
    if not values:
        # 没有更新值
        return update_fields is not None or filtered.exists()
    if self._meta.select_on_save and not forced_update:
        return (
            filtered.exists() and
            (filtered._update(values) > 0 or filtered.exists())
        )
    # 最终调用 QuerySet 对象的_update()方法更新，该方法返回更新成功的记录数
    return filtered._update(values) > 0

def _do_insert(self, manager, using, fields, update_pk, raw):
    # manager 其实是一个 Manager 对象，调用其_insert()方法进行数据插入
    # 而该_insert()方法最终来源于 QuerySet 对象中的_insert()方法
    return manager._insert([self], fields=fields, return_id=update_pk,
                           using=using, raw=raw)
......
```

从 self._do_update()方法和 self._do_insert()方法的源码中可以看到，最终都是调用 Manager 对象的_update()方法和_insert()方法（该 Manager 对象关联了模型类），而这两个方法均来自 QuerySet 类。

继续看在 Model 类中定义的其他方法，相关使用示例如下：

```
>>> from shell_test.models import DjangoBooks
>>> book_obj = DjangoBooks.objects.all().get(book_name='我是奇才')
>>> book_obj
<DjangoBooks: <我是奇才, new_spyinx>>
>>> book_obj.delete()   # 模型类可以直接使用 delete()方法删除记录
(1, {'shell_test.DjangoDetail': 0, 'book_sales.BookSales': 0, 'shell_test.DjangoBooks':
1})
>>> DjangoBooks.objects.all().get(book_name='我是奇才')  # 删除后，无法从数据库中查到该记录
Traceback (most recent call last):
  File "<console>", line 1, in <module>
  File
"/root/.pyenv/versions/django2-core-test/lib/python3.8/site-packages/django/db/model
s/query.py", line 406, in get
    raise self.model.DoesNotExist(
shell_test.models.DjangoBooks.DoesNotExist: DjangoBooks matching query does not exist.
>>> fields = book_obj._meta.fields
>>> book_obj._get_FIELD_display(fields[1])  # 输入的必须是 Field 对象
'我是奇才'
```

至此，关于在 Model 类中定义的方法介绍到这里就结束了，有兴趣的读者可以继续分析和实验后续的方法。

关于 ORM 框架的源码解读到此就结束了，限于篇幅及源码的复杂性，还有很多关于 ORM 框架的源码细节没有介绍，但是本章的内容可以作为读者研究 ORM 框架源码的一个起点，这里也展现

了许多阅读源码的方法，比如带着疑问学习、根据功能现象追踪源码等，希望能对读者阅读源码有所帮助。

## 3.5 小结

本章重点剖析了 Django 内置的 ORM 框架，首先从读取数据库配置信息的代码开始，到 ORM 框架的底层核心，逐层抽丝剥茧，厘清 Django 操作 MySQL 的核心代码。它基于 Python 中的 mysqlclient 模块，并在该模块基础上做进一步的封装，形成了风格独特的数据库操作模式。本章没有深入剖析 Django 是如何将 Query 翻译成 SQL 语句的，有兴趣的读者可以对 SQLCompiler 类中的方法进行深入学习。

# 第4章 Django内置的模板系统

本章将全面解读 Django 内置的模板系统的源码，深入理解 Django 中模板文件及其标签渲染的实现原理。本章的源码主要位于 django/template 目录中。

## 4.1 Django 内置的模板语法

仍以第 1 章创建的 first_django 项目为例进行讲解。Django 自带了一个名为 Django Template Language（DTL）的模板系统，通过该模板系统，我们可以加载模板文件到内存中进行编译并动态插入数据，最后返回渲染后的文本内容。Django 项目的 settings.py 文件中的 TEMPLATES 变量用于配置模板系统的相关信息，包括指定模板引擎类路径、模板文件目录等。

（1）可在 first_django/settings.py 中设置 TEMPLATES 变量的内容，具体如下：

```
# first_django/settings.py
# ……

TEMPLATES = [
    {
        'BACKEND': 'django.template.backends.django.DjangoTemplates',
        # 只添加了下面的数据，指定模板文件所在目录；当创建 Django 项目时，这里为[]
        'DIRS': [os.path.join(BASE_DIR, 'templates')],
        'APP_DIRS': True,
        'OPTIONS': {
            'context_processors': [
                'django.template.context_processors.debug',
                'django.template.context_processors.request',
                'django.contrib.auth.context_processors.auth',
                'django.contrib.messages.context_processors.messages',
            ],
        },
    },
]
```

```
# ......
```

（2）在 first_django 项目的根目录下创建一个名为 templates 的目录，编辑一个简单的模板文件 test.tpl，具体操作如下：

```
(django2-core-test) [root@master first_django]# mkdir templates
(django2-core-test) [root@master first_django]# cat templates/test.tpl
标题: {{ title }}
内容: {{ content }}
```

（3）进入 Django 的 shell 模式，执行如下操作：

```
(django2-core-test) [root@master first_django]# python manage.py shell
Python 3.8.6 (default, Oct 18 2020, 15:33:08)
[GCC 4.8.5 20150623 (Red Hat 4.8.5-39)] on linux
Type "help", "copyright", "credits" or "license" for more information.
(InteractiveConsole)
>>> from django.template.loader import get_template
>>> tf = get_template('test.tpl')
>>> print(tf.render(context={'title': '这是标题', 'content': '这是正文'}))
标题: 这是标题
内容: 这是正文

>>>
```

通过几行简单的命令即可调用 Django 的内置模板系统，并对模板文件内容进行渲染，是不是非常神奇？下面看一些复杂的模板渲染示例。

## 4.1.1　for 标签

for 标签的示例模板文件位于 templates 目录中，内容如下：

(django2-core-test) [root@master first_django]# cat templates/test_for1.tpl

遍历列表：

```
{% for animal in animals %}
{{ animal }}
{% endfor %}
```

遍历字典：

```
{% for key, value in data.items %}
{{ key }}:{{ value }}
{% endfor %}
```

在 Django 的 shell 模式下渲染该模板文件：

```
# 前面执行的python manage.py shell命令部分省略
# ……
>>> from django.template.loader import get_template
>>> tf = get_template('test_for1.tpl')
>>> print(tf.render(context={'animals': ['大象', '狮子', '老虎', '熊猫'], 'data': {'person':
'王某某', 'age': 38, 'sex': '男'}}))
```

遍历列表：

```
大象

狮子

老虎

熊猫


```

遍历字典：

```

person:王某某

age:38

sex:男


>>>
```

可以看到，这里的模板文件已经成功添加了相关信息，但是在每一项下面都有一行空格，如何解决呢？这个问题在分析完 Django 内置的模板系统后就能得到解答。此外，还有一种方法，就是使用 Jinja2 模块作为 Django 的模板引擎，它自带的{%-和-%}能有效去掉空格。

（1）调整 Django 中的 settings 配置，指定 Jinja2 模块作为 Django 的模板引擎：

```
# first_django/settings.py
# ……

TEMPLATES = [
    {
        'BACKEND': 'django.template.backends.jinja2.Jinja2',
        'DIRS': [os.path.join(BASE_DIR, 'templates')],
        'APP_DIRS': True,
        'OPTIONS': {
            'context_processors': [
                'django.template.context_processors.debug',
                'django.template.context_processors.request',
```

```
                'django.contrib.auth.context_processors.auth',
                'django.contrib.messages.context_processors.messages',
            ],
        },
    },
]

# ……
```

(2) 准备一个符合 Jinja2 模块风格的模板文件，并确保在虚拟环境中已安装 Jinja2 模块，如下：

```
(django2-core-test) [root@master first_django]# cat templates/test_for_jinja2.tpl
```

遍历列表：

```
{% for animal in animals -%}
{{ animal }}
{% endfor %}
```

遍历字典：

```
{% for key in data -%}
{{ key }}:{{ data[key] }}
{% endfor -%}
(django2-core-test) [root@master first_django]# pip install jinja2
Looking in indexes: https://pypi.tuna.tsinghua.edu.cn/simple
Requirement already satisfied: jinja2 in
/root/.pyenv/versions/3.8.6/envs/django2-core-test/lib/python3.8/site-packages
(2.11.3)
Requirement already satisfied: MarkupSafe>=0.23 in
/root/.pyenv/versions/3.8.6/envs/django2-core-test/lib/python3.8/site-packages (from
jinja2) (1.1.1)
```

(3) 在 Django 的 shell 模式下渲染语句，结果如下：

```
>>> tf = get_template('test_for_jinja2.tpl')
>>> print(tf.render(context={'animals': ['大象', '狮子', '老虎', '熊猫'], 'data': {'person':
'王某某', 'age': 38, 'sex': '男'}}))
```

遍历列表：

```
大象
狮子
老虎
熊猫
```

遍历字典：

```
person:王某某
age:38
sex:男
```

```
>>>
```

**注意**，带短横线（-）风格的语句不适合 Django 内置的模板引擎，会抛出异常。此外，对于 Django 内置的 for 标签，还有许多变量可以在模板中使用，参见表 4-1。

表 4-1

| 变　量 | 含　义 |
|---|---|
| forloop.counter | 当前循环位置，从 1 开始 |
| forloop.counter0 | 当前循环位置，从 0 开始 |
| forloop.revcounter | 反向循环位置，从 *n* 开始，到 1 结束 |
| forloop.revcounter0 | 反向循环位置，从 *n*-1 开始，到 0 结束 |
| forloop.first | 如果是当前循环的第一位，返回 True |
| forloop.last | 如果是当前循环的最后一位，返回 True |
| forloop.parentloop | 在嵌套 for 循环中，获取上层 for 循环的 forloop |

笔者在 templates 目录下准备了一个 test_for2.tpl 文件，在该文件中使用了表 4-1 中的大部分变量，具体内容如下：

```
(django2-core-test) [root@master first_django]# cat templates/test_for2.tpl
```

遍历列表：

```
<ul>
{% spaceless %}
{% for animal in animals %}
{% if forloop.first %}
<li>第一次:{{ forloop.counter }}:{{ forloop.counter0 }}:{{ animal }}:
{{ forloop.revcounter }}:{{ forloop.revcounter0 }}</li>
{% elif forloop.last %}
<li>最后一次:{{ forloop.counter }}:{{ forloop.counter0 }}:{{ animal  }}:
{{ forloop.revcounter }}:{{ forloop.revcounter0 }}</li>
{% else %}
</li>{{ forloop.counter }}:{{ forloop.counter0 }}:{{ animal }}:{{ forloop.revcounter
 }}:{{ forloop.revcounter0 }}</li>
{% endif %}
{% endfor %}
{% endspaceless %}
</ul>
```

倒序遍历列表：

```
{% spaceless %}
{% for animal in animals reversed %}
<p>{{ animal }}:{{ forloop.revcounter }}</p>
```

```
{% endfor %}
{% endspaceless %}
```

（4）进入 Django 的 shell 模式并调用 get_template()方法进行模板渲染，结果如下：

```
(django2-core-test) [root@master first_django]# python manage.py shell
Python 3.8.6 (default, Oct 18 2020, 15:33:08)
[GCC 4.8.5 20150623 (Red Hat 4.8.5-39)] on linux
Type "help", "copyright", "credits" or "license" for more information.
(InteractiveConsole)
>>> from django.template.loader import get_template
>>> tf = get_template('test_for2.tpl')
>>> print(tf.render(context={'animals': ['大象', '狮子', '老虎', '熊猫']}))
遍历列表:
<ul>
<li>第一次:1:0:大象:4:3</li></li>2:1:狮子:3:2</li></li>3:2:老虎:2:1</li><li>最后一次:4:3:
熊猫:1:0</li>
</ul>

倒序遍历列表:
<p>熊猫:4</p><p>老虎:3</p><p>狮子:2</p><p>大象:1</p>

>>>
```

**注意**，在完成上述操作前，需要将 Django 的模板引擎改为内置的模板引擎，因为只有内置的模板引擎才支持 forloop.counter 等变量。Jinja2 模块支持的 for 循环变量与 Django 的模板引擎支持的 for 循环变量不同。此外，由于 Django 内置的模板引擎不支持带短横线的{%或者%}，为了解决空格问题，这里使用了 HTML 文本配合{% spaceless %}标签的方式去除多余空格。

### 4.1.2　if 标签

一个包含 if 标签的示例模板文件中的内容如下：

```
(django2-core-test) [root@master first_django]# cat templates/test_if.tpl
{% spaceless %}
<label{% if person.sex == 0 %} class="boy" {% endif %}>{{ person.name }}同学，你好!</label>
{% endspaceless %}
```

在 Python 交互模式下，Django 内置的模板系统的渲染效果如下：

```
(django2-core-test) [root@master first_django]# python manage.py shell
Python 3.8.6 (default, Oct 18 2020, 15:33:08)
[GCC 4.8.5 20150623 (Red Hat 4.8.5-39)] on linux
Type "help", "copyright", "credits" or "license" for more information.
(InteractiveConsole)
>>> from django.template.loader import get_template
>>> tf = get_template('test_if.tpl')
>>> print(tf.render(context={'person': {'sex': 0, 'name': 'spyinx'}}))
```

```
<label class="boy" >spyinx 同学，你好!</label>

>>> print(tf.render(context={'person': {'sex': 1, 'name': 'spyinx'}}))
<label>spyinx 同学，你好!</label>

>>>
```

这里的 if 标签和编程中的 if 语句类似，都是通过语句判断来控制流程。

### 4.1.3 csrf_token 标签

一个超级简单的 csrf_token 标签的示例如下：

```
(django2-core-test) [root@master first_django]# cat templates/test_token.tpl
```

测试 csrf_token 标签

```
{% csrf_token %}
(django2-core-test) [root@master first_django]# python manage.py shell
Python 3.8.6 (default, Oct 18 2020, 15:33:08)
[GCC 4.8.5 20150623 (Red Hat 4.8.5-39)] on linux
Type "help", "copyright", "credits" or "license" for more information.
(InteractiveConsole)
>>> from django.template.loader import get_template
>>> tf = get_template('test_token.tpl')
# 如果在 context 中没有 key='csrf_token', {% csrf_token %}标签没有输出
>>> print(tf.render(context={}))
```

测试 csrf_token 标签

```


>>> print(tf.render(context={'csrf_token': '111111'}))
```

测试 csrf_token 标签

```
<input type="hidden" name="csrfmiddlewaretoken" value="111111">

>>>
```

从上面的结果中可以看到，{% csrf_token %}标签最终被渲染成了一个<input>元素，其中，value 的值来自 context 中的 key='csrf_token'对应的值。如果在 context 中没有该 key，则模板标签将没有任何输出。这一点该如何解释呢？同样在分析完模板层的源码后，即可将这一现象解释得一清二楚。

### 4.1.4 with 标签

with 标签可用于对某个变量重新命名并使用。一个简单的带有 with 标签的模板文件的内容如下：

```
(django2-core-test) [root@master first_django]# cat templates/test_with.tpl
{% spaceless %}
{% with name1=person.name %}
<p>{{ name1 }}</p>
{% endwith %}
{% endspaceless %}

{% spaceless %}
{% with person.name as name2 %}
<div>{{ name2 }} </div>
{% endwith %}
{% endspaceless %}
```

下面是 Django 内置的模板系统的渲染效果：

```
(django2-core-test) [root@master first_django]# python manage.py shell
Python 3.8.6 (default, Oct 18 2020, 15:33:08)
[GCC 4.8.5 20150623 (Red Hat 4.8.5-39)] on linux
Type "help", "copyright", "credits" or "license" for more information.
(InteractiveConsole)
>>> from django.template.loader import get_template
>>> tf = get_template('test_with.tpl')
>>> print(tf.render(context={'person': {'sex': 1, 'name': '沈奇才'}}))
<p>沈奇才</p>

<div>沈奇才 </div>

>>>
```

### 4.1.5　cycle 标签

下面是一个 cycle 标签和 for 标签配合使用的示例，其中，测试了不同长度的 for 循环对 cycle 标签的影响：

```
(django2-core-test) [root@master first_django]# cat templates/test_cycle.tpl
<tr class="{% cycle 'test1' 'test2' 'test3' %}">
<td>技能:{{ skill }}</td>
</tr>
{% endfor %}
(django2-core-test) [root@master first_django]# python manage.py shell
Python 3.8.6 (default, Oct 18 2020, 15:33:08)
[GCC 4.8.5 20150623 (Red Hat 4.8.5-39)] on linux
Type "help", "copyright", "credits" or "license" for more information.
(InteractiveConsole)
>>> from django.template.loader import get_template
>>> tf = get_template('test_cycle.tpl')
>>> print(tf.render(context={'person': {'skills': ['数学', '英语', '理科综合']}}))
```

```
<tr class="test1">
<td>技能:数学</td>
</tr>

<tr class="test2">
<td>技能:英语</td>
</tr>

<tr class="test3">
<td>技能:理科综合</td>
</tr>
>>> print(tf.render(context={'person': {'skills': ['数学', '英语', '理科综合', '语文', '
体育']}}))

<tr class="test1">
<td>技能:数学</td>
</tr>

<tr class="test2">
<td>技能:英语</td>
</tr>

<tr class="test3">
<td>技能:理科综合</td>
</tr>

<tr class="test1">
<td>技能:语文</td>
</tr>

<tr class="test2">
<td>技能:体育</td>
</tr>
```

通过上面的例子可以看到，cycle 标签可以配合 for 标签进行渲染。当 for 标签中的循环变量值超过 cycle 标签的值时，cycle 标签的值将反复循环。

### 4.1.6 include 标签

include标签的含义很明确，就是将其他文件包括进来。下面是一个关于include标签的简单示例：

```
(django2-core-test) [root@master first_django]# cat templates/base.tpl
<p>{{ name }}，您好</p>
(django2-core-test) [root@master first_django]# cat templates/test_include.tpl
{% include "base.tpl" with name="火星阁下" %}
<p>吾乃地球骑士，特来一战</p>
(django2-core-test) [root@master first_django]# python manage.py shell
```

```
Python 3.8.6 (default, Oct 18 2020, 15:33:08)
[GCC 4.8.5 20150623 (Red Hat 4.8.5-39)] on linux
Type "help", "copyright", "credits" or "license" for more information.
(InteractiveConsole)
>>> from django.template.loader import get_template
>>> tf = get_template('test_include.tpl')
>>> print(tf.render(context={}))
<p>火星阁下，您好</p>

<p>吾乃地球骑士，特来一战</p>

>>>
```

这里用了一个带变量的模板文件 base.tpl，其中，模板变量名为 name。接着在模板文件 test_include.tpl 中使用 include 标签导入了 base.tpl 文件中的内容，同时传入 name 变量的值。最终渲染的结果如上所示。

## 4.1.7　过滤器标签

Django 中的过滤器标签基本出现在模板变量中，用于对模板变量的多级处理，如首字母大写，或者整个变量大写等。下面是一个简单的多级过滤器示例，如下：

```
(django2-core-test) [root@master first_django]# cat templates/test_filter_tag.tpl
<p>学生技能共{{ person.skills | length }}项:</p>{{ person.skills | join:"、"}}
<p>{{ time|date:'Y/m/d' }}</p>

(django2-core-test) [root@master first_django]# python manage.py shell
Python 3.8.6 (default, Oct 18 2020, 15:33:08)
[GCC 4.8.5 20150623 (Red Hat 4.8.5-39)] on linux
Type "help", "copyright", "credits" or "license" for more information.
(InteractiveConsole)
>>> from datetime import datetime
>>> now_value = datetime.now()
>>> from django.template.loader import get_template
>>> tf = get_template('test_filter_tag.tpl')
>>> print(tf.render(context={'person': {'skills': ['足球', '篮球', '乒乓球', '羽毛球']},
'time': now_value}))
<p>学生技能共 4 项:</p>足球、篮球、乒乓球、羽毛球
<p>2021/02/28</p>



>>>
```

在这面这段代码中，共演示了三个过滤器的使用，分别是不带参数的 length 过滤器，以及带参数的 join 过滤器和 date 过滤器。它们的功能从名称上就能看出来，非常直接。

# 4.2 Django 内置模板引擎源码解读

本节从 get_template()方法的源码开始，一步步深入解析 Django 的模板系统。

## 4.2.1 get_template()方法的源码解析

get_template()方法的源码如下：

```
# 源码位置：django/template/loader.py
# ……

def get_template(template_name, using=None):
    """
    根据给定的名称加载并返回一个 Template 对象。
    当没有对应的 Template 对象时，抛出 TemplateDoesNotExist
    """
    chain = []
    # 获取模板引擎列表
    engines = _engine_list(using)
    # 遍历模板引擎
    for engine in engines:
        try:
            # 如果调用 engine 的 get_template()方法能获得 Template 对象，直接返回
            return engine.get_template(template_name)
        except TemplateDoesNotExist as e:
            chain.append(e)
    # 如果得不到对应的 Template 对象，就直接抛出 TemplateDoesNotExist 异常
    raise TemplateDoesNotExist(template_name, chain=chain)
```

从上面的代码可以看出，Template 对象正是通过调用 engine 的 get_template()方法获得的，而相关的引擎列表则由_engine_list()方法得到。

## 4.2.2 _engine_list()方法的源码解析

_engine_list()方法的源码如下：

```
# 源码位置：django/template/loader.py
# ……

def _engine_list(using=None):
    return engines.all() if using is None else [engines[using]]
```

因为在 4.1 节中演示模板渲染时没有输入 using 参数，所以这里只考虑 using=None 的情况。继续追踪 engines 值：

```
# 源码位置：django/template/__init__.py
# ……

from .engine import Engine
from .utils import EngineHandler

engines = EngineHandler()

# ……
```

从上面的代码可知，engines 值为一个 EngineHandler 对象。

### 4.2.3　EngineHandler 类的源码解析

EngineHandler 类的源码如下：

```
# 源码位置：django/template/utils.py
# ……

class EngineHandler:
    def __init__(self, templates=None):
        # 关键的属性值，可以在初始化时输入，默认为配置文件中的 TEMPLATES 值
        self._templates = templates
        # 支持的模板引擎
        self._engines = {}

    @cached_property
    def templates(self):
        if self._templates is None:
            # 当实例化该类时，如果没有输入 templates 参数值，
            # 则默认的_templates 正好是配置文件中定义的 TEMPLATES 变量
            self._templates = settings.TEMPLATES

        templates = OrderedDict()
        backend_names = []
        for tpl in self._templates:
            try:
                # 获取引擎系统名称
                default_name = tpl['BACKEND'].rsplit('.', 2)[-2]
            except Exception:
                # 处理异常，非常规写法
                # ……

            # 得到一个 tpl 字典
            tpl = {
                'NAME': default_name,
                'DIRS': [],
```

```
                'APP_DIRS': False,
                'OPTIONS': {},
                **tpl,
            }

            templates[tpl['NAME']] = tpl
            backend_names.append(tpl['NAME'])

        # 查看在 backend_names 中元素出现的次数
        counts = Counter(backend_names)
        duplicates = [alias for alias, count in counts.most_common() if count > 1]
        if duplicates:
            # 抛出异常
            # ……

        return templates

    def __getitem__(self, alias):
        try:
            # 从_engine 属性中获取 alias 对应 key 的值
            return self._engines[alias]
        except KeyError:
            try:
                # 如果出现异常，从 templates 值中取对应 key 的模板引擎信息
                params = self.templates[alias]
            except KeyError:
                # 抛出异常
                raise InvalidTemplateEngineError(
                    "Could not find config for '{}' "
                    "in settings.TEMPLATES".format(alias))

            # 如果导入或者初始化模板引擎抛出异常
            # 就表示 self._engines[alias]没有设置，而下面的代码会得到执行
            # 因此，需要保存原来的参数，参见#24265.
            params = params.copy()
            backend = params.pop('BACKEND')
            # 导入模板引擎类
            engine_cls = import_string(backend)
            # 得到模板引擎对象
            engine = engine_cls(params)
            # 把模板引擎对象添加到_engines 属性中
            self._engines[alias] = engine
            return engine

    def __iter__(self):
        return iter(self.templates)
```

```
    def all(self):
        return [self[alias] for alias in self]
```

接下来在 Django 的 shell 模式下演示 EngineHandler 类的使用，同时对部分语句进行简单测试。从上面的代码可以看到：当 EngineHandler 对象的_templates 属性在__init__()函数中没有传入 templates 参数值时，该值会被设置为 settings.py 文件中设置的 TEMPLATES 变量的值。为了能演示相应的效果，下面分别指定 Django 内置的模板引擎和基于 Jinja2 模块的引擎：

```
# 源码位置：first_django/settings.py
# ……

TEMPLATES = [
    {
        'BACKEND': 'django.template.backends.django.DjangoTemplates',
        'DIRS': [os.path.join(BASE_DIR, 'templates')],
        'APP_DIRS': True,
        'OPTIONS': {
            'context_processors': [
                'django.template.context_processors.debug',
                'django.template.context_processors.request',
                'django.contrib.auth.context_processors.auth',
                'django.contrib.messages.context_processors.messages',
            ],
        },
    },
    {
        'BACKEND': 'django.template.backends.jinja2.Jinja2',
        'DIRS': [os.path.join(BASE_DIR, 'jinja2_templates')],
        'APP_DIRS': True,
    },
]

# ……
```

（1）实例化 EngineHandler 类，不传入任何参数。这里既可以手动实例化 EngineHandler 类，也可以直接导入已经实例化的 EngineHandler 对象，即 django/template/__init__.py 文件中的 engines：

```
(django2-core-test) [root@master first_django]# python manage.py shell
Python 3.8.6 (default, Oct 18 2020, 15:33:08)
[GCC 4.8.5 20150623 (Red Hat 4.8.5-39)] on linux
Type "help", "copyright", "credits" or "license" for more information.
(InteractiveConsole)
# 手动实例化 EngineHandler 类
>>> from django.template.utils import EngineHandler
>>> engines = EngineHandler()
# 直接导入实例化的 EngineHandler 对象
>>> from django.template import engines
```

```
>>> engines
<django.template.utils.EngineHandler object at 0x7f917085bf40>
>>>
```

（2）测试 engines.templates 语句中的一些代码，然后给出 engine.templates 的结果，具体操作如下：

```
>>> from django.conf import settings
>>> settings.TEMPLATES
[{'BACKEND': 'django.template.backends.django.DjangoTemplates', 'DIRS':
['/root/django-core-test/first_django/templates'], 'APP_DIRS': True, 'OPTIONS':
{'context_processors': ['django.template.context_processors.debug',
'django.template.context_processors.request',
'django.contrib.auth.context_processors.auth',
'django.contrib.messages.context_processors.messages']}}, {'BACKEND':
'django.template.backends.jinja2.Jinja2', 'DIRS':
['/root/django-core-test/first_django/jinja2_templates'], 'APP_DIRS': True}]
>>> tpl = settings.TEMPLATES[0]
>>> tpl['BACKEND']
'django.template.backends.django.DjangoTemplates'
>>> tpl['BACKEND'].rsplit('.', 2)[-2]
'django'
>>> tpl = settings.TEMPLATES[1]
>>> tpl['BACKEND'].rsplit('.', 2)[-2]
'jinja2'
>>> engines.templates
OrderedDict([('django', {'NAME': 'django', 'DIRS':
['/root/django-core-test/first_django/templates'], 'APP_DIRS': True, 'OPTIONS':
{'context_processors': ['django.template.context_processors.debug',
'django.template.context_processors.request',
'django.contrib.auth.context_processors.auth',
'django.contrib.messages.context_processors.messages']}, 'BACKEND':
'django.template.backends.django.DjangoTemplates'}), ('jinja2', {'NAME': 'jinja2',
'DIRS': ['/root/django-core-test/first_django/jinja2_templates'], 'APP_DIRS': True,
'OPTIONS': {}, 'BACKEND': 'django.template.backends.jinja2.Jinja2'})])
```

上面演示了 templates()方法中的关键语句：获取模板引擎标识名（default_name），最终返回一个有序字典。在字典中保存的正是模板引擎的相关信息。

**注意**，在第 3 章中已多次介绍过魔法函数__getitem__()的含义。engines[alias]语句会调用魔法函数__getitem__()获取模板引擎信息，这里的 alias 对应前面得到的模板引擎标识名。测试语句如下：

```
# 从 templates 值中获取'django'对应的模板引擎信息
>>> params = engines.templates['django']
>>> params = params.copy()
# 获取模板引擎类路径
>>> backend = params.pop('BACKEND')
# 导入该模板引擎类
>>> from django.utils.module_loading import import_string
```

```
>>> engine_cls = import_string(backend)
# 实例化模板引擎类，得到模板引擎对象
>>> engine = engine_cls(params)
>>> engine
<django.template.backends.django.DjangoTemplates object at 0x7f916df42850>
# 上述语句最终会得到 engines['django']，即得到一个模板引擎对象
>>> engines['django']
<django.template.backends.django.DjangoTemplates object at 0x7f916df42730>
```

此外，根据代码可知，engines.all()语句的输出为在配置文件中设置的模板引擎对象列表：

```
>>> engines.all()
[<django.template.backends.django.DjangoTemplates object at 0x7f916df42730>,
<django.template.backends.jinja2.Jinja2 object at 0x7f916df42820>]
```

此时再看_engine_list()函数的代码，是不是非常清楚了？直接运行该函数查看输出结果，如下：

```
>>> from django.template.loader import _engine_list
>>> _engine_list()
[<django.template.backends.django.DjangoTemplates object at 0x7f916df42730>,
<django.template.backends.jinja2.Jinja2 object at 0x7f916df42820>]
>>> _engine_list('django')
[<django.template.backends.django.DjangoTemplates object at 0x7f916df42730>]
>>> _engine_list('jinja2')
[<django.template.backends.jinja2.Jinja2 object at 0x7f916df42820>]
>>>
```

_engine_list()函数中传入的参数也用于指定模板引擎，最终返回的都是列表形式。

### 4.2.4 DjangoTemplates 类的源码解析

接下来将追踪目标转向模板引擎类——DjangoTemplate 类的源码，首先是 Django 内置的模板引擎类，即 DjangoTemplates，其源码如下：

```
# 源码位置：django/template/backends/django.py
# ……

class DjangoTemplates(BaseEngine):

   app_dirname = 'templates'

   def __init__(self, params):
      # params 参数为前面 EngineHandler 对象的 templates[alias]属性值
      params = params.copy()
      options = params.pop('OPTIONS').copy()
      options.setdefault('autoescape', True)
      options.setdefault('debug', settings.DEBUG)
      options.setdefault('file_charset', settings.FILE_CHARSET)
```

```
        libraries = options.get('libraries', {})
        options['libraries'] = self.get_templatetag_libraries(libraries)
        super().__init__(params)
        # 核心语句
        self.engine = Engine(self.dirs, self.app_dirs, **options)

    def from_string(self, template_code):
        return Template(self.engine.from_string(template_code), self)

    def get_template(self, template_name):
        try:
            # 返回 Template 对象
            return Template(self.engine.get_template(template_name), self)
        except TemplateDoesNotExist as exc:
            reraise(exc, self)

    def get_templatetag_libraries(self, custom_libraries):

        libraries = get_installed_libraries()
        libraries.update(custom_libraries)
        return libraries
```

DjangoTemplates 类中的代码并不多，简单浏览上面的代码即可发现 DjangoTemplates 类中的核心类与方法，它们分别为 BaseEngine 类、Engine 类和 Template 类，以及获取模板对象的 get_template() 方法。接下来将依次介绍它们，并通过实战演示其功能。

### BaseEngine 类

BaseEngine 类即模板引擎的父类，其源码如下：

```
# 源码位置：django/template/backends/base.py
# ……

class BaseEngine:

    def __init__(self, params):
        """
        初始化模板引擎，params 参数是关键
        """
        params = params.copy()
        # 模板引擎标识，'django'、'jinja2'等
        self.name = params.pop('NAME')
        # 模板目录
        self.dirs = list(params.pop('DIRS'))
        self.app_dirs = params.pop('APP_DIRS')
        # 如果还有其他参数，抛出异常
        if params:
```

```
            raise ImproperlyConfigured(
                "Unknown parameters: {}".format(", ".join(params)))

    @property
    def app_dirname(self):
        raise ImproperlyConfigured(
            "{} doesn't support loading templates from installed "
            "applications.".format(self.__class__.__name__))

    def from_string(self, template_code):
        """
        通过传入的参数创建并返回一个模板对象，必须在子类中实现
        """
        raise NotImplementedError(
            "subclasses of BaseEngine should provide "
            "a from_string() method")

    def get_template(self, template_name):
        """
        通过传入的模板文件名导入并返回一个模板对象，
        如果对应的模板文件不存在，抛出 TemplateDoesNotExist 异常
        """
        raise NotImplementedError(
            "subclasses of BaseEngine must provide "
            "a get_template() method")

    @cached_property
    def template_dirs(self):
        """
        返回搜索模板文件的目录列表
        """
        template_dirs = tuple(self.dirs)
        if self.app_dirs:
            template_dirs += get_app_template_dirs(self.app_dirname)
        return template_dirs

    def iter_template_filenames(self, template_name):
        for template_dir in self.template_dirs:
            try:
                # 生成迭代对象，根据搜索的模板目录及模板文件名组成全路径
                yield safe_join(template_dir, template_name)
            except SuspiciousFileOperation:
                pass
```

BaseEngine 类中的代码比较简单，只看上面代码中的注释即可理解。由于 BaseEngine 类中的几个核心方法都需要在子类中实现后才能调用,因此这里只演示对 BaseEngine 类的实例化,操作如下：

```
(django2-core-test) [root@master first_django]# python manage.py shell
Python 3.8.6 (default, Oct 18 2020, 15:33:08)
[GCC 4.8.5 20150623 (Red Hat 4.8.5-39)] on linux
Type "help", "copyright", "credits" or "license" for more information.
(InteractiveConsole)
>>> from django.template import engines
>>> params = engines.templates['django']
>>> params
{'NAME': 'django', 'DIRS': ['/root/django-core-test/first_django/templates'],
'APP_DIRS': True, 'OPTIONS': {'context_processors':
['django.template.context_processors.debug',
'django.template.context_processors.request',
'django.contrib.auth.context_processors.auth',
'django.contrib.messages.context_processors.messages']}, 'BACKEND':
'django.template.backends.django.DjangoTemplates'}
>>> from django.template.backends.base import BaseEngine
>>> backend = params.pop('BACKEND')  # 在EngineHandler类的魔法函数__getitem__()中
>>> options = params.pop('OPTIONS')    # 在子类的初始化方法__init__()中
>>> base_engine = BaseEngine(params)    # 子类会调用父类的初始化方法
```

**注意**，在实例化 BaseEngine 类时不能有额外的参数，否则会抛出异常。上述示例得到的 params 参数中包含一些额外的键值，为了能顺利实例化该类，需要移除这些键值。在 Django 中也是这样做的，'BACKEND'键会在 EngineHandler 类的魔法函数__getitem__()中被弹出，'OPTIONS'键会在其子类魔法函数__init__ ()中被弹出。

### Engine 类

Engine 类是真正用于渲染模板内容的引擎类，它的初始化方法如下：

```
# 源码位置：django/template/engine.py
# ……

class Engine:
   # 内置的标签、过滤器路径
   default_builtins = [
      'django.template.defaulttags',
      'django.template.defaultfilters',
      'django.template.loader_tags',
   ]

   def __init__(self, dirs=None, app_dirs=False, context_processors=None,
             debug=False, loaders=None, string_if_invalid='',
             file_charset='utf-8', libraries=None, builtins=None, autoescape=True):
      # 初始化一些变量，避免 None 值影响后续操作
      if dirs is None:
         dirs = []
      if context_processors is None:
```

```
        context_processors = []
    # 关于 loaders 值有一个可能的异常，即输入了 loaders 值且设置 app_dirs 为 True
    if loaders is None:
        # loaders 模块的路径
        loaders = ['django.template.loaders.filesystem.Loader']
        if app_dirs:
            loaders += ['django.template.loaders.app_directories.Loader']
        if not debug:
            loaders = [('django.template.loaders.cached.Loader', loaders)]
    else:
        if app_dirs:
            raise ImproperlyConfigured(
                "app_dirs must not be set when loaders is defined.")
    if libraries is None:
        libraries = {}
    if builtins is None:
        builtins = []

    # 给对象属性赋值
    self.dirs = dirs
    self.app_dirs = app_dirs
    self.autoescape = autoescape
    self.context_processors = context_processors
    self.debug = debug
    self.loaders = loaders
    self.string_if_invalid = string_if_invalid
    self.file_charset = file_charset
    self.libraries = libraries
    # 获取模板库
    self.template_libraries = self.get_template_libraries(libraries)
    # 内置模板标签的路径
    self.builtins = self.default_builtins + builtins
    self.template_builtins = self.get_template_builtins(self.builtins)
```

在上述初始化方法中出现了两个导入模块列表的方法，分别为 get_template_libraries()方法和 get_template_builtins()方法，它们的实现源码如下：

```
# 源码位置：django/template/engine.py
# ……

class Engine:

    # ……

    def get_template_builtins(self, builtins):
        # 根据路径导入相应的模块
        return [import_library(x) for x in builtins]
```

```
    def get_template_libraries(self, libraries):
        loaded = {}
        for name, path in libraries.items():
            loaded[name] = import_library(path)
        return loaded

    # ……
```

上述两个方法的核心为 import_library()方法，它的内容非常简单，就是根据传入的字符串路径，调用 importlib 模块中的 import_module()方法导入相应模块，具体实现如下：

```
# 源码位置：django/template/library.py
# ……

def import_library(name):
    try:
        # 导入指定路径的模块
        module = import_module(name)
    except ImportError as e:
        raise InvalidTemplateLibrary(
            "Invalid template library specified. ImportError raised when "
            "trying to load '%s': %s" % (name, e)
        )
    try:
        # 注意，这里返回的是导入模块的 register 属性值
        return module.register
    except AttributeError:
        raise InvalidTemplateLibrary(
            "Module  %s does not have a variable named 'register'" % name,
        )
```

import_library()方法最后返回的是导入模块的 register 属性值，因此该方法的作用是导入包含 register 属性的模块。如果没有 register 属性，将抛出相关异常，操作示例如下：

```
django2-core-test) [root@master first_django]# python manage.py shell
Python 3.8.6 (default, Oct 18 2020, 15:33:08)
[GCC 4.8.5 20150623 (Red Hat 4.8.5-39)] on linux
Type "help", "copyright", "credits" or "license" for more information.
(InteractiveConsole)
>>> from django.template.engine import Engine
# 注意，传入的 dirs 参数必须是列表
>>> engine = Engine(['/root/django-core-test/first_django/templates'])
>>> engine.builtins
['django.template.defaulttags', 'django.template.defaultfilters',
'django.template.loader_tags']
>>> engine.template_builtins
```

```
[<django.template.library.Library object at 0x7f0ce714d250>,
<django.template.library.Library object at 0x7f0ce71fc610>,
<django.template.library.Library object at 0x7f0ce716c430>]
>>>
```

从上面的代码可以看到，当没有传入 builtins 参数就实例化 Engine 类时，得到的 template_builtins 值为三个 Library 对象组成的列表，该属性值正是在初始化方法__init__()中通过调用 get_template_builtins()方法得到的。

在 Engine 类中有一个静态方法 get_default()，其源码如下：

```
# 源码位置：django/template/engine.py
# ……

class Engine:

   # ……

   @staticmethod
   @functools.lru_cache()
   def get_default():
       # 只能在这里导入，以免造成循环导入问题
       from django.template import engines
       from django.template.backends.django import DjangoTemplates
       # 导入所有的模板引擎系统
       for engine in engines.all():
          # 如果存在 Django 内置的模板引擎系统，直接返回内置引擎对象的 engine 属性
          if isinstance(engine, DjangoTemplates):
             return engine.engine
       raise ImproperlyConfigured('No DjangoTemplates backend is configured.')

   # ……
```

从上面的源码可以看到，静态方法 get_default()的作用是获取 Django 内置模板引擎对象的 engine 属性。从 DjangoTemplates 类的初始化方法中可知，该 engine 属性正是 Engine 对象，具体操作如下：

```
(django2-core-test) [root@master first_django]# python manage.py shell
Python 3.8.6 (default, Oct 18 2020, 15:33:08)
[GCC 4.8.5 20150623 (Red Hat 4.8.5-39)] on linux
Type "help", "copyright", "credits" or "license" for more information.
(InteractiveConsole)
>>> from django.template import engines
>>> engines['django']
<django.template.backends.django.DjangoTemplates object at 0x7f735c8a28e0>
>>> engines['django'].engine
<django.template.engine.Engine object at 0x7f735c8a2a00>
```

接下来讲解 Engine 类中的 template_loaders()方法，该方法用于导入 loaders 中的模块，具体实现

如下：

```
# 源码位置：django/template/engine.py
# ……

class Engine:

    # ……

    @cached_property
    def template_loaders(self):
        return self.get_template_loaders(self.loaders)

    def get_template_loaders(self, template_loaders):
        loaders = []
        for template_loader in template_loaders:
            # 核心，导入 Loader 对象
            loader = self.find_template_loader(template_loader)
            if loader is not None:
                # 加入 loaders 列表中
                loaders.append(loader)
        # 返回列表
        return loaders

    def find_template_loader(self, loader):
        if isinstance(loader, (tuple, list)):
            # 对于元组或者列表，第 1 个为字符串路径，后续为参数
            loader, *args = loader
        else:
            args = []

        if isinstance(loader, str):
            # 使用导入相关的 Loader 类
            loader_class = import_string(loader)
            # 实例化该类
            return loader_class(self, *args)
        else:
            raise ImproperlyConfigured(
                "Invalid value in template loaders configuration: %r" % loader)

    # ……
```

最终调用 import_string()方法，其源码位于 django/utils/module_loading.py 文件中，内容如下：

```
# 源码位置：django/utils/module_loading.py
# ……

def import_string(dotted_path):
    try:
```

```
        # 具体指定到要导入的类，于是按照最后一个.分割成模块路径和导入的类名
        module_path, class_name = dotted_path.rsplit('.', 1)
    except ValueError as err:
        raise ImportError("%s doesn't look like a module path" % dotted_path) from err

    # 导入模块
    module = import_module(module_path)

    try:
        # 获取模块中相应的类
        return getattr(module, class_name)
    except AttributeError as err:
        raise ImportError('Module "%s" does not define a "%s" attribute/class' % (
            module_path, class_name)
        ) from err
```

在理解了上面的 get_default()、template_loaders()和 import_string()这 3 个方法后，使用这些方法进行简单的测试，如下：

```
(django2-core-test) [root@master first_django]# python manage.py shell
Python 3.8.6 (default, Oct 18 2020, 15:33:08)
[GCC 4.8.5 20150623 (Red Hat 4.8.5-39)] on linux
Type "help", "copyright", "credits" or "license" for more information.
(InteractiveConsole)
>>> from django.template.engine import Engine
>>> engine = Engine(['/root/django-core-test/first_django/templates'])
>>> loader =engine.loaders[0]
>>> loader
('django.template.loaders.cached.Loader',
['django.template.loaders.filesystem.Loader'])
# 下面这些语句出自 find_template_loader()方法
>>> isinstance(loader, (tuple, list))
True
>>> loader, *args = loader
>>> loader
'django.template.loaders.cached.Loader'
>>> args
[['django.template.loaders.filesystem.Loader']]
>>> from django.utils.module_loading import import_string
>>> loader_class = import_string(loader)
>>> loader_class
<class 'django.template.loaders.cached.Loader'>
>>> loader = loader_class(engine, *args)
>>> loader
<django.template.loaders.cached.Loader object at 0x7f1bf3525520>
>>>
```

从上面的操作示例可知，在默认情况下，engine. template_loaders 的结果为 Loader 对象组成的列表：

```
# 接着上面的输入，以免重新导入
>>> engine.template_loaders
[<django.template.loaders.cached.Loader object at 0x7f1bf35258b0>]
```

Engine 类中的最后 5 个方法均与模板类（Template）相关，这些方法的源码如下：

```
# 源码位置：django/template/engine.py
# ……

class Engine:

    # ……

    def find_template(self, name, dirs=None, skip=None):
        tried = []
        for loader in self.template_loaders:
            try:
                # 通过 Loader 对象导入 Template 对象，在第一次导入成功后直接返回
                template = loader.get_template(name, skip=skip)
                return template, template.origin
            except TemplateDoesNotExist as e:
                tried.extend(e.tried)
        raise TemplateDoesNotExist(name, tried=tried)

    def from_string(self, template_code):
        """
        根据 template_code 参数返回一个编译过的 Template 对象
        """
        return Template(template_code, engine=self)

    def get_template(self, template_name):
        """
        根据 template_name 参数返回一个编译过的 Template 对象
        """
        template, origin = self.find_template(template_name)
        if not hasattr(template, 'render'):
            template = Template(template, origin, template_name, engine=self)
        return template

    def render_to_string(self, template_name, context=None):
        if isinstance(template_name, (list, tuple)):
            t = self.select_template(template_name)
        else:
            t = self.get_template(template_name)
        if isinstance(context, Context):
```

```
        # 调用模板对象渲染文本
            return t.render(context)
        else:
            return t.render(Context(context))

    def select_template(self, template_name_list):
        """
        给定一系列的模板文件名，返回第一个可被实例化的 Template 对象
        """
        if not template_name_list:
            raise TemplateDoesNotExist("No template names provided")
        not_found = []
        for template_name in template_name_list:
            try:
                # 返回第 1 个可被实例化的 Template 对象
                return self.get_template(template_name)
            except TemplateDoesNotExist as exc:
                if exc.args[0] not in not_found:
                    not_found.append(exc.args[0])
                continue
        # 如果没有满足的模板文件，直接抛出异常
        raise TemplateDoesNotExist(', '.join(not_found))
```

在上述方法中，find_template()、from_string()、get_template()和 select_template()方法均返回 Template 对象，render_to_string()方法则是对模板文件进行渲染，得到最终的内容。这 5 个方法的调用示例如下：

```
(django2-core-test) [root@master first_django]# python manage.py shell
Python 3.8.6 (default, Oct 18 2020, 15:33:08)
[GCC 4.8.5 20150623 (Red Hat 4.8.5-39)] on linux
Type "help", "copyright", "credits" or "license" for more information.
(InteractiveConsole)
>>> from django.template.engine import Engine
# 对应 find_template()方法中的语句
>>> engine = Engine(['/root/django-core-test/first_django/templates'])
>>> engine.template_loaders
[<django.template.loaders.cached.Loader object at 0x7f48a64383d0>]
>>> loader = engine.template_loaders[0]
>>> template = loader.get_template('test.tpl')
# 和 find_template()方法返回的结果一致
>>> template, template.origin
(<django.template.base.Template object at 0x7f48a64387c0>, <django.template.base.Origin
object at 0x7f48a64384f0>)
# 和上面的结果一致，只不过是不同的对象
>>> engine.find_template('test.tpl')
(<django.template.base.Template object at 0x7f48a64387c0>, <django.template.base.Origin
object at 0x7f48a64384f0>)
# 只返回一个 Template 对象
```

```
>>> engine.get_template('test.tpl')
<django.template.base.Template object at 0x7f48a64387c0>
# 只取第 1 个得到的 Template 对象
>>> engine.select_template(['test.tpl', 'test_for1.tpl'])
<django.template.base.Template object at 0x7f48a64387c0>
>>> print(engine.render_to_string('test.tpl', context={'title': '测试', 'content': '这
是一个测试'}))
标题：测试
内容：这是一个测试

>>>
```

从上面的代码可以看到，render_to_string()方法是通过调用模板对象的 render()方法实现的。此外，在 Engine 类中还有一部分代码需要探索，即 find_template() 方法中的 template = loader.get_template(name, skip=skip)语句。该语句中的Template对象是调用Loader对象的get_template()方法得到的，下面继续追踪 Loader 对象的 get_template()方法。根据前面的分析结果可知，在默认情况下得到的 Loader 对象是 django/template/loaders/cached.py 文件中 Loader 类的一个实例，该类的源码如下：

```
# 源码位置：django/template/loaders/cached.py
# ……

class Loader(BaseLoader):

    def __init__(self, engine, loaders):
        self.template_cache = {}
        self.get_template_cache = {}
        self.loaders = engine.get_template_loaders(loaders)
        super().__init__(engine)

    # ……

    def get_template(self, template_name, skip=None):

        # 如果在缓存中有数据，直接返回缓存中的数据
        # ……

        try:
            # 调用父类的 get_templates()方法获得 Template 对象
            template = super().get_template(template_name, skip)
        except TemplateDoesNotExist as e:
            # 抛出异常
            # ……
        else:
            # 在得到 Template 对象后，将其添加到缓存中
            self.get_template_cache[key] = template
```

```
        return template

    # ……
```

从上述代码可知，Loader 对象在调用 get_template()方法后，最终会调用父类的 get_template()方法获得 Template 对象。继续追踪 BaseLoader 类中 get_template()方法的实现，源码如下：

```
# 源码位置：django/template/loaders/base.py
# ……

class Loader:

    def __init__(self, engine):
        self.engine = engine

    def get_template(self, template_name, skip=None):

        tried = []

        for origin in self.get_template_sources(template_name):
            if skip is not None and origin in skip:
                tried.append((origin, 'Skipped'))
                continue

            try:
                # 获取模板文件内容
                contents = self.get_contents(origin)
            except TemplateDoesNotExist:
                tried.append((origin, 'Source does not exist'))
                continue
            else:
                # 根据模板文件内容等参数实例化 Template 对象
                return Template(
                    contents, origin, origin.template_name, self.engine,
                )

        raise TemplateDoesNotExist(template_name, tried=tried)

    def get_template_sources(self, template_name):
        # 由子类实现
        raise NotImplementedError(
            'subclasses of Loader must provide a get_template_sources() method'
        )

# ……
```

从上面的源码可以看到，Loader 对象的 get_template()方法先获取模板文件内容，然后根据内容等参数实例化 Template 对象并返回：

```
(django2-core-test) [root@master first_django]# python manage.py shell
Python 3.8.6 (default, Oct 18 2020, 15:33:08)
[GCC 4.8.5 20150623 (Red Hat 4.8.5-39)] on linux
Type "help", "copyright", "credits" or "license" for more information.
(InteractiveConsole)
>>> from django.template.engine import Engine
>>> engine = Engine(['/root/django-core-test/first_django/templates'])
>>> loader = engine.template_loaders[0]
>>> for origin in loader.get_template_sources('test.tpl'):
...     print(origin)
...     print(type(origin))
...
/root/django-core-test/first_django/templates/test.tpl
<class 'django.template.base.Origin'>
# Loader 对象的 get_contents()方法先传入 Origin 对象，然后返回该对象代表的模板文件内容
>>> for origin in loader.get_template_sources('test.tpl'):
...     print(loader.get_contents(origin))
...
标题：{{ title }}
内容：{{ content }}
```

Origin 对象的实现非常简单，不再赘述。print()语句会打印该对象的结果，即输出其 name 属性值，该值对应模板文件的完整路径。

至此，整个 Engine 类中的方法已经介绍完毕。与 Engine 类关联最深的便是 Template 类，该类的源码位于 django/template/base.py 文件中。

### Template 类

Template 类是核心模板类。这里是模板渲染最核心的地方，前面执行渲染的 render()方法也是在该类中定义的。首先从 Template 类的初始化方法入手，以分析模板文件 test.tpl 为例：

```
# 源码位置：django/template/base.py
# ......

UNKNOWN_SOURCE = '<unknown source>'

# ......

class Template:
    def __init__(self, template_string, origin=None, name=None, engine=None):

        # 如果没有输入 engine 对象，则调用 Engine.get_default()方法获得一个 engine 对象
        if engine is None:
            from .engine import Engine
```

```
            # 前面介绍过该方法
            engine = Engine.get_default()
        if origin is None:
            # 如果没有传入 Origin 对象，则使用默认的
            origin = Origin(UNKNOWN_SOURCE)
        self.name = name
        self.origin = origin
        self.engine = engine
        #注意，这里的 source 属性既可能是模板文件内容，也可能是 Template 对象
        self.source = str(template_string)  # May be lazy.
        # 核心语句
        self.nodelist = self.compile_nodelist()

    # ……
```

在上述初始化方法中有两个需要注意地方，第 1 个是 source 属性，它既可能是模板文件内容，也可能是 Template 对象；第 2 个是 nodelist 属性，操作示例如下：

```
 (django2-core-test) [root@master first_django]# python manage.py shell
Python 3.8.6 (default, Oct 18 2020, 15:33:08)
[GCC 4.8.5 20150623 (Red Hat 4.8.5-39)] on linux
Type "help", "copyright", "credits" or "license" for more information.
(InteractiveConsole)
>>> template_src = "标题: {{ title }}\n 内容: {{ content }}"
>>> from django.template.base import Template
>>> template = Template(template_src)
>>> template.nodelist
[<TextNode: '标题: '>, <Variable Node: title>, <TextNode: '\n 内容: '>, <Variable Node: content>]
>>>
```

结果非常明显，由此可以猜出 compile_nodelist()方法的含义：按照空格切割模板文件内容，同时提取相关的模板变量或者模板标签。下面用一个复杂的模板文件进行二次测试：

```
# 准备一个复杂的模板文件
(django2-core-test) [root@master first_django]# cat templates/test_node.tpl
测试变量: {{ person.name }}
<p {% if person.sex == 1 %}class='boy'{% endif %}>person.sex</p>
<li>
{% for skill in person.skills %}
<li>{{ skill }}</li>
{% endfor %}
(django2-core-test) [root@master first_django]# python manage.py shell
Python 3.8.6 (default, Oct 18 2020, 15:33:08)
[GCC 4.8.5 20150623 (Red Hat 4.8.5-39)] on linux
Type "help", "copyright", "credits" or "license" for more information.
(InteractiveConsole)
>>> f = open('templates/test_node.tpl', 'r+')
```

```
>>> template_src = f.read()
>>> template_src
"测试变量: {{ person.name }}\n<p {% if person.sex == 1 %}class='boy'{%
endif %}>person.sex</p>\n<li>\n{% for skill in person.skills %}\n<li>{{ skill }}</li>\n{%
endfor %}\n"
>>> from django.template.base import Template
>>> template = Template(template_src)
>>> template.nodelist
[<TextNode: '测试变量: '>, <Variable Node: person.name>, <TextNode: '\n<p '>, <IfNode>,
<TextNode: '>person.sex</p>\n<li>\n'>, <ForNode: for skill in person.skills, tail_len:
3>, <TextNode: '\n'>]
```

对比上述两个模板文件的输出结果，即可大致猜出 nodelist 属性值的含义。接下来追踪 compile_nodelist()方法的实现源码，找出 Django 是如何对这些内容进行切割并得到相应 Node 对象的：

```
# 源码位置：django/template/base.py
# ……

class Template:

   # ……

   def compile_nodelist(self):

      # 这里与 3.2.3 节介绍的一样，在默认情况下是非调试的
      if self.engine.debug:
         lexer = DebugLexer(self.source)
      else:
         lexer = Lexer(self.source)

      # 分词
      tokens = lexer.tokenize()
      # 创建解析器对象
      parser = Parser(
         tokens, self.engine.template_libraries, self.engine.template_builtins,
         self.origin,
      )

      try:
         # 调用 parse()方法解析并返回解析结果
         return parser.parse()
      except Exception as e:
         if self.engine.debug:
            e.template_debug = self.get_exception_info(e, e.token)
         raise

   # ……
```

上述源码的逻辑非常简单，首先，创建 Lexer 对象（或者是 DebugLexer 对象）；然后，调用 Lexer 对象的 tokenize()方法对模板文件内容按空格切割，示例如下：

```
(django2-core-test) [root@master first_django]# python manage.py shell
Python 3.8.6 (default, Oct 18 2020, 15:33:08)
[GCC 4.8.5 20150623 (Red Hat 4.8.5-39)] on linux
Type "help", "copyright", "credits" or "license" for more information.
(InteractiveConsole)
>>> f = open('templates/test_node.tpl', 'r+')
>>> template_src = f.read()
>>> from django.template.base import Lexer
>>> lexer = Lexer(template_src)
>>> lexer.tokenize()
[<django.template.base.Token object at 0x7f108733f760>, <django.template.base.Token
object at 0x7f108733f4c0>, <django.template.base.Token object at 0x7f108733f970>,
<django.template.base.Token object at 0x7f108733fa00>, <django.template.base.Token
object at 0x7f108733f820>, <django.template.base.Token object at 0x7f108733f880>,
<django.template.base.Token object at 0x7f108733f640>, <django.template.base.Token
object at 0x7f108733faf0>, <django.template.base.Token object at 0x7f108733f5b0>,
<django.template.base.Token object at 0x7f108733fb80>, <django.template.base.Token
object at 0x7f108733fc40>, <django.template.base.Token object at 0x7f108733fca0>,
<django.template.base.Token object at 0x7f108733f9a0>]
```

当通过手动创建 Lexer 对象并对 test_node.tpl 文件中的内容进行切割时，返回的结果是一个元素为 Token 对象的列表。Token 对象的定义如下：

```
# 源码位置：django/template/base.py
# ……

class Token:
    def __init__(self, token_type, contents, position=None, lineno=None):
        """
        一个 Token 对象代表着模板中的一个字符串

        token_type
            token 的类型，既可以是普通字符、变量或块，也可以是注释等
        contents
            token 表示的字符串内容
        position
            记录 token 在模板内容中的开始位置和结束位置，可选的二元组，用于调试
        lineno
            token 出现在模板内容中的行号，用于追踪信息
        """
        self.token_type, self.contents = token_type, contents
        self.lineno = lineno
        self.position = position

    def __str__(self):
```

```
        token_name = self.token_type.name.capitalize()
        return ('<%s token: "%s...">' %
                (token_name, self.contents[:20].replace('\n', '')))

    # ……

# ……
```

在了解了 Token 对象中的相关属性后，继续在前面的交互模式下输入如下代码：

```
>>> data = {}
>>> for token in lexer.tokenize():
...     data['token_type'] = token.token_type
...     data['contents'] = token.contents
...     print(data)
...
{'token_type': <TokenType.TEXT: 0>, 'contents': '测试变量: '}
{'token_type': <TokenType.VAR: 1>, 'contents': 'person.name'}
{'token_type': <TokenType.TEXT: 0>, 'contents': '\n<p '}
{'token_type': <TokenType.BLOCK: 2>, 'contents': 'if person.sex == 1'}
{'token_type': <TokenType.TEXT: 0>, 'contents': "class='boy'"}
{'token_type': <TokenType.BLOCK: 2>, 'contents': 'endif'}
{'token_type': <TokenType.TEXT: 0>, 'contents': '>person.sex</p>\n<li>\n'}
{'token_type': <TokenType.BLOCK: 2>, 'contents': 'for skill in person.skills'}
{'token_type': <TokenType.TEXT: 0>, 'contents': '\n<li>'}
{'token_type': <TokenType.VAR: 1>, 'contents': 'skill'}
{'token_type': <TokenType.TEXT: 0>, 'contents': '</li>\n'}
{'token_type': <TokenType.BLOCK: 2>, 'contents': 'endfor'}
{'token_type': <TokenType.TEXT: 0>, 'contents': '\n'}
```

将上面的结果与 test_node.tpl 中的内容进行对比，可以看到 tokenize()方法对原模板文件进行了切割，切割单元为定义的三种类型：文本（TEXT）、变量（VAR）和代码块（BLOCK）。继续追踪 tokenize()方法的实现源码，如下：

```
# 源码位置：django/template/base.py
# ……

# 模板符号常量
FILTER_SEPARATOR = '|'
FILTER_ARGUMENT_SEPARATOR = ':'
VARIABLE_ATTRIBUTE_SEPARATOR = '.'
# 模板块标签的开始标识符
BLOCK_TAG_START = '{%'
# 模板块标签的结束标识符
BLOCK_TAG_END = '%}'
# 模板变量标签的开始标识符
VARIABLE_TAG_START = '{{'
# 模板变量标签的结束标识符
```

```
VARIABLE_TAG_END = '}}'
# 注释标签的开始标识符
COMMENT_TAG_START = '{#'
# 注释标签的结束标识符
COMMENT_TAG_END = '#}'
TRANSLATOR_COMMENT_MARK = 'Translators'
# 单括号开始标识符
SINGLE_BRACE_START = '{'
# 单括号结束标识符
SINGLE_BRACE_END = '}'

# ……

# 用于匹配模板变量和模板块标签，并捕获整个标签及其内容
tag_re = (re.compile('(%s.*?%s|%s.*?%s|%s.*?%s)' %
         (re.escape(BLOCK_TAG_START), re.escape(BLOCK_TAG_END),
          re.escape(VARIABLE_TAG_START), re.escape(VARIABLE_TAG_END),
          re.escape(COMMENT_TAG_START), re.escape(COMMENT_TAG_END))))

logger = logging.getLogger('django.template')

# token 的类型，枚举类
class TokenType(Enum):
    TEXT = 0
    VAR = 1
    BLOCK = 2
    COMMENT = 3

# ……

class Lexer:
    def __init__(self, template_string):
        # 模板文件内容
        self.template_string = template_string
        self.verbatim = False

    def tokenize(self):
        # 将模板文件内容切割成一个个 token
        in_tag = False
        lineno = 1
        result = []

        # 按照前面编译的正则表达式进行切割
        for bit in tag_re.split(self.template_string):
            if bit:
                # 生成 Token 对象并将其添加到 result 中
                result.append(self.create_token(bit, None, lineno, in_tag))
            #如果 in_tag=True，则表示 bit 是标签中的内容，否则 bit 为普通文本
```

```
            in_tag = not in_tag
            # 得到下一个 token 的开始行号
            lineno += bit.count('\n')
        return result

    # ……

# ……
```

在上述源码中，最关键的地方是 tag_re 变量，它是一个编译后的正则表达式，用于匹配模板变量（ {{ 表示开始，}} 表示结束）和模板块标签（ {%表示开始，%}表示结束），以及注释标签（{#表示开始，#}表示结束）。此外，在 tokenize()方法中，通过 tag_re.split(self.template_string)语句可以实现对模板文件内容的切割。下面根据 tag_re 变量代表的正则表达式切割模板变量标签、模板块标签及注释标签，操作示例如下：

```
(django2-core-test) [root@master first_django]# python manage.py shell
Python 3.8.6 (default, Oct 18 2020, 15:33:08)
[GCC 4.8.5 20150623 (Red Hat 4.8.5-39)] on linux
Type "help", "copyright", "credits" or "license" for more information.
(InteractiveConsole)
>>> from django.template.base import tag_re
>>> tag_re
re.compile('(\\{%.*?%\\}|\\{\\{.*?\\}\\}|\\{\\#.*?\\#\\})')
>>> f = open('templates/test_node.tpl', 'r+')
>>> template_src = f.read()
>>> tag_re.split(template_src)
['测试变量: ', '{{ person.name }}', '\n<p ', '{% if person.sex == 1 %}', "class='boy'",
'{% endif %}', '>person.sex</p>\n<li>\n', '{% for skill in person.skills %}', '\n<li>',
'{{ skill }}', '</li>\n', '{% endfor %}', '\n']
```

在 tokenize()方法中，首先遍历上面切割的 token 字符串，然后调用 create_token()方法依次将切割的字符串转换为 Token 对象并添加到 result 中，最后返回该 result 值。继续追踪 create_token()方法的实现源码，如下：

```
# 源码位置: django/template/base.py
# ……

class Lexer:

    # ……

    def create_token(self, token_string, position, lineno, in_tag):

        if in_tag and token_string.startswith(BLOCK_TAG_START):
            # 如果是以{%开头的，提取对应块标签中的字符串内容
            block_content = token_string[2:-2].strip()  # 例如，提取{% if %}中的if字符串
            if self.verbatim and block_content == self.verbatim:
```

```
            self.verbatim = False
        if in_tag and not self.verbatim:
            if token_string.startswith(VARIABLE_TAG_START):
                # 如果是以{{开头的，提取模板变量内容，直接生成 Token 对象并返回
                return Token(TokenType.VAR, token_string[2:-2].strip(), position, lineno)
            elif token_string.startswith(BLOCK_TAG_START):
                # 如果是以{%开头的，直接进入这里执行
                if block_content[:9] in ('verbatim', 'verbatim '):
                    self.verbatim = 'end%s' % block_content
                # 最终返回 Token 对象，表示为 TokenType.BLOCK
                return Token(TokenType.BLOCK, block_content, position, lineno)
            elif token_string.startswith(COMMENT_TAG_START):
                content = ''
                if token_string.find(TRANSLATOR_COMMENT_MARK):
                    content = token_string[2:-2].strip()
                return Token(TokenType.COMMENT, content, position, lineno)
        else:
            return Token(TokenType.TEXT, token_string, position, lineno)

# ……
```

如果能假定 self.verbatim 一直为 False，则代码比较容易理解。事实上，self.verbatim 只会在输入块标签且块标签内容以 verbatim 开头时才会重新赋值。而这里的模板文件内容并没有这样的块标签，所以在解析 test_node.tpl 文件时，self.verbatim 将一直保持为 False。此时整个代码的逻辑非常简单，即用 create_token()方法判断传入的字符串是否为模板标签，如果不是模板标签，直接返回 Token 对象，且对应类型为 TokenType.TEXT；如果是模板标签，则根据起始字符串区分模板变量标签（{{）、模板块标签（{%）和注释标签（{#），最终返回对应的 Token 对象。

在解析 tokenize()方法之后，可知这里生成的 Token 对象只分为 4 种字符串，无法区分像{% if %}和{% for %}这样的模板块标签。继续回到 Template 对象的 compile_nodelis()方法中，再次调用 Lexer 对象的 tokenize()方法对模板文件进行切割，会继续生成一个 Parser 对象。调用该对象的 parse()方法解析前面得到的 tokens 列表，最终用 compile_nodelis()方法解析该结果。该结果将对得到的 Token 对象进行进一步解析，这里的区分粒度将更大。先看生成 Parser 对象的语句，如下：

```
parser = Parser(
    tokens, self.engine.template_libraries, self.engine.template_builtins,
    self.origin,
)
```

生成 Parser 对象的 4 个参数都比较容易理解，下面一一进行说明。

- ◎ 参数 Tokens 是调用 Lexer 对象的 tokenize()方法得到的。
- ◎ 参数 self.origin 是一个 Origin 对象，它和模板文件相关。
- ◎ 参数 self.engine.template_builtins 在介绍 Engine 类时详细介绍过，它的默认值如下：

```
>>> from django.template.engine import Engine
>>> Engine.default_builtins
['django.template.defaulttags', 'django.template.defaultfilters',
'django.template.loader_tags']
>>> engine = Engine()
>>> engine.template_builtins
[<django.template.library.Library object at 0x7ff91b9f0250>,
<django.template.library.Library object at 0x7ff91ba9f610>,
<django.template.library.Library object at 0x7ff91ba0d430>]
```

Engine 对象的 template_builtins 属性表示的是内置的模板标签、默认过滤器等。

◎ 参数 self.engine.template_libraries 的默认值如下：

```
>>> from django.template.base import Template, Origin
>>> template_file = "templates/test_node.tpl"
>>> template_src = open(template_file, 'r+').read()
>>> template = Template(template_src, Origin(template_file))
>>> template.engine.template_libraries
{'cache': <django.template.library.Library object at 0x7ff919de95e0>, 'i18n':
<django.template.library.Library object at 0x7ff91ba5f310>, 'l10n':
<django.template.library.Library object at 0x7ff919de98b0>, 'static':
<django.template.library.Library object at 0x7ff91c772c40>, 'tz':
<django.template.library.Library object at 0x7ff919de9d90>, 'admin_list':
<django.template.library.Library object at 0x7ff919743100>, 'admin_modify':
<django.template.library.Library object at 0x7ff919743d00>, 'admin_static':
<django.template.library.Library object at 0x7ff919743d90>, 'admin_urls':
<django.template.library.Library object at 0x7ff91b988a00>, 'log':
<django.template.library.Library object at 0x7ff91974b1f0>, 'staticfiles':
<django.template.library.Library object at 0x7ff91974b2e0>}
```

在分析 Engine 类中的方法时，曾分析过默认的 template_builtins 属性值，它的默认值来自 Engine 类本身的类变量 default_builtins。而默认的 template_libraries 属性值为空，操作如下：

```
>>> from django.template.engine import Engine
>>> engine = Engine()
>>> engine.template_builtins
[<django.template.library.Library object at 0x7ff91b9f0250>,
<django.template.library.Library object at 0x7ff91ba9f610>,
<django.template.library.Library object at 0x7ff91ba0d430>]
>>> engine.template_libraries
{}
```

template.engine.template_libraries 的输出结果是从何处得来的呢？下面看在 Template 类中关于 Engine 对象的生成部分：

```
# 源码位置：django/template/base.py
# ……
```

```
class Template:
    def __init__(self, template_string, origin=None, name=None, engine=None):

        if engine is None:
            from .engine import Engine
            engine = Engine.get_default()

        # ……

# ……
```

前面分析过 Engine 类的静态方法 get_default()，它将获取 DjangoTemplates 对象中的 engine 属性值：

```
# 源码位置：django/template/engine.py
# ……

class Engine:

    # ……

    @staticmethod
    @functools.lru_cache()
    def get_default():

        from django.template import engines
        from django.template.backends.django import DjangoTemplates
        for engine in engines.all():
            if isinstance(engine, DjangoTemplates):
                # 返回的是 DjangoTemplates 对象中的 engine 属性值
                return engine.engine
        raise ImproperlyConfigured('No DjangoTemplates backend is configured.')

    # ……

# ……
```

接下来跳到 DjangoTemplates 类的源码，从中找出 engine 属性的赋值语句，如下：

```
# 源码位置：django/template/backends/django.py
# ……

class DjangoTemplates(BaseEngine):

    app_dirname = 'templates'

    def __init__(self, params):
        # 这个 params 参数正是关于模板引擎的相关参数
        params = params.copy()
```

```
        options = params.pop('OPTIONS').copy()
        options.setdefault('autoescape', True)
        options.setdefault('debug', settings.DEBUG)
        options.setdefault('file_charset', settings.FILE_CHARSET)
        libraries = options.get('libraries', {})
        # 调用 DjangoTemplates 对象的 get_templatetag_libraries()方法获取模板库
        options['libraries'] = self.get_templatetag_libraries(libraries)
        super().__init__(params)
        # 根据 options['libraries']的值，实例化 Engine 对象
        self.engine = Engine(self.dirs, self.app_dirs, **options)

    # ……

    def get_templatetag_libraries(self, custom_libraries):
        libraries = get_installed_libraries()
        libraries.update(custom_libraries)
        return libraries
```

由上述代码可知，DjangoTemplates 对象的 engine 属性值最终来自 get_installed_libraries()方法。在 Django 的 shell 模式下直接调用 get_installed_libraries()方法，结果如下：

```
>>> from django.template.backends.django import get_installed_libraries
>>> get_installed_libraries()
{'cache': 'django.templatetags.cache', 'i18n': 'django.templatetags.i18n', 'l10n':
'django.templatetags.l10n', 'static': 'django.templatetags.static', 'tz':
'django.templatetags.tz', 'admin_list':
'django.contrib.admin.templatetags.admin_list', 'admin_modify':
'django.contrib.admin.templatetags.admin_modify', 'admin_static':
'django.contrib.admin.templatetags.admin_static', 'admin_urls':
'django.contrib.admin.templatetags.admin_urls', 'log':
'django.contrib.admin.templatetags.log', 'staticfiles':
'django.contrib.staticfiles.templatetags.staticfiles'}
```

执行创建 Engine 对象的语句（self.engine = Engine(self.dirs, self.app_dirs, **options)），即可得到前面显示的 self.engine.template_libraries 的结果：

```
>>> from django.template.engine import Engine
>>> from django.template.backends.django import get_installed_libraries
>>> libraries = get_installed_libraries()
>>> engine = Engine()
# 当在 Engine 类中使用 libraries 参数初始化时，效果和下面的语句一样，
# 得到的结果将赋给 template_libraries 属性值
>>> engine.get_template_libraries(libraries)
{'cache': <django.template.library.Library object at 0x7ff919de95e0>, 'i18n':
<django.template.library.Library object at 0x7ff91ba5f310>, 'l10n':
<django.template.library.Library object at 0x7ff919de98b0>, 'static':
<django.template.library.Library object at 0x7ff91c772c40>, 'tz':
```

```
<django.template.library.Library object at 0x7ff919de9d90>, 'admin_list':
<django.template.library.Library object at 0x7ff919743100>, 'admin_modify':
<django.template.library.Library object at 0x7ff919743d00>, 'admin_static':
<django.template.library.Library object at 0x7ff919743d90>, 'admin_urls':
<django.template.library.Library object at 0x7ff91b988a00>, 'log':
<django.template.library.Library object at 0x7ff91974b1f0>, 'staticfiles':
<django.template.library.Library object at 0x7ff91974b2e0>}
```

关于 get_installed_libraries()方法的源码本书不再追踪，其内容比较简单，有兴趣的读者可以自行学习。

在学习上面 4 个参数后，在 Django 的 shell 模式下将生成一个 Parser 对象，下面调用 parse()方法查看最终的解析结果：

```
>>> from django.template.engine import Engine
>>> from django.template.base import Lexer,Origin,Parser
# 模板文件内容
>>> source = open('templates/test_node.tpl', 'r+').read()
>>> lexer = Lexer(source)
# 得到 tokens 参数
>>> tokens = lexer.tokenize()
#在 engine 中实例化 Parser 对象的 libraries 和 builtins 参数
>>> engine = Engine.get_default()
# 得到 origin 参数
>>> origin = Origin('/root/django-core-test/first_django/templates/test_node.tpl')
>>> parser = Parser(tokens, engine.template_libraries, engine.template_builtins, origin)
>>> parser.parse()
[<TextNode: '测试变量:'>, <Variable Node: person.name>, <TextNode: '\n<p '>, <IfNode>,
<TextNode: '>person.sex</p>\n<li>\n'>, <ForNode: for skill in person.skills, tail_len:
3>, <TextNode: '\n'>]
```

从上面的代码可以看到，最终 tokens 会被解析成相应的 Node。比如 TokenType.TEXT 类型的 Token 对象将被直接转换成 TextNode 对象（不包括被标签所包含的 Token 对象，这部分内容将被追加到对应的 Node 对象中），{% if %}标签和{% endif %}标签所包括的 Token 对象将被统一解析成 IfNode 对象，{% for %}标签和{% endfor %}标签所包括的 Token 对象将被统一解析成 ForNode 对象。

下面接着分析 Parser 对象中的 parse()方法。它的实现源码并不复杂，内容如下：

```
# 源码位置：django/template/base.py
# ……

class Parser:
   def __init__(self, tokens, libraries=None, builtins=None, origin=None):
       self.tokens = tokens
       self.tags = {}
       self.filters = {}
       self.command_stack = []
```

```
    if libraries is None:
        libraries = {}
    if builtins is None:
        builtins = []

    self.libraries = libraries
    for builtin in builtins:
        # 注意，这里根据 builtins 得到支持解析的标签及过滤器
        self.add_library(builtin)
    self.origin = origin

def parse(self, parse_until=None):

    if parse_until is None:
        parse_until = []
    nodelist = NodeList()
    while self.tokens:  # 在循环内部，不断弹出 token 元素并解析，直到 tokens 列表为空
        token = self.next_token()  # 弹出最新的一个 Token 对象
        if token.token_type.value == 0:  # TokenType.TEXT
            # 直接添加 TextNode 节点到 nodelist 列表中，并添加额外的属性值
            self.extend_nodelist(nodelist, TextNode(token.contents), token)
        elif token.token_type.value == 1:  # TokenType.VAR
            # 模板变量必须要有变量值，否则抛出异常
            if not token.contents:
                raise self.error(token, 'Empty variable tag on line %d' % token.lineno)
            try:
                # 提取过滤器相关的信息
                filter_expression = self.compile_filter(token.contents)
            except TemplateSyntaxError as e:
                raise self.error(token, e)
            # 得到 VariableNode 对象
            var_node = VariableNode(filter_expression)
            # 将 VariableNode 对象添加到 nodelist 列表中
            self.extend_nodelist(nodelist, var_node, token)
        elif token.token_type.value == 2:  # TokenType.BLOCK
            try:
                command = token.contents.split()[0]
            except IndexError:
                raise self.error(token, 'Empty block tag on line %d' % token.lineno)
            if command in parse_until:
                self.prepend_token(token)
                return nodelist
            # 将 token 添加到命令栈中，如果由于未关闭的块标记而导致解析失败
            # 那么这里的值可用于打印错误信息
            self.command_stack.append((command, token))
            # 获取标签对应的回调方法
            try:
```

```
                compile_func = self.tags[command]
            except KeyError:
                self.invalid_block_tag(token, command, parse_until)

            try:
                # 标签对应的回调方法，标签的核心动作
                compiled_result = compile_func(self, token)
            except Exception as e:
                raise self.error(token, e)
            self.extend_nodelist(nodelist, compiled_result, token)
            # 从命令栈中移除该 token
            self.command_stack.pop()
        if parse_until:
            self.unclosed_block_tag(parse_until)
        return nodelist

    # ……

# ……
```

对于 Parser 类，这里列出了初始化方法__init__()和 parse()方法的实现。在初始化方法__init__()中有一个非常重要的动作，即调用 Parse 对象的 add_library()方法。该方法将设置 Parse 对象的 tags 和 filters 属性，这些是 Parser 对象支持解析的模板标签和过滤器。该方法的实现源码如下：

```
# 源码位置：django/template/base.py
# ……

class Parser:
    # ……

    def add_library(self, lib):
        self.tags.update(lib.tags)
        self.filters.update(lib.filters)

    # ……

# ……
```

查看前文得到的 Parser 对象的 tags 和 filters 属性值，结果如下：

```
# 接着前面得到的 Parser 对象继续执行
>>> parser.tags
{'autoescape': <function autoescape at 0x7f3978c09b80>, 'comment': <function comment at 
0x7f3978c0c820>, 'cycle': <function cycle at 0x7f3978c0c8b0>, 'csrf_token': <function 
csrf_token at 0x7f3978c0c940>, 'debug': <function debug at 0x7f3978c0c9d0>, 'filter': 
<function do_filter at 0x7f3978c0caf0>, 'firstof': <function firstof at 0x7f3978c0ca60>, 
'for': <function do_for at 0x7f3978c0cc10>, 'ifequal': <function ifequal at 
0x7f3978c0cca0>, 'ifnotequal': <function ifnotequal at 0x7f3978c0cd30>, 'if': <function
```

```
do_if at 0x7f3978c12160>, 'ifchanged': <function ifchanged at 0x7f3978c0cdc0>, 'load':
<function load at 0x7f3978c12310>, 'lorem': <function lorem at 0x7f3978c123a0>, 'now':
<function now at 0x7f3978c12430>, 'regroup': <function regroup at 0x7f3978c124c0>,
'resetcycle': <function resetcycle at 0x7f3978c12550>, 'spaceless': <function spaceless
at 0x7f3978c125e0>, 'templatetag': <function templatetag at 0x7f3978c12670>, 'url':
<function url at 0x7f3978c12700>, 'verbatim': <function verbatim at 0x7f3978c12790>,
'widthratio': <function widthratio at 0x7f3978c12820>, 'with': <function do_with at
0x7f3978c12940>, 'block': <function do_block at 0x7f3978c15550>, 'extends': <function
do_extends at 0x7f3978c15670>, 'include': <function do_include at 0x7f3978c15700>}
>>> sorted(parser.tags.keys())
['autoescape', 'block', 'comment', 'csrf_token', 'cycle', 'debug', 'extends', 'filter',
'firstof', 'for', 'if', 'ifchanged', 'ifequal', 'ifnotequal', 'include', 'load', 'lorem',
'now', 'regroup', 'resetcycle', 'spaceless', 'templatetag', 'url', 'verbatim',
'widthratio', 'with']
>>> parser.filters
# 字典形式，略，读者可以自行打印查看
# ……
>>> sorted(parser.filters.keys())
['add', 'addslashes', 'capfirst', 'center', 'cut', 'date', 'default', 'default_if_none',
'dictsort', 'dictsortreversed', 'divisibleby', 'escape', 'escapejs', 'filesizeformat',
'first', 'floatformat', 'force_escape', 'get_digit', 'iriencode', 'join', 'json_script',
'last', 'length', 'length_is', 'linebreaks', 'linebreaksbr', 'linenumbers', 'ljust',
'lower', 'make_list', 'phone2numeric', 'pluralize', 'pprint', 'random', 'rjust', 'safe',
'safeseq', 'slice', 'slugify', 'stringformat', 'striptags', 'time', 'timesince',
'timeuntil', 'title', 'truncatechars', 'truncatechars_html', 'truncatewords',
'truncatewords_html', 'unordered_list', 'upper', 'urlencode', 'urlize', 'urlizetrunc',
'wordcount', 'wordwrap', 'yesno']
```

从上面的结果中不仅可以看到 Django 支持的所有默认标签，如 if、for、cycle 和 csrf_token 等，还可以看到 Django 支持的所有过滤器，如 add、capfirst 和 cut 等。

继续回到 Parse 对象的 parse()方法。该方法的逻辑比较简单，最核心的代码段位于 while 循环中。它将循环检查切割模板文件得到的 tokens 列表，直到 tokens 列表中的元素为空。在 while 循环中，会依次弹出 tokens 列表中的 Token 对象，并进行判断。

(1) 当 token 类型为 TokenType.TEXT 时，直接调用 self.extend_nodelist()方法，将得到的 TextNode 对象添加到 nodelist 中。该方法对应的源码如下：

```
# 源码位置：django/template/base.py
# ……

class Parser:
    # ……

    def extend_nodelist(self, nodelist, node, token):
        if node.must_be_first and nodelist.contains_nontext:
            raise self.error(
```

```
                token, '%r must be the first tag in the template.' % node,
            )
        if isinstance(nodelist, NodeList) and not isinstance(node, TextNode):
            # 只要有非 TextNode，就给 nodelist 标记一个属性值
            nodelist.contains_nontext = True
        # 给相应的节点对象添加相关属性
        node.token = token
        node.origin = self.origin
        # 添加到 nodelist 中
        nodelist.append(node)

    # ……

# ……
```

self.extend_nodelist()方法并不复杂，核心功能就是给相应的节点对象添加相关属性，然后将该节点对象添加到 nodelist 中。

(2)当 token 类型为 TokenType.VAR 时，将调用 compile_filter()方法提取 token 内容中的过滤器，得到 filter_expression。compile_filter()方法的源码如下：

```
# 源码位置：django/template/base.py
# ……

class Parser:
    # ……

    def compile_filter(self, token):
        # 返回 FilterExpression 对象
        return FilterExpression(token, self)

    # ……

# ……
```

compile_filter()方法比较简单，它会返回一个 FilterExpression 对象。对于 FilterExpression 类的源码，这里只需了解初始化方法__init__()即可，如下：

```
# 源码位置：django/template/base.py
# ……

class FilterExpression:
    """
    解析变量 token 和它的过滤器，返回过滤器和参数的元组列表。
    例如[(过滤器 1, 参数 1), (过滤器 2, 参数 2)]这样的形式。
    实例:

        >>> token = 'variable|default:"Default value"|date:"Y-m-d"'
```

```
    # 在官方示例中，这里没有导入相应的标签库和内置过滤器库，因此会报错
    >>> p = Parser('')
    >>> fe = FilterExpression(token, p)
    >>> len(fe.filters)
    2
    >>> fe.var
    <Variable: 'variable'>
"""
def __init__(self, token, parser):
    self.token = token
    # filter_re 是一段用于匹配过滤器的正则表达式
    matches = filter_re.finditer(token)
    var_obj = None
    filters = []
    upto = 0
    for match in matches:
        start = match.start()
        if upto != start:
            raise TemplateSyntaxError("Could not parse some characters: "
                                      "%s|%s|%s" %
                                      (token[:upto], token[upto:start],
                                       token[start:]))
        if var_obj is None:
            # 第一次匹配模板变量，得到 Variable 对象，后续循环匹配的就是过滤器
            var, constant = match.group("var", "constant")
            if constant:
                try:
                    var_obj = Variable(constant).resolve({})
                except VariableDoesNotExist:
                    var_obj = None
            elif var is None:
                raise TemplateSyntaxError("Could not find variable at "
                                          "start of %s." % token)
            else:
                var_obj = Variable(var)
        else:
            # 后续将匹配过滤器名称
            filter_name = match.group("filter_name")
            args = []
            # 匹配参数
            constant_arg, var_arg = match.group("constant_arg", "var_arg")
            if constant_arg:
                args.append((False, Variable(constant_arg).resolve({})))
            elif var_arg:
                args.append((True, Variable(var_arg)))
            # 查找过滤器对应的过滤方法
            filter_func = parser.find_filter(filter_name)
            # 检查过滤方法的参数是否正常
```

```
            self.args_check(filter_name, filter_func, args)
            # 添加（过滤器函数，参数）到过滤器列表中
            filters.append((filter_func, args))
        upto = match.end()
    if upto != len(token):
        raise TemplateSyntaxError("Could not parse the remainder: '%s' "
                              "from '%s'" % (token[upto:], token))
    # 将过滤器列表赋给 filters 属性
    self.filters = filters
    self.var = var_obj

# 忽略其他方法
# ……

def __str__(self):
    return self.token

# ……
```

在官方的注释中给出了关于该类的一个使用示例，不过生成 Parser 对象的语句并不正确，因为 Parser('')语句得到的 Parser 对象没有相应的内置标签库和过滤器库，无法解析后续的过滤器。在 FilterExpression 类的初始化方法__init__()中，最核心的是 filter_re 变量，它是一个由 re 模块编译后的正则表达式，用于匹配变量和过滤器及其参数。注意上述代码中的 for 循环，变量匹配项第 1 个匹配的一定是模板变量，后续匹配的才是相应的过滤器。因此，在正常情况下，代码会先执行 if var_obj is None 语句，得到 Variable 对象并赋值给 var_obj。在接下来的循环中匹配的是过滤器的结果，此时 var_obj 变量不再是 None，因此代码将进行 else 分支，然后获取匹配的过滤器名和过滤器参数。此外，在 else 分支中将调用 Parser 对象的 find_filter()方法，找到对应过滤器名称的过滤器函数和过滤器参数，共同组成一个二元组：(filter_func, args)，然后将该二元组添加到过滤器列表 filters 中。最终整个 for 循环执行后，将得到一个 Variable 对象与一个过滤器函数和过滤器参数的列表，这两个值将分别赋给 FilterExpression 对象的 filters 属性和 var 属性。通过测试上面代码中的部分语句，可以帮助读者理解相关源码的含义，具体操作如下：

```
(django2-core-test) [root@master first_django]# python manage.py shell
Python 3.8.6 (default, Oct 18 2020, 15:33:08)
[GCC 4.8.5 20150623 (Red Hat 4.8.5-39)] on linux
Type "help", "copyright", "credits" or "license" for more information.
(InteractiveConsole)
>>> from django.template.base import filter_re
>>> token = 'variable|default:"Default value"|date:"Y-m-d"'
>>> flag = False
>>> matches = filter_re.finditer(token)
>>> for match in matches:
...     if not flag:
...         print('变量名称及常量:', match.group("var", "constant"))
```

```
...         flag = True
...     else:
...         print('过滤器名称:', match.group("filter_name"))
...         print('过滤器参数:', match.group("constant_arg", "var_arg"))
...
变量名称及常量：('variable', None)
过滤器名称： default
过滤器参数：('"Default value"', None)
过滤器名：date
过滤器参数：('"Y-m-d"', None)
```

上面的代码简单演示了 filter_re 变量的相关功能，即对模板变量和过滤器进行提取，并用一个 flag 模拟初始化方法__init__()中的 if 判断。if 判断与初始化方法__init__()类似，filter_re 变量也是第一次取模板变量信息，接下来取匹配的过滤器和相关参数。继续演示初始化方法__init__()的最终结果：

```
(django2-core-test) [root@master first_django]# python manage.py shell
Python 3.8.6 (default, Oct 18 2020, 15:33:08)
[GCC 4.8.5 20150623 (Red Hat 4.8.5-39)] on linux
Type "help", "copyright", "credits" or "license" for more information.
(InteractiveConsole)
>>> from django.template.engine import Engine
>>> engine = Engine.get_default()
>>> from django.template.base import Parser,FilterExpression
# 得到带内置模板标签库和过滤器库的 Parser 对象
>>> parser = Parser('', engine.template_libraries, engine.template_builtins)
>>> token = 'variable|default:"Default value"|date:"Y-m-d"'
>>> fe = FilterExpression(token, parser)
>>> fe.filters
[(<function default at 0x7f505e626820>, [(False, 'Default value')]), (<function date at 0x7f505e6265e0>, [(False, 'Y-m-d')])]
>>> fe.var
<Variable: 'variable'>
```

上面的代码成功演示了源码注释中给出的示例，compile_filter()方法返回的结果我们已经非常清楚了，不再赘述。最后根据 compile_filter()方法返回的结果对应生成 VariableNode 节点，再调用 self.extend_nodelist()方法将该节点添加到 nodelist 中，这便是针对模板变量做出的处理。

（3）第三种情况比较复杂，主要处理 Django 内置的模板标签，比如{% if xxxx %}和{% for name in names %}等。从前文的代码可以看到，parse()方法会将完整模板标签内的 tokens 压缩成一个 Node 节点。比如下面的 for 标签：

```
{% for skill in person.skills %}
<li>{{ skill }}</li>
{% endfor %}
```

上面的内容将被划分成 5 个 Token 对象，如下：

```
(django2-core-test) [root@master first_django]# python manage.py shell
Python 3.8.6 (default, Oct 18 2020, 15:33:08)
[GCC 4.8.5 20150623 (Red Hat 4.8.5-39)] on linux
Type "help", "copyright", "credits" or "license" for more information.
(InteractiveConsole)
>>> contents = "{% for skill in person.skills %}\n<li>{{ skill }}</li>\n{% endfor %}"
>>> from django.template.base import Lexer
>>> lexer = Lexer(contents)
>>> tokens = lexer.tokenize()
>>> tokens
[<django.template.base.Token object at 0x7feaacb167f0>, <django.template.base.Token
object at 0x7feaacb16580>, <django.template.base.Token object at 0x7feaacb16400>,
<django.template.base.Token object at 0x7feaacb16610>, <django.template.base.Token
object at 0x7feaacb168b0>]
```

根据前面的知识，上述结果不难理解。最终这 5 个 Token 对象将被压缩成一个 ForNode 对象，如下：

```
# 接着上面的操作
>>> from django.template.engine import Engine
>>> engine = Engine.get_default()
>>> from django.template.base import Parser
>>> parser = Parser(tokens, engine.template_libraries, engine.template_builtins)
>>> parser.parse()
[<ForNode: for skill in person.skills, tail_len: 3>]
>>>
```

这部分的处理正是由 parse()方法中 elif token.token_type.value == 2 分支下的代码实现的。下面以前文的 for 标签为例进行说明，逐步介绍该 else 分支下的代码。在 while 循环后，self.tokens 会弹出第 1 个 Token 对象，其类型为 TokenType.BLOCK，值为 for skill in person.skills：

```
>>> tokens[0].token_type
<TokenType.BLOCK: 2>
>>> tokens[0].contents
'for skill in person.skills'
```

此时，parse()方法将进入 elif token.token_type.value == 2 分支。

第 1 步，处理标签名，提取 for 字符串，表明这是一个 for 标签，对应的代码如下：

```
# token 的 contents 属性就是前面得到的{% %}中的内容，即 for skill in person.skills 字符串
command = token.contents.split()[0]
```

command 就是 for 字符串。

第 2 步，判断 command 是否位于 parse_until 列表中。一开始，对 TokenType.BLOCK 类型的 Token 对象来说，parse_until 是 None 值，因此跳过该 if 判断，将 for 字符串推入列表中：

```
self.command_stack.append((command, token))
```

此时的 self.command_stack 是一个空列表，在推入一个二元组(command, token)后，列表中就有了一个元素。

第 3 步的代码非常关键，而且在调试时非常容易出错：

```
try:
   compile_func = self.tags[command]
except KeyError:
   self.invalid_block_tag(token, command, parse_until)
```

上面的代码将从 self.tags 属性中找出对应 command（for 标签）的处理函数，该属性值的结果在前面展示过：

```
>>> parser.tags['for']
<function do_for at 0x7feaae6c9c10>
```

加载 do_for()函数的详细位置并不难找，只需要查看 Engine 类中的 default_builtins 属性值，即可知道 Django 中所有模板标签对应的模块路径。

第 4 步，看最后部分的处理代码，如下：

```
try:
   compiled_result = compile_func(self, token)
except Exception as e:
   raise self.error(token, e)
self.extend_nodelist(nodelist, compiled_result, token)
# Compile success. Remove the token from the command stack.
self.command_stack.pop()
```

注意，这里的代码非常关键，前文找出了 for 标签对应的处理函数，这里需调用该函数。compile_func()方法中传入的第 1 个参数即为该函数所在的 Parser 对象，这一点非常重要，必须牢记！

第 5 步，进入 do_for()函数的源码中：

```
# 源码位置：django/template/defaulttags.py
# ……

@register.tag('for')
def do_for(parser, token):
   """
   源码中有丰富的示例和 for 标签支持的变量及其说明
   """
   bits = token.split_contents() # for skill in person.skills，至少能按空格切成 4 块
   if len(bits) < 4:
      raise TemplateSyntaxError("'for' statements should have at least four"
                           " words: %s" % token.contents)
   # 这里说明了 for 标签中的最后一个字符串可以是'reversed'，即按倒序循环
   is_reversed = bits[-1] == 'reversed'
   # 必须是 for in 语句，如果是 for x in xx，则 in 在-2 索引的位置上
```

```
    # 如果是 for x in xx reversed，则 in 在-3 索引的位置上
    in_index = -3 if is_reversed else -2
    if bits[in_index] != 'in':
        # 如果不是 for in 语句，就抛出异常
        raise TemplateSyntaxError("'for' statements should use the format"
                                  " 'for x in y': %s" % token.contents)

    invalid_chars = frozenset((' ', '"', "'", FILTER_SEPARATOR))
    # 找出循环的变量，对于 for skill in person.skills 来说，loopvars 就是'skill'
    loopvars = re.split(r' *, *', ' '.join(bits[1:in_index]))
    for var in loopvars:
        if not var or not invalid_chars.isdisjoint(var):
            raise TemplateSyntaxError("'for' tag received an invalid argument:"
                                      " %s" % token.contents)
    # 解析循环值，比如对于 for skill in person.skills 来说，
    # bits[in_index + 1]即 person.skills
    sequence = parser.compile_filter(bits[in_index + 1])
    # 核心语句，非常重要，递归调用 parse()方法
    nodelist_loop = parser.parse(('empty', 'endfor',))
    # 取下一个 token，注意这里操作的 token 是同一个 Parser 对象中的 token
    token = parser.next_token()
    if token.contents == 'empty':
        nodelist_empty = parser.parse(('endfor',))
        parser.delete_first_token()
    else:
        nodelist_empty = None
    return ForNode(loopvars, sequence, is_reversed, nodelist_loop, nodelist_empty)
```

do_for()方法的前半本部分比较容易理解，即校验 for 标签的表达式，必须带 in，必须是 for x in xx 或者 for x in xx reversed 形式，否则直接抛出异常。do_for()方法的后半部分代码是它的核心，该方法将提取循环的变量、循环值，并对循环值调用 Parser 对象的 compile_filter()方法，以解析可能出现的过滤器。最核心的一句是调用当前 Parser 对象的 parse()方法，这里实现了一个递归操作（前面从 parse()方法追踪进入 do_for()方法）。不过和上层调用不同的是，这里传递了一个元组('empty', 'endfor',)给 parse_until 变量。

此外，需要注意的是，这里的 Parser 对象和上层的 Parser 对象是同一个。因此当再次进入 parse()方法时，该方法中的 self.tokens 值就只剩下 4 个 Token 对象了（'{% for skill in person.skills %}字符串对应的 Token 对象已经在第一个 parse()方法中被弹出了），但 parse()方法中的 nodelist = NodeList()语句会生成一个新的节点空列表。再次进入 while 循环，此时调用 token = self.next_token()语句将得到内容为\n<li>的 Token 对象，将其加入 nodelist 列表中。接下来是内容为 skill 的 Token 对象（TokenType.Var 类型）和内容为</li>\n 的 Token 对象。再接下来就是类型为 TokenType.BLOCK、内容为'end_for'的 Token 对象，这里将继续进入 elif token.token_type.value == 2 分支下处理。

**注意**，这里再次进入了 elif token.token_type.value == 2 分支，和第一次进入该分支不同的是，

parse_until 的值不再是 None，而是变成了('empty', 'endfor',)。再次执行如下代码段：

```
# 以下是elif token.token_type.value == 2分支下的代码段，在这里，command='end_for',
# 而parse_util=('empty', 'endfor',)
if command in parse_until:
   self.prepend_token(token)
return nodelist
```

当执行到 if 语句时，command='end_for'，而 parse_until=('empty', 'endfor',)，因此语句 command in parse_until 成立，进入该 if 语句。Parser 对象的 prepend_token()方法如下：

```
# 源码位置：django/template/base.py
# ……

class Parser:
   # ……

   def prepend_token(self, token):
      self.tokens.insert(0, token)

   # ……
```

上面的代码非常简单，将内容为 endfor 的 Token 对象插入 self.tokens，接着执行 return 语句，返回 nodelist 变量：

```
# 源码位置：django/template/defaulttags.py
# ……

@register.tag('for')
def do_for(parser, token):
   # 忽略前面的代码
   # ……

   # 对于二次调用parse()方法追踪完毕，返回nodelist
   nodelist_loop = parser.parse(('empty', 'endfor',))
   # 再次弹出内容为endfor的token，此时self.tokens为空
   token = parser.next_token()
   if token.contents == 'empty':
      nodelist_empty = parser.parse(('endfor',))
      parser.delete_first_token()
   else:
      # 内容为endfor
      nodelist_empty = None
   # 最终返回ForNode节点，同时带上标签内的所有信息，即循环变量、循环值和循环体等
   return ForNode(loopvars, sequence, is_reversed, nodelist_loop, nodelist_empty)

# ……
```

由前面的分析可知，nodelist_loop 为 for 标签内的节点列表，具体内容就是两个 TextNode 对象

（内容分别为\n<li>和</li>\n）和一个 VariableNode 对象（内容为 skill）。此外，在 Parser 对象的 tokens 属性中只剩下一个内容为 endfor 的 Token 对象。接下来，do_for()方法将调用 parser.next_token()语句，弹出最后一个 Token 对象。由于该 Token 对象的内容为 endfor，因此进入 else 分支。最终 do_for()方法返回一个 ForNode 对象，该对象中有循环变量、循环值、是否反转标识及循环体等属性，这些属性将用于 for 标签的渲染结果。

继续追踪接下来的执行语句：

```
self.extend_nodelist(nodelist, compiled_result, token)
self.command_stack.pop()
```

上面的代码首先将得到的 ForNode 对象添加到 nodelist 中，然后从 command_stack 列表中弹出之前推入列表中的('for', token)，此时 command_stack 列表为空。根据前面的分析可知，此时 self.tokens 也为空，因此 while 循环结束。跳出循环，继续执行如下代码：

```
if parse_until:
   # 抛出标签，未关闭异常
   self.unclosed_block_tag(parse_until)
return nodelist
```

上面的代码用于判断标签是否有对应的关闭标签，如果没有，直接抛出异常。例如，标签{% for %}需要用标签{% endfor %}才能关闭。下面以关闭模板为例描述抛出异常的过程，最简单的一个模板如下：

```
{% for name in names %}
<li>{{ name }}</li>
```

根据前面的分析过程可知，当处理内容为 for name in names 的 Token 对象时，在 self.command_stack 列表中将压入了一个二元组('for', token)，接着调用 do_for()方法处理 for 标签，然后在 do_for()方法中再次调用 Parser 对象的 parse()方法处理剩余的 tokens。注意，当递归调用 parse()方法时将传入('empty', 'endfor',)值，此时 parse()方法中的 parse_until 将等于('empty', 'endfor',)。在这次调用的 parse()方法中，由于 self.tokens 只剩下三个非标签类的 Token 对象，因此最终 tokens 将被取完，然后跳出循环。此时 parse_until=('empty', 'endfor',)，程序将调用 self.unclosed_block_tag(parse_until)方法抛出异常。程序是如何知道是哪个模板标签没有关闭呢？继续追踪 unclosed_block_tag()方法，如下：

```
# 源码位置：django/template/base.py
# ……

class Parser:
   # ……

   def unclosed_block_tag(self, parse_until):
      command, token = self.command_stack.pop()
```

```
        msg = "Unclosed tag on line %d: '%s'. Looking for one of: %s." % (
            token.lineno,
            command,
            ', '.join(parse_until),
        )
        raise self.error(token, msg)

    # ……
```

**注意**，无论是调用 parse()方法，还是在 do_for()方法中调用 parse()方法，它们对应的 Parser 对象都是同一个。因此，在第一次处理 for 标签时，会将('for', token)元素压入 Parser 对象的 command_stack 属性值中。而在 do_for()方法调用的 parse()方法中，调用的同样是这个 Parser 对象的 command_stack 属性值，并且它里面的元素没有弹出。因此，通过 Parser 对象的 command_stack 属性值就可以知道是哪个标签没有关闭了。

给读者留一个思考题：分析下面代码中 parse()方法的调用过程，该文件中的 if 标签没有对应的结束标签，但同样会抛出异常：

```
{% if person.sex == 0 %}
{% for name in names %}
<li>{{ name }}</li>
{% endfor %}
```

至此，在分析完 parse()方法后，对模板文件的解析过程就全部结束了。本书只以 for 标签为例对模块标签进行了讲解，其他模块标签的分析过程与之类似，读者可以自行分析。

接下来讲解对模板文件的渲染过程，即讲解 Template 类中的 render()方法。在 Template 类中，render()方法的实现源码如下：

```
# 源码位置：django/template/base.py
# ……

class Template:
    # ……

    def render(self, context):
        # 注意，这里的 context 参数为 Context 对象，而非普通的字典
        with context.render_context.push_state(self):
            if context.template is None:
                with context.bind_template(self):
                    context.template_name = self.name
                    return self._render(context)
            else:
                return self._render(context)

    # ……
```

当调用 Template 对象中的 render()方法时，传入的 context 参数必须是 Context 对象，否则 with context.render_context.push_state(self)语句将抛出异常。此时，有的读者可能会有疑问：在 4.1 节中，tf.render()语句传给 context 参数的都是字典形式，为什么不报错？原因很简单，在 tf.render()语句中调用的 render()方法并不是 Template 类中的 render()方法，只需查看一下 tf 的类型即可了解：

```
>>> from django.template.loader import get_template
>>> tf = get_template('test.tpl')
>>> type(tf)
<class 'django.template.backends.django.Template'>
```

追踪这个 Template 对象的 render()方法，如下：

```
# 源码位置：django/template/backends/django.py
# ……

class DjangoTemplates(BaseEngine):
   # ……

   def get_template(self, template_name):
      try:
         # 注意这里得到的 Template 对象是下面定义的类的实例
         # self.engine.get_template()语句得到的是在 base.py 文件中定义的 Template 对象
         return Template(self.engine.get_template(template_name), self)
      except TemplateDoesNotExist as exc:
         reraise(exc, self)

   # ……


class Template:
   # ……

   def render(self, context=None, request=None):
      # 根据输入的字典数据生成 Context 对象
      context = make_context(context, request,
autoescape=self.backend.engine.autoescape)
      try:
         # 最终调用 Template 对象的 render()方法，这里传入的正是 Context 对象
         return self.template.render(context)
      except TemplateDoesNotExist as exc:
         raise(exc, self.backend)
```

从 DjangoTemplates 类的 get_template()方法中可以看到，返回的 Template 对象其实是在该文件中定义的 Template 类的实例。而该对象的 template 属性则是根据 self.engine.get_template()得到的在 django/template/base.py 文件中定义的 Template 对象。测试结果如下：

```
# 接着上面的测试
>>> tf.template
<django.template.base.Template object at 0x7f5703378a90>
```

在 django.py 文件中定义的 Template 类的 render()方法的核心是调用 template 属性的 render()方法，即 base.py 文件中的 Templae 类的 render()方法。不过传入的 context 参数在经过 make_context()方法后变成了 Context 对象。因此，在 base.py 文件中的 Template 类的 render()方法接收的不再是输入的字典数据，而是 Context 对象，正好满足了代码要求。

在解决了上述疑惑后，回到 base.py 文件中 Template 类的 render()方法上来。为了能继续深入学习该方法的渲染逻辑，首先需要理解 Context 对象并掌握其使用方法。与 Context 对象相关的类全部定义在 django/template/context.py 文件中，内容如下：

```
# 源码位置: django/template/context.py
# ……

class ContextDict(dict):
    def __init__(self, context, *args, **kwargs):
        # 实例化的 context 参数是 Context 对象
        super().__init__(*args, **kwargs)
        # 调用 Context 对象的 dicts 属性，添加当前 ContextDict 对象
        context.dicts.append(self)
        self.context = context

    def __enter__(self):
        # with 语句需要实现的方法
        return self

    def __exit__(self, *args, **kwargs):
        # with 语句需要实现的方法
        self.context.pop()


class BaseContext:
    def __init__(self, dict_=None):
        self._reset_dicts(dict_)

    def _reset_dicts(self, value=None):
        builtins = {'True': True, 'False': False, 'None': None}
        self.dicts = [builtins]
        if value is not None:
            #将 value 值添加到 self.dicts 列表中
            self.dicts.append(value)

    def __copy__(self):
        duplicate = copy(super())
        duplicate.dicts = self.dicts[:]
```

```
        return duplicate

    def __repr__(self):
        return repr(self.dicts)

    def __iter__(self):
        return reversed(self.dicts)

    # 其余方法比较简单，不再列出
    # ……

class Context(BaseContext):
    def __init__(self, dict_=None, autoescape=True, use_l10n=None, use_tz=None):
        self.autoescape = autoescape
        self.use_l10n = use_l10n
        self.use_tz = use_tz
        self.template_name = "unknown"
        self.render_context = RenderContext()
        self.template = None
        super().__init__(dict_)

    @contextmanager
    def bind_template(self, template):
        if self.template is not None:
            raise RuntimeError("Context is already bound to a template")
        self.template = template
        try:
            yield
        finally:
            self.template = None

    # ……

# ……

def make_context(context, request=None, **kwargs):
    """
    根据传入的普通字典或者请求的 HttpRequest 对象创建一个 Context 对象
    """
    if context is not None and not isinstance(context, dict):
        # 如果 context 参数不为空，且非普通字典形式，直接抛出异常
        raise TypeError('context must be a dict rather than %s.' %
context.__class__.__name__)
    if request is None:
        # request 为 None，即直接根据 context 实例化 Context 类
        context = Context(context, **kwargs)
    else:
        # 根据 request 实例化 RequestContext 对象 （RequestContext 类为 Context 类的子类）
```

```
        # ……
    return context
```

上面的代码并不复杂，也非常容易理解，主要是在 BaseContext 类中定义的那些魔法函数，相关含义在前面已介绍过多次，此处不再赘述。下面直接使用这些类完成一些简单的测试，以帮助读者理解上下文及其相关类。以下是关于 BaseContext 类的使用示例：

```
Python 3.8.6 (default, Oct 18 2020, 15:33:08)
[GCC 4.8.5 20150623 (Red Hat 4.8.5-39)] on linux
Type "help", "copyright", "credits" or "license" for more information.
(InteractiveConsole)
>>> from django.template.context import BaseContext
>>> base_context = BaseContext({'person': {'name': '沈奇才', 'sex': 0, 'skills': ['足球', '篮球', '乒乓球']}})
>>> base_context.dicts
[{'True': True, 'False': False, 'None': None}, {'person': {'name': '沈奇才', 'sex': 0, 'skills': ['足球', '篮球', '乒乓球']}}]
# 根据魔法函数__iter__()可知，迭代 base_context 将反向迭代 self.dicts 列表
>>> for value in base_context:
...     print(value)
...
{'person': {'name': '沈奇才', 'sex': 0, 'skills': ['足球', '篮球', '乒乓球']}}
{'True': True, 'False': False, 'None': None}
# 以下操作与魔法函数__getitem__()相关
>>> base_context['person']
{'name': '沈奇才', 'sex': 0, 'skills': ['足球', '篮球', '乒乓球']}
>>> base_context['True']
True
# 以下操作与魔法函数__contains__()相关，都是遍历查询 key 是否在 self.dicts 列表的元素中
>>> 'hello' in base_context
False
>>> 'person' in base_context
True
# 下面是调用 get()方法的示例
>>> base_context.get('person')
{'name': '沈奇才', 'sex': 0, 'skills': ['足球', '篮球', '乒乓球']}
>>> base_context.get('None')  # 输出为 None
>>> base_context.get('False')
False
# 调用 setdefault()方法
>>> base_context.setdefault('hello', 'world')
'world'
#给 dicts 属性追加一个字典元素，打印 BaseContext 对象，结果是打印其 dicts 属性值
>>> base_context
[{'True': True, 'False': False, 'None': None}, {'person': {'name': '沈奇才', 'sex': 0, 'skills': ['足球', '篮球', '乒乓球']}, 'hello': 'world'}]
# 以下是 pop()方法和 push()方法，在 self.dicts 的最后位置压入和弹出元素
>>> base_context.push({'hahaha': 'wowowo'})
```

```
{'hahaha': 'wowowo'}
>>> base_context
[{'True': True, 'False': False, 'None': None}, {'person': {'name': '沈奇才', 'sex': 0,
'skills': ['足球', '篮球', '乒乓球']}, 'hello': 'world'}, {'hahaha': 'wowowo'}]
>>> base_context.pop()
{'hahaha': 'wowowo'}
>>> base_context.pop()
{'person': {'name': '沈奇才', 'sex': 0, 'skills': ['足球', '篮球', '乒乓球']}, 'hello':
'world'}
>>> base_context
[{'True': True, 'False': False, 'None': None}]
# flatten()方法会将列表中的所有字典元素合并成一个字典并返回
>>> base_context.flatten()
{'True': True, 'False': False, 'None': None, 'person': {'name': '沈奇才', 'sex': 0, 'skills':
['足球', '篮球', '乒乓球']}, 'hello': 'world'}
# 使用 new()方法不仅可以构建一个 BaseContext 对象,
# 还可以传入字典参数，并将该参数值添加到 self.dicts 中
>>> new_base_context = base_context.new(values={'xyz': 'abc'})
>>> type(new_base_context)
<class 'django.template.context.BaseContext'>
>>> new_base_context
[{'True': True, 'False': False, 'None': None}, {'xyz': 'abc'}]
```

上面的代码不难理解，每一个语句的背后都对应 BaseContext 类中的一个方法。这里只对 set_default()方法进行简单的说明。根据前面给出的 BaseContext 类代码可知，打印 BaseContext 对象将输出其 dicts 属性值，而该值是一个列表，第 1 个元素为一个固定字典{'True': True, 'False': False, 'None': None}，第 2 个元素是实例化 BaseContext 类时传入的字典参数。从上面的测试结果可以看出，当执行 base_context.setdefault('hello', 'world')语句时，在 self.dicts 的最后一个元素中就添加了一个键值对'hello'='world'。setdefault()方法的实现源码如下：

```
# 源码位置：django/template/context.py
# ……

class BaseContext:
    # ……

    def __setitem__(self, key, value):
        # 在 self.dicts 的最后一个元素中添加一个键值对
        self.dicts[-1][key] = value

    # ……

    def __getitem__(self, key):
        for d in reversed(self.dicts):
            if key in d:
                return d[key]
```

```
        # 遍历 self.dicts 中所有元素的键值对，如果没有相应的 key，则抛出 KeyError 异常
        raise KeyError(key)

    # ……

    def setdefault(self, key, default=None):
        try:
            # 如果存在 key，则直接返回对应的 value
            return self[key]
        except KeyError:
            # 如果没有 key，就设置 key，并调用魔法函数__setitem__()
            self[key] = default
        return default

    # ……
```

在看完上面的代码后，读者应该能够理解为何 setdefault()方法会将键值对'hello'='world'设置到 self.dicts 的最后一个元素中了。

在掌握了 BaseContext 类后，学习 Context 类就非常容易了。相比其父类，Context 类多了一些属性，扩展了父类的魔法函数__copy__()，并新增了 bind_template()方法和 update()方法，此处不再赘述。如果想了解其新增方法的具体含义，直接查看其源码即可。

最后测试 make_context()方法的输出，如下：

```
>>> from django.template.context import make_context
>>> context = make_context({'person': {'name': '沈奇才', 'sex': 0, 'skills': ['足球', '篮球', '乒乓球']}})
>>> type(context)
<class 'django.template.context.Context'>
>>> context.dicts
[{'True': True, 'False': False, 'None': None}, {'person': {'name': '沈奇才', 'sex': 0, 'skills': ['足球', '篮球', '乒乓球']}}]
>>> context['person']
{'name': '沈奇才', 'sex': 0, 'skills': ['足球', '篮球', '乒乓球']}
```

有了这些基础之后，继续回到 django/template/base.py 文件的 Template 类中，其 render()方法的第 1 句是一个 with 语句，如下：

```
with context.render_context.push_state(self):
```

Context 对象的 render_context 属性是一个 RenderContext 对象，该对象的 push_state()方法如下：

```
# 源码位置：django/template/context.py
# ……

class RenderContext(BaseContext):
    # ……
```

```
    @contextmanager
    def push_state(self, template, isolated_context=True):
        initial = self.template
        self.template = template
        if isolated_context:
            self.push()
        try:
            yield
        finally:
            self.template = initial
            if isolated_context:
                self.pop()

# ......
```

push_state() 方法中的装饰器 @contextmanager 使其能支持 with 语句。Python 使用装饰器 @contextmanager 来定义上下文管理器，读者可以在其源码中找到关于该装饰器的详细说明，如下：

```
# 源码位置：python3/lib/contextlib.py
# ......

def contextmanager(func):
    """@contextmanager 装饰器.

    典型用法：
        @contextmanager
        def some_generator(<arguments>):
            <setup>
            try:
                yield <value>
            finally:
                <cleanup>

    这使得：

        with some_generator(<arguments>) as <variable>:
            <body>

    等价于下面语句：

        <setup>
        try:
            <variable> = <value>
            <body>
        finally:
            <cleanup>
    """
```

```
    @wraps(func)
    def helper(*args, **kwds):
        return _GeneratorContextManager(func, args, kwds)
    return helper

# ……
```

被@contextmanager 装饰的函数在使用 with 语句时，将使用 with 语句下面的<body>语句替换函数中的 yield 部分。如果是 with A() as <variable>，那么在函数 A()的 yield 语句后面必须加上<value>，然后再用<variable>=<value>和<body>两部分内容替换函数 A()中的 yield 语句中的内容。下面通过一个示例理解装饰器@contextmanager 的作用：

```
(django2-core-test) [root@master first_django]# cat test_contextmanager.py
from contextlib import contextmanager


class Hello:
    def say_hello(self):
        print('您好!')


@contextmanager
def push_state():
    print('开始调用 push_state()方法')
    yield
    print('调用 push_state()方法结束')


@contextmanager
def push_state_with_variable():
    print('开始调用 push_state_with_variable()方法')
    yield Hello()
    print('调用 push_state_with_variable()方法结束')


with push_state():
    print('简单调用')

print('')

def test():
    with push_state_with_variable() as h:
        h.say_hello()
        print('在简单调用后执行 return')
        return  # 这个 return 将在最后执行
    print('最后不会执行')
```

```
test()
```

测试结果如下：

```
(django2-core-test) [root@master first_django]# python test_contextmanager.py
开始调用 push_state()方法
简单调用
调用 push_state()方法结束

开始调用 push_state_with_variable()方法
您好!
在简单调用后执行 return
调用 push_state_with_variable()方法结束
```

继续追踪 render()方法的代码，在 with 语句之后是判断 context.template 值是否为空。首先，通过 make_context()方法将字典数据转成 Context 对象，此时 context.template 为 None：

```
>>> from django.template.context import make_context
>>> context = make_context({'person': {'name': '沈奇才', 'sex': 0}})
>>> context.template is None
True
```

然后，进入 if context.template is None 分支并执行如下语句：

```
with context.bind_template(self):
    context.template_name = self.name
    return self._render(context)
```

又是一个 with 语句，查看 Context 类中的 bind_template()方法，同样是使用装饰器@contextmanager。bind_template()方法的实现源码如下：

```
# 源码位置: django/template/context.py
# ……

class Context(BaseContext):
    # ……

    @contextmanager
    def bind_template(self, template):
        if self.template is not None:
            raise RuntimeError("Context is already bound to a template")
        self.template = template
        try:
            # 替换 with 下面的语句, return 语句在最后执行
            yield
        finally:
            self.template = None
```

从上面的源码可知，在执行 with context.bind_template(self)及其下面的语句后，Context 对象的 template_name 和 template 均被赋值，最后再调用 self._render(context)返回渲染结果。下面完整演示 Template 对象中的 render()方法，其中，self 将通过 Template 对象代替：

```
>>> from django.template.context import make_context
>>> context = make_context({'title': '沈奇才', 'content': '他是电信的一名普通员工'})
>>> from django.template.loader import get_template
>>> tf = get_template('test.tpl')
>>> template = tf.template # base.py 文件中的 Template 对象
>>> template.name
'test.tpl'
# 实现 render()方法中 if 分支下的语句
>>> with context.bind_template(template):
...     context.template_name = template.name
...     print(template._render(context))
...
标题：沈奇才
内容：他是电信的一名普通员工
```

在 render()方法中，在确保 Context 对象有 template 和 template_name 属性后，调用 self._render()方法即可得到渲染结果。_render()方法的实现源码如下：

```
# 源码位置：django/template/base.py
# ……

class Template:
    # ……

    def _render(self, context):
        # 渲染的核心
        return self.nodelist.render(context)

    # ……
```

self.nodelist 正是前面追踪的 Template 对象中的 compile_nodelist()方法的结果，该值是一个 NodeList 对象，它保存了解析模板文件的全部信息。先回顾一下解析模板的最终结果，如下：

```
(django2-core-test) [root@master first_django]# python manage.py shell
Python 3.8.6 (default, Oct 18 2020, 15:33:08)
[GCC 4.8.5 20150623 (Red Hat 4.8.5-39)] on linux
Type "help", "copyright", "credits" or "license" for more information.
(InteractiveConsole)
>>> from django.template.base import Template, Origin
>>> template_file = "templates/test_node.tpl"
>>> template_src = open(template_file, 'r+').read()
>>> template = Template(template_src, Origin(template_file))
>>> template.nodelist
```

```
[<TextNode: '测试变量:'>, <Variable Node: person.name>, <TextNode: '\n<p '>, <IfNode>,
<TextNode: '>person.sex</p>\n<li>\n'>, <ForNode: for skill in person.skills, tail_len:
3>, <TextNode: '\n'>]
```

可以看到，nodelist 值是一个类似列表的结果。nodelist 是一个 NodeList 对象，而 NodeList 继承了列表类 list：

```
# 源码位置：django/template/base.py
# ……

class NodeList(list):
    contains_nontext = False

    def render(self, context):
        bits = []
        # 遍历节点列表
        for node in self:
            if isinstance(node, Node):
                # 如果 node 是 Node 的一个实例，则调用 Node 的 render_annotated()方法
                bit = node.render_annotated(context)
            else:
                bit = node
            # 把每个 node 对应的渲染内容都转成字符串
            bits.append(str(bit))
        # 最后把所有节点渲染的字符串拼接起来，就是最终渲染后的模板文件内容
        return mark_safe(''.join(bits))

    # ……

# ……
```

render()方法的逻辑非常简单：首先遍历 NodeList 对象中的每个节点元素，然后依次渲染这些节点元素得到相应的字符串，最后将所有的字符串拼接起来就得到了渲染结果。因此，这里的核心就变成了渲染单个节点的方法，即上面的 render_annotated()方法。该方法的源码如下：

```
# 源码位置：django/template/base.py
# ……

class Node:
    # ……

    def render(self, context):
        pass

    def render_annotated(self, context):
        try:
            # 每个继承了 Node 的子节点都会实现各自的 render()方法
            return self.render(context)
```

```
        except Exception as e:
            # 处理异常
            # ……

    # ……
```

在 Node 的所有子类中均未定义 render_annotated()方法，只在 Node 类中定义了 render_annotated()方法。该方法最终调用的是节点对象的 render()方法，所有继承了 Node 节点的子类均会自行实现 render()方法。因此，对于从模板文件中解析得到的各种 Node 节点而言，最终都会调用该节点的 render()方法得到对应的渲染节点。

下面先对每个节点进行测试，观察得到的各节点的渲染结果，然后分析源码解释现象。接着前面的交互模式执行如下操作：

```
>>> nodelist = template.nodelist
>>> from django.template.context import make_context
>>> nodelist[0]
<TextNode: '测试变量:'>
>>> nodelist[0].render(make_context({}))
'测试变量:'
>>> nodelist[1]
<Variable Node: person.name>
>>> nodelist[1].render(make_context({'person': {'name': '沈奇才'}}))
'沈奇才'
>>> nodelist[2]
<TextNode: '\n<p '>
>>> nodelist[2].render(make_context({}))
'\n<p '
>>> nodelist[3].render(make_context({'person': {'sex': 0}}))
''
>>> nodelist[3].render(make_context({'person': {'sex': 1}}))
"class='boy'"
# 对于普通文本节点，将直接输出结果 >>> nodelist[4]
<TextNode: '>person.sex</p>\n<li>\n'>
# 演示 ForNode 节点输出
>>> nodelist[5]
<ForNode: for skill in person.skills, tail_len: 3>
>>> nodelist[5].render(make_context({'person': {'skills': ['足球', '篮球', '乒乓球']}}))
'\n<li>足球</li>\n\n<li>篮球</li>\n\n<li>乒乓球</li>\n'
```

从上面的结果可以看出，渲染动作的核心其实在各节点的 render()方法中。下面以测试模板文件中出现的 TextNode、VariableNode 和 IfNode 为例进行说明。

### TextNode：普通文本节点

该节点不需要渲染，直接返回其代表的文本内容即可。在 TextNode 类中定义的 render()方法的内容如下：

```
# 源码位置：django/template/base.py
# ……

class TextNode(Node):
    def __init__(self, s):
        self.s = s

    def __repr__(self):
        return "<%s: %r>" % (self.__class__.__name__, self.s[:25])

    def render(self, context):
        return self.s

# ……
```

### VariableNode：变量节点

从前面的分析结果可知，该节点中包含了 Django 内置的过滤器，变量要经过一系列的过滤器处理后，才能得到最终结果。该节点的 render()方法实现如下：

```
# 源码位置：django/template/base.py
# ……

class VariableNode(Node):
    def __init__(self, filter_expression):
        self.filter_expression = filter_expression

    def __repr__(self):
        return "<Variable Node: %s>" % self.filter_expression

    def render(self, context):
        try:
            # 处理过滤器链
            output = self.filter_expression.resolve(context)
        except UnicodeDecodeError:
            return ''
        return render_value_in_context(output, context)

# ……
```

这里需要回顾一下前文 Parser 对象中 parse()方法的代码。在实例化 VariableNode 类时传入的 filter_expression 其实就是一个 FilterExpression 对象，它表示变量所经过的一系列过滤器表达式。下面测试 resolve()方法的输出：

```
(django2-core-test) [root@master first_django]# python manage.py shell
Python 3.8.6 (default, Oct 18 2020, 15:33:08)
```

```
[GCC 4.8.5 20150623 (Red Hat 4.8.5-39)] on linux
Type "help", "copyright", "credits" or "license" for more information.
(InteractiveConsole)
>>> from django.template.engine import Engine
>>> engine = Engine.get_default()
>>> from django.template.base import Parser,FilterExpression
# 在 Parser 对象中包含了 Django 内置的过滤器
>>> parser = Parser('', engine.template_libraries, engine.template_builtins)
# 待解析的 token
>>> token = 'person.name|default:"Unknown"|capfirst'
>>> fe = FilterExpression(token, parser)
# 解析 token，得到变量和过滤器
>>> fe.filters
[(<function default at 0x7fbd68bee820>, [(False, 'Unknown')]), (<function capfirst at 0x7fbd68be5ca0>, [])]
>>> fe.var
<Variable: 'person.name'>
# 输入 context，得到变量通过一系列过滤器后的值
>>> fe.resolve(context={'person': {'name': ''}})
'Unknown'
>>> fe.resolve(context={'person': {'name': 'spyinx'}})
'Spyinx'
```

上面的代码演示了 FilterExpression 对象调用 resolve()方法的结果，即得到变量经过一系列过滤器后的值。接着查看 resolve()方法的源码，内容如下：

```
# 源码位置：django/template/base.py
# ……

class FilterExpression:

    # ……

    def resolve(self, context, ignore_failures=False):
        if isinstance(self.var, Variable):
            try:
                # self.var 指的是最开始的变量值，在调用 resolve()方法后，
                # 根据输入的 context 值得到变量的值
                obj = self.var.resolve(context)
            except VariableDoesNotExist:
                # 异常处理
                # ……
        else:
            # 一般来说，self.var 是 Variable 对象。如果不是，直接把 self.var 的值赋给 obj
            obj = self.var

        # 在经过一系列过滤器处理后，遍历 self.filters
        for func, args in self.filters:
```

```
        # func 为过滤器函数，args 为默认参数
        arg_vals = []
        for lookup, arg in args:
            if not lookup:
                arg_vals.append(mark_safe(arg))
            else:
                arg_vals.append(arg.resolve(context))

        # 对一些特殊过滤器进行单独处理
        if getattr(func, 'expects_localtime', False):
            obj = template_localtime(obj, context.use_tz)
        if getattr(func, 'needs_autoescape', False):
            new_obj = func(obj, autoescape=context.autoescape, *arg_vals)
        else:
            # 除上述特殊过滤器外，大部分过滤器将直接调用过滤器函数，得到相关结果
            new_obj = func(obj, *arg_vals)

        # 最后判断在过滤器函数中是否有 is_safe 标识；如果有，就需要处理
        if getattr(func, 'is_safe', False) and isinstance(obj, SafeData):
            obj = mark_safe(new_obj)
        else:
            obj = new_obj
    # 最后返回经过一系列过滤器处理的最终结果
    return obj

# ……
```

上面的代码梳理了 resolve()方法的执行过程。该方法的逻辑是先用 context 替换变量的值，得到 obj。接着遍历过滤器，将 obj 传入过滤器函数进行处理，之后再送入下一个过滤器执行，最终返回经过所有过滤器处理后的 obj。继续在上面的交互模式下演示这一过程：

```
#接着前面的交互模式，已经得到了 fe 变量，这里不再重复导入代码
>>> from django.template.base import Variable
>>> isinstance(fe.var, Variable)
True
>>> obj = fe.var.resolve(context={'person': {'name': 'spyinx'}})
>>> obj
'spyinx'
>>> from django.utils.safestring import mark_safe
>>> func, args = fe.filters[0]
>>> func, args
(<function default at 0x7fbd68bee820>, [(False, 'unknown')])
# 在 default 过滤器中，参数中只有一个，所以直接用 append()方法处理一次即可
>>> arg_vals = []
>>> arg_vals.append(mark_safe('unknown'))
>>> arg_vals
['unknown']
# 经过 default 过滤器处理后的结果
```

```
>>> new_obj = func(obj, *arg_vals)
>>> new_obj
'spyinx'
# 将default过滤器处理后的结果赋给obj
>>> obj = new_obj
# 得到第2个过滤器，即capfirst过滤器
>>> func, args = fe.filters[1]
>>> func, args
(<function capfirst at 0x7fbd68be5ca0>, [])
# obj字符串经过capfirst过滤器处理后，首字母变大写，这便是最后的结果
>>> new_obj = func(obj, *[])
>>> new_obj
'Spyinx'
```

在看完上述操作过程后，对 FilterExpression 对象的 resolve()方法是不是非常清楚了？再回到 VariableNode 类的 render()方法中来，在经过一系列过滤器处理后，得到 output 值，最后需要经过 render_value_in_context()方法处理才能得到最终的渲染结果。该方法的实现源码如下：

```
# 源码位置：django/template/base.py
# ……

def render_value_in_context(value, context):
    value = template_localtime(value, use_tz=context.use_tz)
    value = localize(value, use_l10n=context.use_l10n)
    # 传入context参数，这里主要是使用context的autoescape属性值做判断
    if context.autoescape:
        if not issubclass(type(value), str):
            value = str(value)
        return conditional_escape(value)
    else:
        # 直接返回字符串形式
        return str(value)

# ……
```

从上面的源码可以看到，这里已经没有任何需要渲染的操作了，只是对渲染后的字符串做一些时钟、国际化及自动转义等处理。

在前面的演示中，只是输入 fe.var.resolve(context={'person': {'name': 'spyinx'}})即可得到 person.name 变量的结果，并没有分析渲染的过程。下面解释变量是如何根据 context 值渲染得到相应的值的。首先看 Variable 类的初始化方法__init__()，其源码如下：

```
# 源码位置：django/template/base.py
# ……

class Variable:
    """
```

```
    源码中有详细的注释和示例，这里省略了注释相关的内容

    (The example assumes VARIABLE_ATTRIBUTE_SEPARATOR is '.')
    """

    def __init__(self, var):
        self.var = var
        self.literal = None
        self.lookups = None
        self.translate = False
        self.message_context = None

        if not isinstance(var, str):
            # 如果不是字符串，抛出异常
            raise TypeError(
                "Variable must be a string or number, got %s" % type(var))
        try:
            # 首先，尝试将这个变量当作数值进行处理
            if '.' in var or 'e' in var.lower():
                # 判断是否有小数点或者指数符号
                self.literal = float(var)
                # "2." 是无效的，点号不能出现在数值最后
                if var.endswith('.'):
                    raise ValueError
            else:
                #转换成 int 类型
                self.literal = int(var)
        except ValueError:
            # 出现 ValueError 异常，说明 var 不是数值
            if var.startswith('_(') and var.endswith(')'):
                #对于以'_('开头或以')'结尾的，需要把这些去掉
                self.translate = True   # 这里的变量会影响 resolve()方法的处理
                var = var[2:-1]
            # 如果该变量是引号括起来的，则当作普通字符处理
            try:
                self.literal = mark_safe(unescape_string_literal(var))
            except ValueError:
                if var.find(VARIABLE_ATTRIBUTE_SEPARATOR + '_') > -1 or var[0] == '_':
                    raise TemplateSyntaxError("Variables and attributes may "
                                              "not begin with underscores: '%s'" %
                                              var)
                # 按照点号分割，比如对于 person.name 字符串，切割后将得到['person', 'name']
                self.lookups = tuple(var.split(VARIABLE_ATTRIBUTE_SEPARATOR))

# ……
```

在上面的代码中，在 Variable 类进行实例化时（即在初始化方法__init__()中的操作），会对传

入的变量进行一系列处理，处理逻辑如下。

（1）先将字符串变量 var（如果不是字符串，直接抛出异常）当作数值处理。如果包含小数点或者 e'字符，则使用 float()方法将字符串强制转成浮点数，否则使用 int()方法将字符串强制转成整数。如果上述转换没有异常，就直接设置 literal 属性值，初始化过程结束。

（2）如果第(1)步处理出现异常，则把字符串变量 var 当成普通变量处理，此时需要调用 unescape_string_literal()方法和 mark_safe()方法共同判断。如果 var 是表示常规的字符串，直接返回该字符串并赋给 literal 属性值，初始化过程结束；否则继续抛出异常。下面使用 unescape_string_literal()方法进行简单的测试：

```
>>> from django.utils.text import unescape_string_literal
>>> unescape_string_literal('person.name')
Traceback (most recent call last):
  File "<console>", line 1, in <module>
  File
"/root/.pyenv/versions/django2-core-test/lib/python3.8/site-packages/django/utils/fu
nctional.py", line 238, in wrapper
    return func(*args, **kwargs)
  File
"/root/.pyenv/versions/django2-core-test/lib/python3.8/site-packages/django/utils/te
xt.py", line 380, in unescape_string_literal
    raise ValueError("Not a string literal: %r" % s)
ValueError: Not a string literal: 'person.name'
>>> unescape_string_literal('"person.name"')
'person.name'
# 里面有双引号，表明是'xxxxx'
>>> unescape_string_literal('"xxxxx"')
'xxxxx'
>>> unescape_string_literal('xxxxx')
Traceback (most recent call last):
  File "<console>", line 1, in <module>
  File
"/root/.pyenv/versions/django2-core-test/lib/python3.8/site-packages/django/utils/fu
nctional.py", line 238, in wrapper
    return func(*args, **kwargs)
  File
"/root/.pyenv/versions/django2-core-test/lib/python3.8/site-packages/django/utils/te
xt.py", line 380, in unescape_string_literal
    raise ValueError("Not a string literal: %r" % s)
ValueError: Not a string literal: 'xxxxx'
```

看到区别了吗？对于前面测试的模板变量 person.name 而言，这里将继续抛出 ValueError 异常，进行第（3）次判断。

（3）如果第(2)步判断依旧异常，说明它是一个变量，需要根据 context 内容得到。此时在初始化方法__init__()中会根据点号对 var 进行切割，将其转成元组后赋给 lookups 属性，结果如下：

```
>>> from django.template.base import Variable
>>> var = Variable(var='person.name')
>>> var.lookups
('person', 'name')
>>> var.literal is None
True
```

从上面的初始化过程可以知道，对于不用渲染的结果，比如数值字符串、带引号的字符串等，将直接赋给 Variable 对象的 literal 属性。而对于需要根据 context 值进行渲染的，则将相关变量信息保存在 lookup 属性中，最后统一由 resolve()方法渲染得到结果。resolve()方法的实现源码如下：

```
# 源码位置：django/template/base.py
# ……

class Variable:
   # ……

   def resolve(self, context):
      if self.lookups is not None:
         # 继续调用_resolve_lookup()方法渲染变量结果
         value = self._resolve_lookup(context)
      else:
         # 如果 self.lookups 的值为 None，则表示不需要渲染
         value = self.literal
      if self.translate:
         #当 var 字符串以'_('开头且以')'结尾时，self.translate 值会被设置为 True
         # 通常情况下不会执行这里的代码
         is_safe = isinstance(value, SafeData)
         msgid = value.replace('%', '%%')
         msgid = mark_safe(msgid) if is_safe else msgid
         if self.message_context:
            return pgettext_lazy(self.message_context, msgid)
         else:
            return gettext_lazy(msgid)
      # 最终返回 value 结果，即渲染的最终结果
      return value

   # ……

# ……
```

resolve()方法的处理逻辑非常简单，根据前面的分析可知，当 Variable 对象的 lookups 属性值不为空时，需要结合 context 中的 key-value 值渲染模板变量，对于变量或者多级变量的情况，最终调用_resolve_lookup()方法渲染结果。_resolve_lookup()方法的实现源码如下：

```
# 源码位置：django/template/base.py
# ……

class Variable:
    # ……

    def _resolve_lookup(self, context):

        # 将 context 值赋给 current
        current = context
        try:
            # 遍历 self.lookups，对于前面的 person.name 来说，遍历元组('person', 'name')
            for bit in self.lookups:
                try:
                    # 第一种获取方式
                    current = current[bit]
                except (TypeError, AttributeError, KeyError, ValueError, IndexError):
                    try:
                        if isinstance(current, BaseContext) and getattr(type(current), bit):
                            raise AttributeError
                        # 第二种获取方式
                        current = getattr(current, bit)
                    except (TypeError, AttributeError):
                        if not isinstance(current, BaseContext) and bit in dir(current):
                            raise
                        try:
                            # 第三种获取方式
                            current = current[int(bit)]
                        except (IndexError,  # 索引超出范围异常
                                ValueError,  # 调用 int()的异常
                                KeyError,    # current 是一个字典，而该字典没有值为'int(bit) '的键
                                TypeError):  # 不可订阅对象
                            # 仍然找不到对应的属性值
                            raise VariableDoesNotExist("Failed lookup for key "
                                                       "[%s] in %r",
                                                       (bit, current))
                if callable(current):
                    # 如果 current 是可调用的
                    if getattr(current, 'do_not_call_in_templates', False):
                        pass
                    elif getattr(current, 'alters_data', False):
                        current = context.template.engine.string_if_invalid
                    else:
                        try:  # 对于 current 是可调用的情况，不能传入参数
                            current = current()  # 调用 current 代表的函数并将结果赋给 current
                        except TypeError:
                            try:
                                # 再次尝试处理
```

```
                        getcallargs(current)
                    except TypeError:
                        current = context.template.engine.string_if_invalid
                    else:
                        raise
        except Exception as e:
            # 处理异常
            # ……

        return current

# ……
```

_resolve_lookup()方法并不复杂，只是检查错误及对错误的冗余处理较多。在获取普通多级变量时，就是一层层用 current = current[bit]或者 getattr(current, bit)语句，最终获得变量结果。下面看一个简单的示例：

```
>>> from django.template.base import Variable
>>> var = Variable('person.name.first')
>>> var.lookups
('person', 'name', 'first')
>>> context = {'person': {'name': {'first': '沈', 'last': '奇才'}}}
>>> context
{'person': {'name': {'first': '沈', 'last': '奇才'}}}
>>> current = context
>>> current = current['person']
>>> current = current['name']
>>> current = current['first']
>>> current
'沈'
```

变量的简单渲染过程就是如此，还可以通过 getattr()、current[int(bit)]等方式获取变量结果。模板变量的值甚至还可以是函数名，其返回结果将被作为最终渲染结果，示例如下：

```
>>> def hello():
...     return 'hello, 沈奇才'
...
>>> from django.template.base import Variable
>>> var = Variable('hello')
>>> var.resolve(context={'hello': hello})
'hello, 沈奇才'
```

### IfNode：if 标签

if 标签的含义就是控制{% if condition %}和{% endif %}两个块标签内的内容是否显示。在它们内部可能又是一系列的 nodelist，比如有 if 标签、for 标签等。该标签的实现逻辑不难想象，即对于 if 块标签包裹的内容，继续调用 parse()方法得到 nodelist，接着调用 nodelist 的 render()方法得到渲染的文本，最后通过 condition 条件控制是否显示这段渲染的文本。IfNode 类中的 render()方法的实现源码如下：

```
# 源码位置：django/template/defaulttags.py
# ……

class IfNode(Node):

    def __init__(self, conditions_nodelists):
        self.conditions_nodelists = conditions_nodelists

    # ……

    @property
    def nodelist(self):
        return NodeList(self)

    def render(self, context):
        # condition 为 if 条件
        # nodelist 为递归调用 parse()方法解析 if 标签包裹的模板内容得到的 nodelist
        for condition, nodelist in self.conditions_nodelists:

            if condition is not None:          # if / elif 语句
                try:
                    # 得到条件匹配结果
                    match = condition.eval(context)
                except VariableDoesNotExist:
                    match = None
            else:                              # else 语句
                # 条件为空，默认为 True
                match = True

            if match:
                # 如果分支的条件成立，那么渲染的结果为该分支标签下 nodelist 的渲染结果
                return nodelist.render(context)
        # 其余情况返回空字符串
        return ''

# ……
```

在上面的代码中，IfNode 对象中的 render()方法和预想的一致，最关键的是在实例化 IfNode 节点时传入的 conditions_nodelists 参数。这时需要回到生成 IfNode 对象的地方，参考前面的分析过程，对 if 标签的处理是由 do_if()方法完成的，内容如下：

```
# 源码位置：django/template/defaulttags.py
# ……
```

```
@register.tag('if')
def do_if(parser, token):
    # {% if ... %}
    bits = token.split_contents()[1:]  # bits 为条件切割后的结果
    # 调用 TemplateIfParser 对象的 parse()方法解析 if 条件
    condition = TemplateIfParser(parser, bits).parse()
    # 解析该标签下的其余 node，如果遇到 elif、else 等，立即返回
    nodelist = parser.parse(('elif', 'else', 'endif'))
    # 得到初始化 if 条件列表，第 1 个元素为条件和 nodelist 组成的元组
    conditions_nodelists = [(condition, nodelist)]
    # 下一个 token
    token = parser.next_token()

    # 反复解析{% elif ... %}标签
    while token.contents.startswith('elif'):
        bits = token.split_contents()[1:]
        condition = TemplateIfParser(parser, bits).parse()
        nodelist = parser.parse(('elif', 'else', 'endif'))
        conditions_nodelists.append((condition, nodelist))
        token = parser.next_token()

    # 解析{% else %}标签
    if token.contents == 'else':
        # else 标签，接下来只有 endif 标签来结尾
        nodelist = parser.parse(('endif',))
        # 在 else 标签下只有 nodelist，没有相应的条件
        conditions_nodelists.append((None, nodelist))
        token = parser.next_token()

    # 如果接下来的标签不是{% endif %}，直接抛出异常
    if token.contents != 'endif':
        raise TemplateSyntaxError('Malformed template tag at line {0}: "{1}"'.format(
                    token.lineno, token.contents
              ))
    # 最后返回 IfNode 节点
    return IfNode(conditions_nodelists)

# ……
```

上述方法的操作流程整理如下：

（1）从 if 标签进来，提取 if 标签中的条件判断语句，得到 TemplateIfParser 对象。调用该对象的 parse()方法得到 if 标签的条件 condition，继续解析得到 if 标签下的 nodelist，直到遇到 elif、else 或 endif 中的任何一个标签，继续向下执行得到初始化的条件节点列表 conditions_nodelists。

（2）取下一个 token，如果 token 内容以 elif 开始，就需要循环处理。每个 elif 标签都将得到一个(condition, nodelist)元组。condition 为本次分支语句的条件，nodelist 为本次分支下的节点列表，每

次都将得到的元组添加到 conditions_nodelists 列表中。

(3) 继续取下一个 token，如果该 token 的内容为“else”字符串，表明在该分支中没有条件判断，只能得到该分支下的节点列表 nodelist。此时会将(None, nodelist)元组添加到 conditions_nodelists 中。

(4) 继续取下一个 token，如果该 token 内容不是“endif”字符串，直接抛出异常。

(5) 最后返回 IfNode 对象，实例化参数正是前面得到的 conditions_nodelists。

根据上面的分析结果，相信读者已经对得到的 IfNode 对象中的 conditions_nodelists 有了一定的了解。下面将一段包含 if 标签的文本转成 IfNode 对象，并查看 conditions_nodelists 的具体输出结果：

```
(django2-core-test) [root@master first_django]# python manage.py shell
Python 3.8.6 (default, Oct 18 2020, 15:33:08)
[GCC 4.8.5 20150623 (Red Hat 4.8.5-39)] on linux
Type "help", "copyright", "credits" or "license" for more information.
(InteractiveConsole)
>>> contents = "{% if person.sex == 0 %}\n<li>男</li>\n{% elif person.sex == 1 %}\n<li>
女</li>\n{% else %}\n<li>数据不太对</li>{% endif %}"
>>> print(contents)
{% if person.sex == 0 %}
<li>男</li>
{% elif person.sex == 1 %}
<li>女</li>
{% else %}
<li>数据不太对</li>{% endif %}
#获取一个 IfNode 对象
>>> from django.template.base import Lexer
>>> lexer = Lexer(contents)
>>> tokens = lexer.tokenize()
>>> from django.template.engine import Engine
>>> engine = Engine.get_default()
>>> from django.template.base import Parser
>>> parser = Parser(tokens, engine.template_libraries, engine.template_builtins)
#这里的 contents 内容只有一个 if 操作
# 所以最终解析出来只有一个 IfNode 对象，直接取第 1 个进行后续的操作
>>> if_node = parser.parse()
>>> if_node
[<IfNode>]
>>> if_node = if_node[0]
>>> if_node.conditions_nodelists
[((== (literal <django.template.base.FilterExpression object at 0x7fd492c975e0>)
(literal <django.template.base.FilterExpression object at 0x7fd4925742e0>)), [<TextNode:
'\n<li>男</li>\n'>]), ((== (literal <django.template.base.FilterExpression object at
0x7fd492574340>) (literal <django.template.base.FilterExpression object at
0x7fd492574490>)), [<TextNode: '\n<li>女</li>\n'>]), (None, [<TextNode: '\n<li>数据不太
对</li>'>])]
```

再次回到 IfNode 的 render()方法。render()方法遍历 self.conditions_nodelists，即遍历每个 if 分支，通过分支条件判断应渲染哪个分支下的内容。分支条件的判断语句十分明确，即 match = condition.eval(context)语句。下面接着前文的交互模式继续操作：

```
>>> conditions_nodelist = if_node.conditions_nodelists
>>> condition, nodelist = conditions_nodelist[0]  # if 分支及其下面的节点
#当 person.sex=0 时，if 分支条件判断成立
>>> condition.eval({'person': {'sex': 0}})
True
#当 person.sex10 时，if 分支条件判断不成立
>>> condition.eval({'person': {'sex': 1}})
False
# if 下面的内容渲染结果
>>> nodelist.render({})
'\n<li>男</li>\n'
>>> condition, nodelist = conditions_nodelist[1]  # elif 分支及其下面的节点
>>> condition.eval({'person': {'sex': 0}})
False
>>> condition.eval({'person': {'sex': 1}})
True
>>> nodelist.render({})
'\n<li>女</li>\n'
# 在 else 分支下，condition 为 None
>>> _, nodelist = conditions_nodelist[2]  # else 分支及其下面的节点
>>> nodelist.render({})
'\n<li>数据不太对</li>'
```

经上述操作后，对于 if 标签的渲染过程是不是非常清楚了？在追踪 if 标签的渲染过程中，还有一处细节没有深入讲解，即 TemplateIfParser 类。前面展示的 condition 正是该类的一个实例，TemplateIfParser 类中的 parse()方法及 eval()方法都是理解 if 标签实现的细节。限于篇幅，不再赘述，有兴趣的读者可以继续追踪这两个方法的源码，彻底掌握 if 标签。

## 4.3 答疑解惑

通过 4.2 节的详细分析，我们可以完整理解 Django 中内置模板系统对模板文件的解析及渲染过程。其中，单个标签及过滤标签对应的渲染逻辑如下：

（1）if、include、with 等标签的动作函数位于 django/template/defaulttags.py 文件中。比如，if 标签、include 标签和 with 标签对应的处理方法分别为 do_if()、do_include()和 do_with()，它们最终会得到包含该段标签的一个节点，如 IfNode、IncludeNode 和 WithNode。

（2）对于 IfNode、WithNode 等节点而言，最终的渲染由对应节点的 render()方法实现，在 render()方法中会得到这些标签最终的渲染文本，这和前面的标签处理方法（如 do_if()方法等）密切相关。

（3）过滤器方法均在 django/template/defaultfilters.py 文件中。

有了这些知识后，现在尝试解决 4.1 节中提出的一些问题。在 4.1 节中，笔者演示了一些简单的模板标签及过滤器的使用，同时提出了以下几个问题：

（1）在 for 标签实验中，空格是如何产生的？应该如何消除呢？

（2）解释{% csrf_token %}的实验现象。

（3）解释 cycle 标签的实验现象。

对于第 1 个问题，需要通过现象与源码共同配合解释。首先以一个 for 标签为例描述空格的产生，准备一个 for 标签内容并解析得到 tokens，具体操作如下：

```
(django2-core-test) [root@master first_django]# python manage.py shell
Python 3.8.6 (default, Oct 18 2020, 15:33:08)
[GCC 4.8.5 20150623 (Red Hat 4.8.5-39)] on linux
Type "help", "copyright", "credits" or "license" for more information.
(InteractiveConsole)
>>> contents = "{% for name in names %}\n<li>{{ name }}</li>\n{% endfor %}"
>>> print(contents)
{% for name in names %}
<li>{{ name }}</li>
{% endfor %}
>>> from django.template.base import Lexer
>>> lexer = Lexer(contents)
>>> tokens = lexer.tokenize()
>>> tokens[0].contents
'for name in names'
>>> tokens[1].contents
'\n<li>'
>>> tokens[2].contents
'name'
>>> tokens[3].contents
'</li>\n'
>>> tokens[4].contents
'endfor'
```

其次，调用 Parser 对象的 parse()方法解析上述 tokens，得到 ForNode 节点，操作如下：

```
>>> from django.template.engine import Engine
>>> engine = Engine.get_default()
>>> from django.template.base import Parser
>>> parser = Parser(tokens, engine.template_libraries, engine.template_builtins)
>>> for_node = parser.parse()
>>> for_node
[<ForNode: for name in names, tail_len: 3>]
>>> for_node = for_node[0]
```

```
>>> for_node
<ForNode: for name in names, tail_len: 3>
```

先看现象，对该 Node 节点的渲染结果如下：

```
>>> from django.template.context import make_context
>>> for_node.render(context=make_context({'names': ['张三', '李四', '王二麻子']}))
'\n<li>张三</li>\n\n<li>李四</li>\n\n<li>王二麻子</li>\n'
>>>
```

在渲染结果中，每个<li>与</li>与原来相比都多了一个换行符（\n），这就是多出来的空格。那么 render()方法中的哪部分代码产生了多余的换行符，又该如何消除呢？继续看 ForNode 节点的 render()方法，内容如下：

```
# 源码位置：django/template/defaulttags.py
# ……

class ForNode(Node):

    # ……

    def render(self, context):
        # forloop 变量
        if 'forloop' in context:
            parentloop = context['forloop']
        else:
            parentloop = {}
        # 1. with 语句
        with context.push():
            # 2. sequence 属性是循环的对象
            values = self.sequence.resolve(context, ignore_failures=True)
            # 如果 value 为 None，设置为空列表[]
            if values is None:
                values = []
            # 没有__len__属性，强制转成 list 类型
            if not hasattr(values, '__len__'):
                values = list(values)
            # 循环列表的长度
            len_values = len(values)
            # 空循环
            if len_values < 1:
                return self.nodelist_empty.render(context)
            nodelist = []
            # 3. 如果设置了 is_reversed 属性，直接使用 reversed()函数翻转列表
            if self.is_reversed:
                values = reversed(values)
            # 4. loopvars 属性表示循环变量
            num_loopvars = len(self.loopvars)
```

```
        unpack = num_loopvars > 1 # 类似于 for k, v in data 这样的语句
        # 注意，这里是字典。所以 loop_dict 的值将被更新到 context['forloop'] 中
        loop_dict = context['forloop'] = {'parentloop': parentloop}
        for i, item in enumerate(values):
            # 循环变量
            loop_dict['counter0'] = i
            loop_dict['counter'] = i + 1
            # 反向循环变量
            loop_dict['revcounter'] = len_values - i
            loop_dict['revcounter0'] = len_values - i - 1
            # {% for %}标签中的 first 和 last 变量
            loop_dict['first'] = (i == 0)
            loop_dict['last'] = (i == len_values - 1)

            pop_context = False
            if unpack:  # 下面处理多循环变量
                try:
                    len_item = len(item)
                except TypeError: # not an iterable
                    len_item = 1
                if num_loopvars != len_item:
                    raise ValueError(
                        "Need {} values to unpack in for loop; got {}. "
                        .format(num_loopvars, len_item),
                    )
                unpacked_vars = dict(zip(self.loopvars, item))
                pop_context = True  # 需要再次 pop()一次，将下面的 unpacked_vars 弹出去
                context.update(unpacked_vars)
            else:
                # 5. 对于 for k in data 这样的语句，都走该 else 分支
                context[self.loopvars[0]] = item  # 在 context 中导入前面设置的循环变量 item

            # 6. 这里是渲染的重点内容，nodelist_loop 为{% for %}标签包裹的 nodelist
            for node in self.nodelist_loop:
                nodelist.append(node.render_annotated(context))

            if pop_context:
                context.pop()
        # 最后，把渲染结果全部放在 nodelist 列表中
        return mark_safe(''.join(nodelist))

# ……
```

通过分析上面的代码，读者可完整了解 for 标签的渲染过程及在渲染过程中产生的临时变量。例如在 4.1 节中，for 标签中出现的 forloop.counter 或者 forloop.counter0 变量等。上面的代码内容较多，下面对整理出的 6 处重要的代码进行详细讲解，并且都配有详细的示例，以帮助读者更好地理解其含义。

在第 1 处代码中用了一个 with context.push()语句。先看 context.push()的源码，内容如下：

```
# 源码位置：django/template/context.py
# ……

class ContextDict(dict):
    def __init__(self, context, *args, **kwargs):
        super().__init__(*args, **kwargs)

        context.dicts.append(self)
        self.context = context

    # 下面的两个魔法函数对应着 with 语句
    def __enter__(self):
        # 返回 ContextDict 对象本身
        return self

    def __exit__(self, *args, **kwargs):
        # 当离开 with 语句时，在对应的 context 属性中会弹出 dicts 列表中的最后一个元素
        self.context.pop()


class BaseContext:

    # ……

    def push(self, *args, **kwargs):
        dicts = []
        for d in args:
            if isinstance(d, BaseContext):
                dicts += d.dicts[1:]
            else:
                dicts.append(d)
        return ContextDict(self, *dicts, **kwargs)

    # ……
```

在上述代码中，push()方法出现在父类 BaseContext 中，而语句 context.push()将返回一个 ContextDict 对象，该对象可以使用 with 语句。进入该 with 语句及离开该 with 语句的对应动作在 ContextDict 类中定义的魔法函数__enter__()和__exit__()中。下面做一个简单的测试：

```
>>> from django.template.context import make_context
>>> context = make_context({'names':['张三', '李四', '王二麻子']})
>>> context
[{'True': True, 'False': False, 'None': None}, {'names': ['张三', '李四', '王二麻子']}]
>>> with context.push():
...     print(context)
...
```

```
[{'True': True, 'False': False, 'None': None}, {'names': ['张三', '李四', '王二麻子']}, {}]
>>> context
[{'True': True, 'False': False, 'None': None}, {'names': ['张三', '李四', '王二麻子']}]
>>>
```

print(context)打印的其实是 context.dicts 属性，它是一个元素为字典的列表，这些在前文曾介绍过。with context.push()在 context.dicts 属性的最后添加了一个空字典，在离开 with 语句时，在对应的 context 属性中将弹出 dicts 列表中的最后一个元素。那么这个一开始被压入的空字典元素有什么作用呢？继续看下面的分析。

对于第 2 处代码只需掌握一点：ForNode 节点中的 sequence 属性值是循环对。例如，对于 for 标签的语句{% for x in x_list %}而言，sequence 表示的是 x_list，这一点在 4.2 节中曾详细分析过。接着前面得到的 ForNode 节点继续操作：

```
# 参考前面得到 ForNode 对象的语句，这里不再重复导入代码
>>> for_node.sequence
<django.template.base.FilterExpression object at 0x7fd7fed1cf10>
>>> for_node.sequence.token
'names'
>>> for_node.sequence.resolve(context={'names': ['张三', '李四', '王二麻子']})
['张三', '李四', '王二麻子']
```

ForNode 对象的 sequence 属性值其实是一个 FilterExpression 对象，它的 token 属性正是待循环的值。上面还展示了如何调用 resolve()方法得到一个真实场景的 values 值。后续的代码会对得到的 values 值进行判断，确保它是可迭代的，甚至会被强制转成列表。

第 3 处代码比较简单，就是对于{% for x in xxx reversed %}这样的写法，ForNode 对象中的 is_reversed 属性将为 True，然后 render()方法将直接使用 reversed()方法对 values 中的元素进行反转。

在第 4 处代码中，ForNode 对象中的 loopvars 属性表示循环中的变量。对于模板语句{% for x in x_list %}来说，loopvars 表示的是变量 x。在该代码段中有一个非常重要的语句，如下：

```
loop_dict = context['forloop'] = {'parentloop': parentloop}
```

该语句的结果如下：

```
# 接着第 1 处代码中得到的 context 继续执行，这里不再重复导入代码
>>> with context.push():
...     loop_dict = context['forloop'] = {'parentloop': {}}
...     print(context)
...
[{'True': True, 'False': False, 'None': None}, {'names': ['张三', '李四', '王二麻子']},
{'forloop': {'parentloop': {}}}]
>>> context
[{'True': True, 'False': False, 'None': None}, {'names': ['张三', '李四', '王二麻子']}]
```

context['forloop'] = {'parentloop': {}}会在 context.dicts 的最后一个字典元素中加入 forloop 键（关

于 context 的这个特性在 4.2 节中曾详细分析过），而原先的 context 值在调用 push()方法后会在列表的最后添加一个空字典元素。后续向 context 变量中添加的循环变量及其值，如 forloop.couter、forloop.counter()等，都会直接加到 context.dicts（列表）的最后一个元素（字典值）中。在离开 with 语句后，最后包含循环变量的字典元素将被弹出，context 恢复到原来的值。

在第 4 处代码的下面是一个 for 循环，会遍历带循环的列表。在循环过程中将向 context 中加入多个循环变量供渲染时使用。下面简单演示一下该过程，如下：

```
>>> with context.push():
...     values = for_node.sequence.resolve(context, ignore_failures=True)
...     loop_dict = context['forloop'] = {'parentloop': {}}
...     for i, item in enumerate(values):
...         loop_dict['counter0'] = i
...         loop_dict['counter'] = i + 1
...         print('第{}次循环, context={}'.format(i + 1, context))
...
第 1 次循环, context=[{'True': True, 'False': False, 'None': None}, {'names': ['张三', '李四', '王二麻子']}, {'forloop': {'parentloop': {}, 'counter0': 0, 'counter': 1}}]
第 2 次循环, context=[{'True': True, 'False': False, 'None': None}, {'names': ['张三', '李四', '王二麻子']}, {'forloop': {'parentloop': {}, 'counter0': 1, 'counter': 2}}]
第 3 次循环, context=[{'True': True, 'False': False, 'None': None}, {'names': ['张三', '李四', '王二麻子']}, {'forloop': {'parentloop': {}, 'counter0': 2, 'counter': 3}}]
```

接下来，在对 values 的循环中，{% for k, v in names %}这样的模板语句对应着 unpack=True。而普通的 for 循环将执行 context[self.loopvars[0]] = item，即 render()方法中的第 5 处代码。item 即为循环时的变量值。比如对于 names =['张三', '李四', '王二麻子']，每次循环的变量依次为'张三', '李四'和'王二麻子'，也就是对应的 item 值。self.loopvars[0]表示循环的变量名，在这里是'name'，即每次循环将执行 context['name']='xx'这样的语句：

```
# context 和前面得到的方式一致，不再重复导入
>>> with context.push():
...     loop_dict = context['forloop'] = {'parentloop': {}}
...     print(context)
>>> with context.push():
...     loop_dict = context['forloop'] = {'parentloop': {}}
...     for itemin ['张三', '李四', '王二麻子']:
...         context['name'] = item
...         print(context)
...
[{'True': True, 'False': False, 'None': None}, {'names': ['张三', '李四', '王二麻子']}, {'forloop': {'parentloop': {}}, 'name': '张三'}]
[{'True': True, 'False': False, 'None': None}, {'names': ['张三', '李四', '王二麻子']}, {'forloop': {'parentloop': {}}, 'name': '李四'}]
[{'True': True, 'False': False, 'None': None}, {'names': ['张三', '李四', '王二麻子']}, {'forloop': {'parentloop': {}}, 'name': '王二麻子'}]
```

```
>>> context
[{'True': True, 'False': False, 'None': None}, {'names': ['张三', '李四', '王二麻子']}]
```

第 6 处是代码 for 标签渲染的关键，也是本次探查的核心语句。在 ForNode 对象中，属性 nodelist_loop 表示的是循环的主体内容，请看下面的演示语句：

```
# ForNode 节点和前面得到的方式一致，不再重复相关语句
>>> for_node.nodelist_loop
[<TextNode: '\n<li>'>, <Variable Node: name>, <TextNode: '</li>\n'>]
```

node.render_annotated(context)语句则是将上面循环的主体节点根据 context 内容渲染成对应的文本内容：

```
>>> with context.push():
...     loop_dict = context['forloop'] = {'parentloop': {}}
...     nodelist = []
...     for item in ['张三', '李四', '王二麻子']:
...         context['name'] = item
...         for node in for_node.nodelist_loop:
...             nodelist.append(node.render_annotated(context))
# 此时 context 包含变量 name 的值
...
>>> nodelist
['\n<li>', '张三', '</li>\n', '\n<li>', '李四', '</li>\n', '\n<li>', '王二麻子', '</li>\n']
```

这里以非常简单的方式重现了 render()方法的渲染流程。从上面的代码可以看到，渲染中多余的换行符其实来自模板文本中的换行符。Django 中的 for 标签会默认将换行符重复循环，最终在文本全部叠加时，将出现多处换行的现象。

如何消除循环中多余的换行呢？既然空行是在循环'\n<li>'和'</li>\n'的过程中出现的，那么删掉其中一个换行符不就可以了吗？操作如下：

```
# 交互模式下的完整操作，部分导入语句和前面重复
(django2-core-test) [root@master first_django]# python manage.py shell
Python 3.8.6 (default, Oct 18 2020, 15:33:08)
[GCC 4.8.5 20150623 (Red Hat 4.8.5-39)] on linux
Type "help", "copyright", "credits" or "license" for more information.
(InteractiveConsole)
>>> contents = "{% for name in names %}<li>{{ name }}</li>\n{% endfor %}"
 # 去掉<li>前的换行符
>>> print(contents)
{% for name in names %}<li>{{ name }}</li>
{% endfor %}
>>> from django.template.base import Lexer
>>> lexer = Lexer(contents)
>>> tokens = lexer.tokenize()
>>> from django.template.engine import Engine
>>> engine = Engine.get_default()
```

```
>>> from django.template.base import Parser
>>> parser = Parser(tokens, engine.template_libraries, engine.template_builtins)
>>> for_node = parser.parse()[0]
>>> from django.template.context import make_context
>>> print(for_node.render(context=make_context({'names': ['张三', '李四', '王二麻子']})))
<li>张三</li>
<li>李四</li>
<li>王二麻子</li>

>>> contents = "{% for name in names %}\n<li>{{ name }}</li>{% endfor %}"
 # 去掉</li>后的换行符
>>> print(contents)
{% for name in names %}
<li>{{ name }}</li>{% endfor %}
>>> lexer = Lexer(contents)
>>> tokens = lexer.tokenize()
>>> parser = Parser(tokens, engine.template_libraries, engine.template_builtins)
>>> for_node = parser.parse()[0]
>>> print(for_node.render(context=make_context({'names': ['张三', '李四', '王二麻子']})))

<li>张三</li>
<li>李四</li>
<li>王二麻子</li>
>>>
```

继续看第 2 个问题，它比较简单，只需检查一下 csrf_token 标签对应的 render()方法即可。首先找到 csrf_token 标签的处理函数，内容如下：

```
# 源码位置：django/template/defaulttags.py
# ……

@register.tag
def csrf_token(parser, token):
   return CsrfTokenNode()

# ……
```

可以看到，csrf_token 标签最终会被解析成 CsrfTokenNode，而该标签最终的渲染方法如下：

```
# 源码位置：django/template/defaulttags.py
# ……

class CsrfTokenNode(Node):
   def render(self, context):
      # 从 context 中获取 csrf_token 对应的 value 值
      csrf_token = context.get('csrf_token')
      if csrf_token:
         if csrf_token == 'NOTPROVIDED':
            return format_html("")
```

```
            else:
                # 最终 CsrfTokenNode 将被渲染成一个 input 元素，其中，
                # value 属性值正是从 context 中获取的 csrf_token 值
                return format_html('<input type="hidden" name="csrfmiddlewaretoken"
                                        value="{}">', csrf_token)
        else:
            # 无法得到 csrf_token 值，返回空字符串
            return ''

# ……
```

看完上面的代码就不难理解 4.1.3 节中关于 csrf_token 标签的实验现象了。当输入 context={'csrf_token': '111111'}时，{% csrf_token %}将被解析成 CsrfTokenNode 对象，而在渲染 CsrfTokenNode 对象的过程中，将读取 context 中对应'csrf_token'的值。如果该值存在且不是"NOTPROVIDED"字符串，则{% csrf_token %}将被翻译成一段 HTML 文本，否则输出空字符串。

继续追踪第 3 个问题，先看一下对 cycle 标签的处理函数：

```
# 源码位置：django/template/defaulttags.py
# ……

@register.tag
def cycle(parser, token):

    # 一般情况下是按空格切割字符串的，但有一些特殊情况需额外处理，
    # 具体可参考 split_contents()方法的实现源码
    args = token.split_contents()

    if len(args) < 2:
        # 至少有一个参数，需要加上 cycle 标签
        raise TemplateSyntaxError("'cycle' tag requires at least two arguments")

    if len(args) == 2:
        # 处理{% cycle foo %}，foo 是一个列表变量
        name = args[1]
        if not hasattr(parser, '_named_cycle_nodes'):
            raise TemplateSyntaxError("No named cycles in template. '%s' is not defined" %
name)
        if name not in parser._named_cycle_nodes:
            raise TemplateSyntaxError("Named cycle '%s' does not exist" % name)
        return parser._named_cycle_nodes[name]

    as_form = False

    if len(args) > 4:
        # 处理{% cycle ... as foo [silent] %}，倒数第 2 个或者倒数第 3 个字符串是"as"
        if args[-3] == "as":
```

```
        # 如果倒数第 3 个字符串是 as，则最后一个字符串必须是 silent
        if args[-1] != "silent":
            # 抛出异常
            # ……
        as_form = True   # as_form 表示存在"as"语句，和下一个 elif 分支一样
        silent = True    # silent 为 True，和下一个分支不同
        args = args[:-1]   # 将 args 中最后一个 silent 元素去掉，方面后续统一处理
    elif args[-2] == "as":
        as_form = True
        silent = False

if as_form:
    # 对包含 as 语句的 cycle 标签进行处理
    name = args[-1] #  对于含有 as 语句的 cycle 标签而言，最后的元素是循环列表指定的变量名
    # Parse 对象解析循环参数
    values = [parser.compile_filter(arg) for arg in args[1:-2]]
    node = CycleNode(values, name, silent=silent)  # 最终得到 CycleNode 节点
    if not hasattr(parser, '_named_cycle_nodes'):
        parser._named_cycle_nodes = {}
    parser._named_cycle_nodes[name] = node
else:
    # 对普通的 cycle 标签进行处理，既没有 as 语句，也没有变量，在 cycle 标签后面是待循环的列表
    values = [parser.compile_filter(arg) for arg in args[1:]]
    # 最终返回 CycleNode 节点
    node = CycleNode(values)
parser._last_cycle_node = node
return node

# ……
```

对 cycle 标签的处理逻辑非常简单，即解析 cycle 标签的内容，并针对若干情况进行处理。比如对于{% cycle foo %}这样的模板语句，foo 为待循环的列表，在处理后直接返回即可；再比如对于{% cycle ... as foo [silent] %}及{% cycle 'test1' 'test2' 'test3' %}这样的语句，它们最后都将返回一个 CycleNode 节点。这里以普通写法为例进行说明，对于该语句的写法，先进入最后的 else 分支，然后提取循环的每个参数（还会对其中包含的过滤器进行处理），最后统一设置 Parser 对象的 _last_cycle_node 属性，返回得到 CycleNode 节点。为了帮助读者理解代码中的 CycleNode 节点的 values 值，下面给出几个简单的示例，具体如下：

```
>>> from django.template.base import Lexer, Parser
>>> from django.template.engine import Engine
>>> contents = "{% cycle 'test1' 'test2' 'test3' %}"
>>> lexer = Lexer(contents)
>>> tokens = lexer.tokenize()
>>> engine = Engine.get_default()
>>> parser = Parser(tokens, engine.template_libraries, engine.template_builtins)
>>> cycle_node = parser.parse()[0]
```

```
>>> cycle_node
<django.template.defaulttags.CycleNode object at 0x7f7e50d38880>
# 后面在 CycleNode 的初始化方法中可以看到，values 值被传给了 cyclevars 属性
>>> cycle_node.cyclevars
[<django.template.base.FilterExpression object at 0x7f7e5060ce80>,
<django.template.base.FilterExpression object at 0x7f7e5060cca0>,
<django.template.base.FilterExpression object at 0x7f7e5060cee0>]
>>> print(cycle_node.cyclevars[0].token)
'test1'
>>> print(cycle_node.cyclevars[1].token)
'test2'
>>> print(cycle_node.cyclevars[2].token)
'test3'
```

继续追踪 CycleNode 节点的 render()方法，其源码如下：

```
# 源码位置: django/template/defaulttags.py
# ……

class CycleNode(Node):
    def __init__(self, cyclevars, variable_name=None, silent=False):
        # 传过来的循环体
        self.cyclevars = cyclevars
        # 对循环体进行命名，处理成{% cycle ... as xxx %}这样的形式。如果没有 as 语句，该值为 None
        self.variable_name = variable_name
        self.silent = silent

    def render(self, context):
        # context 的 render_context 属性值是一个 RenderContext 对象
        # 该判断将进入 RenderContext 类的魔法函数__contains__()中
        if self not in context.render_context:
            # 得到一个迭代对象
            context.render_context[self] = itertools_cycle(self.cyclevars)
        cycle_iter = context.render_context[self]
        # 每次调用 next()都将获取迭代对象的下一个元素
        value = next(cycle_iter).resolve(context)
        if self.variable_name:
            context.set_upward(self.variable_name, value)
        if self.silent:
            return ''
        return render_value_in_context(value, context)

    # ……

# ……
```

上面的源码和之前的节点渲染不一样，这里出现了一个迭代操作。下面通过示例，帮助读者理解该节点的渲染过程：

```
>>> from django.template.context import make_context
>>> context = make_context({})
>>> context.render_context
[{'True': True, 'False': False, 'None': None}]
```

Context 对象中的 render_context 属性在前文曾介绍过，它是一个 RenderContext 对象。在该类中定义的__iter__()、__contains__()和__getitem__()等魔法函数使得该对象支持 in 判断、迭代及索引等操作：

```
# 源码位置：django/template/context.py
# ……

class RenderContext(BaseContext):

   template = None

   def __iter__(self):
      yield from self.dicts[-1]

   def __contains__(self, key):
      return key in self.dicts[-1]

   def get(self, key, otherwise=None):
      return self.dicts[-1].get(key, otherwise)

   def __getitem__(self, key):
      return self.dicts[-1][key]

   # ……

# ……
```

下面根据 render()方法中的代码语句给出如下操作示例：

```
# 接着前面的操作，主要用到了 CycleNode 对象：cycle_node
>>> context.render_context
[{'True': True, 'False': False, 'None': None}]
>>> from itertools import cycle as itertools_cycle
>>> itertools_cycle(cycle_node.cyclevars)   # 得到一个迭代对象
<itertools.cycle object at 0x7f375350f4c0>
>>> cycle_node in context.render_context
# 模拟第一次调用，cycle_node 不在 RenderContext 对象中
False
>>> context.render_context[cycle_node] = itertools_cycle(cycle_node.cyclevars)
 # 赋值语句
>>> context.render_context
[{'True': True, 'False': False, 'None': None, <django.template.defaulttags.CycleNode
object at 0x7f3753c2d880>: <itertools.cycle object at 0x7f375350f800>}]
```

```
>>> cycle_node in context.render_context
True
>>> cycle_iter = context.render_context[cycle_node]   # 继续取出这个迭代器
>>> next(cycle_iter).resolve(context)                 # 迭代调用
'test1'
>>> next(cycle_iter).resolve(context)
'test2'
>>> next(cycle_iter).resolve(context)
'test3'
>>> next(cycle_iter).resolve(context)          # itertools.cycle 的作用是循环调用
'test1'
>>> next(cycle_iter).resolve(context)
'test2'
```

继续模拟最后返回的文本，操作如下：

```
>>> value = next(cycle_iter).resolve(context)
# 上次已经迭代到“test2”，再迭代就是“test3”了
>>> value
'test3'
>>> from django.template.base import render_value_in_context
>>> render_value_in_context(value, context)      # 最终 render()方法的返回结果
'test3'
```

通过上面的操作不难猜测，是不是多次调用 render()方法就会出现'test1'、'test2'、'test3'……‘test1’、‘test2’、‘test3’这样反复循环的现象呢？继续在上文的交互模式下测试：

```
>>> cycle_node.render(context)
'test1'
>>> cycle_node.render(context)
'test2'
>>> cycle_node.render(context)
'test3'
>>> cycle_node.render(context)
'test1'
>>> cycle_node.render(context)
'test2'
```

至此，一切就非常清楚了。在 4.1 节的实验中，cycle 标签结合 for 标签后，能实现上述不断调用 CycleNode 对象的 render()方法的过程，从而能循环 cycle 标签中的参数。关于 Django 内置模板层的学习方式如下：

（1）先掌握对应模板标签的基本使用，不要求掌握该模板标签的各种用法。

（2）在 defaulttags.py 文件中找到对应标签的处理方法，仔细研究处理逻辑和对应得到的 Node 对象及其实例化参数。从源码中还可以发现关于该标签的更多用法，比如前面介绍的 cycle 标签，通过分析其标签处理方法可以看到其支持多种写法及 as 语句等。此外，根据 cycle()方法中处理循环参

数的语句 values = [parser.compile_filter(arg) for arg in args[1:]]可知，cycle 标签中的循环参数应当支持使用过滤器。下面对 4.1 节中的 template/test_cycle.tpl 文件进行简单改造，内容如下：

```
(django2-core-test) [root@master first_django]# cat templates/test_cycle.tpl
{% for skill in person.skills %}
<tr class="{% cycle 'test1'|upper 'test2'|upper|last ''|default:'unknown' %}">
<td>技能:{{ skill }}</td>
</tr>
{% endfor %}
```

接着运行 4.1 节中的测试语句，模板文件的渲染结果如下：

```
(django2-core-test) [root@master first_django]# python manage.py shell
Python 3.8.6 (default, Oct 18 2020, 15:33:08)
[GCC 4.8.5 20150623 (Red Hat 4.8.5-39)] on linux
Type "help", "copyright", "credits" or "license" for more information.
(InteractiveConsole)
>>> from django.template.loader import get_template
>>> tf = get_template('test_cycle.tpl')
>>> print(tf.render(context={'person': {'skills': ['数学', '英语', '理科综合']}}))

<tr class="TEST1">
<td>技能:数学</td>
</tr>

<tr class="2">
<td>技能:英语</td>
</tr>

<tr class="unknown">
<td>技能:理科综合</td>
</tr>


>>>
```

可见过滤器确实对 cycle 标签中的参数起了作用。

(3) 最后研究该标签对应 Node 节点的 render()方法。从该方法中可以看到 cycle 标签渲染文本的过程。此外，从该方法中还可以找到该模板标签所支持的内置变量，例如，前面多次分析过的 for 标签。

接下来介绍 Django 内置过滤器的相关源码。在 4.2 节中曾详细分析过 Django 内置模板系统解析过滤器的过程，下面再回顾一遍。例如下面的 token：

```
'variable|default:"Default value"|date:"Y-m-d"'
```

该字符串将被 Django 内部解析，提取初始变量名 variable 和两个过滤器 default 和 date，同时保留它们对应的函数及参数信息，然后通过 for 循环对初始变量值用提取的过滤器函数进行处理，最终返回过滤结果。在上述过滤语句中，最核心的就是过滤器函数，这些过滤器函数位于 django/template/defaultfilters.py 文件中。下面以 default 和 date 这两个简单的过滤器函数为例进行说明，其他过滤器的分析过程基本与之相同，不再赘述。

default 过滤器对应的源码如下：

```
# 源码位置：django/template/defaultfilters.py
# ……

@register.filter(is_safe=False)
def default(value, arg):
    """ 如果 value 为 None 或者空字符串、空数组等，则使用默认值 """
    return value or arg

# ……
```

传入 default 过滤器处理的字符串为 value，arg 为过滤器参数，即 default:"Default value"中对应的是“Default value”字符串。其返回逻辑一目了然，当 value 为空（None、空字符串、空数组等）时，返回默认值 arg。

date 过滤器对应的源码如下：

```
# 源码位置：django/template/defaultfilters.py
# ……

@register.filter(expects_localtime=True, is_safe=False)
def date(value, arg=None):
    """ 根据给定的格式来格式化传入的时间"""
    if value in (None, ''):
        return ''
    try:
        # 只有这一条是核心语句
        return formats.date_format(value, arg)
    except AttributeError:
        try:
            return format(value, arg)
        except AttributeError:
            return ''

# ……
```

在 Django 源码中，formats.py 文件封装了许多用于处理时间相关的函数，例如 date_format()函数。date()函数的核心语句是调用 date_format()函数去格式化时间，是不是非常简单？下面简单测试一下该 formats 模块中的一些方法，示例如下：

```
>>> from django.utils import formats
>>> from datetime import datetime
# 对于各种时间的表示，可以参考 Django 官方文档或者从源码中查找，这里 i 表示分钟
>>> formats.date_format(datetime.now(), 'Y-m-d H:i:s')
'2021-03-13 10:36:59'
```

Django 框架提供了这样一套过滤器机制，即只需写一个方法并添加@register.filter 装饰器即可让函数升级为过滤器。对于@register.filter 的工作机制与源码分析这里不再赘述，过程并不复杂，有兴趣的读者可以自行追踪并分析整个过程。

## 4.4 Jinja2 模块封装过程解析

在 4.1 节中笔者展示了以 Jinja2 模块作为模板引擎的案例，当时没有追踪这部分封装代码，只知道它是以 Jinja2 模块作为模板引擎解析和渲染模板文件的，本节将探究这一封装的过程。

下面从在 settings.py 文件中指定的 Jinja2 模块开始分析，对应的路径如下：

```
django.template.backends.jinja2.Jinja2
```

该 Jinja2 类的定义如下：

```
# 源码位置：django/template/backends/jinja2.py
# ……

class Jinja2(BaseEngine):

    app_dirname = 'jinja2'

    def __init__(self, params):
        # 初始化一些属性值
        # ……

    def from_string(self, template_code):
        return Template(self.env.from_string(template_code), self)

    def get_template(self, template_name):
        # 从该方法入手，探究 Jinja2 类的封装过程
        try:
            return Template(self.env.get_template(template_name), self)
        except jinja2.TemplateNotFound as exc:
            raise TemplateDoesNotExist(exc.name, backend=self) from exc
        except jinja2.TemplateSyntaxError as exc:
            # 处理异常
            # ……

    # ……
```

和 Django 的内置模板系统一样，这里定义的 Jinja2 类继承了 BaseEngine 类并实现了获取模板对象的 get_template()方法。为了能理解接下来封装的 Jinja2 模块代码，下面先给出 Jinja2 模块的一个简单示例。以下是在 4.1 节中测试 Jinja2 引擎的模板文件：

```
[root@master first_django]# cat templates/test_for_jinja2.tpl
遍历列表:
{% for animal in animals -%}
{{ animal }}
{% endfor %}
遍历字典:
{% for key in data -%}
{{ key }}:{{ data[key] }}
{% endfor -%}
```

接着按照官网的示例做一次最简单的模板渲染动作，具体如下：

```
>>> from jinja2 import Environment, FileSystemLoader
>>> env =
Environment(loader=FileSystemLoader('/root/django-core-test/first_django/templates')
)
>>> template = env.get_template('test_for_jinja2.tpl')
>>> content = template.render({'animals': ['狮子', '老虎', '大象'], 'data': {'name': '
沈奇才'}})
>>> content
'遍历列表:\n 狮子\n 老虎\n 大象\n\n 遍历字典:\nname:沈奇才\n'
>>> print(content)
遍历列表:
狮子
老虎
大象

遍历字典:
name:沈奇才

>>>
```

使用 Jinja2 模块渲染模板文件的过程就是这么简单。先得到一个 Environment 对象，在实例化 Environment 对象时，在传入其中的 loader 参数中包含了模板文件所在的目录等信息。接着调用 Environment 对象的 get_template()方法获取模板文件的 Template 对象，最后调用该模板对象的 render()方法进行渲染。可以发现，上述过程似乎和前面解析 Django 模板内置系统一致，可以说 Django 实现了一套类 Jinja2 的模板引擎，同时在内部兼容使用 Jinja2 模块作为模板引擎。

继续回到 jinja2.py 文件的源码中来，在 Jinja2 类的初始化方法中出现了一个眼熟的属性——env：

```
# 源码位置: django/template/backends/jinja2.py
# ……
```

```
class Jinja2(BaseEngine):
   # ……

   def __init__(self, params):
      # params 参数其实就是 settings.py 文件中 TEMPLATES 变量的值
      params = params.copy()
      options = params.pop('OPTIONS').copy()
      super().__init__(params)

      self.context_processors = options.pop('context_processors', [])

      environment = options.pop('environment', 'jinja2.Environment')
      environment_cls = import_string(environment)

      if 'loader' not in options:
         options['loader'] = jinja2.FileSystemLoader(self.template_dirs)
      options.setdefault('autoescape', True)
      options.setdefault('auto_reload', settings.DEBUG)
      options.setdefault('undefined',
                  jinja2.DebugUndefined if settings.DEBUG else jinja2.Undefined)
      # 这个属性值就是 Jinja2 模块中的 Environment 对象
      self.env = environment_cls(**options)

   # ……
```

从上面的结果可以看出，self.env 正是 Jinja2 模块中的 Environment 对象，而 loader 参数则被封装在 options 语句中。语句 options['loader'] = jinja2.FileSystemLoader(self.template_dirs)和前面创建 Environment 对象时传入的 loader 参数一模一样。这里的 self.template_dirs 正是在 Django 配置中指定的模板文件所在的目录。下面将里面的代码拆解出来单独执行，具体操作如下：

```
>>> import jinja2
>>> options = {}
>>> options['loader'] =
jinja2.FileSystemLoader('/root/django-core-test/first_django/templates')
>>> environment = options.pop('environment', 'jinja2.Environment')
# 弹出的 environment 值默认是“jinja2.Environment”字符串
>>> environment
'jinja2.Environment'
>>> from django.utils.module_loading import import_string
# 导入 Environment 类
>>> environment_cls = import_string(environment)
>>> environment_cls
<class 'jinja2.environment.Environment'>
# 使用 options 参数实例化 Environment 类
>>> env = environment_cls(**options)
# 得到 Environment 对象
>>> env
```

```
<jinja2.environment.Environment object at 0x7f6aa481fd60>
```

继续回到 Jinja2 类的 get_template()方法中，该方法最终返回的 Template 对象语句如下：

```
Template(self.env.get_template(template_name), self)
```

很明显，self.evn.get_template(template_name)语句得到的是 Jinja2 模块中的 Template 对象，而这里实例化的 Template 类在 jinja2.py 文件中的定义如下：

```
# 源码位置: django/template/backends/jinja2.py
# ……

class Template:

    def __init__(self, template, backend):
        self.template = template
        self.backend = backend
        self.origin = Origin(
            name=template.filename, template_name=template.name,
        )

    def render(self, context=None, request=None):
        from .utils import csrf_input_lazy, csrf_token_lazy
        if context is None:
            context = {}
        # 这部分可暂时不用理会
        if request is not None:
            context['request'] = request
            context['csrf_input'] = csrf_input_lazy(request)
            context['csrf_token'] = csrf_token_lazy(request)
            for context_processor in self.backend.template_context_processors:
                context.update(context_processor(request))
        # 注意最核心的渲染动作
        return self.template.render(context)

# ……
```

根据前面的分析结果可知，当实例化 Template 类时传入的第 1 个参数是 Jinja2 模块中的 Template 对象，操作如下：

```
# 接着前面得到 env 后的操作，避免重复导入
>>> from django.template.backends.jinja2 import Template
# Jinja2 类的 get_template()方法中的语句，最后将返回该结果
>>> template = Template(env.get_template('test_for_jinja2.tpl'), None)
>>> template
<django.template.backends.jinja2.Template object at 0x7f6aa481ff40>
>>> template.template
<Template 'test_for_jinja2.tpl'>
```

在该 Template 对象中，render()方法的核心语句是调用 template 属性的 render()方法渲染模板文件。接着上面的语句继续执行下面的代码：

```
>>> template.render({'animals': ['狮子', '老虎', '大象'], 'data': {'name': '沈奇才'}})
'遍历列表:\n 狮子\n 老虎\n 大象\n\n 遍历字典:\nname:沈奇才\n'
>>> template.template.render({'animals': ['狮子', '老虎', '大象'], 'data': {'name': '沈奇才'}})
'遍历列表:\n 狮子\n 老虎\n 大象\n\n 遍历字典:\nname:沈奇才\n'
```

在阅读了 Jinja2 类及 Template 类的源码后，再看 4.1 节中基于 Jinja2 模块渲染模板文件 test_for_jinja2.tpl 的实验，是不是感觉非常简单？Django 仅仅封装了 Jinja2 模块中的几个基础类，如生成 Environment 对象，调用 Environment 对象的 get_tempate()方法得到 Jinja2 类中的 Template 对象，以及调用前面得到的 Jinja2 类中的 Template 对象的 render()方法完成渲染。

至此，对于 Django 模板系统的源码分析就结束了。限于篇幅，一些和模板层相关的源码并没有介绍，比如 settings.py 配置中模板变量 TEMPLATES 中的其他参数（如 context_processors），以及与视图层之间的交互（从 request 中获取变量）等，有兴趣的读者可以自行追踪和学习相关代码。

## 4.5　小结

本章主要介绍了 Django 的模板系统，深入分析了 django/template 目录下的源码，重点梳理了 Django 内置模板系统对模板文件的解析及渲染过程，通过示例帮助读者更好地理解渲染过程及相关的核心源码。首先在 4.1 节中测试了 Django 的内置模板功能，并根据若干测试现象提出一些问题；然后在 4.2 节中完整分析了 Django 内置模板系统的运作流程；接着在 4.3 节根据前面的分析经验从源码中找到问题的答案；最后探讨了 Django 支持的 Jinja2 模板引擎，分析了封装的引擎代码。至此，对 Django 模板系统的源码分析就告一段落，接下来探索 Django 的核心模块源码。

# 第 5 章
# 解读Django核心模块的源码

本章将完整剖析 Django 源码中 core 目录下的源码文件，并按子目录的功能依次进行解读。本章内容是第 6 章的基础，也是 Django 框架的核心。

## 5.1 core 目录源码一览

本节主要介绍 django/core 目录下的一些源码文件，以及它们在 Django 框架中的用途。在该目录下共有 8 个子目录和一些 Python 文件，它们的内容与用途如下：

◎ cache：该目录中的代码文件实现了 Django 框架的内置缓存引擎，并支持多种缓存模式，比如数据库缓存（db.py）、文件缓存（filebased.py）和本地内存缓存（locmem.py）等。这些缓存模式的实现方式和 Django 内置的 ORM 框架的实现方式十分相似，源码的结构也基本一致，这些现象会在分析完该部分源码后看到。

◎ checks：该目录中的代码文件主要用于 Django 的全局检查，其中定义了非常多的检查函数，有数据库配置检查、URLConf 配置检查、缓存配置检查和请求的安全检查等。此外，该目录中的代码文件还定义了非常多的告警字符串，当出现异常时，用于打印相关告警信息。

◎ files：该目录定义了一些文件相关的类，比如 File 类、ContentFile 类和 UploadedFile 类等。该目录中的部分代码文件与视图层的文件上传功能相关，例如，上传文件的处理器（handler）代码位于 uploadhandler.py 文件中，上传的文件最终会被封装成在 uploadedfile.py 文件中定义的 TemporaryUploadedFile 或者 InMemoryUploadedFile 类对象，然后传入视图函数中进行处理。

◎ handlers：该目录中最重要的一个文件为 wsgi.py。在 wsgi.py 文件中定义的 WSGIRequest 类和 WSGIHandler 类对 Web 服务的 HTTP 请求做了二次封装。

◎ mail：在该目录下，代码文件是 Django 自行封装的一个邮件模块，这里同样内置了不少发送邮件的方式，比如控制台发送消息、基于 SMTP 服务发送邮件等。默认是基于 SMTP 服务发送真实的邮件到邮箱中。

◎ management：该目录下的 commands 目录在第 2 章中已经介绍过，在 commands 目录中的代码文件定义了 Django 支持的所有命令，每个命令都对应一个 Python 文件。比如 python manage.py shell 命令对应的 Python 文件为 shell.py，从该文件中能找出 shell 命令运行的背后逻辑。

◎ serializers：该目录实现了 Django 内置的序列化器与反序列化器，用于将 Django 内置的模型对象与 JSON、Python、XML 或者 YAML 格式进行互相转换。

◎ servers：该目录下的代码主要是对 Python 内置的 wsgiref 模块进行二次封装，用于实现 Django 内置的 WSGI server 服务器。

◎ 其他 Python 文件：这里有非常重要的 wsgi.py 文件、校验器（validators.py）、分页器（paginator.py）、全局异常类（exceptions.py）和安全 URL 处理函数（signing.py），以及定义了 Django 内置信号的 signals.py 文件。

整体来看，core 目录下的子目录结构分明，每个子目录下的代码功能也十分明确。下面将 core 目录下的代码文件按功能分别讲解，并依次讲解其源码实现（management 目录下的代码在第 2 章中曾介绍过，此处不再赘述），最终达到掌握 Django 核心功能的目的。

## 5.2 请求处理

本节从 Django 的 runserver 命令说起。在第 2 章中，我们追踪了 Django 部分命令的执行过程，这里将直接从 runserver 命令调用的核心函数开始。从第 2 章中可知，runserver 命令最终调用的是 django/core/magement/commands/runserver.py 文件中 Command 类下的 handle()方法，其源码实现如下：

```
# 源码位置：django/core/magement/commands/runserver.py
# ……

class Command(BaseCommand):

    default_addr = '127.0.0.1'    # 默认启动监听地址
    default_addr_ipv6 = '::1'
    default_port = '8000'        # 默认启动监听端口
    protocol = 'http'
    server_cls = WSGIServer

    # ……

    def execute(self, *args, **options):
        if options['no_color']:
            os.environ["DJANGO_COLORS"] = "nocolor"
        super().execute(*args, **options)     # 核心是调用父类的 execute()方法

    def get_handler(self, *args, **options):
```

```
        """ 返回当前运行的 WSGI 处理器"""
        return get_internal_wsgi_application()

    def handle(self, *args, **options):
        if not settings.DEBUG and not settings.ALLOWED_HOSTS:
            raise CommandError('You must set settings.ALLOWED_HOSTS if DEBUG is False.')

        # ……
        self.run(**options)

    def run(self, **options):
        """运行 Web 服务"""
        use_reloader = options['use_reloader']  # 设置热加载，即当文件有变化时自动加载

        if use_reloader:
            autoreload.run_with_reloader(self.inner_run, **options)
        else:
            self.inner_run(None, **options)

    def inner_run(self, *args, **options):
        autoreload.raise_last_exception()

        threading = options['use_threading']
        # shutdown_message 是一个隐藏的选项
        shutdown_message = options.get('shutdown_message', '')
        quit_command = 'CTRL-BREAK' if sys.platform == 'win32' else 'CONTROL-C'

        self.stdout.write("Performing system checks...\n\n")
        self.check(display_num_errors=True)
        # 检查迁移文件
        self.check_migrations()
        now = datetime.now().strftime('%B %d, %Y - %X')
        # 输出一些信息
        self.stdout.write(now)
        self.stdout.write((
            "Django version %(version)s, using settings %(settings)r\n"
            "Starting development server at %(protocol)s://%(addr)s:%(port)s/\n"
            "Quit the server with %(quit_command)s.\n"
        ) % {
            "version": self.get_version(),
            "settings": settings.SETTINGS_MODULE,
            "protocol": self.protocol,
            "addr": '[%s]' % self.addr if self._raw_ipv6 else self.addr,
            "port": self.port,
            "quit_command": quit_command,
        })

        try:
```

```
        # handler 变量，追踪源码后可知，该变量为一个 WSGIHandler 对象
        handler = self.get_handler(*args, **options)
        # 调用 django.core.servers.basehttp 模块中的 run()方法，指定监听地址、端口和服务类
        run(self.addr, int(self.port), handler,
            ipv6=self.use_ipv6, threading=threading, server_cls=self.server_cls)
    except socket.error as e:
        # 使用有用的错误信息
        ERRORS = {
            errno.EACCES: "You don't have permission to access that port.",
            errno.EADDRINUSE: "That port is already in use.",
            errno.EADDRNOTAVAIL: "That IP address can't be assigned to.",
        }
        try:
            error_text = ERRORS[e.errno]
        except KeyError:
            error_text = e
        self.stderr.write("Error: %s" % error_text)
        # 需要使用 os 模块退出，因为在线程中 sys.exit 不起作用
        os._exit(1)
    except KeyboardInterrupt:
        # 手动关闭，打印关闭消息并退出
        if shutdown_message:
            self.stdout.write(shutdown_message)
        sys.exit(0)

# ……
```

注意，从 runserver 命令入口可以获取对应执行命令的 Command 类，获取方法是使用 ManagementUtility 类中的 fetch_command()方法。对于 runserver 命令而言，其结果如下：

```
>>> from django.core.management import ManagementUtility
>>> ManagementUtility().fetch_command('runserver')
<django.contrib.staticfiles.management.commands.runserver.Command object at
0x7f20811ee4f0>
>>>
```

注意，runserver 命令、shell 命令及 startproject 命令使用的 Command 类是不同的。runserver 命令使用的 Command 类在 django/contrib/staticfiles/management/commands/runserver.py 文件中，该 Command 类的源码如下：

```
# 源码位置：django/contrib/staticfiles/management/commands/runserver.py
# ……
from django.core.management.commands.runserver import (
    Command as RunserverCommand,
)

class Command(RunserverCommand):
    help = "Starts a lightweight Web server for development and also serves static files."
```

```
    def add_arguments(self, parser):
        # ……

    def get_handler(self, *args, **options):
        # ……
```

在上面的 Command 类中并没有定义 handle()方法，所以最终调用的仍然是 core 目录下的 Command 类中的 handle()方法。

下面从 handle()方法开始追踪，它会继续调用内部的 run()方法。当使用非自动加载模式时，run()方法会调用内部的 inner_run()方法。在 inner_run()方法中，将出现我们十分熟悉的打印信息，如最开始的"Performing system checks...\n\n"字符串。在 inner_run()方法中，比较关键的语句是获取 handler 的 get_handler()方法，以及调用 django/core/servers/basehttp.py 文件中的 run()方法。get_handler()方法会调用 django/core/servers/basehttp.py 文件中定义的 get_internal_wsgi_application()方法。get_internal_wsgi_ application()方法的实现源码如下：

```
# 源码位置：django/core/servers/basehttp.py
# ……

def get_internal_wsgi_application():

    from django.conf import settings
    app_path = getattr(settings, 'WSGI_APPLICATION')
    if app_path is None:
        return get_wsgi_application()

    try:
        # 导入对应的模块
        return import_string(app_path)
    except ImportError as err:
        raise ImproperlyConfigured(
            "WSGI application '%s' could not be loaded; "
            "Error importing module." % app_path
        ) from err

# ……
```

每创建一个 Django 项目，在生成的 settings.py 文件中就会自动设置 WSGI_APPLICATION 变量的值。以前文创建的 first_django 项目为例，它的 WSGI_APPLICATION 变量值如下：

```
(django2-core-test) [root@master first_django]# cat first_django/settings.py | grep
WSGI_APPLICATION
WSGI_APPLICATION = 'first_django.wsgi.application'
```

get_internal_wsgi_application()方法将返回上面路径对应的模块，而该模块的值位于 wsgi.py 文件的最后，如下：

```
(django2-core-test) [root@master first_django]# cat first_django/wsgi.py
import os

from django.core.wsgi import get_wsgi_application

os.environ.setdefault('DJANGO_SETTINGS_MODULE', 'first_django.settings')

# 调用get_internal_wsgi_application()方法导入的模块值
application = get_wsgi_application()
(django2-core-test) [root@master first_django]# python manage.py shell
Python 3.8.6 (default, Oct 18 2020, 15:33:08)
[GCC 4.8.5 20150623 (Red Hat 4.8.5-39)] on linux
Type "help", "copyright", "credits" or "license" for more information.
(InteractiveConsole)
>>> from django.utils.module_loading import import_string
>>> import_string('first_django.wsgi.application')
<django.core.handlers.wsgi.WSGIHandler object at 0x7fda71623fd0>
```

从上面的代码可以看到，application 其实是 get_wsgi_application()方法的返回值。当 app_path 为 None 时，返回的结果也是相同的。get_wsgi_application()方法的实现源码如下：

```
# 源码位置：django/core/wsgi.py
# ……

def get_wsgi_application():

   django.setup(set_prefix=False)
   return WSGIHandler()
```

application 变量是一个 WSGIHandler 对象，因此，从 inner_run()方法中得到的 handler 变量便是 WSGIHandler 对象。在 django/core/servers/basehttp.py 文件中，run()方法的实现源码如下：

```
# 源码位置：django/core/servers/basehttp.py
# ……

def run(addr, port, wsgi_handler, ipv6=False, threading=False, server_cls=WSGIServer):
   server_address = (addr, port)
   if threading:
      httpd_cls = type('WSGIServer', (socketserver.ThreadingMixIn, server_cls), {})
   else:
      httpd_cls = server_cls
   httpd = httpd_cls(server_address, WSGIRequestHandler, ipv6=ipv6)
   if threading:
      httpd.daemon_threads = True
   httpd.set_app(wsgi_handler)
   httpd.serve_forever()
```

为了能理解上述源码，我们需要掌握 Python 的一个内置模块：wsgiref。该模块是基于 WSGI 协议开发的服务模块，主要完成两个功能：按 HTTP 请求协议解析数据和按 HTTP 响应协议组装数据。该模块的使用示例如下：

```
(django2-core-test) [root@master first_django]# cat test_server.py
from wsgiref.simple_server import WSGIServer, WSGIRequestHandler

def application(environ, start_response):
    start_response('200 OK', [('Content-Type', 'text/html')])
    return [b'Hello, world!']

httpd_cls = WSGIServer
# 192.168.26.110 为该虚拟机的内网地址
httpd = httpd_cls(('192.168.26.110', 8001), WSGIRequestHandler)
httpd.set_app(application)
httpd.serve_forever()
```

上述代码和前面的 run()方法非常相似。这里先看一下代码现象，然后分析结果。在一个控制台上运行该 Python 代码：

```
(django2-core-test) [root@master first_django]# python test_server.py
#控制台没有任何输出
```

在执行上述 Python 代码后，从表面上看控制台会卡住不动，实际上该代码是在监听 8001 端口，等待请求的到来。下面另起一个控制台查看该虚拟机的 8001 端口，结果如下：

```
[root@master ~]# sudo netstat -anltp | grep 8001
tcp        0      0 192.168.26.110:8001     0.0.0.0:*               LISTEN      66281/python
```

如果在本机或者其他节点机器上使用 curl 请求 master 节点的 8001 端口，则结果如下：

```
[root@master ~]# curl -I http://192.168.26.110:8001
HTTP/1.0 200 OK
Date: Thu, 25 Mar 2021 07:38:03 GMT
Server: WSGIServer/0.2 CPython/3.8.6
Content-Type: text/html
Content-Length: 13

[root@master ~]# curl http://192.168.26.110:8001
Hello, world![root@master ~]#
[root@master ~]# curl -XPOST http://192.168.26.110:8001
Hello, world![root@master ~]#
```

此时，再回到前面卡住的控制台，即可看到 HTTP 请求及相关的输出信息了：

```
(django2-core-test) [root@master first_django]# python test_server.py
192.168.26.110 - - [25/Mar/2021 15:38:03] "HEAD / HTTP/1.1" 200 13
192.168.26.110 - - [25/Mar/2021 15:38:12] "GET / HTTP/1.1" 200 13
192.168.26.110 - - [25/Mar/2021 15:39:41] "POST / HTTP/1.1" 200 13
```

从上面的结果看，是不是有点像 Django 项目执行 python manage.py runserver 的控制台输出？即同样是控制台阻塞监听相应的端口，等待 HTTP 请求的到来。

上面的示例代码与 django/core/servers/basehttp.py 文件中的 run()方法有哪些不同呢？不同点主要有三处：httpd_cls、WSGIRequestHandler 和 wsgi_handler。下面依次对这些变量进行说明。

### httpd_cls

在上面的示例代码中，httpd_cls 是 wsgiref.simple_server.WSGIServer 类。而对 django/core/servers/basehttp.py 中的 run()方法来说，通常情况下（threading=True）是执行 httpd_cls = type('WSGIServer', (socketserver.ThreadingMixIn, server_cls), {})语句，即得到一个新命名的 WSGIServer 类，该类继承了 ThreadingMixIn 类和在 basehttp.py 文件中定义的 WSGIServer 类。当客户端发送 HTTP 请求时，服务端会创建一个新的线程对 HTTP 请求进行处理。basehttp.py 文件中的 WSGIServer 类的实现如下：

```
# 源码位置：django/core/servers/basehttp.py
# ……

class WSGIServer(simple_server.WSGIServer):
    """BaseHTTPServer 通过继承 simple_server.WSGIServer 类实现了 Python WSGI 协议"""

    request_queue_size = 10

    def __init__(self, *args, ipv6=False, allow_reuse_address=True, **kwargs):
        if ipv6:
            self.address_family = socket.AF_INET6
        self.allow_reuse_address = allow_reuse_address
        super().__init__(*args, **kwargs)

    def handle_error(self, request, client_address):
        if is_broken_pipe_error():
            logger.info("- Broken pipe from %s\n", client_address)
        else:
            super().handle_error(request, client_address)

# ……
```

从上述源码可知，这里的 WSGIServer 类与 wsgiref 模块中的 WSGIServer 类相比，只是简单扩展了初始化方法和处理错误请求的方法，其他功能基本不变。

### WSGIRequestHandler

在前面的示例代码中，WSGIRequestHandler 类是 wsgiref 模块的内置类，用于处理请求。而 run()方法中的 WSGIRequestHandler 类同样继承了 wsgiref 模块中的 WSGIRequestHandler 类，并重写了请求处理的 handle()方法和 handle_one_request()方法：

```
# 源码位置：django/core/servers/basehttp.py
# ……

class WSGIRequestHandler(simple_server.WSGIRequestHandler):
    protocol_version = 'HTTP/1.1'

    # ……

    def handle(self):
        self.close_connection = True
        self.handle_one_request()
        while not self.close_connection:
            self.handle_one_request()
        try:
            self.connection.shutdown(socket.SHUT_WR)
        except (socket.error, AttributeError):
            pass

    def handle_one_request(self):
        """WSGIRequestHandler.handle()的一个拷贝，但不同于 ServerHandler"""
        self.raw_requestline = self.rfile.readline(65537)
        if len(self.raw_requestline) > 65536:
            self.requestline = ''
            self.request_version = ''
            self.command = ''
            self.send_error(414)
            return

        if not self.parse_request():  # 已发送错误代码，直接退出
            return

        handler = ServerHandler(
            self.rfile, self.wfile, self.get_stderr(), self.get_environ()
        )
        handler.request_handler = self     # 用于日志记录和关闭连接的反向指针
        handler.run(self.server.get_app())

# ……
```

从上面的 handle()方法和 handle_one_request()方法中可以看出该请求处理器的逻辑，即循环处理请求直到关闭。handle_one_request()方法是对一次请求进行处理，该方法中的最后一句是 handler.run(self.server.get_app())。其中，self.server.get_app()正是 run()方法中 httpd.set_app(wsgi_handler) 语句的执行结果（通过追踪源码可知，httpd 和 self.server 是同一个 WSGIServer 对象，set_app()方法的作用是设置该对象的 application 属性值，而 get_app()方法的作用是返回该 application 的属性值）。通过反复追踪 ServerHandler 类的父类，即可找到 run()方法的实现源码：

```
# 源码位置：lib/wsgiref/handlers.py
# ……

class BaseHandler:

    # ……

    def run(self, application):
        """Invoke the application"""
        try:
            self.setup_environ()
            # 调用传入的application方法，同时传入self.environ和self.start_response参数
            self.result = application(self.environ, self.start_response)
            self.finish_response()
        except (ConnectionAbortedError, BrokenPipeError, ConnectionResetError):
            return
        except:
            try:
                self.handle_error()
            except:
                self.close()
                raise
```

从上面的源码可以看到，通过 set_app()方法传入的参数最终将被当作函数去处理请求。

### wsgi_handler

在前面的示例中，httpd.set_app()语句传入的参数是处理请求的方法，而在 run()方法中，对应的是 wsgi_handler 变量，该变量值由 run()方法的参数传入，继续向上追踪到 django/core/management/commands/runserver.py 文件的 inner_run()方法中可知，wsgi_handler 变量是 self.get_handler(*args, **options)方法的返回结果。前面分析过，Command 类的 get_handler()方法返回的是一个 WSGIHandler 对象，对于 python manage.py runserver 命令而言，执行到此处得到的 handler 变量一定是这里的 WSGIHandler 对象吗？答案是不一定。运行 runserver 命令最终调用的 Command 对象并不是 core 目录下的，而是 contrib 目录下的，而在对应的 Command 类中刚好有定义 get_handler()方法：

```
# 源码位置：django/contrib/staticfiles/management/commands/runserver.py
# ……
from django.core.management.commands.runserver import (
    Command as RunserverCommand,
)

class Command(RunserverCommand):
    # ……

    def get_handler(self, *args, **options):
        # 通过前面的分析可知，该handler为django.core.handlers.wsgi.WSGIHandler对象
```

```
    handler = super().get_handler(*args, **options)
    # 在默认情况下，use_static_handler的值为True，insecure_serving的值为False
    use_static_handler = options['use_static_handler']
    insecure_serving = options['insecure_serving']
    if use_static_handler and (settings.DEBUG or insecure_serving):
        return StaticFilesHandler(handler)
    return handler
```

在默认情况下，use_static_handler 的值为 True，insecure_serving 的值为 False，因此判断句 if use_static_handler and (settings.DEBUG or insecure_serving)的结果取决于 settings.DEBUG 的值。当 settings.DEBUG 的值为 True 时，上面的判断语句为真，get_handler()方法返回的是一个 StaticFilesHandler 对象。当 settings.DEBUG 的值为 False 时，上面的判断语句为假，get_handler()方法返回的是父类的返回结果，即 django.core.handlers.wsgi.WSGIHandler 对象。继续追踪 StaticFilesHandler 类的实现源码，如下：

```
# 源码位置：django/contrib/staticfiles/handlers.py
# ……
from django.core.handlers.wsgi import WSGIHandler, get_path_info

class StaticFilesHandler(WSGIHandler):

    handles_files = True

    def __init__(self, application):
        self.application = application
        self.base_url = urlparse(self.get_base_url())
        super().__init__()

    # ……

    def __call__(self, environ, start_response):
        if not self._should_handle(get_path_info(environ)):
            return self.application(environ, start_response)
        return super().__call__(environ, start_response)
```

从上面的源码可知，StaticFilesHandler 类继承自 core 目录下的 WSGIHandler 类。从前面的 get_handler()方法可知，在实例化 StaticFilesHandler 类时传入的参数 handler 是一个 WSGIHandler 对象，与得到的 StaticFilesHandler 对象的 application 属性为同一个 WSGIHandler 对象。此时，再看 StaticFilesHandler 类中的魔法函数 __call__()，无论返回的结果是 self.application()还是 super().__call__()，它们都是一样的，都是调用 WSGIHandler 类中的魔法函数__call__()。魔法函数 __call__()的实现源码如下：

```
# 源码位置：django/core/handlers/wsgi.py
# ……
```

```
class WSGIHandler(base.BaseHandler):
    request_class = WSGIRequest

    def __init__(self, *args, **kwargs):
        super().__init__(*args, **kwargs)
        # 载入中间件，将在第 7 章中分析
        self.load_middleware()

    def __call__(self, environ, start_response):
        set_script_prefix(get_script_name(environ))
        signals.request_started.send(sender=self.__class__, environ=environ)
        # 对请求进行封装，得到一个 WSGIRequest 对象
        request = self.request_class(environ)
        # 得到请求结果
        response = self.get_response(request)

        response._handler_class = self.__class__

        # 返回状态码
        status = '%d %s' % (response.status_code, response.reason_phrase)
        # 响应头
        response_headers = [
            *response.items(),
            *(('Set-Cookie', c.output(header='')) for c in response.cookies.values()),
        ]
        # 处理请求
        start_response(status, response_headers)
        if getattr(response, 'file_to_stream', None) is not None and
environ.get('wsgi.file_wrapper'):
            response = environ['wsgi.file_wrapper'](response.file_to_stream)
        # 返回请求响应结果
        return response

# ……
```

在前面的示例代码中，是使用 httpd.set_app(application)语句设置请求响应的处理器的，该示例代码最终会调用 application()函数来处理响应，且该函数要传入两个参数：environ 和 start_response，它们的含义如下。

◎ environ：该参数将包含所有 HTTP 请求信息的 dict 对象。

◎ start_response：该参数为处理 HTTP 请求并返回响应的函数。

在 Django 的 run()方法中，是使用 httpd.set_app(wsgi_handler)语句来设置处理请求的。由于在 WSGIHandler 类中定义了魔法函数__call__()，所以 wsgi_handler 变量可以被当作函数调用，即 wsgi_handler()最终调用的是 WSGIHandler 类中的魔法函数__call__()，且传入的参数刚好和前面调用 application()函数输入的参数吻合。这说明在 Django 中，整个请求处理链的源头正是这里。在第 6 章

中，将以这里为起点，完整剖析视图层的源码，此处不再赘述。

## 5.3 缓存模块

Django 的全局缓存模块位于 core/cache 目录中，下面先看__init__.py 文件，其内容如下：

```
# 源码位置：django/core/cache/__init__.py
# ……

# 默认的缓存模块别名
DEFAULT_CACHE_ALIAS = 'default'


def _create_cache(backend, **kwargs):
    try:
        # 从 settings.py 文件中获取后端缓存模块路径
        try:
            conf = settings.CACHES[backend]
        except KeyError:
            try:
                # 尝试导入一次，假设输入的 backend 是完整的模块路径
                import_string(backend)
            except ImportError as e:
                raise InvalidCacheBackendError("Could not find backend '%s': %s" % (
                    backend, e))
            location = kwargs.pop('LOCATION', '')
            params = kwargs
        else:
            # 获取 settings.py 文件中缓存的配置参数信息
            params = {**conf, **kwargs}
            # 使用后端引擎模块路径
            backend = params.pop('BACKEND')
            # 数据保存位置，默认为空
            location = params.pop('LOCATION', '')
        # 导入缓存模块
        backend_cls = import_string(backend)
    except ImportError as e:
        raise InvalidCacheBackendError(
            "Could not find backend '%s': %s" % (backend, e))
    return backend_cls(location, params)  # 返回缓存对象


class CacheHandler:

    def __init__(self):
        self._caches = local()  # 全局对象
```

```
    def __getitem__(self, alias):
        try:
            # 如果能在缓存中找到该键对应的值，直接返回
            return self._caches.caches[alias]
        except AttributeError:
            self._caches.caches = {}
        except KeyError:
            pass

        # 如果 alias 不是 settings.CACHES 中的键值，直接抛出异常
        if alias not in settings.CACHES:
            raise InvalidCacheBackendError(
                "Could not find config for '%s' in settings.CACHES" % alias
            )
        # 创建缓存系统对象
        cache = _create_cache(alias)
        # 加入对象缓存中
        self._caches.caches[alias] = cache
        return cache

    def all(self):
        # 返回全部缓存对象结果
        return getattr(self._caches, 'caches', {}).values()

# CacheHandler 对象
caches = CacheHandler()


class DefaultCacheProxy:
    """
    访问默认缓存对象的属性代理类，允许旧缓存对象使用新的 caches，API 是线程安全的
    """
    def __getattr__(self, name):
        return getattr(caches[DEFAULT_CACHE_ALIAS], name)

    def __setattr__(self, name, value):
        return setattr(caches[DEFAULT_CACHE_ALIAS], name, value)

    def __delattr__(self, name):
        return delattr(caches[DEFAULT_CACHE_ALIAS], name)

    def __contains__(self, key):
        return key in caches[DEFAULT_CACHE_ALIAS]

    def __eq__(self, other):
        return caches[DEFAULT_CACHE_ALIAS] == other
```

```
cache = DefaultCacheProxy()

# ......
```

上面的内容和在 3.3 节中分析 Django 中模型层的代码十分类似。django/db/__init__.py 文件中的 ConnectionHandler 类对应这里的 CacheHandler 类，DefaultConnectionProxy 类对应这里的 DefaultCacheProxy 类。它们的实现代码及调用方式都十分相似。从_create_cache()函数的源码可知，该方法从 settings.CACHES[backend]中获取缓存的信息，而在默认生成的 Django 项目的 settings.py 文件中没有定义 CACHES 变量，因此需要查找默认的全局配置文件 django/conf/global_settings.py，其中，CACHES 变量的定义如下：

```
# 源码位置：django/conf/global_settings.py
# ......

CACHES = {
    'default': {
        'BACKEND': 'django.core.cache.backends.locmem.LocMemCache',
    }
}
```

从这里可以看到 Django 默认的缓存模块路径，该模块的源码会在后面重点分析。继续看__init__.py 文件中的代码，CacheHandler 类和 DefaultCacheProxy 类的内容都比较简单，主要需要理解魔法函数的相关含义。接下来在 Python 交互模式下对 CacheHandler 类、DefaultCacheProxy 类和_create_cache()函数进行测试，操作如下：

```
(django2-core-test) [root@master first_django]# python manage.py shell
Python 3.8.6 (default, Oct 18 2020, 15:33:08)
[GCC 4.8.5 20150623 (Red Hat 4.8.5-39)] on linux
Type "help", "copyright", "credits" or "license" for more information.
(InteractiveConsole)
>>> from django.core.cache import caches # 导入 caches
>>> caches._caches
<_thread._local object at 0x7fd995e36bd0>
>>> caches['default'] # 调用在 CacheHandler 类中定义的魔法函数__getitem__()
<django.core.cache.backends.locmem.LocMemCache object at 0x7fd98f388460>
>>> caches._caches.caches
{'default': <django.core.cache.backends.locmem.LocMemCache object at 0x7fd98f388460>}
>>> caches['xxx']
Traceback (most recent call last):
  File "<console>", line 1, in <module>
  File
"/root/.pyenv/versions/django2-core-test/lib/python3.8/site-packages/django/core/cac
he/__init__.py", line 75, in __getitem__
    raise InvalidCacheBackendError(
```

```
django.core.cache.backends.base.InvalidCacheBackendError: Could not find config for
'xxx' in settings.CACHES
>>> caches.all()
dict_values([<django.core.cache.backends.locmem.LocMemCache object at
0x7fd98f388460>])
```

从上面的代码可以看到，caches['default']的核心逻辑是先从缓存中查找名为 default 的缓存对象。如果没有找到，就调用_create_cache(alias)得到对应的对象并加入缓存中，以便下次查找。因此_create_cache()函数的使用就非常简单了，示例如下：

```
>>> _create_cache('default')
<django.core.cache.backends.locmem.LocMemCache object at 0x7fd98f388ac0>
```

接着操作 DefaultCacheProxy 类，从其源码中可以看到，其实就是操作 CacheHandler 对象中默认的缓存对象（即上面的 LocMemCache 对象），操作示例如下：

```
>>> from django.core.cache import cache   # cache 是一个 DefaultCacheProxy 对象
>>> cache
<django.core.cache.DefaultCacheProxy object at 0x7fd995e378e0>
>>> cache.pickle_protocol
5
>>> cache.add   # 得到的是 LocMemCache 对象的 add 属性（在 LocMemCache 中定义了 add()方法）
<bound method LocMemCache.add of <django.core.cache.backends.locmem.LocMemCache object
at 0x7fd98f388460>>
>>> cache.get
<bound method LocMemCache.get of <django.core.cache.backends.locmem.LocMemCache object
at 0x7fd98f388460>>
>>>
```

上述代码中导入的 cache 为一个 DefaultCacheProxy 对象，表面是获取 cache 的属性值，实际上是获取配置的默认引擎（即前面得到的 LocMemCache 对象）的对应属性，例如，上面的 add()方法和 get()方法。

从 Django 默认的全局配置可知，其内置的缓存系统是本地缓存，对应的缓存类路径为 django.core.cache.backends.locmem.LocMemCache，源码实现如下：

```
# 源码位置：django/core/cache/backends/locmem.py
# ……

class LocMemCache(BaseCache):
   pickle_protocol = pickle.HIGHEST_PROTOCOL

   def __init__(self, name, params):
      super().__init__(params)
      self._cache = _caches.setdefault(name, OrderedDict())
      self._expire_info = _expire_info.setdefault(name, {})
      self._lock = _locks.setdefault(name, Lock())
```

```
    # ……
```

从上面的源码可以看到，LocMemCache 对象继承了 BaseCache 类，即 Django 定义的缓存基础类。LocMemCache 对象的初始化方法如下：

```
# 源码位置：django/core/cache/backends/locmem.py
# ……

_caches = {}
_expire_info = {}
_locks = {}


class LocMemCache(BaseCache):
    pickle_protocol = pickle.HIGHEST_PROTOCOL

    def __init__(self, name, params):
        super().__init__(params)
        self._cache = _caches.setdefault(name, OrderedDict()) # self._cache 为一个有序字典
        # self._expire_info 为一个空字典
        self._expire_info = _expire_info.setdefault(name, {})
        self._lock = _locks.setdefault(name, Lock())  # self._lock 为 Lock 对象

    # ……

# ……
```

在上面的初始化方法__init__()中设置的_cache、_expire_info 和_locks 属性都非常重要。它们分别表示该缓存对象实际缓存数据的字典、key 过期信息及线程锁。所有写入这些属性的信息都会被全局的字典值（最上面定义的三个字典_caches、_expire_info 和_locks）访问。下面给出一个简单的示例，操作如下：

```
>>> from django.core.cache.backends.locmem import _caches, _expire_info, _locks
>>> from django.core.cache.backends.locmem import LocMemCache
>>> loc = LocMemCache('test', {})  # 创建 LocMemCache 对象
# 下面展示 LocMemCache 对象中的三个属性值
>>> loc._cache
OrderedDict()
>>> loc._expire_info
{}
>>> loc._lock
<unlocked _thread.lock object at 0x7fbda7466840>
# 下面展示全局的_caches、 _expire_info 和_locks 属性值
>>> _caches
{'test': OrderedDict()}
>>> _expire_info
{'test': {}}
```

```
>>> _locks
{'test': <unlocked _thread.lock object at 0x7fbda7466840>}
# 再次创建一个新的 LocMemCache 对象
>>> loc2 = LocMemCache('test2', {})
# 从代码中可以看出，相关属性信息在全局变量中同样可以看到
>>> _caches
{'test': OrderedDict(), 'test2': OrderedDict()}
>>> _expire_info
{'test': {}, 'test2': {}}
>>> _locks
{'test': <unlocked _thread.lock object at 0x7fbda7466840>, 'test2': <unlocked
_thread.lock object at 0x7fbda744e270>}
```

Django 中的这种写法比较有意思，它能在全局中保存相关的缓存信息，且每个对象之间的缓存数据互不干扰。接着上面的交互模式，开始向缓存对象写入数据，操作如下：

```
>>> loc.set('hello', 'world')
>>> loc2.set('hello2','world')
>>> _caches
{'test1': OrderedDict([(':1:hello',
b'\x80\x05\x95\t\x00\x00\x00\x00\x00\x00\x00\x8c\x05world\x94.')]), 'test2':
OrderedDict([(':1:hello2',
b'\x80\x05\x95\t\x00\x00\x00\x00\x00\x00\x00\x8c\x05world\x94.')])}
>>> _expire_info
{'test1': {':1:hello': 1615984001.81032}, 'test2': {':1:hello2': 1615984006.0741954}}
>>> import time
# 由于输入需要时间，可能有一定的误差，1615984001-1615983709=292，按整数来算应该是 300
>>> time.time()
1615983709.2988918
```

调用缓存对象的 set()方法向这个缓存对象中分别写入一个 key-value 数据，相关的数据均记录在全局缓存字典_caches 中。此外，这两个 key 还分别对应着一个过期时间。由于输入语句之间有误差，所以无法从上述结果中看到准确的时间差。继续输入下面的语句，即可得到一个准确的默认过期时间：

```
>>> loc.set('xxx',
'yyy');now_time=time.time();gap=_expire_info['test1'][':1:xxx']-now_time;print(gap)
299.9999952316284
```

由于执行 Python 语句也需要时间，所以在 LocMemCache 类中设置缓存数据的默认过期时间应该为 300 秒。300 这个数值可以从父类 BaseCache 的初始化方法__init__()中看到：

```
# 源码位置：django/core/cache/backends/base.py
# ……

class BaseCache:
    def __init__(self, params):
        timeout = params.get('timeout', params.get('TIMEOUT', 300))
```

```
        if timeout is not None:
            try:
                timeout = int(timeout)
            except (ValueError, TypeError):
                timeout = 300
        self.default_timeout = timeout

        # 根据 params 参数值设置其他属性
        # ……

# ……
```

从 BaseCache 类的初始化方法__init__()中可以看到，当参数 params 中没有 timeout 键时，将使用默认的 300 设置缓存对象的 default_timeout 属性值：

```
>>> loc.default_timeout
300
```

不妨大胆猜测一下，这里实现的是一个缓存系统，提供了 set()、get()这样的方法，同时还有过期时间，是不是和 Redis 的过期时间一样？在超过过期时间后，是不是就无法通过 get()方法获取缓存值了？为了测试上面的想法，在上述的交互模式下继续执行下面的代码：

```
>>> loc3 = LocMemCache('test3', {'timeout': 10})  # 设置过期时间
>>> loc3.set('hello', '你好')
>>> loc3.get('hello')
'你好'
>>> loc3.get('hello')
'你好'
>>> loc3.get('hello')    # 大约 10 秒后再操作，已经无法返回相应的结果了
>>> loc3.get('hello')
>>>
```

果然，上面的操作和 Redis 十分类似，那么 Django 是如何实现缓存数据过期这个效果的呢？这就要从源码中寻找答案了：

```
# 源码位置：django/core/cache/backends/locmem.py
# ……

class LocMemCache(BaseCache):
    pickle_protocol = pickle.HIGHEST_PROTOCOL
    # ……

    def get(self, key, default=None, version=None):
        # make_key()得到存储在字典中的 key，即前面出现的':1:hello'这样的字符串
        key = self.make_key(key, version=version)
        # 校验 key
        self.validate_key(key)
        # 用线程锁操作缓存对象中的数据
```

```
        with self._lock:
            # 先检测数据是否过期
            if self._has_expired(key):
                # 删除缓存值
                self._delete(key)
                # 如果数据已过期，返回传入的默认值；如果没有传入的默认值，则为 None
                return default
            # 如果没有数据过期，从缓存中取数据
            pickled = self._cache[key]
            # 设置 last=False，表示将当前的键移动到最前端
            self._cache.move_to_end(key, last=False)
        # 将结果反序列化后返回，
        # 如 b'\x80\x05\x95\t\x00\x00\x00\x00\x00\x00\x00\x8c\x05world\x94.'
        return pickle.loads(pickled)

    def _set(self, key, value, timeout=DEFAULT_TIMEOUT):
        # 检查缓存是否已满
        if len(self._cache) >= self._max_entries:
            # 按一定的规则从缓存中弹出一些数据，腾出空间
            self._cull()
        # 将 key-value 值设置到_cache 属性值中
        self._cache[key] = value
        # 将该值移动到有序字典的最前端
        self._cache.move_to_end(key, last=False)
        # 设置过期时间
        self._expire_info[key] = self.get_backend_timeout(timeout)

    def set(self, key, value, timeout=DEFAULT_TIMEOUT, version=None):
        key = self.make_key(key, version=version)
        self.validate_key(key)
        # 将 value 值序列化，转成字节流
        pickled = pickle.dumps(value, self.pickle_protocol)
        with self._lock:
            # 设置 key-value 值，同时添加过期时间
            self._set(key, pickled, timeout)
    # ……

# ……
```

设置 key-value 值和获取 key 的逻辑都非常简单。LocMemCache 对象中的 set()方法，先通过 make_key()方法生成要保存的 key，再校验 key，然后使用 pickle 模块将 value 序列化，最后加上线程锁，并调用_set()方法设置 key-value 值。在_set()方法中会先校验缓存是否已满，然后将 key-value 值添加到_cache 字典中，同时将该 key-value 值移到有序字典的最前端。最后在_expire_info 属性值中设置该 key 的过期时间。这便是整个 set()方法的执行逻辑。

get()方法同样是调用 make_key()方法来得到要获取的 key 并校验该 key，接着使用 with 语句调

用加锁操作。在加锁期间会调用_has_expired()方法判断对应的 key 是否已经过期。如果已经过期，从缓存中删除该 key-value 值，并返回默认值（如果没有传入的默认值，则默认为 None）。如果该 key 没有过期，则从缓存（self._cache）中读取对应的 key，并将该 key 移到缓存字典（有序字典）的最前端，最后调用 pickle 模块反序列化，返回得到的结果。

在上述源码中用到了三个常用方法，分别是 make_key()、validate_key()和_has_expired()，下面分别介绍这三个方法的实现源码。

（1）make_key()方法的含义很明确，对于传入的 key，将生成一个新的 key 作为缓存的键。它由父类实现，内容如下：

```
# 源码位置：django/core/cache/backends/base.py
# ……

def default_key_func(key, key_prefix, version):
   # 三个参数组成新的 key
   return '%s:%s:%s' % (key_prefix, version, key)


def get_key_func(key_func):
   # 获取生成新 key 的方法
   if key_func is not None:
      if callable(key_func):
         return key_func
      else:
         # 支持字符串路径导入
         return import_string(key_func)
   # 默认方法
   return default_key_func


class BaseCache:
   def __init__(self, params):
         # ……

      # key_prefix 前缀需要从 params 参数中获取
      self.key_prefix = params.get('KEY_PREFIX', '')
      # version 属性也需要从 params 参数中获取
      self.version = params.get('VERSION', 1)
      # 获取新 key 的方法也从 params 参数中获取，默认为 default_key_func()方法
      self.key_func = get_key_func(params.get('KEY_FUNCTION'))
   # ……

   def make_key(self, key, version=None):

      if version is None:
```

```
        version = self.version
    # 调用对象的key_func()方法生成新的key，这里传入self.key_prefix参数和version参数
    new_key = self.key_func(key, self.key_prefix, version)
    return new_key
  # ……

# ……
```

make_key()方法比较简单，它的核心是调用对象的 key_func()方法。key_func()方法是从 params 参数中获取的，一同组成 key 的 key_prefix 同样是从 params 参数中获取的。如果没有传入生成 key 的函数，默认使用 default_key_func()方法生成 key。下面在交互模式下进行测试，具体操作如下：

```
(django2-core-test) [root@master first_django]# python
Python 3.8.6 (default, Oct 18 2020, 15:33:08)
[GCC 4.8.5 20150623 (Red Hat 4.8.5-39)] on linux
Type "help", "copyright", "credits" or "license" for more information.
>>> from django.core.cache.backends.locmem import LocMemCache
>>> loc = LocMemCache('test', {'KEY_PREFIX': 'qicai'})
>>> loc.make_key('xxx')
'qicai:1:xxx'
>>> loc.make_key('xxx', version=2)
'qicai:2:xxx'
>>> def new_key_func(key, key_prefix, version):
...     return '%s#%s#%s' % (key_prefix, version, key)
...
>>> loc = LocMemCache('test', {'KEY_PREFIX': 'qicai', 'KEY_FUNCTION': new_key_func,
'VERSION': 3})
>>> loc.make_key('xyz')
'qicai#3#xyz'
```

（2）validate_key()方法的作用是校验 key 的合法性。在 LocMemCache 类中并没有重写 validate_key()方法，因此在调用 LocMemCache 对象的 validate_key()方法时将进入父类的 validate_key()方法中执行。以下是父类 BaseCache 中 validate_key()方法的源码：

```
# 源码位置：django/core/cache/backends/base.py
# ……

class BaseCache:   # ……

   def validate_key(self, key):

       for warning in memcache_key_warnings(key):
           # 如果存在需要告警的信息，打印即可
           warnings.warn(warning, CacheKeyWarning)
   # ……

def memcache_key_warnings(key):
```

```
    # 使用 yield 语句得到一个迭代对象，支持 for 循环
    if len(key) > MEMCACHE_MAX_KEY_LENGTH:
        yield (
            'Cache key will cause errors if used with memcached: %r '
            '(longer than %s)' % (key, MEMCACHE_MAX_KEY_LENGTH)
        )
    for char in key:
        # ord()方法返回 char 字符对应的 ASCII 码
        if ord(char) < 33 or ord(char) == 127:
            yield (
                'Cache key contains characters that will cause errors if '
                'used with memcached: %r' % key
            )
            break
```

从上面真正校验 key 值的 memcache_key_warnings()方法中可以看到，主要存在两个告警信息：

- ◎ 如果 key 的长度超过默认值（250），产生告警信息。
- ◎ 如果字符的 ASCII 码小于 33 或者等于 127，产生告警信息。这样的缓存 key 在使用 memcached 时将导致一些异常。

下面简单测试一下 memcache_key_warnings()方法，操作如下：

```
>>> from django.core.cache.backends.base import memcache_key_warnings
>>> from django.core.cache.backends.base import CacheKeyWarning
>>> ord('\n')
10
>>> key = "x" * 250 + '\n'
>>> import warnings
>>> for warning in memcache_key_warnings(key):
...     warnings.warn(warning, CacheKeyWarning)
...
<stdin>:2: CacheKeyWarning: Cache key will cause errors if used with memcached:
'xxxxxxxxxxxxxxxxxxxxxxxxxxxxxxxxxxxxxxxxxxxxxxxxxxxxxxxxxxxxxxxxxxxxxxxxxxxxxxxxxxx
xxxxxxxxxxxxxxxxxxxxxxxxxxxxxxxxxxxxxxxxxxxxxxxxxxxxxxxxxxxxxxxxxxxxxxxxxxxxxxxxxxxx
xxxxxxxxxxxxxxxxxxxxxxxxxxxxxxxxxxxxxxxxxxxxxxxxxxxxxxxxxxxxxxxxxxxxxxxxxxxxxxxxxxx\
n' (longer than 250)
<stdin>:2: CacheKeyWarning: Cache key contains characters that will cause errors if used
with memcached:
'xxxxxxxxxxxxxxxxxxxxxxxxxxxxxxxxxxxxxxxxxxxxxxxxxxxxxxxxxxxxxxxxxxxxxxxxxxxxxxxxxxx
xxxxxxxxxxxxxxxxxxxxxxxxxxxxxxxxxxxxxxxxxxxxxxxxxxxxxxxxxxxxxxxxxxxxxxxxxxxxxxxxxxxx
xxxxxxxxxxxxxxxxxxxxxxxxxxxxxxxxxxxxxxxxxxxxxxxxxxxxxxxxxxxxxxxxxxxxxxxxxxxxxxxxxxx\n'
```

在上面的操作中，我们使用了一个异常的 key，即该 key 不仅长度异常，而且存在坏字符，因此会生成两个告警信息。

（3）_has_expired()方法的源码如下：

```
# 源码位置: django/core/cache/backends/locmem.py
# ……

class LocMemCache(BaseCache):

    # ……

    def _has_expired(self, key):
        # 获取该 key-value 值保存在内存中的过期时间
        exp = self._expire_info.get(key, -1)
        # 如果超过了这个时间，则返回 None
        return exp is not None and exp <= time.time()

    # ……
```

上面的源码非常清晰，在把 key-value 值保存到内存中的同时，还会在_expire_info 属性中保存该 key 的过期时间。之后调用_has_expired()方法，即可将对应 key 的过期时间与当前时间进行对比。如果 exp 非 None 且过期时间已到，则返回 True（已过期）；否则返回 False：

```
# 源码位置: django/core/cache/backends/locmem.py
# ……

class LocMemCache(BaseCache):
    # ……

    def add(self, key, value, timeout=DEFAULT_TIMEOUT, version=None):
        key = self.make_key(key, version=version)  # 生成缓存 key
        self.validate_key(key)                     # 校验 key
        pickled = pickle.dumps(value, self.pickle_protocol)  # 序列化
        with self._lock:
            if self._has_expired(key):  # 如果已过期
                self._set(key, pickled, timeout)  # 设置 key 并返回 True
                return True
            return False                  # 否则返回 False，但不设置缓存

    # ……

    def touch(self, key, timeout=DEFAULT_TIMEOUT, version=None):
        key = self.make_key(key, version=version)
        with self._lock:
            # 如果已过期，直接返回 False
            if self._has_expired(key):
                return False
            # 否则重新设置过期时间并返回 True
            self._expire_info[key] = self.get_backend_timeout(timeout)
            return True

    def incr(self, key, delta=1, version=None):
```

```
        key = self.make_key(key, version=version)
        self.validate_key(key)
        with self._lock:
            if self._has_expired(key):  # 如果已过期，删除并抛出异常
                self._delete(key)
                raise ValueError("Key '%s' not found" % key)
            pickled = self._cache[key]   # 从缓存中获取数据
            value = pickle.loads(pickled) # 反序列化
            new_value = value + delta   # 计算新值
            pickled = pickle.dumps(new_value, self.pickle_protocol) # 序列化
            self._cache[key] = pickled   # 保存新值
            self._cache.move_to_end(key, last=False)  # 移动 key-value 值到最前端
        return new_value

    # ……

    def _delete(self, key):
        try:
            # 删除 key-value 值
            del self._cache[key]
            del self._expire_info[key]
        except KeyError:
            pass

    def delete(self, key, version=None):
        key = self.make_key(key, version=version)
        self.validate_key(key)
        with self._lock:
            # 加锁并执行
            self._delete(key)

    def clear(self):
        with self._lock:
            # 清除所有的 key-value 值
            self._cache.clear()
            self._expire_info.clear()
```

上面的代码给出了 LocMemCache 类中剩余的几个方法及其实现源码。例如，add()方法只有在对应 key 过期时才会重新设置新 key 并返回 True，其余情况下则直接返回 False。touch()方法在 key 过期时直接返回 False，其余情况下则向_expire_info 属性值中写入该 key 的过期信息并返回 True。这些逻辑很容易从方法的实现源码中找到。接下来，在交互模式下测试这几个方法：

```
(django2-core-test) [root@master ~]# python
Python 3.8.6 (default, Oct 18 2020, 15:33:08)
[GCC 4.8.5 20150623 (Red Hat 4.8.5-39)] on linux
Type "help", "copyright", "credits" or "license" for more information.
```

```
>>> from django.core.cache.backends.locmem import LocMemCache
>>> lc = LocMemCache('test', {'KET_PERFIX': 'qicai', 'timeout': 20})
>>> lc.set('hello', 'world')    # 第一次设置'hello'值
>>> lc.get('hello')         # 快速调用 get()方法获取 key='hello'对应的 value 值
'world'
>>> lc.add('hello', 'world1')  # 调用 add()方法，但是 key='hello'没有过期，因此返回 False
False
>>> lc.add('hello', 'world1')  # 在 20 秒后，key='hello'过期，此时再次调用 add()方法，返回 True
True
>>> lc.get('hello')         # 在缓存中成功设置了 key='hello'的新值
'world1'
>>> lc1 = LocMemCache('test1', {}) # 此时过期时间是默认的 300 秒
>>> lc1.set('hello', 100)
>>> lc1.incr('hello', 12)
112
>>> lc1.get('hello')
112
# 测试 delete()方法和 clear()方法，前者删除对应的 key-value 值，
# 后者清除缓存对象中的所有 key-value 值
>>> from django.core.cache.backends.locmem import _caches
>>> _caches
{'test': OrderedDict([(':1:hello',
b'\x80\x05\x95\n\x00\x00\x00\x00\x00\x00\x00\x8c\x06world1\x94.')]), 'test1':
OrderedDict([(':1:hello', b'\x80\x05Kp.')])}
>>> lc1.delete('hello')
>>> _caches
{'test': OrderedDict([(':1:hello',
b'\x80\x05\x95\n\x00\x00\x00\x00\x00\x00\x00\x8c\x06world1\x94.')]), 'test1':
OrderedDict()}
>>> lc.clear()
>>> _caches
{'test': OrderedDict(), 'test1': OrderedDict()}
```

至此，对默认的缓存系统的分析就结束了，有兴趣的读者可以自行学习 Django 提供的其他缓存系统，比如基于数据库保存的缓存系统(db.py)、基于 memcached 服务保存的缓存系统(memcached.py)等。它们都提供了统一的 add()、set()、get()、touch()、incr()、delete()、clear()等方法，这些方法的含义类似，只是保存缓存数据的方式不同。默认的缓存系统会把缓存数据保存到内存中，而 db.py 文件中的缓存类会把数据保存到数据库中（具体的数据库可根据 settings.py 文件中的 DATABASES 变量确定），memcached.py 文件中的缓存类会把数据保存到 memcached 服务中。

从 django/core/cache/__init__.py 文件提供的 caches 和 cache 变量可知，它们是 Django 对外提供缓存的入口。在 Django 的源码中全局搜索“from django.core.cache import caches”字符串，结果如图 5-1 所示。

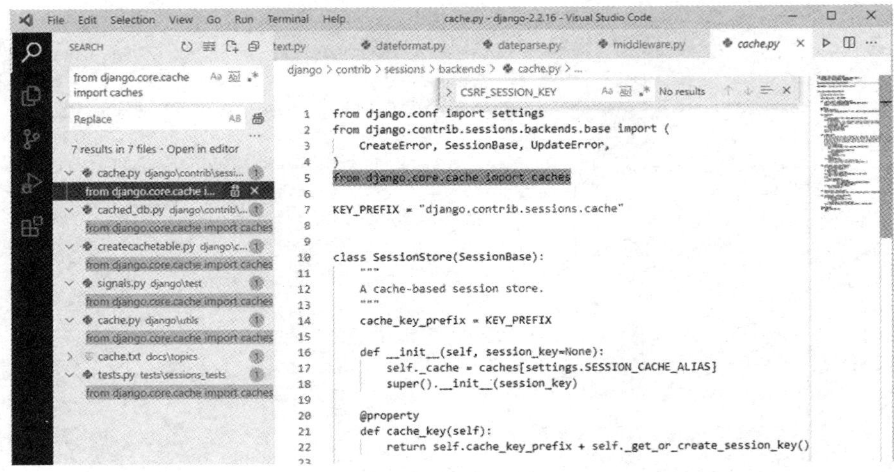


图 5-1

从图 5-1 的搜索结果可以看出，Django 的缓存系统代码主要在三个地方被调用：

（1）Session 相关的代码主要用于缓存和操作 Session 信息。

（2）命令行操作（createcachetable 命令）主要用于创建缓存数据表。

（3）辅助代码主要用于 Django 全局的缓存操作。

## 5.4 检查模块

检查模块的源码位于 django/core/checks 目录中，其中定义了非常多的检查函数。下面解读该目录下的核心源码，并通过实战演示其功能。

### 5.4.1 messages.py 文件的源码解析

首先看依赖最少的 messages.py 文件，其源码如下：

```
# 源码位置：django/core/checks/messages.py

# Levels
DEBUG = 10
INFO = 20
WARNING = 30
ERROR = 40
CRITICAL = 50
```

```
class CheckMessage:

    def __init__(self, level, msg, hint=None, obj=None, id=None):
        # 第 1 个参数必须是日志级别、整型
        assert isinstance(level, int), "The first argument should be level."
        self.level = level
        self.msg = msg
        self.hint = hint
        self.obj = obj
        self.id = id

    def __eq__(self, other):
        return (
            isinstance(other, self.__class__) and
            all(getattr(self, attr) == getattr(other, attr)
                for attr in ['level', 'msg', 'hint', 'obj', 'id'])
        )

    def __str__(self):
        from django.db import models

        if self.obj is None:
            obj = "?"
        elif isinstance(self.obj, models.base.ModelBase):
            obj = self.obj._meta.label
        else:
            obj = str(self.obj)
        id = "(%s) " % self.id if self.id else ""
        hint = "\n\tHINT: %s" % self.hint if self.hint else ''
        return "%s: %s%s%s" % (obj, id, self.msg, hint)

    def __repr__(self):
        return "<%s: level=%r, msg=%r, hint=%r, obj=%r, id=%r>" % \
            (self.__class__.__name__, self.level, self.msg, self.hint, self.obj, self.id)

    def is_serious(self, level=ERROR):
        return self.level >= level

    def is_silenced(self):
        from django.conf import settings
        return self.id in settings.SILENCED_SYSTEM_CHECKS


class Debug(CheckMessage):
    def __init__(self, *args, **kwargs):
        super().__init__(DEBUG, *args, **kwargs)
```

```
class Info(CheckMessage):
    def __init__(self, *args, **kwargs):
        super().__init__(INFO, *args, **kwargs)


class Warning(CheckMessage):
    def __init__(self, *args, **kwargs):
        super().__init__(WARNING, *args, **kwargs)


class Error(CheckMessage):
    def __init__(self, *args, **kwargs):
        super().__init__(ERROR, *args, **kwargs)


class Critical(CheckMessage):
    def __init__(self, *args, **kwargs):
        super().__init__(CRITICAL, *args, **kwargs)
```

CheckMessage 类有点类似日志，实例化的第 1 个参数为日志级别，共有 5 种；第 2 个参数是消息（msg），后续的 DEBUG、INFO、WARNING、ERROR 和 CRITICAL 类均继承自该类，只是对应的日志级别不同。调用示例如下：

```
>>> from django.core.checks.messages import CheckMessage
>>> from django.core.checks.messages import ERROR
>>> cm = CheckMessage(ERROR, '这是错误消息')
>>> cm # 调用魔法函数__repr__()
<CheckMessage: level=40, msg='这是错误消息', hint=None, obj=None, id=None>
>>> print(cm)  # 调用魔法函数__str__()
?: 这是错误消息
>>> from django.core.checks.messages import DEBUG, INFO, ERROR, CRITICAL
>>> cm.is_serious(DEBUG)   # ERROR 比 DEBUG 严重
True
>>> cm.is_serious(CRITICAL)  # ERROR 没有 CRITICAL 严重
False
>>> cm2 = CheckMessage(ERROR, '这是错误消息')
>>> cm == cm2   # 所有的属性值都相同，因而返回 True
True
>>> cm3 = CheckMessage(ERROR, '这是错误消息 2')
>>> cm == cm3   # 属性值 msg 不同，因而返回 False
>>> from django.core.checks.messages import Error
>>> err = Error('错误消息')
>>> err.level
40
>>> err.msg
'错误消息'
>>> err.is_serious(INFO)
True
```

以上是对 CheckMessage 类和 ERROR 类的一些调用示例,以帮助读者更好地理解这些类的使用。

### 5.4.2　registry.py 文件的源码解析

registry.py 文件是一个非常重要的代码文件，在该文件中定义了 Django 内置的标签类和用于注册检查函数的 CheckRegistry 类，它们的源码如下：

```
# 源码位置：django/core/checks/registry.py
from itertools import chain

from django.utils.itercompat import is_iterable


class Tags:
    """
    内部检查的内置标签
    """
    admin = 'admin'
    caches = 'caches'
    compatibility = 'compatibility'
    database = 'database'
    models = 'models'
    security = 'security'
    signals = 'signals'
    templates = 'templates'
    translation = 'translation'
    urls = 'urls'


class CheckRegistry:

    def __init__(self):
        # 注册的检查函数集合
        self.registered_checks = set()
        self.deployment_checks = set()

    def register(self, check=None, *tags, **kwargs):
        """
        可以用作函数或者装饰器， 注册给被装饰的函数并指定标签 tags。
        该函数接收**kwargs 参数，并返回一系列检查错误或者告警信息

        示例：

            registry = CheckRegistry()
            @registry.register('mytag', 'anothertag')
            def my_check(apps, **kwargs):
```

```
            # ... perform checks and collect `errors` ...
            return errors
        # or
        registry.register(my_check, 'mytag', 'anothertag')
        """
        def inner(check):
            # 将 tags 值赋给函数 check 的 tags 属性
            check.tags = tags
            # 根据函数中传入的 deploy 参数值，将 check 函数加入不同的 set 中
            checks = self.deployment_checks if kwargs.get('deploy') else
self.registered_checks
            checks.add(check)
            return check

        if callable(check):
            # 如果 check 可以被调用，就调用 inner()函数的结果
            return inner(check)
        else:
            if check:
                tags += (check,)
            return inner

    def run_checks(self, app_configs=None, tags=None, include_deployment_checks=False):
        """
        运行所有注册的检查函数并返回错误信息和告警信息的列表
        """
        errors = []
        checks = self.get_checks(include_deployment_checks)  # 获取对应的函数集

        if tags is not None:
            checks = [check for check in checks if not set(check.tags).isdisjoint(tags)]
        else:
            checks = [check for check in checks if Tags.database not in check.tags]

        for check in checks:
            new_errors = check(app_configs=app_configs)
            # 确保 check()函数返回的是一个列表结果
            assert is_iterable(new_errors), (
                "The function %r did not return a list. All functions registered "
                "with the checks registry must return a list." % check)
            # 将检查的错误列表追加到 errors 列表中
            errors.extend(new_errors)
        return errors

    def tag_exists(self, tag, include_deployment_checks=False):
        # 判断 tag 是否存在
        return tag in self.tags_available(include_deployment_checks)
```

```
    def tags_available(self, deployment_checks=False):
        # 支持的可用标签
        return set(chain.from_iterable(
            check.tags for check in self.get_checks(deployment_checks)
        ))

    def get_checks(self, include_deployment_checks=False):
        # 检查所有注册的检查函数
        checks = list(self.registered_checks)
        if include_deployment_checks:
            checks.extend(self.deployment_checks)
        # 返回检查的函数集,
        # 如果include_deployment_checks为True，则包括deployment_checks集
        return checks


registry = CheckRegistry()
register = registry.register
run_checks = registry.run_checks
tag_exists = registry.tag_exists
```

上述源码并不复杂，下面依次解读这些源码的含义。Tags 类比较基础，就是定义 Django 内部的一些检查标签，用于给每个检查函数打上标签，表明它属于哪类检查。比如，Tags.cache 表示的是缓存配置的检查，Tags.urls 表示的是与 URL 配置相关的检查等。CheckRegistry 类非常关键，在该类中共定义了 6 个方法，它们的含义如下：

（1）__init__()：对象的初始化方法，在这里只给两个属性值（registered_checks 和 deployment_checks）设置初始值，均为空集合。

（2）register()：注册方法。该方法在 Django 中主要被用作装饰器，将被装饰的函数添加到对应的属性集合中。当在 register()方法的参数中有 deploy 参数且为 True 时，register()方法将被加入 deployment_checks 属性集合中；否则 register()方法将被加入 registered_checks 属性集合中。下面是 @register 装饰器的两种使用示例：

```
# 源码位置：django/core/checks/urls.py
# ……

@register(Tags.urls)
def check_url_config(app_configs, **kwargs):
    if getattr(settings, 'ROOT_URLCONF', None):
        from django.urls import get_resolver
        resolver = get_resolver()
        return check_resolver(resolver)
return []

# 源码位置：django/core/checks/security/base.py
```

```
# ......

@register(Tags.security, deploy=True)
def check_security_middleware(app_configs, **kwargs):
   passed_check = _security_middleware()
   return [] if passed_check else [W001]
```

很明显，在上述两个@register 装饰器中，除标签参数不同外，一个没有 deploy 参数，另一个 deploy 参数为 True。因 check_url_config()方法将被注册到 registr 变量（即 CheckRegistry 对象）的 registered_checks 属性集合中，而 check_security_middleware()方法将被注册到 registry 变量的 deployment_checks 属性集合中，示例如下：

```
(django2-core-test) [root@master first_django]# python manage.py shell
Python 3.8.6 (default, Oct 18 2020, 15:33:08)
[GCC 4.8.5 20150623 (Red Hat 4.8.5-39)] on linux
Type "help", "copyright", "credits" or "license" for more information.
(InteractiveConsole)
>>> from django.core.checks.registry import registry
>>> 'check_url_config' in [check.__name__ for check in registry.registered_checks]
True
>>> 'check_url_config' in [check.__name__ for check in registry.deployment_checks]
False
>>> 'check_security_middleware' in [check.__name__ for check in
registry.registered_checks]
False
>>> 'check_security_middleware' in [check.__name__ for check in
registry.deployment_checks]
True
```

继续看 register()方法内部的 inner()方法，从代码中可以看到，传给@register 装饰器的 tags 参数值将被赋给方法的 tags 属性：

```
>>> [check.tags for check in registry.registered_checks]
[('models',), ('templates',), ('caches',), ('urls',), ('admin',), ('admin',),
('models',), ('models',), ('templates',), ('models',), ('urls',), ('models',),
('staticfiles',), ('models',), ('translation',), ('database',), ('urls',)]
>>> [check.tags for check in registry.deployment_checks]
[('security',), ('security',), ('security',), ('security',), ('security',),
('security',), ('security',), ('security',), ('security',), ('security',),
('security',), ('security',), ('security',), ('security',), ('security',),
('security',)]
```

（3）get_checks()：该方法的功能是获取所有检查函数，在默认情况下，即当 include_deployment_checks=False 时，获取 registered_checks 属性值的列表形式。如果 include_deployment_checks=True，将包含 deployment_checks 中的所有函数集合。

（4）run_checks()：该方法将运行所有的检查函数（同样由 include_deployment_checks 参数控制

是否加入 deployment_checks 属性集合中），并返回检查的错误列表。

（5）tags_available()：该方法将返回所有在注册函数中使用的属性集合。下面简单测试一下该方法的返回结果：

```
>>> from django.core.checks.registry import registry
>>> registry.tags_available()
{'templates', 'admin', 'staticfiles', 'database', 'translation', 'urls', 'caches',
'models'}
# 可以看到，deployment_checks 属性集合中的检查函数的 tags 都是 security
>>> registry.tags_available(True)
{'templates', 'admin', 'staticfiles', 'database', 'security', 'translation', 'urls',
'caches', 'models'}
```

从上面的结果对比可知，对 deployment_checks 属性集合中的所有检查函数而言，它们的 tags 都是 security，因而会出现 registry.tags_available(True)比 registry.tags_available()多了一个'security'元素的现象。

（6）tag_exists()：该方法有两个参数，即 tag 和 deployment_checks。第 2 个参数将被传给 tags_available()方法以获取对应的 tag 集合，返回结果为判断传入的 tag 是否在前面得到的 tag 集合中，示例如下：

```
>>> from django.core.checks.registry import registry
>>> registry.tag_exists('security')
False
>>> registry.tag_exists('security', True)
True
```

从 tags_available()方法的测试结果可以看到，当 deployment_checks 为 False 时，得到的 tags 集合并不包含 security 字符串，所以 registry.tag_exists('security')的结果为 False。当 deployment_checks 为 True 时，得到的 tags 集合将包含 security 字符串，因此 registry.tag_exists('security')的结果为 True。

在掌握了 CheckRegistry 类后，再来看在 registry.py 文件中最后定义的四个变量。registry 变量是一个全局的 CheckRegistry 对象，而 register、run_checks 和 tag_exists 变量表示该 CheckRegistry 对象的方法名。其中，register 变量是在所有的检查函数上进行装饰，用于将该检查函数注册到 CheckRegistry 对象的属性中。下面用一个简单的示例进行说明，其余文件中相应检查函数的分析过程基本与之一致，不再赘述。

继续分析 django/core/checks/urls.py 文件中的 check_url_config()方法，其源码如下：

```
# 源码位置：django/core/checks/urls.py
# ……

@register(Tags.urls)
def check_url_config(app_configs, **kwargs):
```

```
    if getattr(settings, 'ROOT_URLCONF', None):
        from django.urls import get_resolver
        resolver = get_resolver()
        return check_resolver(resolver)
    return []


def check_resolver(resolver):
    """
    检查 resolver 变量
    """
    check_method = getattr(resolver, 'check', None) # 获取 check 属性
    if check_method is not None:
        return check_method()  # 直接调用对应 check 属性的方法
    elif not hasattr(resolver, 'resolve'):  # 如果没有 resolve 属性，返回告警信息
        return get_warning_for_invalid_pattern(resolver)
    else:
        return []

# ……
```

首先看该检查函数的内容，这里会用到 Django 项目的配置变量 settings.ROOT_URLCONF。对于第 2 章中的 first_django 项目而言，其值为'first_django.urls'。因此调用 check_url_config()方法会“走”到 if 分支下，并调用 get_resolver()方法得到 resolver 变量，最后调用 check_resolver()方法返回检查结果。check_resolver()方法的逻辑比较简单：如果 resolver 变量中有 check 属性，返回该属性的结果；如果 resolver 变量中没有 resolver 属性，则调用该文件中定义的 get_warning_for_invalid_pattern()方法返回错误信息列表；否则返回空列表。因此，该校验函数（check_url_config()）的核心在于得到的 resolver 变量。继续追踪得到该变量的 get_resolver()方法，其源码如下：

```
# 源码位置：django/urls/resolvers.py
# ……

@functools.lru_cache(maxsize=None)
def get_resolver(urlconf=None):
    if urlconf is None:
        urlconf = settings.ROOT_URLCONF
    # 对 first_django 项目而言，urlconf 的结果为'first_django.urls'
    return URLResolver(RegexPattern(r'^/'), urlconf)

# ……
```

在上面的代码中，又涉及了 2 个新类，这些类会在第 6 章中详细分析，这里只重点剖析用到的方法。首先看 URLResolver 类，它的源码如下：

```
# 源码位置：django/urls/resolvers.py
# ……
```

```
class URLResolver:
    # ……

    def check(self):
        # 在检查中被调用的方法
        messages = []
        for pattern in self.url_patterns:
            # 这个 check_resolver()方法正是在 urls.py 中定义的 check_resolver()方法
            messages.extend(check_resolver(pattern))
        messages.extend(self._check_custom_error_handlers())
        # self.pattern 就是前面实例化的第 1 个参数 RegexPattern(r'^/'),
        # 如果 messages 为空，继续调用 RegexPattern 对象的 check()方法
        return messages or self.pattern.check()

    # ……

    @cached_property
    def url_patterns(self):
        # urlconf_module might be a valid set of patterns, so we default to it
        patterns = getattr(self.urlconf_module, "urlpatterns", self.urlconf_module)
        try:
            iter(patterns)
        except TypeError:
            # 异常处理
            # ……
        return patterns

    # ……
```

对于上述源码，首先看 url_patterns()方法。从前面可知，urlconf 的结果为字符串“first_django.urls”，该模块中的源码如下：

```
[root@master first_django]# cat first_django/urls.py
"""first_django URL Configuration

忽略注释
"""
from django.contrib import admin
from django.urls import path

urlpatterns = [
    path('admin/', admin.site.urls),
]
```

从上面的源码可以看到，这里定义的是 Django 项目中 URL 与视图函数的映射关系列表 urlpatterns。而 URLResolver 类中的属性方法 url_patterns 刚好可以获取这个映射关系列表：

```
>>> from django.urls.resolvers import URLResolver
>>> from django.urls.resolvers import RegexPattern
>>> urlconf = "first_django.urls"
>>> resolver = URLResolver(RegexPattern(r'^/'), urlconf)
>>> resolver.url_patterns
[<URLResolver <URLPattern list> (admin:admin) 'admin/'>]
>>> from first_django.urls import urlpatterns
>>> resolver.url_patterns is urlpatterns
True
```

再来看 URLResolver 类中的 check()方法，它的逻辑同样比较简单，即遍历 url_patterns 属性值，并依次调用 check_resolver()方法获取检查的错误列表。由于在 first_django.urls 中只有一行默认的 URLConf 配置，因此直接使用第 1 个元素，查看它的类型及 url_patterns 属性值，结果如下：

```
>>> type(resolver.url_patterns[0])
<class 'django.urls.resolvers.URLResolver'>
>>> resolver.url_patterns[0].pattern
<django.urls.resolvers.RoutePattern object at 0x7f020778d0a0>
>>> resolver.url_patterns[0].urlconf_name
[<URLPattern '' [name='index']>, <URLPattern 'login/' [name='login']>, <URLPattern
'logout/' [name='logout']>, <URLPattern 'password_change/' [name='password_change']>,
<URLPattern 'password_change/done/' [name='password_change_done']>, <URLPattern
'jsi18n/' [name='jsi18n']>, <URLPattern 'r/<int:content_type_id>/<path:object_id>/'
[name='view_on_site']>, <URLResolver <URLPattern list> (None:None) 'auth/group/'>,
<URLResolver <URLPattern list> (None:None) 'auth/user/'>, <URLPattern
'^(?P<app_label>auth)/$' [name='app_list']>]
>>> resolver.url_patterns[0].url_patterns
[<URLPattern '' [name='index']>, <URLPattern 'login/' [name='login']>, <URLPattern
'logout/' [name='logout']>, <URLPattern 'password_change/' [name='password_change']>,
<URLPattern 'password_change/done/' [name='password_change_done']>, <URLPattern
'jsi18n/' [name='jsi18n']>, <URLPattern 'r/<int:content_type_id>/<path:object_id>/'
[name='view_on_site']>, <URLResolver <URLPattern list> (None:None) 'auth/group/'>,
<URLResolver <URLPattern list> (None:None) 'auth/user/'>, <URLPattern
'^(?P<app_label>auth)/$' [name='app_list']>]
```

从上面的源码可以看到，resolver.url_patterns[0]是一个 URLResolver 对象，因此对它调用 check_resolver()方法又会回到该类的 check()方法中，而此时该对象的 url_patterns 则是上面显示的列表（这些是 admin 模块设置的，具体细节读者可自行追踪）。列表中的元素都是 URLPattern 对象，当对象再次调用 check_resolver()方法对这些 URLPattern 进行检查时，又会调用该对象的 check()方法，相关的实现源码如下：

```
# 源码位置：django/urls/resolvers.py
# ……

class URLPattern:
    # ……
```

```
    def check(self):
        warnings = self._check_pattern_name()
        warnings.extend(self.pattern.check())
        return warnings

    def _check_pattern_name(self):
        """
        检查在模式名称中是否包含冒号，如果包含冒号，产生告警信息
        """
        if self.pattern.name is not None and ":" in self.pattern.name:
            warning = Warning(
                "Your URL pattern {} has a name including a ':'. Remove the colon, to "
                "avoid ambiguous namespace references.".format(self.pattern.describe()),
                id="urls.W003",
            )
            return [warning]
        else:
            return []

    # ……

# ……
```

check()方法会调用_check_pattern_name()方法来检查在 URL 的模式名称中是否包含冒号，如果包含冒号，则产生告警信息。此外，在 URLResolver 类中，check()方法的最后一句是：

```
return messages or self.pattern.check()
```

除检查 url_patterns 属性外，还会调用 patterns 属性的 check()方法检查错误。如何才能让 checkl 方法中报错呢？由于 admin.site.urls 得到的 URLPattern 对象列表都符合要求（即在 pattern 字符串中没有冒号），所以只能考虑从 self.patterns 入手。从前面 resolver.url_patterns[0]的结果来看，它是一个 RoutePattern 对象：

```
>>> resolver.url_patterns[0].pattern
<django.urls.resolvers.RoutePattern object at 0x7fa16fa0e2e0>
>>> resolver.url_patterns[0].pattern._route
'admin/'
>>> resolver.url_patterns[0].pattern.check()
[]
```

从上面的代码可以看到，该 RoutePattern 对象的 check()方法返回的结果为空列表，检查正常。继续看该 check()方法的实现源码，如下：

```
# 源码位置：django/urls/resolvers.py
# ……

class RoutePattern(CheckURLMixin):
    regex = LocaleRegexDescriptor('_route')
```

```
    def __init__(self, route, name=None, is_endpoint=False):
        self._route = route
        # ……

    # ……

    def check(self):
        # 调用父类的self._check_pattern_startswith_slash()方法进行检查
        warnings = self._check_pattern_startswith_slash()
        route = self._route
        # 产生告警信息的条件是包含(?P<，并且以^开头或者以$结尾
        if '(?P<' in route or route.startswith('^') or route.endswith('$'):
            warnings.append(Warning(
                "Your URL pattern {} has a route that contains '(?P<', begins "
                "with a '^', or ends with a '$'. This was likely an oversight "
                "when migrating to django.urls.path().".format(self.describe()),
                id='2_0.W001',
            ))
        return warnings

    # ……
```

从 RoutePattern 类的 check()方法可以看到，产生告警信息的条件是：在 self._route 字符串中包含(?P<，并且以^开头或者以$结尾。此外还有一个产生告警信息的地方，即调用父类的_check_pattern_startswith_slash()方法，它的实现源码如下：

```
# 源码位置：django/urls/resolvers.py
# ……

class CheckURLMixin:
    # ……

    def _check_pattern_startswith_slash(self):
        """
        检查pattern，不能以/开头
        """
        regex_pattern = self.regex.pattern
        if not settings.APPEND_SLASH:
            return []
        if regex_pattern.startswith(('/', '^/', '^\\/')) and not
regex_pattern.endswith('/'):
            warning = Warning(
                "Your URL pattern {} has a route beginning with a '/'. Remove this "
                "slash as it is unnecessary. If this pattern is targeted in an "
                "include(), ensure the include() pattern has a trailing '/'.".format(
                    self.describe()
                ),
```

```
            id="urls.W002",
        )
        return [warning]
    else:
        return []
# ……
```

从上面的源码可以看到，另一个产生告警信息的方式，即 self.regex.pattern 如果以/、 ^/或^\/中的任意一个开头且不以/结尾，将产生告警信息。对于 self.regex.pattern 的结果，参见源码或者示例结果：

```
>>> resolver.url_patterns[0].pattern.regex.pattern
'^admin/'
```

有了前面的铺垫后，就可以进行代码测试了。调整 first_django/urls.py 中的 urlpatterns 值如下：

```
# 源码位置：first_django/urls.py
# ……

urlpatterns = [
    path('admin/', admin.site.urls),        # 正常不产生告警信息
    path('^admin/', admin.site.urls),       # 以^开头，产生告警信息，id=2_0.W001
    path('admin$', admin.site.urls),        # 以$结尾，产生告警信息，id=2_0.W001
    path('/admin/', admin.site.urls),       # 以/开头，同时以/结尾，不产生告警信息
    path('/admin', admin.site.urls),        # 产生告警信息，id=urls.W002
]
```

再次进入 Django 的 shell 模式并运行 check_url_config()方法，结果如下：

```
(django2-core-test) [root@master first_django]# python manage.py shell
Python 3.8.6 (default, Oct 18 2020, 15:33:08)
[GCC 4.8.5 20150623 (Red Hat 4.8.5-39)] on linux
Type "help", "copyright", "credits" or "license" for more information.
(InteractiveConsole)
>>> from django.core.checks.urls import check_url_config
>>> warnings = check_url_config(None)
>>> len(warnings)
3
>>> warnings
[<Warning: level=30, msg="Your URL pattern '^admin/' has a route that contains '(?P<', begins with a '^', or ends with a '$'. This was likely an oversight when migrating to django.urls.path().", hint=None, obj=None, id='2_0.W001'>, <Warning: level=30, msg="Your URL pattern 'admin$' has a route that contains '(?P<', begins with a '^', or ends with a '$'. This was likely an oversight when migrating to django.urls.path().", hint=None, obj=None, id='2_0.W001'>, <Warning: level=30, msg="Your URL pattern '/admin' has a route beginning with a '/'. Remove this slash as it is unnecessary. If this pattern is targeted in an include(), ensure the include() pattern has a trailing '/'.", hint=None, obj=None, id='urls.W002'>]
```

在上述结果中出现了 3 个告警信息，它们对应着 3 个 URL 的映射配置。虽然能直接从 URL 的表达式中判断是否会产生告警信息（比如最后一行 URL 配置，总有人会将字符串/admin 带入_check_pattern_startswith_slash()方法中进行判断），但实际上 id=urls.W002 的告警信息并不是通过该 URL 的表达式进行判断的。继续在交互模式下查看最后一个 URLPattern 对象的 regex.pattern 值，结果如下：

```
>>> from django.urls import get_resolver
>>> resolver = get_resolver()
>>> resolver.url_patterns[4]
<URLResolver <URLPattern list> (admin:admin) '/admin'>
>>> resolver.url_patterns[4].pattern._route
'/admin'
>>> resolver.url_patterns[4].pattern.regex.pattern
'^/admin'
```

可以看到，相比原来的/admin 字符串，通过 regex.pattern 最终得到的是^/admin，前面多加了一个^。问题又来了，为什么会多一个^？先来看看其他几个 URL 匹配的结果：

```
>>> resolver.url_patterns[0].pattern.regex.pattern
'^admin/'
>>> resolver.url_patterns[1].pattern.regex.pattern
'^\\^admin/'
>>> resolver.url_patterns[2].pattern.regex.pattern
'^admin\\$'
>>> resolver.url_patterns[3].pattern.regex.pattern
'^/admin/'
>>> resolver.url_patterns[4].pattern.regex.pattern
'^/admin'
```

可以看到，确实是在所有的 URL 匹配表达式的最前面都加上了^字符串。这些内容是在哪里添加的呢？追踪 URLPattern 类中定义 regex 属性的源码，如下：

```
# 源码位置：django/urls/resolvers.py
# ……

class RoutePattern(CheckURLMixin):
regex = LocaleRegexDescriptor('_route')

# ……
```

继续追踪 LocaleRegexDescription 的结果，如下：

```
# 源码位置：django/urls/resolvers.py
# ……

class LocaleRegexDescriptor:
   def __init__(self, attr):
      self.attr = attr
```

```
    def __get__(self, instance, cls=None):
        """
        基于当前语言返回一个编译好的正则表达式
        """
        if instance is None:
            return self
        pattern = getattr(instance, self.attr)
        if isinstance(pattern, str):
            # 核心语句
            instance.__dict__['regex'] = instance._compile(pattern)
            return instance.__dict__['regex']
        language_code = get_language()
        if language_code not in instance._regex_dict:
            instance._regex_dict[language_code] = instance._compile(str(pattern))
        return instance._regex_dict[language_code]
```

上述源码实现了一个属性描述符（即实现了魔法函数 __get__()、__set__()、__delete__()中的一个），其中，regex 就是 RoutePattern 类中的属性。为了能理解上述源码，下面给出上述源码的一个简化版本，如下：

```
(django2-core-test) [root@master first_django]# cat test_descriptor.py
import re


class LocaleRegexDescriptor():
    def __init__(self, attr):
        self.attr = attr

    def __get__(self, instance, cls=None):
        print('进入魔法函数__get__(): instance={}, addr={}'.format(instance,
id(instance)))
        pattern = getattr(instance, self.attr)
        return re.compile(pattern)


class RoutePattern():
    regex = LocaleRegexDescriptor('_route')

    def __init__(self, route):
        self._route = route


route_pattern = RoutePattern('admin/')
print(id(route_pattern))
regex = route_pattern.regex
print(regex, regex.pattern)
```

当执行到 pattern.regex 语句时，程序会进入魔法函数__get__()中执行，此时 instance 为当前的 RoutePattern 对象，即 route_pattern 值。接着得到的 pattern 为该 RoutePattern 对象的_routes 属性值，即 admin/字符串，最后返回使用 re 模块对该 pattern 编译的结果。以下是该源码的运行结果：

```
(django2-core-test) [root@master first_django]# python test_descriptor.py
140207422302384
进入魔法函数__get__(): instance=<__main__.RoutePattern object at 0x7f849599e8b0>,
addr=140207422302384
re.compile('admin/') admin/
```

在理解上述结果后，就不难理解 RoutePattern 对象访问 regex 属性语句了，它将进入 LocaleRegexDescriptor 类的魔法函数__get__()，得到的 pattern 变量和本书给出的源码一致，但是在 Django 源码中则多了一步处理，即 instance._compile(pattern)。该语句的结果正是 regex 属性值。只需再次追踪该方法的源码，一切就非常清楚了：

```
# 源码位置：django/urls/resolvers.py
# ……

_PATH_PARAMETER_COMPONENT_RE = re.compile(
    r'<(?:(?P<converter>[^>:]+):)?(?P<parameter>\w+)>'
)

def _route_to_regex(route, is_endpoint=False):
    """
    转换路径模式为一个普通的正则表达式。
    例如，将'foo/<int:pk>' 字符串转换成 '^foo\\/(?P<pk>[0-9]+)'
    和 {'pk': <django.urls.converters.IntConverter>}.
    """
    original_route = route
    # 注意，这里的第 1 个元素为^
        parts = ['^']
    converters = {}
    while True:
        match = _PATH_PARAMETER_COMPONENT_RE.search(route)
        if not match:
            # 不匹配，直接将 route 字符串转义后添加到 parts 中并返回
            parts.append(re.escape(route))
            break
        # 处理匹配情况，类似于'foo/<int:pk>'这样的字符串
        # ……
    if is_endpoint:
        parts.append('$')
    return ''.join(parts), converters


class RoutePattern(CheckURLMixin):
    regex = LocaleRegexDescriptor('_route')
```

```
    # ……

    def _compile(self, route):
        return re.compile(_route_to_regex(route, self._is_endpoint)[0])

    # ……
```

对于一些复杂情况的处理，我们留到下一节去分析，这里只介绍简单情况。_route_to_regex()方法在传入/admin 字符串时，正则表达式_PATH_PARAMETER_COMPONENT_RE 无法匹配，所以对/admin 字符串进行转义后直接将其添加到列表 parts 中，而列表 parts 中的第 1 个元素为^。最终返回的结果为二元组(''.join(parts), converters)，而_compile()方法只取第 1 个元素并返回。从这里即可搞清楚在语句 resolver.url_patterns[4].pattern.regex.pattern 中多出一个^的原因。下面我们简单模拟这一过程：

```
>>> from django.urls.resolvers import _PATH_PARAMETER_COMPONENT_RE
>>> route = '/admin'
>>> parts = ['^']
>>> match = _PATH_PARAMETER_COMPONENT_RE.search(route)
>>> match is None
True
>>> import re
>>> parts.append(re.escape(route))
>>> "".join(parts)
'^/admin'
>>> route = '^admin/'
>>> parts = ['^']
>>> parts.append(re.escape(route))
>>> parts
['^', '\\^admin/']
>>> ''.join(parts)   # 刚好是 resolver.url_patterns[1].pattern.regex.pattern 的结果
'^\\^admin/'
```

最后，所有的检查代码将在 Django 的 check 命令中被调用执行，执行 check 命令，结果如下：

```
(django2-core-test) [root@master first_django]# python manage.py check
System check identified some issues:

WARNINGS:
?: (2_0.W001) Your URL pattern '^admin/' has a route that contains '(?P<', begins with a '^', or ends with a '$'. This was likely an oversight when migrating to django.urls.path().
?: (2_0.W001) Your URL pattern 'admin$' has a route that contains '(?P<', begins with a '^', or ends with a '$'. This was likely an oversight when migrating to django.urls.path().
?: (urls.W002) Your URL pattern '/admin' has a route beginning with a '/'. Remove this slash as it is unnecessary. If this pattern is targeted in an include(), ensure the include() pattern has a trailing '/'.
```

```
?: (urls.W005) URL namespace 'admin' isn't unique. You may not be able to reverse all
URLs in this namespace

System check identified 4 issues (0 silenced).
```

可以看到，前三个告警信息正是 check_url_config()方法的检查结果。通过最后一个告警信息可知，其 tags 依旧是 urls。在 django/core/checks/urls.py 中简单浏览一下即可很快发现这条告警信息的出处了，正是该文件中 check_url_namespaces_unique()方法的检查结果。

对 python manage.py check 命令的完整追踪过程这里不再演示，有兴趣的读者可以自行学习，整体来看和第 2 章的分析基本一致，且更加简单。

此外，对 checks 目录下的其他检查函数的分析过程基本可以参考上面的示例，本书不再赘述。

# 5.5 序列化

Django 具备基本的序列化与反序列化功能，能够将 QuerySet 序列化为 JSON、XML 或 YAML 等格式的数据，或者是将对应格式的数据反序列化为 Django 的模型对象。下面看一个简单的示例：

```
(django2-core-test) [root@master first_django]# python manage.py shell
Python 3.8.6 (default, Oct 18 2020, 15:33:08)
[GCC 4.8.5 20150623 (Red Hat 4.8.5-39)] on linux
Type "help", "copyright", "credits" or "license" for more information.
(InteractiveConsole)
>>> from django.core import serializers
>>> from shell_test.models import DjangoBooks
>>> DjangoBooks.objects.all().filter(sex=1)
<QuerySet [<DjangoBooks: <Python 自动化运维, test>>, <DjangoBooks: <小郭历险记, mmmm>>,
<DjangoBooks: <精通 Django 框架, zzzzz>>, <DjangoBooks: <Django 实战, test>>]>
>>> data = serializers.serialize("json", DjangoBooks.objects.all().filter(sex=1))
>>> data
'[{"model": "shell_test.djangobooks", "pk": 3, "fields": {"book_name":
"Python\\u81ea\\u52a8\\u5316\\u8fd0\\u7ef4", "author": "test", "sex": 1, "price": 79.0,
"isbn": "333333", "publish_date": "2021-02-04T21:43:37Z"}}, {"model":
"shell_test.djangobooks", "pk": 6, "fields": {"book_name":
"\\u5c0f\\u90ed\\u5386\\u9669\\u8bb0", "author": "mmmm", "sex": 1, "price": 39.0, "isbn":
"222221", "publish_date": "2021-02-14T10:23:00Z"}}, {"model": "shell_test.djangobooks",
"pk": 8,
 "fields": {"book_name": "\\u7cbe\\u901aDjango\\u6846\\u67b6", "author": "zzzzz", "sex":
1, "price": 129.0, "isbn": "1234321", "publish_date": "2020-01-18T10:27:00Z"}}, {"model":
"shell_test.djangobooks",
"pk": 9, "fields": {"book_name": "Django\\u5b9e\\u6218", "author": "test", "sex": 1,
"price": 89.0, "isbn": "1111112", "publish_date": "2009-02-28T11:37:00Z"}}]'
>>>
# 使用 Python 序列化器
```

```
>>> data = serializers.serialize("python", DjangoBooks.objects.all().filter(sex=1))
>>> data
[OrderedDict([('model', 'shell_test.djangobooks'), ('pk', 3), ('fields',
OrderedDict([('book_name', 'Python 自动化运维'), ('author', 'test'), ('sex', 1), ('price',
79.0), ('isbn', '333333'), ('publish_date',
datetime.datetime(2021, 2, 4, 21, 43, 37, tzinfo=<UTC>))]))]), OrderedDict([('model',
'shell_test.djangobooks'), ('pk', 6), ('fields', OrderedDict([('book_name', '小郭历险记
'), ('author', 'mmmm'), ('sex', 1), ('price', 39.0), ('isbn', '222221'), ('publish_date',
datetime.datetime(2021, 2, 14, 10, 23, tzinfo=<UTC>))]))]), OrderedDict([('model',
'shell_test.djangobooks'), ('pk', 8), ('fields', OrderedDict([('book_name', '精通 Django
框架'), ('author', 'zzzzz'), ('sex', 1), ('price', 129.0), ('isbn', '1234321'),
('publish_date', datetime.datetime(2020, 1, 18, 10, 27, tzinfo=<UTC>))]))]),
OrderedDict([('model', 'shell_test.djangobooks'), ('pk', 9), ('fields',
OrderedDict([('book_name', 'Django 实战'), ('author', 'test'), ('sex', 1), ('price', 89.0),
('isbn', '1111112'), ('publish_date', datetime.datetime(2009, 2, 28, 11, 37,
tzinfo=<UTC>))]))])]
# 反序列化
>>> objects = list(serializers.deserialize("python", data))
>>> objects
[<DeserializedObject: shell_test.DjangoBooks(pk=3)>, <DeserializedObject:
shell_test.DjangoBooks(pk=6)>, <DeserializedObject: shell_test.DjangoBooks(pk=8)>,
<DeserializedObject: shell_test.DjangoBooks(pk=9)>]
>>> type(objects[0])
<class 'django.core.serializers.base.DeserializedObject'>
>>> objects[0].object
<DjangoBooks: <Python 自动化运维, test>>
>>> type(objects[0].object)
<class 'shell_test.models.DjangoBooks'>
>>> objects[0].object.book_name
'Python 自动化运维'
```

下面通过几个问题探索这部分源码：

（1）在上面的源码中，序列化的完整过程是怎样的？

（2）为什么 JSON 序列化器得到的中文结果为乱码 Unicode 字符，而 Python 序列化器得到的中文结果正常显示呢？如何修正这一问题？

## 5.5.1　serialize()方法的源码解析

与 Django 序列化相关的代码均位于 django/core/serializers 目录中。为了追踪序列化操作的底层逻辑，首先从 serialize()方法开始追踪，该方法的源码如下：

```
# 源码位置：django/core/serializers/__init__.py
# ……

def serialize(format, queryset, **options):
```

```
    """
    使用一个确定的序列化器来序列化一个 queryset（或者其他返回模型对象的迭代器）
    """
    s = get_serializer(format)()     # 得到一个序列化器对象
    s.serialize(queryset, **options)  # 使用对应的序列化器对象序列化该 queryset
    return s.getvalue()            # 返回最后的序列化结果

# ……
```

serialize()方法的内容非常简单，首先根据 format 字符串调用 get_serializer()方法来得到序列化器类，然后实例化得到对应的序列化器对象。接着调用该对象的 serialize()方法序列化该 queryset，最后调用序列化器对象的 getvalue()方法获取序列化后的结果。

get_serializer()方法的源码如下：

```
# 源码位置：django/core/serializers/__init__.py
# ……

def get_serializer(format):
    # 如果_serializers 为空，调用_load_serializers()加载序列化器
    if not _serializers:
        _load_serializers()
    if format not in _serializers:
    # 如果 format 字符串不是_serializers 中的 key，抛出异常
        raise SerializerDoesNotExist(format)
    # 返回对应的序列化器类
    return _serializers[format].Serializer

# ……
```

该方法中出现的_serializers 变量可以在 django/core/serializers/__init__.py 文件中找到：

```
# 源码位置：django/core/serializers/__init__.py
# ……

_serializers = {}
```

可以看到，在开始时_serializers 为一个空字典，因此会调用_load_serializers()方法获得 Django 内置的序列化器：

```
# 源码位置：django/core/serializers/__init__.py
# ……

# Built-in serializers
BUILTIN_SERIALIZERS = {
    "xml": "django.core.serializers.xml_serializer",
    "python": "django.core.serializers.python",
    "json": "django.core.serializers.json",
    "yaml": "django.core.serializers.pyyaml",
```

```
}

# ......

def _load_serializers():

    global _serializers   # 前面定义的全局变量
    serializers = {}    # 局部变量，位于该方法内
    for format in BUILTIN_SERIALIZERS:
        # 注册序列化器，其实就是写入 serializers 变量
        register_serializer(format, BUILTIN_SERIALIZERS[format], serializers)
    if hasattr(settings, "SERIALIZATION_MODULES"):
        # 如果在 Django 的全局配置中由 SERIALIZATION_MODULES 变量指定序列化器，
        # 则设置进一步添加外部的序列化器
        for format in settings.SERIALIZATION_MODULES:
            register_serializer(format, settings.SERIALIZATION_MODULES[format],
                                serializers)
    _serializers = serializers

# ......
```

从上面的源码可以看到，Django 内置了 4 种序列化器，同时支持在全局配置文件中指定自定义的序列化器，即支持自定义，这是框架中的常见方式。上述代码的核心语句是调用 register_serializer() 方法：

```
# 源码位置：django/core/serializers/__init__.py
# ......

def register_serializer(format, serializer_module, serializers=None):

    if serializers is None and not _serializers:
        _load_serializers()

    try:
        # 导入对应字符串地址的模块
        module = importlib.import_module(serializer_module)
    except ImportError as exc:
        bad_serializer = BadSerializer(exc)

        module = type('BadSerializerModule', (), {
            'Deserializer': bad_serializer,
            'Serializer': bad_serializer,
        })

    if serializers is None:
        _serializers[format] = module
    else:
        # 如果传入了 serializers，就设置到 serializers 中
```

```
        serializers[format] = module

# ……
```

从上面的源码可以看到，调用 register_serializer()方法其实就是导入传入的模块字符串，并将对应的名称（即 format 字符串值）保存到传入的 serializers 字典中（如果该值为 None，则保存到全局变量_serializers 中）。在掌握了上述方法的源码后，就可以简单测试一下这些方法了，操作示例如下：

```
>>> from django.core.serializers import BUILTIN_SERIALIZERS, register_serializer
>>> serializers = {}
>>> register_serializer('json', BUILTIN_SERIALIZERS['json'], serializers)
>>> serializers
{'json': <module 'django.core.serializers.json' from
'/root/.pyenv/versions/django2-core-test/lib/python3.8/site-packages/django/core/ser
ializers/json.py'>}
>>> register_serializer('xml', BUILTIN_SERIALIZERS['xml'], serializers)
>>> serializers
{'json': <module 'django.core.serializers.json' from
'/root/.pyenv/versions/django2-core-test/lib/python3.8/site-packages/django/core/ser
ializers/json.py'>, 'xml': <module 'django.core.serializers.xml_serializer' from
'/root/.pyenv/versions/django2-core-test/lib/python3.8/site-packages/django/core/ser
ializers/xml_serializer.py'>}
```

接下来测试一个简单的错误操作，Python 初学者很可能会遇到这个问题。前面在介绍_load_serializers()方法时提到，该方法会将 Django 内置的序列化器和自定义的序列化器模块一同导入全局变量_serializers。因此，在早期调试 Django 源码时测试了如下语句：

```
>>> from django.core.serializers import _serializers
>>> _serializers
{}
>>> from django.core.serializers import _load_serializers
>>> _load_serializers()
>>> _serializers
{}
```

在调用_load_serializers()方法后，这里的全局变量_serializer()依然是空字典。这是 Python 中的变量引用问题，只需再次查看_load_serializers()方法的源码，就能理解其中的原因了：

```
# 源码位置：django/core/serializers/__init__.py
# ……

def _load_serializers():

    global _serializers   # 前面定义的全局变量
    serializers = {}      # 局部变量，位于该方法内
    # ……

    _serializers = serializers
```

```
# ……
```

_load_serializers()方法一开始就用 global 关键字指定了使用前面定义的全局定量_serializers，接着在注册相应的序列化器后得到变量 serializers，最后将变量 serializers 赋给变量_serializers。为了帮助读者理解这个问题，下面给出关于该变量转换前后的一个示意图，如图 5-2 所示。

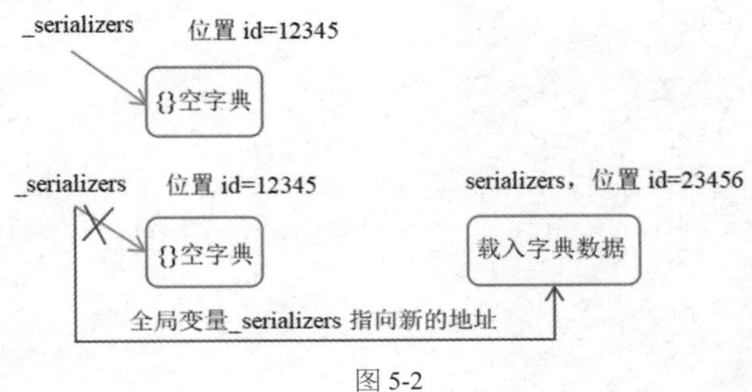


图 5-2

在理解了全局变量_serializers 的赋值操作后，就清楚为何在执行导入序列化器后打印的_serializers 值依旧为空了。因为再次输入的全局变量_serializers 仍旧是被替换前的变量，所以指向的仍旧是空字典地址。如何才能得到更新后有数据的地址呢？非常简单，再导入一次全局变量_serializers 即可，示例如下：

```
>>> from django.core.serializers import _serializers
>>> _serializers
{}
>>> id(_serializers)
140568936534912
>>> from django.core.serializers import _load_serializers
>>> _load_serializers()  # 调用载入方法
>>> id(_serializers)    # 全局变量_serializers 对应的位置没有变，它仍表示前面导入的变量地址
140568936534912
>>> _serializers       #打印该变量仍旧为空字典
{}
>>> from django.core.serializers import _serializers   # 再次导入全局变量_serializers
>>> id(_serializers)    # 可以看到，id 值已经发生了变化
140568887650176
>>> _serializers       # 新的 id 值对应的正是导入数据后的位置
{'xml': <module 'django.core.serializers.xml_serializer' from
'/root/.pyenv/versions/django2-core-test/lib/python3.8/site-packages/django/core/ser
ializers/xml_serializer.py'>, 'python': <module 'django.core.serializers.python' from
'/root/.pyenv/versions/django2-core-test/lib/python3.8/site-packages/django/core/ser
ializers/python.py'>, 'json': <module 'django.core.serializers.json' from
'/root/.pyenv/versions/django2-core-test/lib/python3.8/site-packages/django/core/ser
ializers/json.py'>, 'yaml': <class 'django.core.serializers.BadSerializerModule'>}
```

继续回到 get_serializer()方法，正常情况下它返回的结果如下：

```
_serializers[format].Serializer
```

当 register_serializer()方法中的 format 参数等于'json'时，上述结果为在 django/core/serializers/json.py 文件中定义的 Serializer 类，其余情况均是如此。也就是说，在所有的序列化模块中都必须定义 Serializer 类，包括外部自定义的序列化模块。接下来分析 JSON 序列化器的实现源码，其余类型的序列化器的分析方式基本与之一致：

```
# 源码位置：django/core/serializers/json.py
# ……

from django.core.serializers.python import (
    Deserializer as PythonDeserializer, Serializer as PythonSerializer,
)

# ……

class Serializer(PythonSerializer):
    """将一个 queryset 转换成 JSON 数据"""
    internal_use_only = False

    def _init_options(self):
        self._current = None
        self.json_kwargs = self.options.copy()
        self.json_kwargs.pop('stream', None)
        self.json_kwargs.pop('fields', None)
        if self.options.get('indent'):
            self.json_kwargs['separators'] = (',', ': ')
        self.json_kwargs.setdefault('cls', DjangoJSONEncoder)

    # ……

    def getvalue(self):
        # 调用 PythonSerializer 的父类（即 base.Serializer）的 getvalue()方法
        return super(PythonSerializer, self).getvalue()

# ……
```

从上面的源码可以看到，JSON 序列化器中的 Serializer 类继承自 Python 序列化器中的 Serializer 类。在该类中并没有定义 serialize()方法，此外 getvalue()方法仅调用了父类的 getvalue()方法并返回其结果。继续追踪在 django/core/serializers/python.py 文件中定义的 Serializer 类，其源码如下：

```
# 源码位置：django/core/serializers/python.py
# ……

class Serializer(base.Serializer):
```

```
    """
    将一个 QuerySet 对象序列化成基本的 Python 对象
    """

    # ……

    def getvalue(self):
        return self.objects

# ……
```

这里没有列出该 Serializer 类中的其他方法，但是在 Python 序列化器的 Serializer 类中仍旧找不到 serialize()方法，且其 getvalue()方法也异常简单，只返回了序列化对象的 objects 属性。继续深入该类的父类 base.Serializer 的实现源码，可以看到 serialize()方法，内容如下：

```
# 源码位置：django/core/serializers/base.py
# ……

class Serializer:
    """
    序列化器的抽象基类
    """

    # 表明该序列化器是否仅供 Django 内部使用
    internal_use_only = False
    progress_class = ProgressBar
    stream_class = StringIO

    def serialize(self, queryset, *, stream=None, fields=None,
use_natural_foreign_keys=False, use_natural_primary_keys=False, progress_output=None,
object_count=0, **options):

        self.options = options
        # 默认就是一个 StringIO 对象
        self.stream = stream if stream is not None else self.stream_class()
        self.selected_fields = fields   # 选定需要序列化的字段
        # 用于处理外键的序列化
        self.use_natural_foreign_keys = use_natural_foreign_keys
        self.use_natural_primary_keys = use_natural_primary_keys
        # ProgressBar 对象，序列化进度
        progress_bar = self.progress_class(progress_output, object_count)

        self.start_serialization()  # 调用 start_serialization()方法开始序列化
        self.first = True           # 第一次为 True，后续在被设置为 False 后，将一直为 False
        for count, obj in enumerate(queryset, start=1):
            self.start_object(obj)  # obj 变量就是一个模型对象，关联一条查询记录
            concrete_model = obj._meta.concrete_model
            if self.use_natural_primary_keys:
```

```
                pk = concrete_model._meta.pk
                pk_parent = pk if pk.remote_field and pk.remote_field.parent_link else None
            else:
                # 没有关联键的将执行这里的语句
                pk_parent = None
            # concrete_model._meta.local_fields 正是模型类的属性字段
            for field in concrete_model._meta.local_fields:
                if field.serialize or field is pk_parent:
                    if field.remote_field is None:
                        # 通过传入的 fields 参数选定需要序列化的字段
                        if self.selected_fields is None or field.attname in
self.selected_fields:
                            self.handle_field(obj, field)
                    else:
                        if self.selected_fields is None or field.attname[:-3] in
self.selected_fields:
                            self.handle_fk_field(obj, field)

            # 处理多对多字段
            for field in concrete_model._meta.many_to_many:
                if field.serialize:
                    if self.selected_fields is None or field.attname in self.selected_fields:
                        self.handle_m2m_field(obj, field)  # 处理多对多字段序列化的核心
            self.end_object(obj)
            progress_bar.update(count)
            self.first = self.first and False
        self.end_serialization()   # 调用 end_serialization()方法，结束序列化过程
        return self.getvalue()

    # ……

    def getvalue(self):

        if callable(getattr(self.stream, 'getvalue', None)):
            return self.stream.getvalue()

# ……
```

在上述代码中，如果把处理关联字段的代码删掉，则整个 serialize()方法看起来就非常简单了。除此之外，还需要理解几个变量的含义，比如第 1 个 for 循环中的 obj 和 concrete_model 变量的含义。obj 变量的含义非常清楚，它指代的就是 QuerySet 类中对应的一个个模型对象，示例如下：

```
(django2-core-test) [root@master first_django]# python manage.py shell
Python 3.8.6 (default, Oct 18 2020, 15:33:08)
[GCC 4.8.5 20150623 (Red Hat 4.8.5-39)] on linux
Type "help", "copyright", "credits" or "license" for more information.
(InteractiveConsole)
>>> from shell_test.models import DjangoBooks
```

```
>>> DjangoBooks.objects.all().filter(sex=1)
<QuerySet [<DjangoBooks: <Python 自动化运维, test>>, <DjangoBooks: <小郭历险记, mmmm>>,
<DjangoBooks: <精通 Django 框架, zzzzz>>, <DjangoBooks: <Django 实战, test>>]>
>>> queryset = DjangoBooks.objects.all().filter(sex=1)
# serialize()方法中的第 1 个 for 循环语句
>>> for count, obj in enumerate(queryset, start=1):
...     print(count, obj, type(obj))
...
1 <Python 自动化运维, test> <class 'shell_test.models.DjangoBooks'>
2 <小郭历险记, mmmm> <class 'shell_test.models.DjangoBooks'>
3 <精通 Django 框架, zzzzz> <class 'shell_test.models.DjangoBooks'>
4 <Django 实战, test> <class 'shell_test.models.DjangoBooks'>
```

对于 concrete_model 变量的含义，可以通过如下语句进行理解：

```
>>> from shell_test.models import DjangoBooks
>>> obj = DjangoBooks.objects.all().filter(sex=1).first()
>>> obj
<DjangoBooks: <Python 自动化运维, test>>
>>> concrete_model = obj._meta.concrete_model
>>> type(concrete_model)
<class 'django.db.models.base.ModelBase'>
>>> concrete_model._meta.local_fields
[<django.db.models.fields.AutoField: id>, <django.db.models.fields.CharField:
book_name>, <django.db.models.fields.CharField: author>,
<django.db.models.fields.SmallIntegerField: sex>, <django.db.models.fields.FloatField:
price>, <django.db.models.fields.CharField: isbn>,
<django.db.models.fields.DateTimeField: publish_date>]
>>> concrete_model._meta.local_fields[0]
<django.db.models.fields.AutoField: id>
>>> type(concrete_model._meta.local_fields[0])
<class 'django.db.models.fields.AutoField'>
```

从上面的演示结果可以看到，concrete_model 变量为一个 ModelBase 对象，通过语句 concrete_model._meta.local_fields 可以得到对应模型类的全部字段。此时，再来看 serialize()方法的代码就非常清楚了，它的处理逻辑总结如下：

（1）为变量赋值，然后调用对象的 start_serialization()方法开始序列化。

（2）用 for 循环遍历 QuerySet 对象代表的查询结果，obj 变量为查询集中的某个记录，concrete_model 变量则是根据 obj 变量得到的一个 ModelBase 对象，接下来通过该对象得到对应模型的全部字段。

（3）继续用 for 循环得到模型类的全部字段，对于普通字段，当 serialize 属性为 True 时，调用 Serializer 对象的 handle_field()方法处理该字段；对于外键字段，则调用 Serializer 对象的 handle_fk_field()方法处理该字段；对于多对多字段，在满足条件后调用 handle_m2m_field()方法处理该字段。

（4）在序列化结束后，依次调用该 Serializer 对象的 end_serialization()方法和 getvalue()方法，返回 getvalue()方法的结果。

此外，从代码中可知，serialize()方法中的 fields 参数可用于对指定的字段进行序列化，下面给出一个简单的示例：

```
>>> from shell_test.models import DjangoBooks
>>> from django.core.serializers.json import Serializer
>>> serializer = Serializer()
>>> serializer.serialize(DjangoBooks.objects.all().filter(sex=1), fields=['book_name',
'author', 'price'])
'[{"model": "shell_test.djangobooks", "pk": 3, "fields": {"book_name":
"Python\\u81ea\\u52a8\\u5316\\u8fd0\\u7ef4", "author": "test", "price": 79.0}},
{"model": "shell_test.djangobooks", "pk": 6, "fields": {"book_name":
"\\u5c0f\\u90ed\\u5386\\u9669\\u8bb0",
"author": "mmmm", "price": 39.0}}, {"model": "shell_test.djangobooks", "pk": 8, "fields":
{"book_name": "\\u7cbe\\u901aDjango\\u6846\\u67b6", "author": "zzzzz", "price": 129.0}},
{"model": "shell_test.djangobooks", "pk": 9, "fields": {"book_name":
"Django\\u5b9e\\u6218", "author": "test",
"price": 89.0}}]'
```

可以看到，在 serialize()方法中传入 fields=['book_name', 'author', 'price']后，序列化的结果就被限定为指定字段。

从上面总结的流程来看，在 Django 中，序列化的核心语句就是 handle_field()、handle_fk_field()和 handle_m2m_field()这三个方法。此外，还有一些辅助的方法，比如开始序列化和停止序列化的 start_serialization()和 end_serialization()方法，再比如开始处理模型对象和停止处理模型对象的 start_object()和 end_object()方法。

先看 handle_field()方法，对于 JSON 序列化器来说，在它的 Serializer 类中并没有定义处理字段的方法，而是全部调用父类（Python 序列化器中的 Serializer 类）的 handle_field()方法，其实现源码如下：

```
# 源码位置：django/core/serializers/python.py
# ……

class Serializer(base.Serializer):

   # ……

   def _value_from_field(self, obj, field):
      value = field.value_from_object(obj)
      # 如果字段为保护类型，例如 None、 numbers、dates 或者 Decimals 等 ，则保持不变；
      # 如果字段为其他类型，则先转换成字符串
```

```
        return value if is_protected_type(value) else field.value_to_string(obj)

    def handle_field(self, obj, field):
        self._current[field.name] = self._value_from_field(obj, field)

    # ……

# ……
```

从上述源码来看，handle_field()方法似乎异常简单，只是将模型对象的相关属性值转换成 key-value 值，并保存到_current 属性值中。接下来一步步演示 JSON 序列化器的序列化过程。

## 5.5.2 JSON 序列化器的底层逻辑

先考虑简单情况，在 DjangoBooks 模型类的字段中没有外键或者多对多的字段。此外，还限定了序列化字段，只序列化 book_name 和 author 两个字段。首先，创建一个 JSON 序列化器的 Serializer 对象，操作如下：

```
(django2-core-test) [root@master first_django]# python manage.py shell
Python 3.8.6 (default, Oct 18 2020, 15:33:08)
[GCC 4.8.5 20150623 (Red Hat 4.8.5-39)] on linux
Type "help", "copyright", "credits" or "license" for more information.
(InteractiveConsole)
>>> from django.core.serializers.json import Serializer
>>> serializer = Serializer()
>>>
```

（1）在 serialize()方法的开头部分初始化一些属性值，比如 stream 属性值和 selected_fields 属性值等，模拟操作如下：

```
>>> serializer.stream = serializer.stream_class()
# 在默认情况下，stream 属性值是一个 StringIO 对象
>>> serializer.stream
<_io.StringIO object at 0x7f3d13c64160>
>>> serializer.stream.getvalue()  # 一开始没有任何写入，所以是空字符串
''
>>> serializer.selected_fields = ['book_name', 'author']
# 指定只序列化模型对象中的特定字段
>>>
```

（2）调用 start_serialization()方法，在 JSON 序列化器的 Serialize 类中，开始序列化和停止序列化的源码如下：

```
# 源码位置：django/core/serializers/json.py
# ……


class Serializer(PythonSerializer):
```

```
    # ……

    def _init_options(self):
        self._current = None
        self.json_kwargs = self.options.copy()
        self.json_kwargs.pop('stream', None)
        self.json_kwargs.pop('fields', None)
        if self.options.get('indent'):
            # Prevent trailing spaces
            self.json_kwargs['separators'] = (',', ': ')
        self.json_kwargs.setdefault('cls', DjangoJSONEncoder)

    def start_serialization(self):
        # 初始化选项值
        self._init_options()
        # 在 StringIO 对象中写入 "[" 字符串，类似于列表的开始
        self.stream.write("[")

    def end_serialization(self):
        if self.options.get("indent"):
            self.stream.write("\n")
        # 在 StringIO 对象中写入 "]" 字符串，类似于列表的结束
        self.stream.write( "]" )
        if self.options.get("indent"):
            self.stream.write("\n")

    # ……

# ……
```

从上述源码可以看到，对于默认情况，JSON 序列化器的 start_serialization()方法会初始化一些属性值，并调用 stream 属性的 write()方法写入一个“[”字符串。接着前面的交互模式继续输入下面的语句：

```
>>> serializer.options = {}
>>> serializer.start_serialization()
#对 serializer.stream 的操作就是调用 write()方法写入 "[" 字符串
>>> serializer.stream.getvalue()
'['
>>> serializer.json_kwargs
{'cls': <class 'django.core.serializers.json.DjangoJSONEncoder'>}
```

（3）上述结果可以直接从源码中得到，接下来是最核心的两个 for 循环语句：

```
>>> serializer.use_natural_foreign_keys = False
>>> serializer.use_natural_primary_keys = False
>>> serializer.first = True
>>> for count, obj in enumerate(queryset, start=1):
```

```
    # 在 python.py 文件中，Serializer 类的 start_object()方法中的内容
...     serializer._current = OrderedDict()
...     concrete_model = obj._meta.concrete_model
...     pk_parent = None
...     for field in concrete_model._meta.local_fields:
...        if field.remote_field is None and field.attname in serializer.selected_fields:
...           serializer.handle_field(obj, field)  # 处理指定的字段
...     serializer.end_object(obj)
...     serializer.first = serializer.first and False
...
# 模拟 JSON 序列化器中 Serializer 类的 end_serialization()方法
>>> serializer.stream.write("]")
1
>>> serializer.stream.getvalue()
'[{"model": "shell_test.djangobooks", "pk": 1, "fields": {"book_name":
"Django\\u6e90\\u7801\\u7b14\\u8bb0", "author": "spyinx"}}, {"model":
"shell_test.djangobooks", "pk": 2, "fields": {"book_name":
"Ansible\\u6838\\u5fc3\\u6e90\\u7801\\u5256\\u6790\\u4e0e\\u9879\\u76ee\\u5b9e\\u621
8", "author": "spyinx"}}]'
```

在上述代码中，我们简化了原来的两层 for 循环代码，实现了和 serialize()方法一样的效果，即删掉了外键字段和多对多字段的处理语句，另外，还删掉了一些额外的判断。JOSN 序列化器的输出结果其实是由 stream 属性的 getvalue()方法得到的，这一点从该序列化器的 Serializer 类的 getvalue()方法的源码中可以看到：

```
# 源码位置：django/core/serializers/json.py
# ……

class Serializer(PythonSerializer):
   # ……

   def getvalue(self):
      # 调用父类 PythonSerializer 的 getvalue()方法
      return super(PythonSerializer, self).getvalue()

# ……

# 源码位置：django/core/serializers/base.py
# ……

class Serializer:
   # ……

   def getvalue(self):
      if callable(getattr(self.stream, 'getvalue', None)):
         return self.stream.getvalue()
```

```
# ……
```

默认的 stream 是一个 StringIO 对象，它的使用非常简单：使用 write()方法写入字符串，然后用 getvalue()方法输出全部写入的值。关于该对象的简单操作示例如下：

```
>>> from io import StringIO
>>> stream = StringIO()
>>> stream.write('[')
1
>>> stream.write('你好，地球骑士')
7
>>> stream.write(']')
1
>>> stream.getvalue()
'[你好，地球骑士]'
```

从 JSON 序列化器定义的 start_serialization()和 end_serialization()方法中可以看到，serialize()方法在序列化之前和序列化之后分别调用 stream 属性的 write()方法写入一个“[”和“]”字符串，表示一个列表字符串的首尾。接着在循环 QuerySet 对象时，每处理一个模型对象，在处理的开始和结束位置就分别调用一次 start_object()方法和 end_object()方法。以下是在 json.py 文件中定义的 end_object()方法的实现源码：

```
# 源码位置：django/core/serializers/json.py
# ……

class Serializer(PythonSerializer):
    # ……

    def end_object(self, obj):
        # 在该对象的_current 属性中保存了字段数据
        indent = self.options.get("indent")  # 默认为 None
        if not self.first:   # 如果 first 属性值为 False，则执行下面的语句
            self.stream.write(",") # 在第一次序列化模型对象时没有逗号和空字符串，在之后的处理中才有
            if not indent:
                self.stream.write(" ")  # 输出空字符串
        if indent:
            self.stream.write("\n")
        # 使用 json.dump()方法将 obj 值转成字符串，然后写入 self.stream 中
        json.dump(self.get_dump_object(obj), self.stream, **self.json_kwargs)
        self._current = None

    # ……

# ……
```

**注意**，对于 JSON 序列化器而言，使用的 start_object()方法并不是在 json.py 文件中定义的，而是在 python.py 文件中定义的。该方法的实现非常简单，仅仅是初始化该 Serializer 对象的_current 属

性为一个有序字典。

上述源码非常关键，在调用 end_object()方法之前会循环处理模型对象中的字段（调用 handle_field()方法处理字段数据），而 end_object()方法只是简单保存字段数据到_current 属性中。此外，这里用到了 first 属性值，其含义也非常简单，只有在处理第 1 个模型对象时为 True，其余时候为 False。因此在处理完第 2 个模型对象后，才会调用 stream 的 write()方法添加逗号。接着是调用该 Serializer 对象的 get_dump_object()方法将模型对象的结果转成字典形式，最后通过 json.dump()方法将该字典转成字符串形式写入 self.stream 中。get_dump_object()方法的实现源码位于 django/core/serializers/python.py 文件中，具体如下：

```
# 源码位置：django/core/serializers/python.py
# ……

class Serializer(PythonSerializer):
    # ……

    def get_dump_object(self, obj):
        data = OrderedDict([('model', str(obj._meta))])
        if not self.use_natural_primary_keys or not hasattr(obj, 'natural_key'):
            # 没有外键，设置 pk 主键值
            data["pk"] = self._value_from_field(obj, obj._meta.pk)
        # 在处理一次模型对象后，_current 中保存的 key-value 值正是当前模型对象的数据
        data['fields'] = self._current
        return data
    # ……

# ……
```

上面的结果非常清楚，已经有前面结果的影子了。首先对于每个模型对象，序列化的结果都是对一个字典结果进行 json.dump()操作。而字典结果有三个 key：pk、model 和 fields。下面以一个简单的模型对象演示上述结果，操作如下：

```
>>> from shell_test.models import DjangoBooks
>>> queryset = DjangoBooks.objects.all().filter(author='spyinx')
>>> obj = queryset[0]
>>> obj
<DjangoBooks: <Django 源码笔记, spyinx>>
>>> from collections import OrderedDict
>>> data = OrderedDict([('model', str(obj._meta))])
>>> from django.core.serializers.json import Serializer
>>> serializer = Serializer()
>>> concrete_model = obj._meta.concrete_model
# 模拟得到 Serializer 对象的_current 属性值
>>> current = {}
# 这里分别是 serialize()和 handle_field()方法中的语句
```

```
>>> for field in concrete_model._meta.local_fields:
...     current[field.name] = serializer._value_from_field(obj, field)
...
>>> current
{'id': 1, 'book_name': 'Django 源码笔记', 'author': 'spyinx', 'sex': 0, 'price': 109.0,
'isbn': '11111', 'publish_date': datetime.datetime(2021, 2, 4, 21, 42, 43, tzinfo=<UTC>)}
>>> not hasattr(obj, 'natural_key')
True
>>> data["pk"] = serializer._value_from_field(obj, obj._meta.pk)
>>> data['fields'] = current
# 模拟得到源码中的 data 值
>>> data
OrderedDict([('model', 'shell_test.djangobooks'), ('pk', 1), ('fields', {'id': 1,
'book_name': 'Django 源码笔记', 'author': 'spyinx', 'sex': 0, 'price': 109.0, 'isbn':
'11111', 'publish_date': datetime.datetime(2021, 2, 4, 21, 42,
43, tzinfo=<UTC>)})])
```

最后，测试 json.dump()语句的效果，如下：

```
# 接着上面得到的 data 值
>>> from io import StringIO
>>> stream = StringIO()
>>> stream.write('[')   # 模拟 start_serizalization()方法
1
>>> from django.core.serializers.json import DjangoJSONEncoder
>>> json_kwargs = {}
>>> json_kwargs['cls'] = DjangoJSONEncoder
>>> import json
>>> json.dump(data, stream, **json_kwargs)
# 这是第 1 个序列化的结果，后续的序列化的对象结果需要加一个逗号与上一个数据分隔开
>>> stream.getvalue()
'[{"model": "shell_test.djangobooks", "pk": 1, "fields": {"id": 1, "book_name":
"Django\\u6e90\\u7801\\u7b14\\u8bb0", "author": "spyinx", "sex": 0, "price": 109.0,
"isbn": "11111", "publish_date": "2021-02-04T21:42:43Z"}}'
>>> stream.write(']')  # 模拟 end_serizalization()方法
>>> stream.getvalue()
'[{"model": "shell_test.djangobooks", "pk": 1, "fields": {"id": 1, "book_name":
"Django\\u6e90\\u7801\\u7b14\\u8bb0", "author": "spyinx", "sex": 0, "price": 109.0,
"isbn": "11111", "publish_date": "2021-02-04T21:42:43Z"}}]'
```

上面的代码一步步展示了 serialize()方法的实现全过程。如果单独使用 json.dump()方法将字典数据转成字符串形式，则将抛出一个异常，因为在正常情况下，时间数据是无法使用 JSON 序列化的，具体如下：

```
>>> json.dump(data, stream)
Traceback (most recent call last):
 File "<console>", line 1, in <module>
 File "/root/.pyenv/versions/3.8.6/lib/python3.8/json/__init__.py", line 179, in dump
```

```
    for chunk in iterable:
  File "/root/.pyenv/versions/3.8.6/lib/python3.8/json/encoder.py", line 431, in
_iterencode
    yield from _iterencode_dict(o, _current_indent_level)
  File "/root/.pyenv/versions/3.8.6/lib/python3.8/json/encoder.py", line 405, in
_iterencode_dict
    yield from chunks
  File "/root/.pyenv/versions/3.8.6/lib/python3.8/json/encoder.py", line 405, in
_iterencode_dict
    yield from chunks
  File "/root/.pyenv/versions/3.8.6/lib/python3.8/json/encoder.py", line 438, in
_iterencode
    o = _default(o)
  File "/root/.pyenv/versions/3.8.6/lib/python3.8/json/encoder.py", line 179, in default
    raise TypeError(f'Object of type {o.__class__.__name__} '
TypeError: Object of type datetime is not JSON serializable
```

因此，Django 在 json.py 文件中专门实现了一个 DjangoJSONEncoder 类用于序列化时间数据，对于该方法的实现源码本书不再介绍，有兴趣的读者可以自行学习。至此，JSON 序列化器的底层逻辑就分析完毕了。下面解决中文乱码的问题，请看如下示例语句：

```
>>> import json
>>> data = {}
>>> data['hello'] = '你好'
>>> data['x'] = 'xxx'
>>> import sys
>>> json.dump(data, sys.stdout)
{"hello": "\u4f60\u597d", "x": "xxx"}>>>
```

中文乱码的问题非常容易解决，只需在 dump()方法中加入参数 ensure_ascii=False 即可：

```
>>> json.dump(data, sys.stdout, ensure_ascii=False)
{"hello": "你好", "x": "xxx"}>>>
```

因此，如果想让 JSON 序列化器的序列化结果正常显示中文，只需想办法在 Serializer 类的 json_kwargs 属性中加上参数 ensure_ascii=False 即可，操作如下：

```
>>> from django.core import serializers
>>> from shell_test.models import DjangoBooks
>>> queryset = DjangoBooks.objects.all().filter(author='spyinx')
# 注意，在 Django3 中，笔者发现 ensure_ascii=False 已经被默认设置了，所以不会有中文乱码问题
>>> data = serializers.serialize("json", queryset, ensure_ascii=False)
>>> data
'[{"model": "shell_test.djangobooks", "pk": 1, "fields": {"book_name": "Django 源码笔记
", "author": "spyinx", "sex": 0, "price": 109.0, "isbn": "11111", "publish_date":
"2021-02-04T21:42:43Z"}}, {"model": "shell_test.djangobooks", "pk": 2, "fields":
{"book_name": "Ansible 核心源码剖析与项目实战", "author": "spyinx", "sex": 0, "price": 119.0,
"isbn": "22222", "publish_date": "2021-02-04T21:43:14Z"}}]'
>>>
```

上述解决问题的思路非常简单，将 ensure_ascii=False 作为位置参数传给 options 变量，而 options 变量又会被进一步传到对应 Serializer 对象的 serialize()方法中。serialize()方法的第一行语句就是将 options 变量赋给该 Serializer 对象的 options 属性：

```
# 源码位置：django/core/serializers/base.py
# ……

class Serializer:
   # ……

   def serialize(self, queryset, *, stream=None, fields=None,
use_natural_foreign_keys=False, use_natural_primary_keys=False, progress_output=None,
object_count=0, **options):

      self.options = options
      # ……

# ……
```

注意，JSON 序列化器的 Serializer 类中的 start_serialization()方法会调用_init_options()方法。在_init_options()方法中，会将 Serializer 对象的 options 属性值拷贝一份到 json_kwargs 属性值中，而 json_kwargs 属性值正是前面使用 json.dump()序列化数据时传入的额外参数。因此，最终在 serializers.serialize()方法中传入的 ensure_ascii=False 参数会被传入对应 Serializer 对象的 json_kwargs 属性值中，从而使 json.dump()输出的中文可以正常显示。

## 5.5.3 简单分析 Python 序列化器的输出结果

根据前文的分析可知，序列化的核心方法是 end_object()方法，在 Python 序列化器中，该方法的实现源码如下：

```
# 源码位置：django/core/serializers/python.py
# ……

class Serializer:
   # ……

   def end_object(self, obj):
      # 将得到的字典结果添加到 self.objects 中
      self.objects.append(self.get_dump_object(obj))
      self._current = None

   def get_dump_object(self, obj):
      data = OrderedDict([('model', str(obj._meta))])
      if not self.use_natural_primary_keys or not hasattr(obj, 'natural_key'):
         data["pk"] = self._value_from_field(obj, obj._meta.pk)
```

```
        data['fields'] = self._current
        return data

    # ……

    def getvalue(self):
        # 最后返回 self.objects 的值
        return self.objects

# ……
```

从上面的源码可以看到，Python 序列化器在将模型对象转成字典数据（在 JSON 序列化器中分析过）后，直接将该数据追加到对应 Serializer 对象的 objects 属性值中。该属性值是一个列表，保存着查询集对应的所有模型对象的字典形式，最后序列化的结果即返回该属性值。

## 5.6 文件模块

本节主要剖析 django/core/file 目录下的源码，这部分源码主要涉及 Django 内置的文件模块，包括对文件处理的二次封装、文件操作等。

### 5.6.1 uploadedfile.py 文件

在视图层的文件上传流程中，便用到了在 uploadedfile.py 文件中定义的相关类——Mixin 类：

```
# 源码位置：django/core/files/utils.py
# ……

class FileProxyMixin:
    """
    Mixin 类的使用示例：

        class FileProxy(FileProxyMixin):
            def __init__(self, file):
                self.file = file
    """

    encoding = property(lambda self: self.file.encoding)
    # 各种属性
    # ……

    @property
    def closed(self):
        # 检查是否关闭
        return not self.file or self.file.closed
```

```
    def readable(self):
        # 检查是否可读
        if self.closed:
            return False
        if hasattr(self.file, 'readable'):
            return self.file.readable()
        return True

    def writable(self):
        # 检查是否可写
        if self.closed:
            return False
        if hasattr(self.file, 'writable'):
            return self.file.writable()
        return 'w' in getattr(self.file, 'mode', '')

    def seekable(self):
        if self.closed:
            return False
        if hasattr(self.file, 'seekable'):
            return self.file.seekable()
        return True

    def __iter__(self):
        return iter(self.file)
```

FileProxyMixin 类是一个 Mixin 类，在视图层中曾介绍过该类型的类，它用于实现一个特定功能。在该类的源码中曾给出了相关的使用示例，继承该 Mixin 类的必须要有 file 属性，才能执行该 Mixin 类中的各种操作，比如检查文件是否关闭（closed）、检查文件是否可读（readable）等。

### File 类源码解析

下面学习继承该 Mixin 类的一个重要类——File 类：

```
# 源码位置：django/core/files/base.py
# ……

class File(FileProxyMixin):
    # 默认的分片大小为 64KB
    DEFAULT_CHUNK_SIZE = 64 * 2 ** 10

    def __init__(self, file, name=None):
        self.file = file
        if name is None:
            # 如果没有传入文件名，则从 file 属性中获取
            name = getattr(file, 'name', None)
```

```
        self.name = name
        if hasattr(file, 'mode'):
            # 获取权限
            self.mode = file.mode

    def __str__(self):
        return self.name or ''

    def __repr__(self):
        return "<%s: %s>" % (self.__class__.__name__, self or "None")

    def __bool__(self):
        return bool(self.name)

    def __len__(self):
        return self.size

    @cached_property
    def size(self):
        if hasattr(self.file, 'size'):
            # 如果有 size 属性，则直接返回该属性值
            return self.file.size
        if hasattr(self.file, 'name'):
            try:
                # 通过 os 模块获取文件大小，如果成功，则直接返回
                return os.path.getsize(self.file.name)
            except (OSError, TypeError):
                pass
        # 另一种获取文件大小的方式
        if hasattr(self.file, 'tell') and hasattr(self.file, 'seek'):
            pos = self.file.tell()
            self.file.seek(0, os.SEEK_END)
            size = self.file.tell()
            self.file.seek(pos)
            return size
        # 如果无法获取文件大小，则抛出异常
        raise AttributeError("Unable to determine the file's size.")

    def chunks(self, chunk_size=None):
        """
        读取文件内容，每次调用都返回 chunk_size 字节内容
        """
        chunk_size = chunk_size or self.DEFAULT_CHUNK_SIZE  # 分片大小
        try:
            self.seek(0)  # 文件指针指向内容的最开始处
        except (AttributeError, UnsupportedOperation):
            pass
```

```
    while True:
        data = self.read(chunk_size)  # 每次都读取 chunk_size 大小的内容
        if not data:
            break   # 如果读不到任何内容，直接跳出循环
        yield data

def multiple_chunks(self, chunk_size=None):
    """
    如果有多个分片，则返回 True，只需判断该文件大小与分片大小即可
    """
    return self.size > (chunk_size or self.DEFAULT_CHUNK_SIZE)

def __iter__(self):
    # 按行迭代类似文件的对象
    buffer_ = None
    for chunk in self.chunks():
        # chunk 读的是分片内容，默认每片大小是 64KB
        for line in chunk.splitlines(True):
            if buffer_:
                if endswith_cr(buffer_) and not equals_lf(line):
                    yield buffer_
                else:
                    line = buffer_ + line
                # buffer_已处理，清掉它
                buffer_ = None

            if endswith_lf(line):
                yield line
            else:
                buffer_ = line

    if buffer_ is not None:
        yield buffer_

# 以下两个魔法函数用于支持 File 类的 with 语句
def __enter__(self):
    return self

def __exit__(self, exc_type, exc_value, tb):
    self.close()

def open(self, mode=None):
    # 其实就是封装了 Python 中的 open()函数，用于打开文件句柄
    if not self.closed:
        self.seek(0)
    elif self.name and os.path.exists(self.name):
        self.file = open(self.name, mode or self.mode)
    else:
```

```
            raise ValueError("The file cannot be reopened.")
        return self

    def close(self):
        self.file.close()

# ……
```

上面是 File 类的核心语句，它集成了 FileProxyMixin 类，所以有 file 属性。而 file 属性通常是 Python 中的 open()函数打开的一个文件句柄。通过 file 属性值，以及 FileProxyMixin 类源码，可知该 File 对象有非常多的属性，操作示例如下：

```
(django2-core-test) [root@master first_django]# cat test.txt
dajdalj
这是一个测试文本
来自地球的沈奇才先生

宇宙万物，地球万岁
all in china
(django2-core-test) [root@master first_django]# python manage.py shell
Python 3.8.6 (default, Oct 18 2020, 15:33:08)
[GCC 4.8.5 20150623 (Red Hat 4.8.5-39)] on linux
Type "help", "copyright", "credits" or "license" for more information.
(InteractiveConsole)
>>> from django.core.files.base import File
>>> fp = open('/root/django-core-test/first_django/test.txt', 'r+')
# 以下属性都来自 FileProxyMixin 类中的定义
>>> f = File(fp)
>>> f.encoding
'UTF-8'
>>> f.fileno()  # fileno 是 fp 中的 fileno 属性，该属性是一个函数对象，用于获取打开的文件描述符
3
>>> f.closed
False
>>> f.readable()
True
>>> f.writable()
True
>>> f.seekable()
True
>>> f.read()    # 在 FileProxyMixin 类中获取 file 属性中的 read 属性值
'dajdalj\n 这是一个测试文本\n 来自地球的沈奇才先生\n\n 宇宙万物，地球万岁\nall in china\n'
```

接着，调用在 File 对象中定义的相关方法，示例如下：

```
>>> print(f)    # 调用魔法函数__str__()
/root/django-core-test/first_django/test.txt
>>> f       # 调用魔法函数__repr__()
```

```
<File: /root/django-core-test/first_django/test.txt>
>>> bool(f)        # 调用魔法函数__bool__()
True
>>> len(f)         # 调用魔法函数__len__()
106
>>> f.size
106
>>> f.close()      # 关闭文件
>>> f.closed       # 检查文件句柄是否关闭
True
>>> f.fileno()     # 在关闭状态下查看文件描述符，抛出异常
Traceback (most recent call last):
  File "<console>", line 1, in <module>
ValueError: I/O operation on closed file
>>> f.open()       # 调用 open()方法打开文件
<File: /root/django-core-test/first_django/test.txt>
>>> f.closed       # 再次检查关闭状态
False
>>> f.fileno()
3
# 通过分片读取文本内容，这里指定每次读取 10 字节内容
>>> for chunk in f.chunks(10):
...     print('本次读取切片内容:\n{}'.format(chunk))
...
本次读取切片内容:
dajdalj
这是
本次读取切片内容:
一个测试文本
来自地
本次读取切片内容:
球的沈奇才先生

宇
本次读取切片内容:
宙万物，地球万岁
a
本次读取切片内容:
ll in chin
本次读取切片内容:
a

>>> f.seek(0)    # 经上述操作后，文件指针已经指向文件内容的末尾，需要设置指针到开头
0
>>> data = {}
# 在上述演示中，用 print()方法打印了换行符，这里先将内容写到字典上再打印
>>> for chunk in f.chunks(10):
...     data['10 个字符'] = chunk
```

```
...    print(data, len(chunk))
...
{'10 个字符': 'dajdalj\n 这是'} 10
{'10 个字符': '一个测试文本\n 来自地'} 10
{'10 个字符': '球的沈奇才先生\n\n 宇'} 10
{'10 个字符': '宙万物, 地球万岁\na'} 10
{'10 个字符': 'll in chin'} 10
{'10 个字符': 'a\n'} 2
>>> f.seek(0)
0
>>> f = File(fp)
>>> for line in f:    # 调用魔法函数__iter__(), 每次迭代获取文件中的一行
...    print('本行内容:{}'.format(line))
...
本行内容:dajdalj

本行内容:这是一个测试文本

本行内容:来自地球的沈奇才先生

本行内容:

本行内容:宇宙万物, 地球万岁

本行内容:all in china

>>>
```

### ContentFile 类源码解析

经上述操作后，读者是否已经完全掌握了在 Django 中定义的这个 File 类呢？接下来继承操作该 File 类的子类 ContentFile：

```
# 源码位置: django/core/files/base.py
# ……

class ContentFile(File):

   def __init__(self, content, name=None):
      stream_class = StringIO if isinstance(content, str) else BytesIO  #
      super().__init__(stream_class(content), name=name)
      self.size = len(content) # 直接获取 content 的大小

   def __str__(self):
      return 'Raw content'

   def __bool__(self):
```

```
        return True

    def open(self, mode=None):
        self.seek(0)
        return self

    def close(self):
        pass

    def write(self, data):
        self.__dict__.pop('size', None)  # 清除计算的大小值
        return self.file.write(data)
```

从 ContentFile 类的源码可知，当实例化该类时，需要输入文件的内容，而不是输入文件的句柄。

注意，在调用父类的初始化方法时，是使用 stream_class(content)作为 file 参数的，而 stream_class 为 StringIO 类或者 BytesIO 类。read()和 write()方法的主要作用是在内存缓冲区中进行读写，示例如下：

```
[GCC 4.8.5 20150623 (Red Hat 4.8.5-39)] on linux
Type "help", "copyright", "credits" or "license" for more information.
>>> from django.core.files.base import ContentFile
>>> fp = ContentFile('奇才，你好')
>>> fp.read()
'奇才，你好'
>>> fp.file.getvalue()   # file 的 file 属性其实是一个 StringIO 对象，所以有 getvalue()方法
'奇才，你好'
>>> fp.write('。感谢你拯救了地球')  # 指针已经指向字符串末尾
9
>>> fp.read()
''
>>> fp.seek(0)
0
>>> fp.read()
'奇才，你好。感谢你拯救了地球'
>>>
```

## 5.6.2 images.py 文件

django/core/files/images.py 文件中的 ImageFile 类同样继承自 File 类，其源码实现如下：

```
# 源码位置：django/core/files/images.py
# ……

class ImageFile(File):
    """
    一个 Mixin 类，提供了处理图片文件的额外功能
    """
    @property
```

```
    def width(self):
        return self._get_image_dimensions()[0]

    @property
    def height(self):
        return self._get_image_dimensions()[1]

    def _get_image_dimensions(self):
        if not hasattr(self, '_dimensions_cache'):
            close = self.closed
            # 打开文件句柄
            self.open()
            # 获取图片的宽和高
            self._dimensions_cache = get_image_dimensions(self, close=close)
        return self._dimensions_cache


def get_image_dimensions(file_or_path, close=False):
    """
    返回给定路径图片的宽和高
    """
    from PIL import ImageFile as PillowImageFile  # 使用第三方模块

    p = PillowImageFile.Parser()
    if hasattr(file_or_path, 'read'):
        file = file_or_path
        file_pos = file.tell()
        file.seek(0)
    else:
        file = open(file_or_path, 'rb')
        close = True
    try:
        chunk_size = 1024
        while 1:
            data = file.read(chunk_size)
            if not data:
                break
            try:
                p.feed(data)
            except zlib.error as e:
                # 异常处理
                # ……
            except struct.error:
                pass
            except RuntimeError:
                pass
            if p.image:
                return p.image.size
```

```
            chunk_size *= 2
        return (None, None)
    finally:
        if close:
            file.close()
        else:
            file.seek(file_pos)
```

上面的代码非常简单，对于 ImageFile 类而言，其额外的特性是提供了 width()、height()和_get_image_dimensions()方法去获取图片文件的宽和高。而_get_image_dimensions()方法是通过调用get_image_dimensions()方法获取图片相关信息的。从 get_image_dimensions()方法的源码可知，它使用了 Python 的第三方图像处理库来操作图片文件。在 Python 3 中，对应的模块是 pillow 模块，需要额外安装。下面用一个名为 test.jpg 的图片在 Python 交互模式下演示 ImageFile 类的使用，操作示例如下：

```
(django2-core-test) [root@master ~]# pip install pillow -i
https://pypi.mirrors.ustc.edu.cn/simple/
# ……
(django2-core-test) [root@master first_django]# python
Python 3.8.6 (default, Oct 18 2020, 15:33:08)
[GCC 4.8.5 20150623 (Red Hat 4.8.5-39)] on linux
Type "help", "copyright", "credits" or "license" for more information.
>>> from django.core.files.images import get_image_dimensions
>>> get_image_dimensions('/root/django-core-test/first_django/test.jpg')
(1024, 768)
>>> from django.core.files.images import ImageFile
>>> f = open('/root/django-core-test/first_django/test.jpg', 'rb')
>>> imf = ImageFile(f)
>>> imf.width
1024
>>> imf.height
768
>>> imf._get_image_dimensions()
(1024, 768)
```

### 5.6.3 locks.py 文件

在 django/core/files/locks.py 文件中定义了对文件的加锁和解锁操作，另外，在该文件的注释中有对锁方法的使用示例，具体内容如下：

```
# 源码位置：django/core/files/locks.py
# ……

"""
辅助的文件锁
```

```
使用示例::

   >>> from django.core.files import locks
   >>> with open('./file', 'wb') as f:
      ...   locks.lock(f, locks.LOCK_EX)
      ...   f.write('Django')
"""
import os

__all__ = ('LOCK_EX', 'LOCK_SH', 'LOCK_NB', 'lock', 'unlock')


def _fd(f):
   """获取文件描述符，输入可以是文件或者是文件描述符本身"""
   return f.fileno() if hasattr(f, 'fileno') else f


if os.name == 'nt':
   # 在 Windows 系统下定义加锁函数和解锁函数及相应的锁标识
   # ……
else:
   try:
      import fcntl
      # 共享锁，所有进程都没有写访问权限，即使是加锁进程也没有。所有进程都有读访问权限
      LOCK_SH = fcntl.LOCK_SH
      # 非阻塞锁，如果指定此参数，那么函数不能获得文件锁就立即返回;
      # 如果不指定此参数，函数会等待获得文件锁
      LOCK_NB = fcntl.LOCK_NB
      # 排他锁，除加锁进程外，其他进程没有对已加锁文件的读或写访问权限
      LOCK_EX = fcntl.LOCK_EX
   except (ImportError, AttributeError):
      # 如果不支持文件锁，则自定义锁变量值为 0
      LOCK_EX = LOCK_SH = LOCK_NB = 0

      # 笨办法，返回 True 或 False，但什么也不做
      def lock(f, flags):
         # File is not locked
         return False

      def unlock(f):
         return True
   else:
      def lock(f, flags):
         ret = fcntl.flock(_fd(f), flags)  # 加锁操作
         return ret == 0  # 判断加锁是否成功

      def unlock(f):
         ret = fcntl.flock(_fd(f), fcntl.LOCK_UN)  # 解锁操作
```

```
            return ret == 0  # 判断的锁是否成功
```

在上述代码中，主要演示了加锁操作和解锁操作。此外，对于 LOCK_SH、LOCK_NB 和 LOCK_EX，在注释中也进行了相应的说明，此处不再深入介绍。

## 5.6.4 temp.py 文件

django/core/files/temp.py 文件的内容非常简单，对于 Linux 系统而言，只是给 tempfile 模块中的两个变量分别取一个别名，即 NamedTemporaryFile 和 gettempdir：

```
# 源码位置：django/core/files/temp.py

import os
import tempfile

from django.core.files.utils import FileProxyMixin

__all__ = ('NamedTemporaryFile', 'gettempdir',)


if os.name == 'nt':
   # 在 Windows 系统下定义 TemporaryFile 类，同时设置 NamedTemporaryFile 变量值
   # ……
else:
   NamedTemporaryFile = tempfile.NamedTemporaryFile

gettempdir = tempfile.gettempdir
```

下面简单演示一下 tempfile 模块中的 NamedTemporaryFile 类和 gettempdir()方法的使用示例，代码如下：

```
>>> import tempfile
>>> NamedTemporaryFile = tempfile.NamedTemporaryFile
>>> gettempdir = tempfile.gettempdir
>>> gettempdir()  # 获取临时文件目录
'/tmp'
>>> tf = NamedTemporaryFile()
>>> tf.name
'/tmp/tmp_5m1ofrg'
>>> tf.write(b'hello, world')
12
>>> tf.read()
b''
>>> tf.seek(0)
0
>>> tf.read()
b'hello, world'
```

```
>>> import os
>>> os.path.exists(tf.name)   # 临时文件存在
True
>>> tf.close()            # 在关闭临时文件句柄后，临时文件自动被删除
>>> os.path.exists(tf.name)
False
# 在加上 deletet=False 后，再次关闭临时文件句柄时就不会删除该临时文件了
>>> tf = NamedTemporaryFile(delete=False)
>>> tf.name
'/tmp/tmpjokybe9h'
>>> tf.close()
>>> os.path.exists(tf.name)
True
```

### 5.6.5　move.py 文件

在 django/core/files/move.py 文件中定义了两个函数：_samefile()和 file_move_safe()。前者表示比较输入的两个文件路径是否指向相同的文件，后者则在函数内部调用了前者，并以最安全的方式移动文件到另一个位置：

```
# 源码位置：django/core/files/move.py

"""
以最安全的方式移动文件：

   >>> from django.core.files.move import file_move_safe
   >>> file_move_safe("/tmp/old_file", "/tmp/new_file")
"""

import errno
import os
from shutil import copystat

from django.core.files import locks

__all__ = ['file_move_safe']


def _samefile(src, dst):
   # mac 系统和 UNIX 系统
   if hasattr(os.path, 'samefile'):
      try:
         # 调用 os.path.samefile()方法进行判断
         return os.path.samefile(src, dst)
      except OSError:
         return False
```

```
    # 其他平台： 检查文件路径是否相同
    return (os.path.normcase(os.path.abspath(src)) ==
            os.path.normcase(os.path.abspath(dst)))


def file_move_safe(old_file_name, new_file_name, chunk_size=1024 * 64,
allow_overwrite=False):

    # 如果是相同文件，无须移动
    if _samefile(old_file_name, new_file_name):
        return

    try:
        # 如果新地址不能覆盖或者不可访问，则直接抛出异常
        if not allow_overwrite and os.access(new_file_name, os.F_OK):
            raise IOError("Destination file %s exists and allow_overwrite is False" %
new_file_name)

        # 尝试调用 os.rename()方法
        os.rename(old_file_name, new_file_name)
        return
    except OSError:
        # 调用 os.rename()方法抛出 OSError 异常，可能是移动到另一个文件系统，
        # 或者是在特定系统上移动一个打开的文件
        pass

    # 首先，打开旧文件，确保它不会丢失
    with open(old_file_name, 'rb') as old_file:
        # 接着，打开新文件，确保允许覆盖，即指定相关标识
        fd = os.open(new_file_name, (os.O_WRONLY | os.O_CREAT | getattr(os, 'O_BINARY',
0) | (os.O_EXCL if not allow_overwrite else 0)))
        try:
            # 使用排它锁加锁
            locks.lock(fd, locks.LOCK_EX)
            current_chunk = None
            # 循环读取旧文件，并将旧文件中的内容并写入新文件中
            while current_chunk != b'':
                current_chunk = old_file.read(chunk_size)
                os.write(fd, current_chunk)
        finally:
            # 写完后解锁，并关闭新文件句柄
            locks.unlock(fd)
            os.close(fd)

    try:
        # 拷贝旧文件的元数据信息
        copystat(old_file_name, new_file_name)
```

```
    except PermissionError as e:
        if e.errno != errno.EPERM:
            raise

    try:
        # 删除旧文件
        os.remove(old_file_name)
    except PermissionError as e:
        if getattr(e, 'winerror', 0) != 32:
            raise
```

这两个函数的使用示例如下：

```
(django2-core-test) [root@master first_django]# ln -s test.txt link_file
(django2-core-test) [root@master first_django]# ls -l link_file
lrwxrwxrwx 1 root root 8 Mar 23 22:35 link_file -> test.txt
>>> from django.core.files.move import _samefile
# 比较不同的文件
>>> _samefile('/root/django-core-test/first_django/test.txt',
'/root/django-core-test/first_django/test.file')
False
# 比较相同的文件，只不过后者是链接到前者的链接文件
>>> _samefile('/root/django-core-test/first_django/test.txt',
'/root/django-core-test/first_django/link_file')
True
>>> import os
>>> os.path.exists('/root/test.txt')
False
>>> from django.core.files.move import file_move_safe
# 移动文件到/root 目录下
>>> file_move_safe('/root/django-core-test/first_django/test.txt', '/root/test.txt')
# 新文件路径存在
>>> os.path.exists('/root/test.txt')
True
# 旧文件路径已经不存在
>>> os.path.exists('/root/django-core-test/first_django/test.txt')
False
```

### 5.6.6　storage.py 文件

在 django/core/files/storage.py 文件中主要定义了一个基类 Storage 及其子类 FileSystemStorage。其源码实现非常简单，限于篇幅，不再展示其源码，而是通过实际调用演示其功能，具体操作如下：

```
>>> from django.core.files.storage import default_storage
>>> default_storage._wrapped
<object object at 0x7fa85c1490e0>
>>> default_storage.exists('test.txt')
True
```

```
>>> default_storage._wrapped
<django.core.files.storage.FileSystemStorage object at 0x7fa85451ceb0>
>>> default_storage.location
'/root/test'
>>> default_storage.listdir('/root/test')
(['django'], ['django_pm0lQkb', 'test.txt'])
>>> default_storage.open('/root/test/test.txt')
<File: /root/test/test.txt>
>>> default_storage.open('/root/test/test.txt').size
27
>>> default_storage.get_accessed_time('/root/test/test.txt')
datetime.datetime(2021, 4, 28, 16, 7, 4, 467722)
```

最后，在 django/core/files/uploadedfile.py 文件中将看到在 Django 框架的文件上传功能中涉及的两个文件类：TemporaryUploadedFile 和 InMemoryUploadedFile。它们都继承自 UploadedFile 类，而 UploadedFile 类继承自 File 类：

```
# 源码位置：django/core/files/uploadedfile.py
# ……

class UploadedFile(File):

    def __init__(self, file=None, name=None, content_type=None, size=None, charset=None,
content_type_extra=None):
        super().__init__(file, name)
        self.size = size        # 文件大小
        self.content_type = content_type
        self.charset = charset    # 文件编码
        self.content_type_extra = content_type_extra

    def __repr__(self):
        return "<%s: %s (%s)>" % (self.__class__.__name__, self.name, self.content_type)

    def _get_name(self):
        return self._name

    def _set_name(self, name):
        if name is not None:
            name = os.path.basename(name)

            # 当文件名称的长度超过 255 个字符时会出现问题
            if len(name) > 255:
                name, ext = os.path.splitext(name)  # 分割成文件名（name）和后缀（ext）
                ext = ext[:255]                     # 避免后缀长度超过 255 个字符
                name = name[:255 - len(ext)] + ext  # 确保加上后缀也不超过 255 个字符

        self._name = name
```

```
    name = property(_get_name, _set_name)


class TemporaryUploadedFile(UploadedFile):
    """
    上传的文件位于临时路径中
    """
    def __init__(self, name, content_type, size, charset, content_type_extra=None):
        _, ext = os.path.splitext(name)
        # 使用 tempfile 模块的 NamedTemporaryFile()方法创建一个临时文件
        file = tempfile.NamedTemporaryFile(suffix='.upload' + ext,
dir=settings.FILE_UPLOAD_TEMP_DIR)
        # 把临时文件 file 作为 file 属性初始化父类
        super().__init__(file, name, content_type, size, charset, content_type_extra)

    def temporary_file_path(self):
        """ 返回该文件的完整路径 """
        return self.file.name

    def close(self):
        try:
            return self.file.close()
        except FileNotFoundError:
            pass


class InMemoryUploadedFile(UploadedFile):
    """
    上传文件保存在内存中
    """
    def __init__(self, file, field_name, name, content_type, size, charset,
content_type_extra=None):
        super().__init__(file, name, content_type, size, charset, content_type_extra)
        self.field_name = field_name

    def open(self, mode=None):
        # 文件指针指向内容开头
        self.file.seek(0)
        return self

    def chunks(self, chunk_size=None):
        self.file.seek(0)  # 文件指针指向内容开头
        yield self.read()  # 读取内存中的全部数据

    def multiple_chunks(self, chunk_size=None):
        # 由于文件数据全部在内存中，所以永远不会有分片
        return False
```

上述类的源码实现与 File 类区别不大，只是零星重写了一些方法并增加了一些方法，但如果想要理解上述类的作用，则必须和对应处理的 Handler 类一起分析。

## 5.6.7 uploadhandler.py 文件

Handler 类的定义位于 django/core/files/uploadhandler.py 文件中，其源码实现如下：

```
# 源码位置：django/core/files/uploadhandler.py
# ……

class FileUploadHandler:
    """
    文件流上传处理器的基类
    """
    chunk_size = 64 * 2 ** 10  # 默认的分片大小是 64KB

    def __init__(self, request=None):
        self.file_name = None
        self.content_type = None
        self.content_length = None
        self.charset = None
        self.content_type_extra = None
        self.request = request

    def handle_raw_input(self, input_data, META, content_length, boundary,
encoding=None):
        """
        处理客户端的原始输入
        """
        pass

    def new_file(self, field_name, file_name, content_type, content_length, charset=None,
content_type_extra=None):
        """
        新文件开始上传的信号
        """
        self.field_name = field_name
        self.file_name = file_name
        self.content_type = content_type
        self.content_length = content_length
        self.charset = charset
        self.content_type_extra = content_type_extra

    def receive_data_chunk(self, raw_data, start):
        """
        接收数据流，start 表示分片文件的位置，需要用子类实现
```

```
        """
        raise NotImplementedError('...')

    def file_complete(self, file_size):
        """
        文件上传完成的信号，文件的大小即所有分片的真实累加值。
        子类需要返回一个有效的 UploadedFile 对象（也可以是 UploadedFile 对象的子类对象）
        """
        raise NotImplementedError('...')

    def upload_complete(self):
        """
        上传过程结束的信号，子类需要在此做一些收尾工作
        """
        pass


class TemporaryFileUploadHandler(FileUploadHandler):
    """
    上传文件处理器：将上传文件的数据流保存到临时文件中
    """
    def new_file(self, *args, **kwargs):
        super().new_file(*args, **kwargs)
        # 创建临时上传文件类对象
        self.file = TemporaryUploadedFile(self.file_name, self.content_type, 0, 
self.charset, self.content_type_extra)

    def receive_data_chunk(self, raw_data, start):
        # 调用 write()方法将上传的数据分片写入临时文件中
        self.file.write(raw_data)

    def file_complete(self, file_size):
        self.file.seek(0)       # 重置文件指针
        self.file.size = file_size  # 设置文件大小
        return self.file


class MemoryFileUploadHandler(FileUploadHandler):
    """
    上传文件处理器：将上传的数据流保存到内存中（适合对小文件的处理）
    """

    def handle_raw_input(self, input_data, META, content_length, boundary, 
encoding=None):
        # 如果上传文件超过设定值，则无法使用内存保存上传文件的方式
        self.activated = content_length <= settings.FILE_UPLOAD_MAX_MEMORY_SIZE

    def new_file(self, *args, **kwargs):
```

```
        super().new_file(*args, **kwargs)
        if self.activated:
            self.file = BytesIO()
            raise StopFutureHandlers()

    def receive_data_chunk(self, raw_data, start):
        if self.activated:
            # 将数据添加到内存中，self.file 为一个 BytesIO 对象，用来操作内存中数据的读写
            self.file.write(raw_data)
        else:
            return raw_data

    def file_complete(self, file_size):
        if not self.activated:
            # 如果 self.activated 为 False，则表示不符合用内存保存上传文件的方式
            return

        self.file.seek(0)
        return InMemoryUploadedFile(
            file=self.file,
            field_name=self.field_name,
            name=self.file_name,
            content_type=self.content_type,
            size=file_size,
            charset=self.charset,
            content_type_extra=self.content_type_extra
        )


def load_handler(path, *args, **kwargs):
    """
    根据路径选择上传文件的处理器,并返回该处理器的一个实例

    E.g.::
        >>> from django.http import HttpRequest
        >>> request = HttpRequest()
        >>> load_handler('django.core.files.uploadhandler.TemporaryFileUploadHandler',
request)
        <TemporaryFileUploadHandler object at 0x...>
    """
    return import_string(path)(*args, **kwargs)  # 使用前面熟悉的 import_string()方法
```

上面两个 Handler 类是 Django 处理上传文件的处理器，load_handler()方法是导入这两个类的一个对外接口。只需在 VSCode 中简单搜索一下就能知道是哪里调用了这里的上传文件处理器，搜索结果如图 5-3 所示。

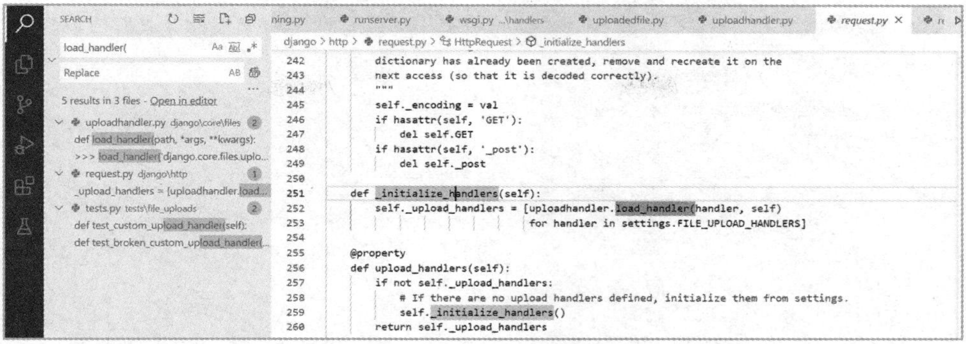

图 5-3

很明显，只有 resquest.py 文件中的语句符合追踪条件，test.py 文件中的语句是测试用的。调用该加载器（load_handler()）的方法如下：

```
# 源码位置：django/http/request.py
# ......

class HttpRequest:
    """一个基本的 HTTP 请求"""

    # ......

    def _initialize_handlers(self):
        # 载入上传文件处理器对象
        self._upload_handlers = [uploadhandler.load_handler(handler, self)
                                 for handler in settings.FILE_UPLOAD_HANDLERS]

    # ......
```

从上述源码可知，默认上传文件处理器是在 settings.FILE_UPLOAD_HANDLERS 中定义的，其值位于 django/conf/global_settings.py 文件中：

```
# 源码位置：django/conf/global_settings.py
# ......

# 上传处理器类的列表
FILE_UPLOAD_HANDLERS = [
    'django.core.files.uploadhandler.MemoryFileUploadHandler',
    'django.core.files.uploadhandler.TemporaryFileUploadHandler',
]

# ......
```

由此可知，_initialize_handlers()其实就是导入前面介绍的两个文件处理器。至于该文件处理器如何被调用，用到了哪些方法，则需要从 HttpRequest 对象的_upload_handlers 属性入手（内置的文件处理器被赋给该属性值）。查看 HttpRequest 类的源码，可以看到调用该属性（或者该属性的另一个变体 upload_handlers）的方法如下：

```
# 源码位置：django/http/request.py
# ……

class HttpRequest:
    # ……

    @property
    def upload_handlers(self):
        if not self._upload_handlers:
            # 如果该属性一开始为空，则调用_initialize_handlers()方法导入内置的文件处理器集
            self._initialize_handlers()
        return self._upload_handlers

    @upload_handlers.setter
    def upload_handlers(self, upload_handlers):
        if hasattr(self, '_files'):
            raise AttributeError("...")
        self._upload_handlers = upload_handlers

    def parse_file_upload(self, META, post_data):
        """返回(POST QueryDict, FILES MultiValueDict)这样的元组数据"""
        self.upload_handlers = ImmutableList(   # 进行类型转换，不可变列表
            self.upload_handlers,
            warning="You cannot alter upload handlers after the upload has been processed."
        )
        # 上传文件处理器集 self.upload_handlers，用于实例化 MultiPartParser 对象
        parser = MultiPartParser(META, post_data, self.upload_handlers, self.encoding)
        return parser.parse()

    # ……

# ……
```

在 HttpRequest 类中定义了一个属性方法 upload_handlers()，通过其源码可以看到，该方法返回的是_upload_handlers 属性值。如果_upload_handlers 属性值为空，就调用 self._initialize_handlers()方法给_upload_handlers 属性赋值后再返回。由此可知，self.upload_handlers 语句其实就是内置的上传文件处理器集：

```
(django2-core-test) [root@master first_django]# python manage.py shell
Python 3.8.6 (default, Oct 18 2020, 15:33:08)
[GCC 4.8.5 20150623 (Red Hat 4.8.5-39)] on linux
Type "help", "copyright", "credits" or "license" for more information.
(InteractiveConsole)
>>> from django.http.request import HttpRequest
```

```
>>> request = HttpRequest()
>>> request.upload_handlers
[<django.core.files.uploadhandler.MemoryFileUploadHandler object at 0x7f7d267a4400>,
<django.core.files.uploadhandler.TemporaryFileUploadHandler object at 0x7f7d267a46a0>]
```

从 parse_file_upload()方法的源码可知，self.upload_handlers 被用于实例化 MultiPartParser 类，最终返回 MultiPartParser 对象的 parse()方法。继续追踪 MultiPartParser 类的源码，如下：

```
# 源码位置：django/http/MultiPartParser
# ……

class MultiPartParser:

    def __init__(self, META, input_data, upload_handlers, encoding=None):
        # ……

        self._upload_handlers = upload_handlers

    def parse(self):
        # ……

        # 将文件处理器集赋给变量 handlers
        handlers = self._upload_handlers

        # ……

        # 1. 循环调用文件处理器的 handle_raw_input()方法
        for handler in handlers:
            result = handler.handle_raw_input(
                self._input_data,
                self._meta,
                self._content_length,
                self._boundary,
                encoding,
            )
            # 检查是否已经处理过。如果已经处理过，则直接返回结果，默认的处理器均返回 None
            if result is not None:
                return result[0], result[1]

        # 最后该方法会返回这两个属性值
        self._post = QueryDict(mutable=True)  # 请求上传的数据
        self._files = MultiValueDict()        # 上传的文件列表

        # ……

        try:
            for item_type, meta_data, field_stream in Parser(stream, self._boundary):
                if old_field_name:
```

```
        # 2. 循环调用文件处理器的 file_complete()方法，更新 self._files 属性值
        self.handle_file_complete(old_field_name, counters)
        old_field_name = None

    try:
        disposition = meta_data['content-disposition'][1]
        # 对于上传文件来说，就是获取上传文件名
        field_name = disposition['name'].strip()
    except (KeyError, IndexError, AttributeError):
        continue

    # ……

    if item_type == FIELD:
        # ……
    elif item_type == FILE:
        # 处理文件上传
        file_name = disposition.get('filename')
        # ……

        try:
            # 3. 循环调用文件处理器的 new_file()方法，创建相应的文件对象
            for handler in handlers:
                try:
                    handler.new_file(
                        field_name, file_name, content_type,
                        content_length, charset, content_type_extra,
                    )
                except StopFutureHandlers:
                    break

            for chunk in field_stream:
                # ……

                # 4. 循环调用文件处理器的 receive_data_chunk()方法
                for i, handler in enumerate(handlers):
                    chunk_length = len(chunk)
                    chunk = handler.receive_data_chunk(chunk, counters[i])
                    # 在 counters[i]中记录了前 i 次写入的总字节数
                    counters[i] += chunk_length
                    if chunk is None:
                        # 当文件处理器收到的数据为 None 时，跳出循环，处理完毕
                        break

        except SkipFile:
            self._close_files()
            exhaust(field_stream)
        else:
            # Handle file upload completions on next iteration.
            # 处理完的文件名
```

```
                old_field_name = field_name

            else:
                # 既不是FIELD，也不是FILE，耗尽数据流（对于空读不用处理，一直到读完为止）
                exhaust(stream)

    except StopUpload as e:
        self._close_files()
        if not e.connection_reset:
            exhaust(self._input_data)
    else:
        # 确保请求数据已经全部输入
        exhaust(self._input_data)

    # 5. 调用文件处理器的upload_complete()方法
    any(handler.upload_complete() for handler in handlers)
    self._post._mutable = False
    return self._post, self._files
```

MultiPartParser 类中的内容较多，这里对 parse()方法进行了简化，省略了许多处理细节。此外，我们在 parse()方法中标注了 5 处调用文件处理器的方法，它们的含义如下。

（1）handle_raw_input()：处理原始的数据流。对于默认的两个文件处理器而言，在它们的 handle_raw_input()方法中均没有返回值，所以得到的 result 为 None。

（2）file_complete()：在一个文件流处理完毕后再调用文件处理器的 file_complete()方法，返回文件处理器对应的上传文件对象（MemoryFileUploadHandler 处理器对应 InMemoryUploadedFile 对象，TemporaryFileUploadHandler 处理器对应 TemporaryUploadedFile 对象）。该方法是在 self.handle_file_complete()方法中被调用的。每处理完一个上传文件，都会将其添加到该 MultiPartParser 对象的_files 属性值中：

```
# 源码位置：django/http/MultiPartParser
# ……

class MultiPartParser:

    # ……

    def handle_file_complete(self, old_field_name, counters):
        for i, handler in enumerate(self._upload_handlers):
            # 两个文件处理器会返回对应的上传文件对象，
            # 即TemporaryUploadedFile对象和InMemoryUploadedFile对象
            file_obj = handler.file_complete(counters[i])
            if file_obj:
                # 对于相同的文件，只添加一次
                self._files.appendlist(force_text(old_field_name, self._encoding,
errors='replace'), file_obj)
```

```
            break

    # ……
```

(3) new_file()：开始接收一个新文件。在上传文件处理器的代码中，首先初始化相关的属性值，其中最核心的是初始化 file 属性：

```
# 源码位置：django/core/files/uploadhandler.py
# ……

class FileUploadHandler:
    # ……

    def new_file(self, field_name, file_name, content_type, content_length, charset=None,
content_type_extra=None):

        self.field_name = field_name
        self.file_name = file_name
        self.content_type = content_type
        self.content_length = content_length
        self.charset = charset
        self.content_type_extra = content_type_extra

    # ……

class TemporaryFileUploadHandler(FileUploadHandler):
    # ……

    def new_file(self, *args, **kwargs):
        super().new_file(*args, **kwargs)
        # 创建文件对象，当文件数据传过来时将其添加到该对象中
        self.file = TemporaryUploadedFile(self.file_name, self.content_type, 0,
self.charset, self.content_type_extra)

    # ……

class MemoryFileUploadHandler(FileUploadHandler):
    # ……

    def new_file(self, *args, **kwargs):
        super().new_file(*args, **kwargs)
        if self.activated:
            # 初始化 file 属性为 BytesIO 对象
            self.file = BytesIO()
            raise StopFutureHandlers()

    # ……
```

（4）receive_data_chunk()：循环接收数据流中的数据，如果直接接收的 chunk 为 None，则退出该循环。这里是获取文件数据的关键方法，各 Handler 类中的 receive_data_chunk()方法的源码如下：

```
# 源码位置：django/core/files/uploadhandler.py
# ……

class TemporaryFileUploadHandler(FileUploadHandler):
    # ……

    def receive_data_chunk(self, raw_data, start):
        # self.file 为 TemporaryUploadedFile 对象，即把数据写入临时文件中，start 为写入的开始位置
        self.file.write(raw_data)

    # ……

class MemoryFileUploadHandler(FileUploadHandler):

    # ……

    def receive_data_chunk(self, raw_data, start):
        """Add the data to the BytesIO file."""
        if self.activated:
            # 把数据写入内存中，由于 self.file 为 BytesIO 对象，所以不需要 start 值
            self.file.write(raw_data)
        else:
            return raw_data

    # ……
```

对于临时文件对象，由于要写入临时文件中，所以需要记录前面写入的总字节数，即需要 start 值。如果是写入 BytesIO 对象中，则不需要 start 值，直接添加数据到 BytesIO 对象中即可。

（5）upload_complete()：表示上传过程结束，内置的 handlers 对上传结束的方法没有做任何处理。

至此，对 django/core/files 目录下的源码分析就结束了。本节简单追踪了在文件上传过程中经过文件处理器的过程及相关的类方法，并没有深入分析文件上传功能的细节，有兴趣的读者可以自行分析。

## 5.7　发送邮件

为了能理解 Django 内置的邮件系统，下面先展示一段 Python 发送邮件的代码，内容如下：

```
(django2-core-test) [root@master first_django]# cat test_email.py
"""
发送邮件测试
"""
import smtplib
```

```
from email.mime.text import MIMEText
from email.header import Header


def send_email(msg_content):
    mail_from = '2894577759@qq.com'  # 发送方邮箱
    passwd = '你的登录密码'          # 在 QQ 邮箱中设置开启 POP3/SMTP 服务

    msg = MIMEText(msg_content, 'plain', 'utf-8')

    #邮件主题
    msg['Subject'] = "这是一封来自奇才的邀请信!"
    msg['From'] = mail_from

    try:
        s = smtplib.SMTP_SSL("smtp.qq.cn", 465)
        # 登录邮箱
        s.login(mail_from, passwd)
        # 发送邮件：发送方，收件方，要发送的消息
        s.sendmail(mail_from, ['2894577759@qq.com'], msg.as_string())
        print("发送邮件成功")
    except s.SMTPException as e:
        print("发送邮件失败:{}，请检查".format(str(e)))
    finally:
        s.quit()

send_email('欢迎有才之士加盟媒体存储小组')
```

在上述代码中，passwd 变量是在对应邮箱中设置开启 POP3/SMTP 服务时对应得到的一个密码，用于登录邮箱后发送邮件。例如，在 QQ 邮箱中设置开启 POP3/SMTP 服务，如图 5-4 所示。

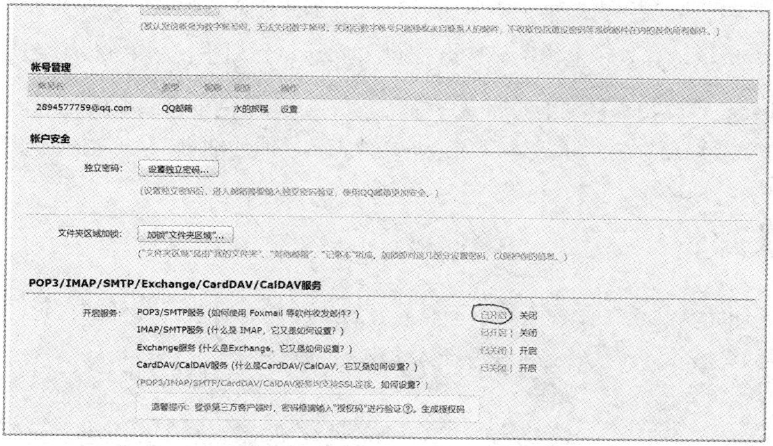

图 5-4

在 Linux 系统中执行该脚本，几秒后就可以收到一封自己发来的邮件了，如图 5-5 所示。

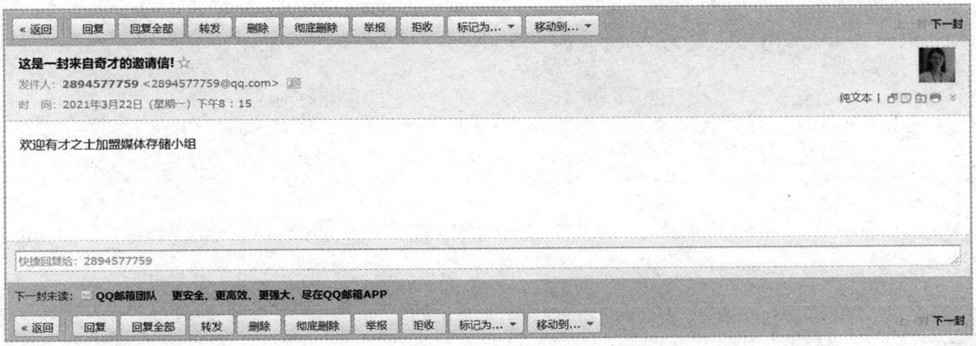

图 5-5

接下来演示 Django 中的邮件系统，通过调用 Django 代码来完成相同的邮件实验：

```
[GCC 4.8.5 20150623 (Red Hat 4.8.5-39)] on linux
Type "help", "copyright", "credits" or "license" for more information.
(InteractiveConsole)
>>> from django.core.mail import get_connection, send_mail
>>> connection = get_connection(username='2894577759@qq.com', password='密码',
host='smtp.qq.com')
>>> send_mail('这是一封来自奇才的邀请信(Django 的邮件系统)!', '欢迎有才之士加盟媒体存储小组',
'2894577759@qq.com', ['2894577759@qq.com'], connection=connection)
1
>>>
```

在执行完上述操作后，就可以在 QQ 邮箱中收到调用 Django 代码发送过来的邮件了，如图 5-6 所示。

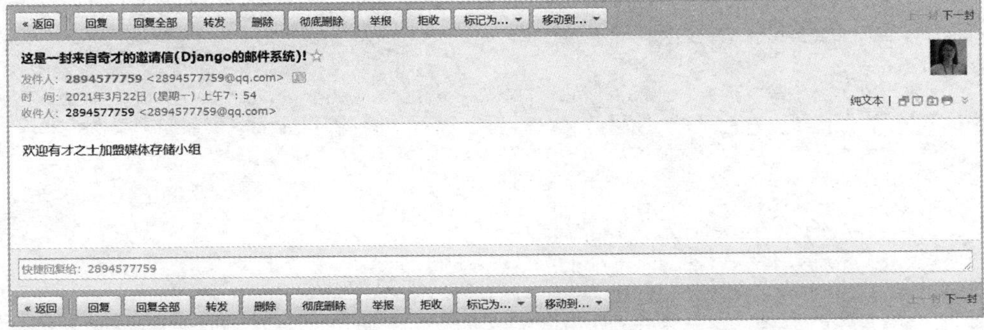


图 5-6

接下来从 Django 的 get_connection()方法和 send_mail()方法入手，探究 Django 中邮件功能的具体实现源码：

```
# 源码位置：django/core/mail/__init__.py
# ……

def get_connection(backend=None, fail_silently=False, **kwds):

    klass = import_string(backend or settings.EMAIL_BACKEND)
    return klass(fail_silently=fail_silently, **kwds)

# ……
```

这里又出现了全局的配置变量：EMAIL_BACKEND。该变量的默认值同样可以从 django/conf/global_settings.py 文件中找到，定义如下：

```
# 源码位置：django/conf/global_settings.py
# ……

EMAIL_BACKEND = 'django.core.mail.backends.smtp.EmailBackend'

# ……
```

由此可知，get_connection()方法只是根据输入的 backend 值（如果没有传入，就取默认的全局变量 settings.EMAIL_BACKEND）导入对应的邮件系统类，最后返回该类的一个对象值。

```
>>> from django.core.mail import get_connection
>>> get_connection()
<django.core.mail.backends.smtp.EmailBackend object at 0x7fae951bc4f0>
```

为了节省篇幅，这里只分析默认的邮件系统，其余邮件系统分析方法基本与之相同：

```
# 源码位置：django/core/mail/backends/smtp.py
# ……

class EmailBackend(BaseEmailBackend):
    """
    管理 SMTP 网络连接的一个封装类
    """
    def __init__(self, host=None, port=None, username=None, password=None,
                 use_tls=None, fail_silently=False, use_ssl=None, timeout=None,
                 ssl_keyfile=None, ssl_certfile=None,
                 **kwargs):
        super().__init__(fail_silently=fail_silently)
        # 关于邮件的很多参数都可以在项目的 settings.py 文件中设置
        self.host = host or settings.EMAIL_HOST
        self.port = port or settings.EMAIL_PORT
        self.username = settings.EMAIL_HOST_USER if username is None else username
        self.password = settings.EMAIL_HOST_PASSWORD if password is None else password
        self.use_tls = settings.EMAIL_USE_TLS if use_tls is None else use_tls
        self.use_ssl = settings.EMAIL_USE_SSL if use_ssl is None else use_ssl
        self.timeout = settings.EMAIL_TIMEOUT if timeout is None else timeout
```

```
        self.ssl_keyfile = settings.EMAIL_SSL_KEYFILE if ssl_keyfile is None else
ssl_keyfile
        self.ssl_certfile = settings.EMAIL_SSL_CERTFILE if ssl_certfile is None else
ssl_certfile
        if self.use_ssl and self.use_tls:
            raise ValueError(
                "EMAIL_USE_TLS/EMAIL_USE_SSL are mutually exclusive, so only set "
                "one of those settings to True.")
        # 核心属性
        self.connection = None
        self._lock = threading.RLock()  # 多重锁

    @property
    def connection_class(self):
        # 关键语句，默认的邮件服务都是 https 访问，所以通常得到 smtplib.SMTP_SSL 类
        return smtplib.SMTP_SSL if self.use_ssl else smtplib.SMTP

    def open(self):

        if self.connection:
            # 如果连接已经打开，直接返回
            return False

        # 如果本地名称没有确定，则通过 socket.getfqdn()语句获取
        # 为了性能，Django 会将上述数据缓存起来
        connection_params = {'local_hostname': DNS_NAME.get_fqdn()}
        if self.timeout is not None:
            connection_params['timeout'] = self.timeout
        if self.use_ssl:
            connection_params.update({
                'keyfile': self.ssl_keyfile,
                'certfile': self.ssl_certfile,
            })
        # 得到一些参数
        try:
            # 核心语句，这里的 self.host 是 SMTP 服务地址，self.port 是该服务开放的端口
            self.connection = self.connection_class(self.host, self.port,
**connection_params)

            if not self.use_ssl and self.use_tls:
                self.connection.starttls(keyfile=self.ssl_keyfile,
certfile=self.ssl_certfile)

            if self.username and self.password:
                # 登录动作，参考前文 Python 代码实现的邮件发送
                self.connection.login(self.username, self.password)
            return True
        except (smtplib.SMTPException, socket.error):
```

```
            if not self.fail_silently:
                raise

    def close(self):
        """关闭和邮件服务器的连接"""
        if self.connection is None:
            return
        try:
            try:
                # 主动关闭与邮件服务的连接
                self.connection.quit()
            except (ssl.SSLError, smtplib.SMTPServerDisconnected):
                self.connection.close()
            except smtplib.SMTPException:
                if self.fail_silently:
                    return
                raise
        finally:
            self.connection = None

    def send_messages(self, email_messages):
        """
        发送一个或者多个 EmailMessage 对象，并返回发送的邮件消息数
        """
        if not email_messages:
            return 0
        # 在发送邮件时加锁
        with self._lock:
            # 连接邮件服务器
            new_conn_created = self.open()
            if not self.connection or new_conn_created is None:
                # 连接失败，直接返回 0
                return 0
            # 记录发送邮件数
            num_sent = 0
            for message in email_messages:
                # 发送邮件
                sent = self._send(message)
                if sent:
                    # 如果发送成功，加 1
                    num_sent += 1
            if new_conn_created:
                # 在全部发送完毕后，关闭连接
                self.close()
        # 返回成功发送的邮件数
        return num_sent

    def _send(self, email_message):
```

```
        """ 真实发送邮件的方法 """
        if not email_message.recipients():
            return False
        # 邮件编码
        encoding = email_message.encoding or settings.DEFAULT_CHARSET
        # 邮件发送方
        from_email = sanitize_address(email_message.from_email, encoding)
        # 邮件抄送对象
        recipients = [sanitize_address(addr, encoding) for addr in
email_message.recipients()]
        # 消息主体
        message = email_message.message()
        try:
            # 发送邮件
            self.connection.sendmail(from_email, recipients,
message.as_bytes(linesep='\r\n'))
        except smtplib.SMTPException:
            if not self.fail_silently:
                # 当设置 self.fail_silently 为 False 时，如果出现异常则抛出；否则返回 False
                raise
            return False
        return True
```

通过阅读上述代码，可以发现其核心是 connection_class()方法，它将返回一个 smtplib.SMTP_SSL 类或者 smtplib.SMTP 类。而 EmailBackend 类的 open()方法的作用是在得到 smtplib.SMTP_SSL（或者 smtplib.SMTP）的一个对象后，调用其 login()方法连接指定的邮件服务。此外，_send()方法是发送邮件的核心方法，该方法的核心是调用 smtplib.SMTP_SSL（或者 smtplib.SMTP）对象的 sendmail()方法发送邮件。整体来看，EmailBackend 类其实就是对前面给出的邮件代码进行了拆分，将其封装成了一个邮件系统类，其核心依旧是调用 smtplib.SMTP_SSL 对象的 login()方法登录邮件系统，调用 sendmail()方法发送具体的邮件。

此外，从 send_messages()方法的源码可以看到，调用该方法可以多次发送邮件。该方法的参数 email_messages 其实是一个 EmailMessage 对象列表，而在 EmailMessage 对象列表中包含了待发送邮件的内容，比如邮件抄送对象、邮件主题、邮件内容，甚至附件等。EmailMessage 类的定义位于 django/core/mail/message.py 文件中，源码如下：

```
# 源码位置：django/core/mail/message.py
# ……

class EmailMessage:
    """封装邮件信息的类"""
    content_subtype = 'plain'
    mixed_subtype = 'mixed'
    encoding = None
```

```
def __init__(self, subject='', body='', from_email=None, to=None, bcc=None,
          connection=None, attachments=None, headers=None, cc=None,
          reply_to=None):
    """
    初始化单个邮件消息
    """
    if to:
        if isinstance(to, str):
            raise TypeError('"to" argument must be a list or tuple')
        # 收件人，转成列表
        self.to = list(to)
    else:
        self.to = []
    if cc:
        # 抄送人
        if isinstance(cc, str):
            raise TypeError('"cc" argument must be a list or tuple')
        self.cc = list(cc)
    else:
        self.cc = []
    if bcc:
        if isinstance(bcc, str):
            raise TypeError('"bcc" argument must be a list or tuple')
        self.bcc = list(bcc)
    else:
        self.bcc = []
    if reply_to:
        if isinstance(reply_to, str):
            raise TypeError('"reply_to" argument must be a list or tuple')
        self.reply_to = list(reply_to)
    else:
        self.reply_to = []
    # 发件人
    self.from_email = from_email or settings.DEFAULT_FROM_EMAIL
    # 邮件主题
    self.subject = subject
    # 邮件内容
    self.body = body or ''
    # 附件
    self.attachments = []
    if attachments:
        for attachment in attachments:
            if isinstance(attachment, MIMEBase):
                self.attach(attachment)
            else:
                self.attach(*attachment)
    self.extra_headers = headers or {}
    self.connection = connection
```

```
def get_connection(self, fail_silently=False):
    from django.core.mail import get_connection
    if not self.connection:
        self.connection = get_connection(fail_silently=fail_silently)
    # 返回对应邮件系统的 EmailBackend 对象
    return self.connection

def message(self):
    encoding = self.encoding or settings.DEFAULT_CHARSET
    # 发送邮件的消息体，SafeMIMEText 继承自 MIMEText 类
    msg = SafeMIMEText(self.body, self.content_subtype, encoding)
    msg = self._create_message(msg)
    msg['Subject'] = self.subject
    msg['From'] = self.extra_headers.get('From', self.from_email)
    self._set_list_header_if_not_empty(msg, 'To', self.to)
    self._set_list_header_if_not_empty(msg, 'Cc', self.cc)
    self._set_list_header_if_not_empty(msg, 'Reply-To', self.reply_to)

    header_names = [key.lower() for key in self.extra_headers]
    if 'date' not in header_names:
        msg['Date'] = formatdate(localtime=settings.EMAIL_USE_LOCALTIME)
    if 'message-id' not in header_names:
        msg['Message-ID'] = make_msgid(domain=DNS_NAME)
    for name, value in self.extra_headers.items():
        if name.lower() != 'from':  # From is already handled
            msg[name] = value
    return msg

def recipients(self):
    # 抄送人
    return [email for email in (self.to + self.cc + self.bcc) if email]

def send(self, fail_silently=False):
    """ 发送邮件 """
    if not self.recipients():
        return 0
    # 调用 EmailBackend 对象的 send_messages()方法发送自身
    return self.get_connection(fail_silently).send_messages([self])

def attach(self, filename=None, content=None, mimetype=None):
    """ 添加附件 """
    if isinstance(filename, MIMEBase):
        assert content is None
        assert mimetype is None
        self.attachments.append(filename)
    else:
        assert content is not None
```

```
        mimetype = mimetype or mimetypes.guess_type(filename)[0] or
DEFAULT_ATTACHMENT_MIME_TYPE
        basetype, subtype = mimetype.split('/', 1)

        if basetype == 'text':
            if isinstance(content, bytes):
                try:
                    content = content.decode()
                except UnicodeDecodeError:
                    mimetype = DEFAULT_ATTACHMENT_MIME_TYPE

        self.attachments.append((filename, content, mimetype))

    def attach_file(self, path, mimetype=None):
        """
        系统上的文件附件
        """
        path = Path(path)
        with path.open('rb') as file:
            content = file.read()
            self.attach(path.name, content, mimetype)

    def _create_message(self, msg):
        return self._create_attachments(msg)

    # ……

# ……
```

从 EmailMessage 类的 message()方法中可以看出，该方法封装了邮件消息的主体内容，包括邮件主题、发送方、抄送者等，还有一些日期信息。它返回的 msg 是一个 SafeMIMEText 对象；而该对象继承自 MIMEText 类，同时还继承了 MIMEMixin 类（添加了 as_string()和 as_bytes()方法）。

此外，EmailMessage 类中的 get_connection()方法和 django/core/mail/__init__.py 文件中的 get_connection()方法的功能一致，均可返回指定邮件系统中的一个 EmailBackend 对象。其中，send()方法是调用 get_connection()方法得到的一个 EmailBackend 对象，然后调用该对象的 send_messages()方法发送自身邮件内容。从这里的语句可以看到，EmailBackend 类中的 send_messages()方法的参数是一个 EmailMessage 对象的列表。EmailMessage 类的示例如下：

```
(django2-core-test) [root@master first_django]# python manage.py shell
Python 3.8.6 (default, Oct 18 2020, 15:33:08)
[GCC 4.8.5 20150623 (Red Hat 4.8.5-39)] on linux
Type "help", "copyright", "credits" or "license" for more information.
(InteractiveConsole)
>>> from django.core.mail.message import EmailMessage
>>> from django.core.mail import get_connection
```

```
>>> connection = get_connection(username='2894577759@qq.com', password='密码',
host='smtp.qq.com')
>>> message = EmailMessage('这是使用 EmailMessage 类封装的邮件主题', '这是使用 EmailMessage
类封装的邮件内容', from_email='2894577759@qq.com', to=['2894577759@qq.com'],
cc=['2894577759@qq.com'], connection=connection)
>>> message.send()
1
>>>
```

在执行完该操作后，就能收到对应的邮件了，如图 5-7 所示。

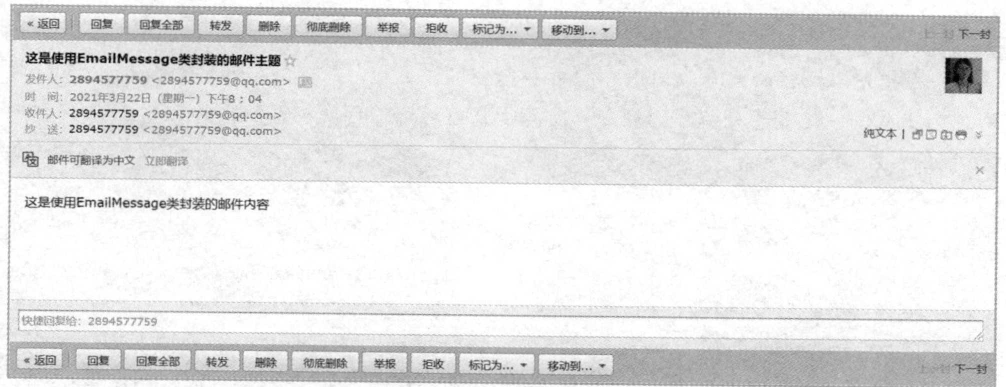


图 5-7

再次回到 EmailBackend 类中，使用 send_messages()方法一次发送多封邮件，操作示例如下：

```
>>> messages = []
>>> for i in range(10):
...     subject = "多封邮件同时发送-标题"
...     content = "多封邮件同时发送-内容"
...     message = EmailMessage(subject, content, from_email='2894577759@qq.com',
to=['2894577759@qq.com'], cc=['2894577759@qq.com'])
...     messages.append(message)
...
>>> from django.core.mail.backends.smtp import EmailBackend
>>> mail = EmailBackend(host='smtp.qq.com', username='2894577759@qq.com',
password='xusmcykiuxqudhcd')
>>> mail.open()
True
>>> mail.send_messages(messages)
10
>>>
```

在执行完上述代码后，从 send_messages()方法的返回结果中可以看出，该语句成功发送了 10 封邮件，QQ 邮箱收到的邮件如图 5-8 所示。

今天 (11 封)

| | | |
|---|---|---|
| 2894577759 | 多封邮件同时发送-标题 - 多封邮件同时发送-内容 | 1秒前 |
| 2894577759 | 多封邮件同时发送-标题 - 多封邮件同时发送-内容 | 3秒前 |
| 2894577759 | 多封邮件同时发送-标题 - 多封邮件同时发送-内容 | 3秒前 |
| 2894577759 | 多封邮件同时发送-标题 - 多封邮件同时发送-内容 | 4秒前 |
| 2894577759 | 多封邮件同时发送-标题 - 多封邮件同时发送-内容 | 5秒前 |
| 2894577759 | 多封邮件同时发送-标题 - 多封邮件同时发送-内容 | 6秒前 |
| 2894577759 | 多封邮件同时发送-标题 - 多封邮件同时发送-内容 | 7秒前 |
| 2894577759 | 多封邮件同时发送-标题 - 多封邮件同时发送-内容 | 9秒前 |
| 2894577759 | 多封邮件同时发送-标题 - 多封邮件同时发送-内容 | 10秒前 |
| 2894577759 | 多封邮件同时发送-标题 - 多封邮件同时发送-内容 | 11秒前 |
| 2894577759 | 这是使用EmailMessage类封装的邮件主题 - 这是使用EmailMessage类封装的邮件内容 | 4分钟前 |

图 5-8

**注意**，当想用上述代码发送不同内容的文本或者邮件主题时，即传入带变量的文本，QQ 邮箱服务可能会抛出异常，这并不是程序问题，而是 QQ 服务对疑似群发邮件做出的反应：

```
>>> mail.send_messages(messages)
Traceback (most recent call last):
  File "<console>", line 1, in <module>
  File
"/root/.pyenv/versions/django2-core-test/lib/python3.8/site-packages/django/core/mai
l/backends/smtp.py", line 110, in send_messages
    sent = self._send(message)
  File
"/root/.pyenv/versions/django2-core-test/lib/python3.8/site-packages/django/core/mai
l/backends/smtp.py", line 126, in _send
    self.connection.sendmail(from_email, recipients, message.as_bytes(linesep='\r\n'))
  File "/root/.pyenv/versions/3.8.6/lib/python3.8/smtplib.py", line 892, in sendmail
    raise SMTPDataError(code, resp)
smtplib.SMTPDataError: (550, b'Mail content denied.
http://service.mail.qq.com/cgi-bin/help?subtype=1&&id=20022&&no=1000726
[MO1cXDLvIp/f550KcgtGca+5ejWclaP4tzihEpKYnMww9uM+nmUHxwa5TyUG/pd0bA== IP:
183.40.221.62]')
```

最后再来看调用发送邮件的 send_mail()方法，其源码实现如下：

```
# 源码位置：django/core/mail/__init__.py
# ……

def send_mail(subject, message, from_email, recipient_list,
              fail_silently=False, auth_user=None, auth_password=None,
              connection=None, html_message=None):

    # 获得指定邮件系统的一个 EmailBackend 对象
    connection = connection or get_connection(
        username=auth_user,
        password=auth_password,
        fail_silently=fail_silently,
    )
```

```
    # EmailMultiAlternatives 类继承自 EmailMessage 类，表示邮件信息，包括邮件主题、内容、附件等
    mail = EmailMultiAlternatives(subject, message, from_email, recipient_list,
connection=connection)
    if html_message:
        mail.attach_alternative(html_message, 'text/html')

    # 调用 EmailMultiAlternatives 对象的 send()方法，其实是调用 EmailMessage 类中的 send()方法
    return mail.send()

# ……
```

在上述代码中，connection 是指定的邮件系统（默认是 django/core/mail/backends/smtp.py 文件）中的一个 EmailBackend 对象，EmailMultiAlternatives 类继承自 EmailMessage 类，因此变量 mail 就表示一个邮件的内容体，包括邮件主题、内容、附件等。最终 send_mail()方法调用 EmailBackend 对象的 send()方法发送邮件。而 send()方法在 EmailMultiAlternatives()中并未定义，所以调用的是其父类 EmailMessage 中的 send()方法。

## 5.8　小结

本章主要介绍了 django/core 目录下的源码，并对部分源码进行了总结与划分，整理出若干核心模块并依次进行讲解和说明。对于部分核心目录，本书并没有介绍里面全部的函数与类，而是选取了若干典型代码进行完整且详细的剖析，为后续分析其他类似代码做了示范。此外，一些函数细节与第 6 章中的内容相同，所以对于这部分代码只进行了简单的介绍，详细内容见第 6 章。

# 第 6 章
# Django的视图层

本章介绍 Django 框架中另一个核心模块的源码：视图层（View），从中可以看到 Django 框架对 HTTP 请求处理的全过程，揭示请求路径、视图函数和视图类映射的本质。本章源码导读部分依赖第 5 章部分内容，因此请务必在学习第 5 章后再学习本章内容。

## 6.1 视图层实战

本节首先基于第 1 章创建的 first_django 项目完成 5 个简单的视图层实验，然后根据实验内容依次提出相关的问题。

### 6.1.1 实验 1：Django 中的“hello, world”

最简单的视图实验内容在第 1 章已经介绍过了，只是当时的视图函数是直接定义在 first_django/urls.py 文件中的，下面使用 include 语句将 URLConf 配置写到新建的应用目录下，这也是项目开发中最常见的做法。

（1）新建一个名为 test_view 的应用：

```
(django2-core-test) [root@master first_django]# django-admin startapp test_view
(django2-core-test) [root@master first_django]# ls test_view/
admin.py  apps.py  __init__.py  migrations  models.py  tests.py  views.py
```

在 first_django 项目的全局配置中添加 test_view 应用，同时新建 urls.py 文件：

```
(django2-core-test) [root@master first_django]# cat first_django/settings.py | grep
INSTALLED_APPS -A 10  # -A 选项为显示从匹配行开始往后的 10 行内容
INSTALLED_APPS = [
    'django.contrib.admin',
    'django.contrib.auth',
    'django.contrib.contenttypes',
    'django.contrib.sessions',
    'django.contrib.messages',
```

```
    'django.contrib.staticfiles',
    'shell_test',
    'book_sales',
    'test_view', # 新增应用
]
(django2-core-test) [root@master first_django]# touch test_view/urls.py
```

（2）编写一个简单的视图函数，内容如下：

```
(django2-core-test) [root@master first_django]# cat test_view/views.py
from django.shortcuts import render
from django.http import HttpResponse
# Create your views here.

def hello_view(request, *args, **kwargs):
    return HttpResponse('hello, world!')
```

（3）编写 URLConf 配置，给该视图函数添加一个匹配的 URL 路径：

```
(django2-core-test) [root@master first_django]# cat test_view/urls.py
from django.urls import path
from django.conf.urls import url
from .views import hello_view

urlpatterns = [
    url('hello/', hello_view, name='hello_view')
]
```

**注意**，这个 urls.py 文件是笔者自己创建的，在 Django 默认的应用文件中并没有它。Django 是如何知道从这里获取映射关系的呢？实际上，在 first_django/settings.py 文件中有整个项目的 URL 映射关系的总入口：

```
(django2-core-test) [root@master first_django]# cat first_django/settings.py | grep
ROOT_URLCONF
ROOT_URLCONF = 'first_django.urls'
```

为了让 Django 能够顺利找到应用目录下的 URLConf 配置，还需要在 first_django/urls.py 文件中添加如下配置：

```
(django2-core-test) [root@master first_django]# cat first_django/urls.py
"""first_django URL Configuration
Function views
    1. Add an import:  from my_app import views
    2. Add a URL to urlpatterns:  path('', views.home, name='home')
Class-based views
    1. Add an import:  from other_app.views import Home
    2. Add a URL to urlpatterns:  path('', Home.as_view(), name='home')
Including another URLconf
    1. Import the include() function: from django.urls import include, path
```

```
   2. Add a URL to urlpatterns:  path('blog/', include('blog.urls'))
"""
from django.contrib import admin
from django.urls import path
from django.conf.urls import include, url

urlpatterns = [
   path('admin/', admin.site.urls),         # 使用path()方法
   url('test_view/', include('test_view.urls')) # 使用url()方法，使用path()方法也可以
]
```

在阅读源码时，一定要看相关的注释。first_django/urls.py 文件中的注释也是 Django 提供的，其中给出了添加 URLConf 的三种常见方式：

- ◎ 函数视图：将类似 path('', views.home, name='home')这样的语句添加到 urlpatterns 中，其中，views.home 为自定义的一个函数。
- ◎ 类视图：将类似 path('', Home.as_view(), name='home')这样的语句添加到 urlpatterns 中，其中，Home 为 Django 内置 View 类的子类。
- ◎ 包含其他 URLConf 配置：将类似 path('blog/', include('blog.urls'))这样的语句添加到 urlpatterns 中，其中，blog.urls 指向 blog 应用下的 urls.py 文件（目录层级可以任意扩展，只要能通过模块导入的方式找到该模块中的 urlpatterns 即可，甚至可以更改 urls.py 文件名）。

这里使用的是 url()方法，与使用 path()方法的效果相同。接下来，启动该工程并进行测试：

```
(django2-core-test) [root@master first_django]# python manage.py runserver
192.168.26.110:8000
Performing system checks...

System check identified no issues (0 silenced).

You have 17 unapplied migration(s). Your project may not work properly until you apply
the migrations for app(s): admin, auth, contenttypes, sessions.
Run 'python manage.py migrate' to apply them.

April 01, 2021 - 21:43:08
Django version 2.2.16, using settings 'first_django.settings'
Starting development server at http://192.168.26.110:8000/
Quit the server with CONTROL-C.
```

此时，服务已经成功启动，并且监听 8000 端口。

（4）打开另一个 xshell 窗口，连接 master 节点，执行如下命令：

```
[root@master ~]# sudo netstat -anltp | grep 8000
tcp        0      0 192.168.26.110:8000     0.0.0.0:*           LISTEN      39307/python
[root@master ~]# curl -I http://192.168.26.110:8000
HTTP/1.1 404 Not Found
```

```
Date: Thu, 01 Apr 2021 21:48:12 GMT
Server: WSGIServer/0.2 CPython/3.8.6
Content-Type: text/html
X-Frame-Options: SAMEORIGIN
Content-Length: 77

[root@master ~]# curl -I http://192.168.26.110:8000/test_view/hello/
HTTP/1.1 200 OK
Date: Thu, 01 Apr 2021 21:48:24 GMT
Server: WSGIServer/0.2 CPython/3.8.6
Content-Type: text/html; charset=utf-8
X-Frame-Options: SAMEORIGIN
Content-Length: 13

[root@master ~]# curl http://192.168.26.110:8000/test_view/hello/
hello, world![root@master ~]#
```

这样，第 1 个简单的实验就完成了，这也是所有初学者都会经历的第一步。如果你是第一次完成上述实验，那么你一定会有诸多疑问：

◎ 在 Django 中是如何实现这样的映射关系的？为什么在配置 urls.py 文件中的 urlpatterns 后，就可以让对应路径的 HTTP 请求进入相关的函数去处理？

◎ path()方法和 url()方法有何区别？

◎ include()函数的原理是什么？为何最后匹配的 URL 路径会加上 test_view/路径作为开头？

### 6.1.2　实验 2：Django 中的视图类

（1）定义一个视图类 HelloView，该类继承自 Django 内部的 View 类：

```
(django2-core-test) [root@master first_django]# cat test_view/views.py
from django.http import HttpResponse
from django.views.decorators.csrf import csrf_exempt
from django.views import View

# ……

class HelloView(View):

   def get(self, request, *args, **kwargs):
      return HttpResponse('hello, get\n')

   def post(self, request, *args, **kwargs):
      return HttpResponse('hello, post\n')

   def put(self, request, *args, **kwargs):
      return HttpResponse('hello, put\n')
```

```
    def delete(self, request, *args, **kwargs):
        return HttpResponse('hello, delete\n')

    @csrf_exempt
    def dispatch(self, request, *args, **kwargs):
        return super(HelloView, self).dispatch(request, *args, **kwargs)
```

（2）添加该视图类的 URL 配置，内容如下：

```
(django2-core-test) [root@master first_django]# cat test_view/urls.py
from django.urls import path
from django.conf.urls import url
from .views import hello_view, HelloView

urlpatterns = [
    url('hello/', hello_view, name='hello_view'),
    url('hello_class/', HelloView.as_view(), name='hello_view_class')  # 新增的 URL 配置
]
```

（3）再次运行 first_django 项目，同时打开另一个连接窗口，使用 curl 命令进行测试：

```
[root@master ~]# curl http://192.168.26.110:8000/test_view/hello_class/
hello, get
[root@master ~]# curl -XPOST http://192.168.26.110:8000/test_view/hello_class/
hello, post
[root@master ~]# curl -XPUT http://192.168.26.110:8000/test_view/hello_class/
hello, put
[root@master ~]# curl -XDELETE http://192.168.26.110:8000/test_view/hello_class/
hello, delete
```

通常，在第一次接触这种视图类写法时，你肯定会有诸多疑问，具体如下：

- ◎ 为什么 Django 能够把 HTTP 中的 GET、POST 及 PUT 等请求自动映射到对应的方法上？
- ◎ 为什么在对应的 URLConf 配置中必须要用 HelloView.as_view()语句而不能直接使用 HelloView？
- ◎ 在 HelloView 中出现的装饰器@csrf_exempt 有什么作用，删掉可以吗？

在学完本章内容后，即可知道这些问题的答案。

### 6.1.3 实验 3：Django 中的请求传参

在 Django 中有 2 种常见的传参方式，分别为动态 URL 传参和 HTTP 请求传参。其中，HTTP 请求传参方式是几乎所有 Django 工程都会用到的。下面先准备 4 个新的视图函数：

```
(django2-core-test) [root@master first_django]# cat test_view/views.py
from django.http import HttpResponse
from django.views.decorators.csrf import csrf_exempt
```

```
from django.views import View

# ……

def article_dynamic(request, year, month, **kwargs):
   return HttpResponse('接收参数:year={}, month={}, kwargs={}\n'.format(year, month,
kwargs))


def article_dynamic_regex(request, year, month, title, **kwargs):
   return HttpResponse('接收参数:year={}, month={}, title={}, kwargs={}\n'.format(
year, month, title, kwargs))


def year_archive(request, year, **kwargs):
   return HttpResponse('接收参数:year={}, kwargs={}\n'.format(year, kwargs))


@csrf_exempt
def check_params(request, *args, **kwargs):

   params = "request.GET={}\n".format(request.GET)
   params += "request.POST={}\n".format(request.POST)
   # 在上传文件时，需要注释掉下面这行，否则会抛出异常
   params += "request.body={}\n".format(request.body)
   params += "request.FILES={}\n".format(request.FILES)

   return HttpResponse(params)
```

接着，准备 URLConf 配置，将这些视图函数映射到具体的请求路径上：

```
(django2-core-test) [root@master first_django]# cat test_view/urls.py
from django.urls import path
from django.conf.urls import url
from django.urls import re_path
from . import views

urlpatterns = [
    url('hello/', views.hello_view, name='hello_view'),
    url('hello_class/', views.HelloView.as_view(), name='hello_view_class'),
    # 动态传参
    path('articles/<int:year>/<int:month>/', views.article_dynamic),
    # 动态正则表达式

   re_path('articles_regex/(?P<year>[0-9]{4})/(?P<month>0[1-9]|1[0-2])/(?P<title>[a-
zA-Z0-9-_]+)/', views.article_dynamic_regex),
    # URLconf 传递额外参数
```

```
    re_path('articles_regex_extra/(?P<year>[0-9]{4})/', views.year_archive, {'hello':
'world'}),
    # 请求参数
    url('check_params/', views.check_params)
]
```

在上述配置中，我们分别使用了动态 URL 路径和正则表达式的 URL 路径。接下来，启动和测试这些路径下的 HTTP 请求：

```
[root@master ~]# curl
http://192.168.26.110:8000/test_view/articles_regex/1991/05/report/
接收参数:year=1991, month=05, title=report, kwargs={}
[root@master ~]# curl http://192.168.26.110:8000/test_view/articles/1991/05/
接收参数:year=1991, month=5, kwargs={}
[root@master ~]# curl
http://192.168.26.110:8000/test_view/articles_regex/1991/05/report/
接收参数:year=1991, month=05, title=report, kwargs={}
[root@master ~]# curl
http://192.168.26.110:8000/test_view/articles_regex_extra/1991/05/
接收参数:year=1991, kwargs={'hello': 'world'}
[root@master ~]# curl http://192.168.26.110:8000/test_view/articles_regex_extra/1991/
接收参数:year=1991, kwargs={'hello': 'world'}
[root@master ~]# curl
"http://192.168.26.110:8000/test_view/check_params/?xx=1&yy=2&zz=test"
request.GET=<QueryDict: {'xx': ['1'], 'yy': ['2'], 'zz': ['test']}>
request.POST=<QueryDict: {}>
request.body=b''
request.FILES=<MultiValueDict: {}>
[root@master ~]# curl -d "data=xyz&x=123"
http://192.168.26.110:8000/test_view/check_params/
request.GET=<QueryDict: {}>
request.POST=<QueryDict: {'data': ['xyz'], 'x': ['123']}>
request.body=b'data=xyz&x=123'
request.FILES=<MultiValueDict: {}>
# 当使用 curl 上传文件时，请求传参及指定本地文件需要使用-F 选项
[root@master ~]# curl -F "file_name=@hosts" -F "username=d1" -F password=d2
http://192.168.26.110:8000/test_view/check_params/
request.GET=<QueryDict: {}>
request.POST=<QueryDict: {'username': ['d1'], 'password': ['d2']}>
request.FILES=<MultiValueDict: {'file_name': [<InMemoryUploadedFile: hosts
(application/octet-stream)>]}>
```

注意，在测试文件上传的请求中，需要注释掉 params += "request.body={}\n".format(request.body) 语句，否则会报错：

```
[root@master ~]# curl -F "file_name=@hosts" -F "username=d1" -F password=d2
http://192.168.26.110:8000/test_view/check_params/
<h1>Server Error (500)</h1>[root@master ~]#
```

如果想查看错误详情，可以在 first_django/settings.py 文件中设置 DEBUG=True，这样在请求时即可在服务端打印出异常详情，以便定位问题。下面是笔者打开 DEBUG 后，执行文件上传时的报错内容，可以看到错误信息指向了 params += "request.body={}\n".format(request.body)语句：

```
# ……
Starting development server at http://192.168.26.110:8000/
Quit the server with CONTROL-C.
Internal Server Error: /test_view/check_params/
Traceback (most recent call last):
  File
"/root/.pyenv/versions/django2-core-test/lib/python3.8/site-packages/django/core/han
dlers/exception.py", line 34, in inner
    response = get_response(request)
  File
"/root/.pyenv/versions/django2-core-test/lib/python3.8/site-packages/django/core/han
dlers/base.py", line 115, in _get_response
    response = self.process_exception_by_middleware(e, request)
  File
"/root/.pyenv/versions/django2-core-test/lib/python3.8/site-packages/django/core/han
dlers/base.py", line 113, in _get_response
    response = wrapped_callback(request, *callback_args, **callback_kwargs)
  File
"/root/.pyenv/versions/django2-core-test/lib/python3.8/site-packages/django/views/de
corators/csrf.py", line 54, in wrapped_view
    return view_func(*args, **kwargs)
  File "/root/django-core-test/first_django/test_view/views.py", line 46, in
check_params
    params += "request.body={}\n".format(request.body)
  File
"/root/.pyenv/versions/django2-core-test/lib/python3.8/site-packages/django/http/req
uest.py", line 281, in body
    raise RawPostDataException("You cannot access body after reading from request's data
stream")
django.http.request.RawPostDataException: You cannot access body after reading from
request's data stream
[02/Apr/2021 02:21:21] "POST /test_view/check_params/ HTTP/1.1" 500 72144
```

在本次实验中我们看到了几个不一样的 URL 路径写法，还遇到了一个报错，因而可以思考以下几个问题：

◎ 在 Django 中，如何实现动态 URL 路径传参？正则表达式如何匹配 URL 路径？

◎ 在 Django 中，对 HTTP 请求参数是如何处理的？这些参数是如何传递到 request.GET 和 request.POST 等变量上的？

◎ 在上传请求中，为何使用 request.body 变量会报错？

### 6.1.4 实验 4：Django 中的文件上传演示

前面我们演示了上传文件，但是在视图函数中并没有把该文件保存到本地。下面将完整实现一个文件上传功能。

（1）准备处理上传文件的视图函数：

```
(django2-core-test) [root@master first_django]# cat test_view/views.py
import os

# 忽略之前实验的视图函数
# ……

def handle_uploaded_file(f):
    save_dir = "/root/test/django"
    if not os.path.exists(save_dir):
        os.makedirs(save_dir)  # 如果目录不存在，则创建该目录
    save_path = os.path.join(save_dir, f.name)
    with open(save_path, 'wb+') as fp:
        for chunk in f.chunks():
            fp.write(chunk)


@csrf_exempt
def file_upload(request, *args, **kwargs):
    handle_uploaded_file(request.FILES['file'])
    return HttpResponse('上传成功')
```

（2）准备对应的 URLConf 配置：

```
(django2-core-test) [root@master first_django]# cat test_view/urls.py
# ……

urlpatterns = [
    # ……
    # 文件上传
    url('file_upload/', views.file_upload)
]
```

从对应的视图函数中可以看到，上传文件将被保存到本地的/root/test/Django 目录下。如果该目录不存在，则直接创建一个。启动 first_django 工程，在另一个 xshell 窗口中测试如下命令：

```
[root@master ~]# ls /root/test     # 在主机（master）上确认/root/test 目录是否存在
ls: cannot access /root/test: No such file or directory
[root@master ~]# su - store
Last login: Fri Apr  2 12:39:57 CST 2021 on pts/2
[store@master ~]$ ssh ceph-1     # 登录 ceph-1 主机
Last login: Thu Mar 25 14:24:41 2021 from master
```

```
[store@ceph-1 ~]$ ls -l basic_check.sh    # 查看 ceph-1 的待上传文件 basic_check.sh
-rw-rw-r-- 1 store store 1271 Mar  3 16:56 basic_check.sh
# 请求 master 上 first_django 服务的上传文件地址
[store@ceph-1 ~]$ curl -F "file=@basic_check.sh"
http://192.168.26.110:8000/test_view/file_upload/
上传成功[store@ceph-1 ~]$
[store@ceph-1 ~]$ exit
logout
Connection to ceph-1 closed.
[store@master ~]$ exit
logout
[root@master ~]# ls -l /root/test/django/    # 该目录已创建，对应的文件已上传成功
total 4
-rw-r--r-- 1 root root 1271 Apr  2 12:41 basic_check.sh
```

上面的代码实现了一个简单的文件上传功能。除此之外，还要注意上传文件的大小，这部分内容在 5.6.1 节分析文件上传时介绍过，有兴趣的读者可以自行尝试通过该接口上传较大的文件。如果某个文件大小被限制了，则应该调整哪个配置的值呢？

接下来展示一种基于 Django 数据库模型类的文件上传方式。

（1）定义相应上传文件的模型类，为了简单起见，只定义两个字段，即上传文件名称和上传文件路径：

```
(django2-core-test) [root@master first_django]# cat test_view/models.py
from django.db import models


def upload_dir(instance, filename):
    # 文件将被上传到 MEDIA_ROOT/django/<filename>路径
    return "django/{}".format(filename)

# 创建模型类
class FileModel(models.Model):
    name = models.CharField('上传文件名', max_length=20)
    # FileField 为字段类型
    upload_path = models.FileField(upload_to=upload_dir)

    class Meta:
        # 通过 db_table 自定义数据表名
        db_table = 'upload_file_record'
```

**注意**，upload_to 参数和 settings.py 文件中的 MEDIA_ROOT 属性值可以一起确定上传文件的目录，形式有很多种，比如写成 upload_to='django/%Y/%m/%d'等。此外，upload_to 参数可以接收方法名，该方法返回的是上传文件的目录。我们在 first_django/settings.py 文件中设置 MEDIA_ROOT 的值为/root/test。

（2）使用 makemigrations 命令和 migrate 命令将该模型类映射到数据库中，建立相应的表。具体操作如下：

```
(django2-core-test) [root@master first_django]# python manage.py makemigrations
test_view
Migrations for 'test_view':
  test_view/migrations/0001_initial.py
    - Create model FileModel
(django2-core-test) [root@master first_django]# python manage.py migrate test_view
System check identified some issues:

# 告警信息，忽略
# ……
Operations to perform:
  Apply all migrations: test_view
Running migrations:
  Applying test_view.0001_initial... OK
```

（3）准备一个简单的视图类，内容如下：

```
(django2-core-test) [root@master first_django]# cat test_view/views.py
# ……

@csrf_exempt
def file_upload2(request, *args, **kwargs):
    upload_file = request.FILES['file']
    FileModel.objects.create(name=upload_file.name, upload_file=upload_file)
    return HttpResponse('上传成功')
```

（4）设置 URLConf 配置，相应的映射语句如下：

```
(django2-core-test) [root@master first_django]# cat test_view/urls.py
# ……

urlpatterns = [
    # ……
    url('file_upload2/', views.file_upload2)
]
```

（5）启动服务并打开另一个窗口（和前面一样，不再重复），使用新的上传文件接口：

```
[root@master django]# ls /root/test/django   # 在该目录中只有第一次上传成功的文件
basic_check.sh
[root@master django]# su - store
Last login: Fri Apr  2 13:42:44 CST 2021 on pts/2
[store@master ~]$ ssh ceph-1
Last login: Sun Apr  4 22:42:40 2021 from master
[store@ceph-1 ~]$ ls -l bird6-1.6.8-1.el7.x86_64.rpm
-rw-rw-r-- 1 store store 307964 Jun 11  2020 bird6-1.6.8-1.el7.x86_64.rpm
# 使用新的上传文件接口
```

```
[store@ceph-1 ~]$ curl -F "file=@bird6-1.6.8-1.el7.x86_64.rpm"
http://192.168.26.110:8000/test_view/file_upload2/
上传成功[store@ceph-1 ~]$

# ……
[root@master django]# ls -l /root/test/django   #文件上传成功
total 308
-rw-r--r-- 1 root root   1271 Apr  2 12:41 basic_check.sh
-rw-r--r-- 1 root root 307964 Apr  2 13:43 bird6-1.6.8-1.el7.x86_64.rpm
```

从上面的代码可以看到，使用新的上传文件接口也能成功上传文件。此外，在 MySQL 数据库中查看表 upload_file_record，可以看到其中多了一条上传记录。

在完成上述实验后，请思考在 Django 中，上传文件的完整流程是什么？第 1 个实验的大致过程在 5.6 节中介绍过，而在第 2 个实验中，似乎没有看到保存文件的动作，为什么在对数据库操作时就能成功保存上传文件？还有 upload_to 参数，它还有哪些其他用法？

### 6.1.5　实验 5：在 Django 中操作 Session

在 Django 的默认配置中可以设置将 Session 保存到数据库中，为了简单起见，先修改默认的 Session 数据保存配置，将其修改为在内存中保存，即修改项目的全局文件 settings.py，具体如下：

```
(django2-core-test) [root@master first_django]# cat first_django/settings.py | grep
SESSION
SESSION_ENGINE = 'django.contrib.sessions.backends.cache'
SESSION_CACHE_ALIAS = 'default'
```

接着准备 Session 的视图类，其中包含 get()方法和 post()方法，分别用于处理 GET 请求和 POST 请求：

```
(django2-core-test) [root@master first_django]# cat test_view/views.py
# ……

class TestLoginSessionView(View):
    def get(self, request, *args, **kwargs):
        # 如果在 Session 中含有 has_login 字段且为 True，则提示已成功登录
        if request.session.get('has_login', False):
            return HttpResponse('已成功登录，无须再次登录')
        return HttpResponse('请先登录')


    def post(self, request, *args, **kwargs):
        # 获取登录账号和密码
        name = request.POST.get('username')
        password = request.POST.get('password')
        if password != '123456':
```

```
            return HttpResponse('用户名或密码不正确' )

        # 登录成功
        request.session['has_login'] = True
        # 设置 10s 后过期
        request.session.set_expiry(10)
        return HttpResponse('登录成功，信息已保存到 Session 中')
```

准备 URLConf 配置，内容如下：

```
(django2-core-test) [root@master first_django]# cat test_view/urls.py
# ……

urlpatterns = [
    # ……
    url('login_session/', views.TestLoginSessionView.as_view())
]
```

最后，在 ceph-1 上准备一段测试该接口的 Python 代码，内容如下：

```
[store@ceph-1 ~]$ cat test_session.py
#  -*- coding: utf-8  -*-

import time
from datetime import datetime
import requests

def get_current_time():
    return datetime.now().strftime("%Y-%m-%d %H:%M:%S")

r = requests.session()
params = {
    'username': 'spyinx',
    'password': '123456'
}
result = r.post("http://192.168.26.110:8000/test_view/login_session/", data=params)
print("[{}]第一次请求结果:{}".format(get_current_time(), result.content))
time.sleep(2)
result = r.get("http://192.168.26.110:8000/test_view/login_session/")
print("[{}]GET 请求/test_view/login_session/路径结果:{}".format(get_current_time(),
result.content))

time.sleep(10)

result = r.get("http://192.168.26.110:8000/test_view/login_session/")
print("[{}]Session 过期之后再次请求:{}".format(get_current_time(), result.content))
```

在上述代码中，是使用 requests.session()语句来维持一个 Session 会话的，这在网络爬虫中十分常用。最后，启动 master 节点上的 first_django 服务，执行上述 Python 代码，结果如下：

```
[store@ceph-1 ~]$ python test_session.py
[2021-04-05 22:07:58]第一次请求结果:登录成功，信息已保存到 Session 中
[2021-04-05 22:08:00]GET 请求/test_view/login_session/路径结果:已成功登录，无须再次登录
[2021-04-05 22:08:10]Session 过期之后再次请求:请先登录
```

从上面的结果可以看出，第一次登录请求成功后设置了 Session 信息，在维持会话的基础上，在 2 秒后发起 GET 请求，会被识别为“已成功登录，无须再次登录”。在 10s 后再次发起 GET 请求，由于 Session 信息已过期，所以返回“请先登录”。

在完成上述实验后，请思考一个问题，Django 中的 Session 有多种存储方式，如数据库存储、内存存储等，这些存储数据的方式究竟是怎样的？如何判断 Session 中的信息是否过期了呢？

## 6.2　请求与响应

在正式分析视图层源码之前，先介绍两个非常重要的类，即 HttpRequest 类和 HttpResponse 类。这两个类依赖的基础类是 CaseInsensitiveMapping、MultiValueDict、QueryDict 和 HttpHeaders：

```
# 源码位置：django/utils/datastructures.py
# ……

class CaseInsensitiveMapping(Mapping):

    def __init__(self, data):
        if not isinstance(data, Mapping):
            data = {k: v for k, v in _destruct_iterable_mapping_values(data)}
        # 保存到_store 属性中，key 统一小写
        self._store = {k.lower(): (k, v) for k, v in data.items()}

    def __getitem__(self, key):
        # 对输入的 key 取小写，然后再去_store 属性值中获取对应 key 的值
        return self._store[key.lower()][1]

    def __len__(self):
        return len(self._store)

    def __eq__(self, other):
        # 对象判定完全相等，除 Mapping 实例外，要求所有的 k-v 值均相同
        return isinstance(other, Mapping) and {
            k.lower(): v for k, v in self.items()
        } == {
            k.lower(): v for k, v in other.items()
        }

    def __iter__(self):
        return (original_key for original_key, value in self._store.values())
```

```
    def __repr__(self):
        return repr({key: value for key, value in self._store.values()})

    def copy(self):
        return self

# ……
```

（1）CaseInsensitiveMapping 类继承自 Mapping 类，可以将其看成是一个类映射的类。与 Python 中的 Mapping 类相比，CaseInsensitiveMapping 类在获取 value 值时对 key 的大小不敏感，即 xxx['key'] 和 xxx['KEY']的结果是一样的，这一点可以从魔法函数__getitem__()中看到：

```
>>> from django.utils.datastructures import CaseInsensitiveMapping
>>> m = CaseInsensitiveMapping({'hello': 'world', 'XYZ': 'xxxx'})
>>> m['xyz']       # 对应魔法函数__getitem__()
'xxxx'
>>> m['HELLO']
'world'
>>> len(m)       # 对应魔法函数__len__()
2
>>> m          # 对应魔法函数_repr__()
{'hello': 'world', 'XYZ': 'xxxx'}
>>> for key in m:            # 对应魔法函数__iter__()
...     print(key, ':', m[key])
...
hello : world
XYZ : xxxx
>>> m._store     # 打印_store 属性值
{'hello': ('hello', 'world'), 'xyz': ('XYZ', 'xxxx')}
```

（2）MultiValueDict 类是 dict 类的一个子类，用于处理一个 key 对应多个值的情况。对于该类，我们不直接列出其全部源码并进行分析，而是通过实战演示该类的使用方法，并选择性地解读对应的源码。具体操作如下：

```
>>> from django.utils.datastructures import MultiValueDict
>>> d = MultiValueDict({'name': ['Adrian', 'Simon'], 'position': ['Developer'], 'sex': 
'男'})
>>> d['sex']
'男'
>>> d['name']   # 通常来说，在获取 key 对应的 value 值时，只能得到 value 列表中的最后一个 value 值
'Simon'
>>> d.get('name')
'Simon'
>>> d['position']
'Developer'
>>> d.getlist('name')  # getlist()方法可以完整获取对应的 value 列表
```

```
['Adrian', 'Simon']
>>> d.getlist('does-not-exist', ['hello', 'world'])
# 从 d 中获取 does-not-exist 对应的列表。
# 如果有，则直接返回 does-not-exist；如果没有，则返回默认值['hello', 'world']
['hello', 'world']
>>> d.get('parents', '没有')  # 和字典一样的 GET 操作，当 key 不存在时，使用默认值
'没有'
>>> d.setlist('parents', ['Tom', 'Lucy'])  # 设置 key，支持设置列表元素
>>> d.getlist('parents')
['Tom', 'Lucy']
>>>
```

MultiValueDict 类的调用主要与其源码中定义的一些方法和它集成的字典相关。例如，通过查看源码即可了解，MultiValueDict 对象中的 getlist()方法获取列表中最后一个元素的原因，内容如下：

```
# 源码位置：django/utils/datastructures.py
# ……
class MultiValueDictKeyError(KeyError):
    pass

class MultiValueDict(dict):

    # ……

    def __getitem__(self, key):

        try:
            # 使用父类字典获取 key
            list_ = super().__getitem__(key)
        except KeyError:
            # 如果异常，抛出对应的错误
            raise MultiValueDictKeyError(key)
        try:
            # 返回列表中的最后一个元素
            return list_[-1]
        except IndexError:
            return []

    # ……
```

再来看 MultiValueDict 类的 getlist()方法，从前面的操作示例可以看到，该方法在获取对应 key 的 value 值的同时，可以将该 value 值转成列表形式，源码如下：

```
class MultiValueDict(dict):

    # ……

    def _getlist(self, key, default=None, force_list=False):
```

```
    try:
        # 使用dict方式获取对应key的value值
        values = super().__getitem__(key)
    except KeyError:
        # 当出现异常时，如果没有默认值，返回[]；如果有默认值，返回默认值
        if default is None:
            return []
        return default
    else:
        # 正常情况
        if force_list:
            #把force_list参数设置为True，将结果转成list类型
            values = list(values) if values is not None else None
        return values

def getlist(self, key, default=None):
    return self._getlist(key, default, force_list=True)

# ……
```

从上面的源码可以看到，_getlist()方法中的 force_list 参数的作用是是否将结果强制转换成 list 类型。当在 getlist()方法中调用_getlist()方法时，默认设置 force_list 参数为 True。下面手动调用_getlist()方法测试 force_list=False，操作如下：

```
>>> d.getlist('sex')
['男']
>>> d._getlist('sex', force_list=False)
'男'
```

此外，在 MultiValueDict 类中定义了许多方法，并重写了字典中的一些方法，如 items()、values()和 update()等，这些方法大部分是在 dict 类的源码上稍加扩展得到的，并不复杂，十分适合初学者阅读。

（3）QueryDict 类的源码继承了前面介绍的 MultiValueDict 类：

```
# 源码位置：django/http/request.py
# ……

class QueryDict(MultiValueDict):

    _mutable = True      # 可变性
    _encoding = None     # 编码

    def __init__(self, query_string=None, mutable=False, encoding=None):
        super().__init__()
        self.encoding = encoding or settings.DEFAULT_CHARSET
        query_string = query_string or ''
        parse_qsl_kwargs = {
```

```
        'keep_blank_values': True,
        'fields_limit': settings.DATA_UPLOAD_MAX_NUMBER_FIELDS,
        'encoding': self.encoding,
    }
    if isinstance(query_string, bytes):
        # query_string 通常包含 URL 编码的数据，以及 ASCII 码的一个子集
        try:
            query_string = query_string.decode(self.encoding)
        except UnicodeDecodeError:
            query_string = query_string.decode('iso-8859-1')
    for key, value in limited_parse_qsl(query_string, **parse_qsl_kwargs):
        self.appendlist(key, value)
    # 默认的 mutable 为 False，即标识不可变
    self._mutable = mutable

@classmethod
def fromkeys(cls, iterable, value='', mutable=False, encoding=None):
    # 实例化一个 QueryDict 对象
    q = cls('', mutable=True, encoding=encoding)
    # 添加(key, value)对
    for key in iterable:
        q.appendlist(key, value)
    # 根据 mutable 设置可变性，在实例化中已设置为 True
    if not mutable:
        q._mutable = False
    # 返回该 QueryDict 对象
    return q

@property
def encoding(self):
    # 返回编码格式
    if self._encoding is None:
        self._encoding = settings.DEFAULT_CHARSET
    return self._encoding

@encoding.setter
def encoding(self, value):
    # 设置编码
    self._encoding = value

def _assert_mutable(self):
    if not self._mutable:
        # 如果_mutable 属性值为 False，即不可变，则抛出异常
        raise AttributeError("This QueryDict instance is immutable")

def __setitem__(self, key, value):
    # 如果_mutable 属性值为 False，即不可变，则无法设置
    self._assert_mutable()
```

```
        # 字节转文本，指定编码
        key = bytes_to_text(key, self.encoding)
        value = bytes_to_text(value, self.encoding)
        # 调用父类的设置，即调用 MultiValueDict 类的魔法函数__setitem__()
        # 调用 dict 类的赋值方法，类似字典的赋值形式：a['x'] = 'xxx'
        super().__setitem__(key, value)

    def __delitem__(self, key):
        # 同理，对于删除的魔法函数，也需要判断该类的可变性（即_mutable 属性）
        self._assert_mutable()
        super().__delitem__(key)

    def __copy__(self):
        # 得到一个该类的对象，为了能复制该 QueryDict 对象，需要设置 mutable=True
        result = self.__class__('', mutable=True, encoding=self.encoding)
        # 复制 key-value 值，并将其赋给新的 QueryDict 对象
        for key, value in self.lists():
            result.setlist(key, value)
        return result

    def __deepcopy__(self, memo):
        # 深拷贝，利用 copy 模块中的 deepcopy()方法实现
        result = self.__class__('', mutable=True, encoding=self.encoding)
        memo[id(self)] = result
        for key, value in self.lists():
            result.setlist(copy.deepcopy(key, memo), copy.deepcopy(value, memo))
        return result

    def setlist(self, key, list_):
        # 检查可变性
        self._assert_mutable()
        key = bytes_to_text(key, self.encoding)
        list_ = [bytes_to_text(elt, self.encoding) for elt in list_]
        # 调用父类的 setlist()方法
        super().setlist(key, list_)

    def setlistdefault(self, key, default_list=None):
        # 检查可变性
        self._assert_mutable()
        # 调用父类的 setlistdefault()方法
        return super().setlistdefault(key, default_list)

    def appendlist(self, key, value):
        # 将(key, value)对添加到 QueryDict 对象中
        self._assert_mutable()
        # 全部转成文本
        key = bytes_to_text(key, self.encoding)
        value = bytes_to_text(value, self.encoding)
```

```
        super().appendlist(key, value)

    def pop(self, key, *args):
        self._assert_mutable()
        # 调用父类的 pop()方法，其实就是 dict 类中的 pop()方法
        return super().pop(key, *args)

    def popitem(self):
        self._assert_mutable()
        # 调用父类的 popitem()方法
        return super().popitem()

    def clear(self):
        self._assert_mutable()
        # 字典中的 clear()方法
        super().clear()

    def setdefault(self, key, default=None):
        # 设置默认值
        self._assert_mutable()
        # 全部转成文本类型
        key = bytes_to_text(key, self.encoding)
        default = bytes_to_text(default, self.encoding)
        return super().setdefault(key, default)

    def copy(self):
        """ 返回当前对象的一个可变副本 """
        return self.__deepcopy__({})

    def urlencode(self, safe=None):
        """
        返回所有查询参数的一个编码后的字符串

        示例:

            >>> q = QueryDict(mutable=True)
            >>> q['next'] = '/a&b/'
            >>> q.urlencode()
            'next=%2Fa%26b%2F'
            >>> q.urlencode(safe='/')
            'next=/a%26b/'
        """
        output = []
        if safe:
            # 字符串安全编码
            safe = safe.encode(self.encoding)

            def encode(k, v):
```

```
            return '%s=%s' % ((quote(k, safe), quote(v, safe)))
    else:
        def encode(k, v):
            return urlencode({k: v})
    for k, list_ in self.lists():
        output.extend(
            encode(k.encode(self.encoding), str(v).encode(self.encoding))
            for v in list_
        )
    # 最后使用'&'连接
    return '&'.join(output)
```

QueryDict 类的源码略多但并不复杂，很多方法都是直接调用父类（MultiValueDict 类）的方法实现的。而其父类的源码在前面曾使用并分析过，都非常简单。因此，这里只需不断地调用和实践 QueryDict 类中的方法即可，操作示例如下：

```
>>> from django.http.request import QueryDict
>>> q = QueryDict('a=xxx&b=xyz&a=haha')  # 自动解析 URL 请求中的参数值
>>> q
<QueryDict: {'a': ['xxx', 'haha'], 'b': ['xyz']}>
>>> fq = q.fromkeys(['a', 'b'], value='xxxx', mutable=True)   # 设置_mutable 属性值为 True
>>> fq
<QueryDict: {'a': ['xxxx'], 'b': ['xxxx']}>
>>> q._mutable
False
>>> q['test'] = 'hello'
# 由于_mutable 属性值为 False，所以在执行 self._assert_mutable()语句时将抛出异常
Traceback (most recent call last):
File "<console>", line 1, in <module>
File
"/root/.pyenv/versions/django2-core-test/lib/python3.8/site-packages/django/http/req
uest.py", line 459, in __setitem__
self._assert_mutable()
File
"/root/.pyenv/versions/django2-core-test/lib/python3.8/site-packages/django/http/req
uest.py", line 456, in _assert_mutable
raise AttributeError("This QueryDict instance is immutable")
AttributeError: This QueryDict instance is immutable
>>> fq._mutable
True
>>> fq['test'] = 'hello'   # 由于_mutable 属性值为 True，所以可以修改 key-value 值
>>> fq
<QueryDict: {'a': ['xxxx'], 'b': ['xxxx'], 'test': ['hello']}>
>>> fq.setlist('test_setlist', 'xx')  # 对字符串使用 for 语句，字符串将被拆分成单个字符
>>> fq
<QueryDict: {'a': ['xxxx'], 'b': ['xxxx'], 'test': ['hello'], 'test_setlist': ['x', 'x']}>
>>> fq.setlist('test_setlist2', ['xx'])  # 直接设置列表值
>>> fq
```

```
<QueryDict: {'a': ['xxxx'], 'b': ['xxxx'], 'test': ['hello'], 'test_setlist': ['x', 'x'],
'test_setlist2': ['xx']}>
>>> fq.setlistdefault('test_setlistdefault', ['default'])
# 设置默认值
['default']
>>> fq
<QueryDict: {'a': ['xxxx'], 'b': ['xxxx'], 'test': ['hello'], 'test_setlist': ['x', 'x'],
'test_setlist2': ['xx'], 'test_setlistdefault': ['default']}>
>>> del fq['test_setlistdefault']  # 删除对应的键值
>>> fq
<QueryDict: {'a': ['xxxx'], 'b': ['xxxx'], 'test': ['hello'], 'test_setlist': ['x', 'x'],
'test_setlist2': ['xx']}>
>>> fq.pop('test_setlist')   # 弹出 key 为 test_setlist 的键值对
['x', 'x']
>>> fq
<QueryDict: {'a': ['xxxx'], 'b': ['xxxx'], 'test': ['hello'], 'test_setlist2': ['xx']}>
>>> fq.popitem()   # 弹出该对象中的一个(key, value)对
('test_setlist2', ['xx'])
>>> fq
<QueryDict: {'a': ['xxxx'], 'b': ['xxxx'], 'test': ['hello']}>
>>> fq.appendlist('appendlist', 'v1')     # appendlist()方法示例
>>> fq
<QueryDict: {'a': ['xxxx'], 'b': ['xxxx'], 'test': ['hello'], 'appendlist': ['v1']}>
>>> fq.appendlist('appendlist', 'v2')
>>> fq
<QueryDict: {'a': ['xxxx'], 'b': ['xxxx'], 'test': ['hello'], 'appendlist': ['v1', 'v2']}>
>>> fq.urlencode()
'a=xxxx&b=xxxx&test=hello&appendlist=v1&appendlist=v2'
>>> fq.clear()
>>> fq
<QueryDict: {}>
```

上面的代码演示了 QueryDict 类中大部分方法的使用方式。注意，必须将 QueryDict 对象中的_mutable 属性值设置为 True 后才能进行增删改等操作。在每次操作之前，都需要调用_assert_mutable()方法去校验_mutable 属性值，如果是 False，则抛出异常。

（4）HttpHeaders 类的实现源码如下：

```
# 源码位置：django/http/request.py
# ……

class HttpHeaders(CaseInsensitiveMapping):
    HTTP_PREFIX = 'HTTP_'
    # 以下两个头信息还没有准备好，所以以 HTTP_开头
    UNPREFIXED_HEADERS = {'CONTENT_TYPE', 'CONTENT_LENGTH'}

    def __init__(self, environ):
        headers = {}
```

```
        # 获取头信息（key-value）
        for header, value in environ.items():
            # 解析头字段，例如，CONTENT_TYPE 会被转换成 CONTENT-TYPE，
            # 其余以 HTTP_开头的头参数，将在去掉 HTTP_开头后得到参数名
            name = self.parse_header_name(header)
            if name:
                headers[name] = value   # 保存最终的 key-value 值
        super().__init__(headers)

    @classmethod
    def parse_header_name(cls, header):
        if header.startswith(cls.HTTP_PREFIX):          # 以 HTTP_开头
            header = header[len(cls.HTTP_PREFIX):]      # 删掉以 HTTP_开头的部分
        elif header not in cls.UNPREFIXED_HEADERS:
            # 如果不是 UNPREFIXED_HEADERS 中的字符串，并且不以 HTTP_开头，则直接返回 None
            return None
        #将 header 变量值中的下画线（_）替换成短横线（-）
        return header.replace('_', '-').title()

# ......
```

HttpHeaders 类的内容比较简单，且源码逻辑十分清晰。首先，它继承自 CaseInsensitiveMapping 类，因此 HttpHeaders 类是一个对 key 大小写不敏感的字典。从 parse_header_name()方法的源码可以看到，对于传入的 header 变量，如果其值以 HTTP_开头，则 header 将删掉以 HTTP_开头的部分。如果该值不以 HTTP_开头，并且不在 UNPREFIXED_HEADERS 数组中，则直接返回 None。对于其他情况，则将 header 变量值中的下画线（_）替换成短横线（-），调用字符串的 title()方法，将该值转换成首字母大写其余部分小写的形式，并返回该值。HttpHeaders 类的初始化方法可从传入的 environ 信息中解析出 HTTP 头信息，并将 HTTP 头信息保存在 headers 中。该类的操作示例如下：

```
>>> from django.http.request import HttpHeaders
>>> HttpHeaders.parse_header_name('HTTP_HELLO')
'Hello'
>>> HttpHeaders.parse_header_name('XXXXX') is None
True
>>> HttpHeaders.parse_header_name('CONTENT_TYPE')
'Content-Type'
>>> h = HttpHeaders({'HTTP_HELLO': 'wrold', 'CONTENT_TYPE': 'text/plain', 'XXX': 'NO',
'CONTENT-LEN': 194, 'HTTP_XXX': '参数'})
>>> h._store   # 从 CaseInsensitiveMapping 中可知，最终字典的值保存在了_store 属性中
{'hello': ('Hello', 'wrold'), 'content-type': ('Content-Type', 'text/plain'), 'xxx':
('Xxx', '参数')}
>>> h['content-type']   # 对 key 大小写不敏感
'text/plain'
>>> h['xxx']
'参数'
```

接下来，正式学习 HttpRequest 类和 HttpResponse 类的源码。

## 6.2.1　HttpRequest 类的源码

这里将 HttpRequest 类中的源码分成以下三部分进行讲解。

（1）获取请求信息的基本方法，如_get_raw_host()、get_host()、get_port()、get_full_path()、get_raw_uri()和 build_absolute_uri()等，主要用于获取客户端的请求信息。这些方法的源码实现都较为简单，我们通过简单的示例演示这些方法的输出结果即可。修改本章中实验 1 的视图函数 hello_view()，使该函数的返回结果如下：

```
# 源码位置： first_django/test_view/views.py
# ……

def hello_view(request, *args, **kwargs):
   content = 'request.headers={}\n'.format(request.headers)
   content += 'request._get_raw_host()结果={}\n'.format(request._get_raw_host())
   content += 'request.get_host()结果={}\n'.format(request.get_host())
   content += 'request.get_port()结果={}\n'.format(request.get_port())
   content += 'request.get_full_path()结果={}\n'.format(request.get_full_path())
   content += 'request.get_raw_uri()结果={}\n'.format(request.get_raw_uri())
   content += 'request.build_absolute_uri()结果={}\n'.format(
                       request.build_absolute_uri())
   content += 'request._current_scheme_host={}\n'.format(
                       request._current_scheme_host)
   content += 'request.scheme={}\n'.format(request.scheme)

return HttpResponse(content)

# ……
```

启动 first_django 服务（python manage.py runserver 192.168.26.110:8000），在 ceph-1 节点上使用 curl 命令模拟 HTTP 请求，同时使用-H 选项模拟请求头参数，结果如下：

```
[store@ceph-1 ~]$ curl http://192.168.26.110:8000/test_view/hello/ -H "Accept-Language:
zh-cn" -H 'test: hello, world'
request.headers={'Content-Length': '', 'Content-Type': 'text/plain', 'User-Agent':
'curl/7.29.0', 'Host': '192.168.26.110:8000', 'Accept': '*/*', 'Accept-Language':
'zh-cn', 'Test': 'hello, world'}
request._get_raw_host()结果=192.168.26.110:8000
request.get_host()结果=192.168.26.110:8000
request.get_port()结果=8000
request.get_full_path()结果=/test_view/hello/
request.get_raw_uri()结果=http://192.168.26.110:8000/test_view/hello/
request.build_absolute_uri()结果=http://192.168.26.110:8000/test_view/hello/
request._current_scheme_host=http://192.168.26.110:8000
```

```
request.scheme=http
```

从上面的结果不难看出 get_host()等方法的含义。此外，通过阅读 get_host()方法的源码也可以看出，设置 ALLOWED_HOSTS 变量（位于项目的配置文件 settings.py 中）可以限制请求来源的地址，内容如下：

```
# 源码位置：django/http/request.py
# ……

class HttpRequest:

    # ……

    def get_host(self):
        host = self._get_raw_host()

        allowed_hosts = settings.ALLOWED_HOSTS  # 从配置文件中获取允许请求的 hosts
        if settings.DEBUG and not allowed_hosts:
            # 如果设置 DEBUG=True，并且 allowed_hosts 为空，则需要设置默认的 allowed_hosts 值
            allowed_hosts = ['localhost', '127.0.0.1', '[::1]']

        # 获取请求
        domain, port = split_domain_port(host)
        if domain and validate_host(domain, allowed_hosts):
            # 校验通过，返回正常的 host
            return host
        else:
            # 校验不通过，抛出 DisallowedHost 异常
            msg = "Invalid HTTP_HOST header: %r." % host
            if domain:
                msg += " You may need to add %r to ALLOWED_HOSTS." % domain
            else:
                msg += " The domain name provided is not valid according to RFC 1034/1035."
            raise DisallowedHost(msg)

    # ……
```

（2）处理 cookie 相关的方法。get_signed_cookie()方法可以获取请求中加密 cookie 中某个 key 对应的 value 值，而该加密 cookie 一般是通过 HttpResponse 类中的 set_signed_cookie()方法返回给客户端的。get_signed_cookie()方法的代码如下：

```
# 源码位置：django/http/request.py
# ……

class HttpRequest:

    # ……
```

```
    def get_signed_cookie(self, key, default=RAISE_ERROR, salt='', max_age=None):

        try:
            # 从请求中获取加密 cookie
            cookie_value = self.COOKIES[key]
        except KeyError:
            # 如果不是 RAISE_ERROR，则返回默认值 default，否则抛出异常
            if default is not RAISE_ERROR:
                return default
            else:
                raise
        try:
            # 从加密的 cookie 中获取 key 对应的 value 值
            value = signing.get_cookie_signer(salt=key + salt).unsign(
                cookie_value, max_age=max_age)
        except signing.BadSignature:
            if default is not RAISE_ERROR:
                return default
            else:
                raise
        # 返回获取的 value 值
        return value

    # ……
```

后续我们将在 HttpResponse 中给出一个关于 cookies 数据签名和解密的实验，以帮助读者更好地理解该部分源码。

（3）处理与上传文件相关的方法，如_initialize_handlers()（初始化文件上传处理器）、upload_handlers()（文件上传处理器）和 parse_file_upload()（解析上传的文件）等。相关方法的源码如下：

```
# 源码位置：django/http/request.py
# ……

class HttpRequest:

    # ……

    def _initialize_handlers(self):
        # 导入文件上传处理器，相关模块路径在 django/conf/global_settings.py 文件中设置
        self._upload_handlers = [uploadhandler.load_handler(handler, self)
                                 for handler in settings.FILE_UPLOAD_HANDLERS]

    @property
    def upload_handlers(self):
```

```
        if not self._upload_handlers:
            # 如果没有定义文件上传处理器，则从 Django 的配置模块中导入
            self._initialize_handlers()
        # 返回导入的文件上传处理器
        return self._upload_handlers

    @upload_handlers.setter
    def upload_handlers(self, upload_handlers):
        # 设置文件上传处理器
        if hasattr(self, '_files'):
            raise AttributeError("...")
        self._upload_handlers = upload_handlers

    def parse_file_upload(self, META, post_data):
        # 将 self.upload_handlers 转成不可变列表
        self.upload_handlers = ImmutableList(
            self.upload_handlers,
            warning="You cannot alter upload handlers after the upload has been processed."
        )
        parser = MultiPartParser(META, post_data, self.upload_handlers, self.encoding)
        # 解析上传的文件
        return parser.parse()

    @property
    def body(self):
        if not hasattr(self, '_body'):
            if self._read_started:
                # 在开始读取请求的数据流之前，无法获取 body 属性值
                raise RawPostDataException("...")

            # 限制请求数据的大小，这些数据将在内存中处理，相关配置变量已经非常清楚
            if (settings.DATA_UPLOAD_MAX_MEMORY_SIZE is not None and
                    int(self.META.get('CONTENT_LENGTH') or 0) >
settings.DATA_UPLOAD_MAX_MEMORY_SIZE):

                raise RequestDataTooBig('...')

            try:
                self._body = self.read()
            except IOError as e:
                raise UnreadablePostError(*e.args) from e
            self._stream = BytesIO(self._body)
        return self._body

    def _mark_post_parse_error(self):
        # 分别给_post 和_files 属性设置初始对象
        self._post = QueryDict()
        self._files = MultiValueDict()
```

```
    def _load_post_and_files(self):
        # 设置_post 和_files 属性值，获得 POST 请求参数和请求文件数据
        if self.method != 'POST':
            # 如果不是 POST 请求，则直接返回
            self._post, self._files = QueryDict(encoding=self._encoding),
MultiValueDict()
            return
        if self._read_started and not hasattr(self, '_body'):
            # 如果读操作已经开始，但是请求对象没有_body 属性，则标记异常并返回
            self._mark_post_parse_error()
            return

        if self.content_type == 'multipart/form-data':
            # 对于 content-type 类型为 multipart/form-data 的处理
            if hasattr(self, '_body'):
                # 使用已经读取的数据
                data = BytesIO(self._body)
            else:
                data = self
            try:
                self._post, self._files = self.parse_file_upload(self.META, data)
            except MultiPartParserError:
                self._mark_post_parse_error()
                raise
        elif self.content_type == 'application/x-www-form-urlencoded':
            self._post, self._files = QueryDict(self.body, encoding=self._encoding),
MultiValueDict()
        else:
            self._post, self._files = QueryDict(encoding=self._encoding),
MultiValueDict()

    # ……
```

在上面的源码中，对文件上传处理器的部分已经在 5.6 节中介绍过，不再赘述。

## 6.2.2　HttpResponse 类的源码

由于 HttpResponse 类继承自 HttpResponseBase 类，因此先学习 HttpResponseBase 类。HttpResponseBase 类的实现源码如下：

```
# 源码位置：django/http/response.py
# ……

class HttpResponseBase:
    """
    HTTP 响应的基类，将请求头转换成字典类型进行处理。
```

```
    该类并不会直接处理请求的数据内容，而是由继承的子类（如 HttpResponse
    和 StreamingHttpResponse）实现
    """

    status_code = 200

    def __init__(self, content_type=None, status=None, reason=None, charset=None):
        self._headers = {}  # 将请求头转成字典格式保存在_headers 属性中
        self._closable_objects = []
        self._handler_class = None
        self.cookies = SimpleCookie()  # 返回的 cookies 信息
        self.closed = False
        if status is not None:
            try:
                self.status_code = int(status)   # 设置返回状态码
            except (ValueError, TypeError):
                raise TypeError('HTTP status code must be an integer.')

            if not 100 <= self.status_code <= 599:   # 如果不是正常的状态码，则抛出异常
                raise ValueError('HTTP status code must be an integer from 100 to 599.')
        self._reason_phrase = reason
        self._charset = charset
        if content_type is None:    # 设置响应中的 content-type 参数
            content_type = '%s; charset=%s' % (settings.DEFAULT_CONTENT_TYPE,
                                               self.charset)
        self['Content-Type'] = content_type

    # ……

    def serialize_headers(self):
        def to_bytes(val, encoding):
            return val if isinstance(val, bytes) else val.encode(encoding)

        # 将变量响应头中的 key-value 值转换成字节形式
        headers = [
            (to_bytes(key, 'ascii') + b': ' + to_bytes(value, 'latin-1'))
            for key, value in self._headers.values()
        ]
        return b'\r\n'.join(headers)   # 使用'\r\n'将请求头元素连接起来

    __bytes__ = serialize_headers

    # ……

    def __setitem__(self, header, value):
        # 魔法函数，支持 self[key] = 'xxx'形式
        header = self._convert_to_charset(header, 'ascii')
        value = self._convert_to_charset(value, 'latin-1', mime_encode=True)
```

```
        self._headers[header.lower()] = (header, value)

    def __delitem__(self, header):
        self._headers.pop(header.lower(), False)

    def __getitem__(self, header):
        # 魔法函数，支持使用 self[key]语句查询结果
        return self._headers[header.lower()][1]

    def has_header(self, header):
        return header.lower() in self._headers

    __contains__ = has_header  # 魔法函数，支持 for header in self 语句

    def items(self):
        return self._headers.values()

    def get(self, header, alternate=None):
        return self._headers.get(header.lower(), (None, alternate))[1]

    def set_cookie(self, key, value='', max_age=None, expires=None, path='/',
                   domain=None, secure=False, httponly=False, samesite=None):
        '''设置 cookie 信息'''
        self.cookies[key] = value
        if expires is not None:
            if isinstance(expires, datetime.datetime):
                # 如果设置的 expires 是 datetime.datetime 类型的实例
                if timezone.is_aware(expires):
                    expires = timezone.make_naive(expires, timezone.utc)
                delta = expires - expires.utcnow()   # 计算和当前时间的差值
                delta = delta + datetime.timedelta(seconds=1)
                expires = None
                # 计算时间差代表的秒数，86400=24×60×60
                max_age = max(0, delta.days * 86400 + delta.seconds)
            else:
                # 其余情况，直接设置 self.cookies[key]['expires']值
                self.cookies[key]['expires'] = expires
        else:
            #当 expires 未设置时，将 self.cookies[key]['expires']值设置为空
            self.cookies[key]['expires'] = ''
        if max_age is not None:
            # 直接设置 self.cookies[key]['max-age']值
            self.cookies[key]['max-age'] = max_age
            # IE requires expires, so set it if hasn't been already.
            if not expires:
                # 如果 expires 未设置，则根据 max-age 值计算过期时间
                self.cookies[key]['expires'] = http_date(time.time() + max_age)
        # 根据位置参数设置 self.cookies
```

```
        if path is not None:
            self.cookies[key]['path'] = path
        if domain is not None:
            self.cookies[key]['domain'] = domain
        if secure:
            self.cookies[key]['secure'] = True
        if httponly:
            self.cookies[key]['httponly'] = True
        if samesite:
            if samesite.lower() not in ('lax', 'strict'):
                raise ValueError('samesite must be "lax" or "strict".')
            self.cookies[key]['samesite'] = samesite

    def setdefault(self, key, value):
        if key not in self:  # 当 key 不是 self._headers 字典中的键时，设置对应的键
            self[key] = value

    def set_signed_cookie(self, key, value, salt='', **kwargs):
        # 设置带签名的 cookie 信息
        value = signing.get_cookie_signer(salt=key + salt).sign(value)
        return self.set_cookie(key, value, **kwargs)

    def delete_cookie(self, key, path='/', domain=None, samesite=None):
        secure = key.startswith(('__Secure-', '__Host-'))
        self.set_cookie(
            key, max_age=0, path=path, domain=domain, secure=secure,
            expires='Thu, 01 Jan 1970 00:00:00 GMT', samesite=samesite,
        )

    # 子类中使用的一些公共方法
    # ……
```

在上述源码中，HttpResponseBase 类的核心是操作_headers 属性和 cookies 属性，它们分别代表响应头及返回给客户端的 cookies 信息。为了能直观地看到上述方法的输出，我们在本章实验 2 的基础上进行改造。

（1）调整对应的视图类 HelloView 的代码如下：

```
# 源码位置：test_view/views.py
# ……

class HelloView(View):

    def get(self, request, *args, **kwargs):
        content = ''
        response = HttpResponse(content='', content_type='text/html', status=201, 
charset='utf-8')
        # 调用 HttpResponseBase 类中的魔法函数__setitem__()，设置_headers 属性
```

```
        response['TEST_header'] = 'value1'
        content += '赋值前:{}\n'.format(response._headers)
        content += '获取"test_header"的键:{}\n'.format(response['test_header'])
        content += '判断键是否在:{}\n'.format('test_header' in response)
        del response['test_header']
        content += '删除头后:{}\n'.format('test_header' in response)

        # 请求后查看响应头数据
        response['Hello'] = 'world'
        # 设置的另一种方式，如果 test2 不在_header 的键中，则设置成功
        response.setdefault('test1', 'value1')

        # 设置签名后的 cookies
        response.set_signed_cookie('user', 'spyinx', max_age=10)
        content += '设置 cookies 后:{}\n'.format(response.cookies)
        response.content = content
        return response
    # ……

# ……
```

在上述代码中，我们使用 HttpResponse 类代替 HttpResponseBase 类完成了相关测试。实际上，在 HttpResponse 类中并没有定义操作_headers 和 cookies 的魔法函数及相关方法，因此上述代码中的相关语句最终都是调用父类（即 HttpResponseBase 类）中的相关代码实现的。

（2）准备一段测试用的 Python 代码，使用 requests 模块模拟 HTTP 请求，请求路径为 /test_view/hello_class/，代码如下：

```
(django2-core-test) [root@master first_django]# cat test_response.py
from datetime import datetime

import requests

response = requests.get('http://192.168.26.110:8000/test_view/hello_class/')
response.encoding = 'utf-8'
print(response.text)
print('当前时间:', datetime.now().strftime('%Y-%m-%d %H:%M:%S'))
print('headers:{}'.format(response.headers))
print('cookies:{}'.format(response.cookies))
print('带签名的用户:', response.cookies.get('user'))
```

在 Django 中，默认的时区为 TIME_ZONE = 'America/Chicago'（在 Django 的源码文件 django/conf/global_settings.py 中），为了保证过期时钟正常，需要在 first_django/settings.py 文件中调整 TIME_ZONE 的值为 Asia/Shanghai：

```
# 源码位置：first_django/settings.py
# ……
```

```
TIME_ZONE = 'Asia/Shanghai'

# ……
```

启动 first_django 服务，执行 test_response.py 中的代码（应使用 Python 3 执行，如果使用的是 Python 2，则会存在编码问题，需要调整代码），结果如下：

```
(django2-core-test) [root@master first_django]# python manage.py runserver
192.168.26.110:8000
# ……


# 打开另一个终端窗口进行测试
(django2-core-test) [root@master first_django]# python test_response.py
赋值前：{'content-type': ('Content-Type', 'text/html'), 'test_header': ('TEST_header',
'value1')}
获取"test_header"的键：value1
判断键是否在：True
删除头后：False
设置 cookies 后：Set-Cookie: user=spyinx:1laAdA:ZAuCgYMb7tXKfcAkwZFkWqXcp90; expires=Sat,
24 Apr 2021 05:15:10 GMT; Max-Age=10; Path=/

当前时间：2021-04-24 13:15:00
headers:{'Date': 'Sat, 24 Apr 2021 05:15:00 GMT', 'Server': 'WSGIServer/0.2 CPython/3.8.6',
'Content-Type': 'text/html', 'Hello': 'world', 'test1': 'value1', 'X-Frame-Options':
'SAMEORIGIN', 'Content-Length': '314', 'Set-Cookie':
'user=spyinx:1laAdA:ZAuCgYMb7tXKfcAkwZFkWqXcp90; expires=Sat, 24 Apr 2021 05:15:10 GMT;
Max-Age=10; Path=/'}
cookies:<RequestsCookieJar[<Cookie user=spyinx:1laAdA:ZAuCgYMb7tXKfcAkwZFkWqXcp90 for
192.168.26.110/>]>
带签名的用户：spyinx:1laAdA:ZAuCgYMb7tXKfcAkwZFkWqXcp90
(django2-core-test) [root@master first_django]# date
Sat Apr 24 13:15:09 CST 2021
```

**注意**，结果中返回的 cookies 的过期时间比当前时间慢了 8 小时（添加的 10s 先忽略），这是因为格林尼治时间（GMT）与北京时间相差 8 小时。如果想要获取正确的结果，应如何处理呢？在 HttpResponseBase 类的 set_cookie()方法中有这样一个逻辑，即如果没有设置过期时间参数，但设置了过期的最大秒数（max_age），则执行如下语句计算过期时间：

```
# 在 set_cookie()方法中，设置如下语句
self.cookies[key]['expires'] = http_date(time.time() + max_age)
```

从上面的代码可以看到，cookies 中的过期时间是由 http_date()方法得到的。http_date()方法的实现源码如下：

```
# 源码位置：django/utils/http.py
# ……
```

```
from email.utils import formatdate  # 内置模块
# ……

def http_date(epoch_seconds=None):
    return formatdate(epoch_seconds, usegmt=True)

# ……
```

从上面的结果可以看到，http_date()方法会调用 Python 3 内置的 formatdate()函数转换时间戳，usegmt 参数负责将北京时间戳转换成格林尼治时间，因此会出现前面的 8 小时时差。操作示例如下：

```
>>> import time
>>> from email.utils import formatdate
>>> from datetime import datetime
>>> timestamp = int(time.time())  # 获取当前时间戳
>>> datetime.fromtimestamp(timestamp).strftime('%Y-%m-%d %H:%M:%S')  # 打印北京时间
'2021-04-24 14:14:00'
>>> formatdate(timestamp, usegmt=True)  # 打印格林威治时间
'Sat, 24 Apr 2021 06:14:00 GMT'
```

如何让 formatdate()函数打印本地时间呢？只需进一步追踪 formatdate()函数支持的参数即可：

```
# Python 3 标准库下的位置：lib/email/utils.py
# ……

def formatdate(timeval=None, localtime=False, usegmt=False):
    if timeval is None:
        timeval = time.time()
    if localtime or usegmt:
        dt = datetime.datetime.fromtimestamp(timeval, datetime.timezone.utc)
    else:
        dt = datetime.datetime.utcfromtimestamp(timeval)
    if localtime:
        dt = dt.astimezone()
        usegmt = False
    return format_datetime(dt, usegmt)

# ……
```

从上面的代码可以看到，在 formatdate()函数中有一个 localtime 参数，默认为 False。继续测试该参数，如下：

```
>>> formatdate(timestamp, localtime=True)
'Sat, 24 Apr 2021 14:14:00 +0800'
>>> formatdate(timestamp, localtime=True, usegmt=True)
'Sat, 24 Apr 2021 14:14:00 +0800'
```

从上面的代码可以看到，只需设置 localtime 参数为 True 即可得到和本地一样的时间。为了得到和本地一样的时间，修改 http_date()方法的源码如下：

```
# 源码位置：django/utils/http.py
# ……

def http_date(epoch_seconds=None):
    return formatdate(epoch_seconds, True)
    # 使用上面的语句，屏蔽原 return 语句
    # return formatdate(epoch_seconds, usegmt=True)

# ……
```

为了验证修改后的效果，需要同步修改虚拟环境中的源码，然后启动 first_django 服务，再次测试 test_response.py 文件中的代码，即可看到正确的过期时间：

```
# 修改虚拟环境中的源码
(django2-core-test) [root@master first_django]# vim
~/.pyenv/versions/django2-core-test/lib/python3.8/site-packages/django/utils/http.py
(django2-core-test) [root@master first_django]# python manage.py runserver
192.168.26.110:8000
# ……

# 打开另一个终端，进入虚拟环境后运行 test_response.py 文件中的代码
(django2-core-test) [root@master first_django]# python test_response.py
# ……

设置 cookies 后:Set-Cookie: user=spyinx:1laBqo:CdGGOKffONmzD5CeRPcFKzZHx7M; expires=Sat,
24 Apr 2021 14:33:20 +0800; Max-Age=10; Path=/

当前时间: 2021-04-24 14:33:10
headers:{'Date': 'Sat, 24 Apr 2021 06:33:10 GMT', 'Server': 'WSGIServer/0.2 CPython/3.8.6',
'Content-Type': 'text/html', 'Hello': 'world', 'test1': 'value1', 'X-Frame-Options':
'SAMEORIGIN', 'Content-Length': '316', 'Set-Cookie':
'user=spyinx:1laBqo:CdGGOKffONmzD5CeRPcFKzZHx7M; expires=Sat, 24 Apr 2021 14:33:20 +0800;
Max-Age=10; Path=/'}
cookies:<RequestsCookieJar[<Cookie user=spyinx:1laBqo:CdGGOKffONmzD5CeRPcFKzZHx7M for
192.168.26.110/>]>
带签名的用户: spyinx:1laBqo:CdGGOKffONmzD5CeRPcFKzZHx7M
```

接下来通过一个简单的示例，帮助读者更好地理解在 HttpRequest 类和 HttpResponseBase 类中操作 cookies 的方法。

### 6.2.3 HttpRequest 类和 HttpResponseBase 类的操作示例

（1）准备一个视图类，该类中的 get()方法会校验请求中的 cookies。如果在解析签名的 cookies 并提取用户信息时出现异常，则返回需要登录的信息。post()方法会在用户名和密码验证通过后返回相应的 cookies，并在 cookies 中放入签名版的用户信息。具体内容如下：

```
# 源码位置：test_view/views.py
# ……

class TestLoginCookieView(View):

    def get(self, request, *args, **kwargs):
        success = False
        if hasattr(request, 'COOKIES'):
            print('cookie=', request.COOKIES)
        # 从请求中获取 cookies
        user = request.get_signed_cookie('user', default='anonymous', salt=default_salt,
max_age=10)
        if user and user != 'anonymous':
            return HttpResponse('登录成功，欢迎{}'.format(user))
        return HttpResponse('您好，请先登录')

    def post(self, request, *args, **kwargs):
        # 获取请求参数
        name = request.POST.get('username')
        password = request.POST.get('password')
        # 校验密码
        if password != 'SPYinx123456':
            return HttpResponse('用户名或密码不正确')

        response = HttpResponse(content='登录成功')
        # 设置带签名的 user 值，将其保存到相应的 cookies 中
        response.set_signed_cookie('user', name, salt=default_salt, max_age=10)
        return response

# ……
```

（2）Python 测试代码如下：

```
(django2-core-test) [root@master first_django]# cat test_cookies.py
import time
import requests

result = requests.get("http://192.168.26.110:8000/test_view/login_cookie/")
print('未登录，请求:', result.text)

params = {
    'username': 'spyinx',
    'password': 'SPYinx123456'
}
result = requests.post("http://192.168.26.110:8000/test_view/login_cookie/",
data=params)
print('得到的 cookie:', result.cookies)
cookies = result.cookies
```

```
result = requests.get("http://192.168.26.110:8000/test_view/login_cookie/",
cookies=cookies)
print('第一次请求:', result.text)

time.sleep(12)

result = requests.get("http://192.168.26.110:8000/test_view/login_cookie/",
cookies=cookies)
print('12s 之后再次请求:', result.text)
```

（3）启动 first_django 项目，在另一个连接终端上测试上述代码，结果如下：

```
(django2-core-test) [root@master first_django]# python test_cookies.py
未登录，请求：您好，请先登录
得到的 cookie: <RequestsCookieJar[<Cookie
user=spyinx:1laCJr:-WuLGDiOZ9Z-kfWTlrvCRzs48RA for 192.168.26.110/>]>
第一次请求：登录成功，欢迎 spyinx
12s 之后再次请求：您好，请先登录
```

可以看到，在 first_django 项目中已成功模拟了基于 cookies 登录的简单过程。继续学习 HttpResponse 类的源码内容：

```
class HttpResponse(HttpResponseBase):
    """
    一个带字符串内容的 HTTP 响应类
    """

    streaming = False

    def __init__(self, content=b'', *args, **kwargs):
        super().__init__(*args, **kwargs)
        # Content is a bytestring.
        self.content = content

    def __repr__(self):
        return '<%(cls)s status_code=%(status_code)d%(content_type)s>' % {
            'cls': self.__class__.__name__,
            'status_code': self.status_code,
            'content_type': self._content_type_for_repr,
        }

    def serialize(self):
        """返回完整的 HTTP 信息"""
        return self.serialize_headers() + b'\r\n\r\n' + self.content

    __bytes__ = serialize

    @property
```

```
    def content(self):
        return b''.join(self._container)

    @content.setter
    def content(self, value):
        if hasattr(value, '__iter__') and not isinstance(value, (bytes, str)):
            content = b''.join(self.make_bytes(chunk) for chunk in value)
            if hasattr(value, 'close'):
                try:
                    value.close()
                except Exception:
                    pass
        else:
            content = self.make_bytes(value)  # 将 content 转换成字节
        self._container = [content]

    def __iter__(self):
        return iter(self._container)  # 迭代_container 属性，它是一个列表

    def write(self, content):
        # 将 content 转换成字节后，添加到_container 属性列表中
        self._container.append(self.make_bytes(content))

    def tell(self):
        return len(self.content)  # 返回 content 属性值的长度

    def getvalue(self):
        return self.content  # 返回 content 属性值

    def writable(self):
        return True  # 一直返回 True

    def writelines(self, lines):
        for line in lines:
            self.write(line)

# ……
```

HttpResponse 类中定义的相关方法都比较简单，下面通过实际操作演示这些方法的功能：

```
>>> from django.http.response import HttpResponse
>>> response = HttpResponse('这是一个测试')
>>> response  # 执行魔法函数__repr__()
<HttpResponse status_code=200, "text/html; charset=utf-8">
>>> response.serialize()   # 打印响应头及响应内容信息，字节化
b'Content-Type: text/html;
charset=utf-8\r\n\r\n\xe8\xbf\x99\xe6\x98\xaf\xe4\xb8\x80\xe4\xb8\xaa\xe6\xb5\x8b\xe
8\xaf\x95'
```

```
>>> response.__bytes__()  # 因为__bytes__ = serialize，所以__bytes__()即serialize()
b'Content-Type: text/html;
charset=utf-8\r\n\r\n\xe8\xbf\x99\xe6\x98\xaf\xe4\xb8\x80\xe4\xb8\xaa\xe6\xb5\x8b\xe8\xaf\x95'
>>> response.content  # 查看对象的content属性，注意@content.setter下面的方法
b'\xe8\xbf\x99\xe6\x98\xaf\xe4\xb8\x80\xe4\xb8\xaa\xe6\xb5\x8b\xe8\xaf\x95'
>>> response.content.decode('utf-8')  # 将字节内容解码成字符串
'这是一个测试'
>>> response.tell()  # 获取content长度，共有18字节
18
>>> response.getvalue().decode('utf-8')
'这是一个测试'
>>> response.write('新的一行')  #添加内容到_container列表中
>>> response.getvalue().decode('utf-8')
'这是一个测试，新的一行'
>>> [content.decode('utf-8') for content in response._container]
['这是一个测试', '新的一行']
```

至此，对 HTTP 请求和响应封装的相关核心类就介绍完毕了，接下来深入分析在 Django 项目中一个常规 HTTP 请求的完整处理过程。

## 6.3 视图层核心源码解读

本节探索在 Django 项目中一个常规 HTTP 请求的完整处理过程，本节内容将以 6.1 节中的实验 1 为例进行展开。从 5.2 节的最后部分继续追踪 Django 源码，HTTP 请求的处理入口源码如下：

```
# 源码位置：django/core/handlers/wsgi.py
# ……

class WSGIHandler(base.BaseHandler):
    request_class = WSGIRequest

    def __init__(self, *args, **kwargs):
        super().__init__(*args, **kwargs)
        # 载入中间件
        self.load_middleware()

    def __call__(self, environ, start_response):
        # ……
        request = self.request_class(environ)
        # 得到响应结果
        response = self.get_response(request)

        # ……

        # 返回请求响应结果
```

```
        return response

    # ……
```

从上面的源码可以看到，request 变量其实是一个 WSGIRequest 对象，它将在后续的 HTTP 处理请求中进行传递。最核心的语句是 self.get_response()，该语句可以得到最后的响应结果。继续追踪 get_response()方法的实现源码，它在父类 base.BaseHandler 中，内容如下：

```
# 源码位置：django/core/handlers/base.py
# ……

class BaseHandler:

    # ……

    def get_response(self, request):
        """根据给定的 HttpRequest 对象，返回一个 HttpResponse 对象"""

        set_urlconf(settings.ROOT_URLCONF)     # 为该线程设置默认的 URL 解析器
        response = self._middleware_chain(request) # 核心语句是 self._middleware_chain
        # 下面是处理响应
        response._closable_objects.append(request)
        if response.status_code >= 400:
            log_response(
                '%s: %s', response.reason_phrase, request.path,
                response=response,
                request=request,
            )
        return response

    # ……
```

可以看到，在 get_response()方法方法中，首先，使用 set_urlconf()函数保存了 URLConf 配置的入口地址；然后，执行 self._middleware_chain(request)语句获得对 HTTP 请求的响应。其中，set_urlconf()函数的功能非常简单，只是用来设置本地变量_urlconfs 的 value 属性值：

```
# 源码位置：django/urls/base.py
# ……

_urlconfs = local()

# ……

def set_urlconf(urlconf_name):
    if urlconf_name:
        _urlconfs.value = urlconf_name  # 保存 urlconf_name
    else:
```

```
        if hasattr(_urlconfs, "value"):
            del _urlconfs.value        # 删除对应的属性值

# ……
```

下面分析 self._middleware_chain 的属性值。通过查看 base.BaseHandler 类的源码可知，BaseHandler 对象的_middleware_chain 属性值是在 load_middleware()方法中赋值的，如下：

```
# 源码位置：django/core/handlers/base.py
# ……

class BaseHandler:
    # ……
    _middleware_chain = None

    def load_middleware(self):

        self._view_middleware = []
        self._template_response_middleware = []
        self._exception_middleware = []

        handler = convert_exception_to_response(self._get_response)
        for middleware_path in reversed(settings.MIDDLEWARE):
            middleware = import_string(middleware_path)
            try:
                mw_instance = middleware(handler)
            except MiddlewareNotUsed as exc:
                # 加载某个中间件异常，如果设置了 DEBUG=True，则打印异常日志
                # ……
                continue

            # 如果 mw_instance 为 None，则直接抛出异常
            # ……

            # 将各中间件中的方法添加到对应的属性中
            if hasattr(mw_instance, 'process_view'):
                self._view_middleware.insert(0, mw_instance.process_view)
            if hasattr(mw_instance, 'process_template_response'):
                self._template_response_middleware.append(
                        mw_instance.process_template_response)
            if hasattr(mw_instance, 'process_exception'):
                self._exception_middleware.append(mw_instance.process_exception)

            handler = convert_exception_to_response(mw_instance)

        # 注意，_middleware_chain 属性值等于 handler
        self._middleware_chain = handler
```

```
    # ……
```

关于中间件的处理细节将在第 7 章中进行分析，这里只需知道一点，即_middleware_chain 属性值代表着一个方法名，是由装饰函数 convert_exception_to_response()多次处理后得到的：

```
# 源码位置：django/core/handlers/exception.py
# ……

def convert_exception_to_response(get_response):
    @wraps(get_response)
    def inner(request):
        try:
            # 执行传入的方法，得到响应
            response = get_response(request)
        except Exception as exc:
            # 处理异常
            response = response_for_exception(request, exc)
        # 返回响应
        return response
    # 返回 inner 函数名
    return inner

# ……
```

而 self._middleware_chain()语句最终将执行_get_response()方法来处理 HTTP 请求（至于为何调用_get_response()方法将在第 7 章中介绍），继续追踪_get_response()方法的实现源码：

```
# 源码位置：django/core/handlers/base.py
# ……

class BaseHandler:
    # ……
    def _get_response(self, request):
        response = None

        if hasattr(request, 'urlconf'):
            urlconf = request.urlconf
            set_urlconf(urlconf)
            resolver = get_resolver(urlconf)
        else:
            # 通常情况下，执行这里的语句将得到 URLConf 解析器
            resolver = get_resolver()

        # 根据请求路径匹配 URLConf 值
        resolver_match = resolver.resolve(request.path_info)
        # 获得该路径对应的处理函数及函数参数
        callback, callback_args, callback_kwargs = resolver_match
        request.resolver_match = resolver_match
```

```
        # 依次进入中间件
        for middleware_method in self._view_middleware:
            response = middleware_method(request, callback, callback_args,
callback_kwargs)
            if response:
                break

        if response is None:
            # 最终得到处理对应 URL 路径的视图函数，是在文件 urls.py 中定义的
            wrapped_callback = self.make_view_atomic(callback)
            try:
                # 执行对应的视图函数，注意视图函数的三个参数
                response = wrapped_callback(request, *callback_args, **callback_kwargs)
            except Exception as e:
                response = self.process_exception_by_middleware(e, request)

        if response is None:
            # 如果 HTTP 请求处理出现异常，则直接抛出异常信息
            # ……

        elif hasattr(response, 'render') and callable(response.render):
            # 由模板层文件进行渲染，需要额外处理
            for middleware_method in self._template_response_middleware:
                response = middleware_method(request, response)
                # 如果得到的 response 为 None，则直接抛出异常
                # ……

            try:
                # 进行渲染，最终得到响应结果
                response = response.render()
            except Exception as e:
                response = self.process_exception_by_middleware(e, request)

        # 得到初步的请求响应并返回
        return response

# ……
```

通过阅读_get_response()方法的源码，可以看到 Django 框架处理 HTTP 请求的大致过程。首先，得到 URLConf 解析器，即一个 URLResolver 对象。其次，根据请求的 URL 路径及在 Django 项目中所有设置的 URL 映射关系表(由各应用下的 urls.py 文件得到)匹配出符合该路径的 URLConf 配置，从中可以得到处理该请求路径的视图函数 callback()。最后，Django 项目将调用该视图函数处理 HTTP 请求，并得到 HttpResponse 对象。如果在该响应中返回了待渲染的模板文件，则还需要调用 response.render()做进一步处理，最终返回标准的响应数据。

### 6.3.1　HTTP 请求路径的匹配过程

（1）从 get_resolver()函数开始介绍（在 5.4 节中介绍过），该函数的实现源码如下：

```
# 源码位置：django/urls/resolvers.py
# ……

@functools.lru_cache(maxsize=None)
def get_resolver(urlconf=None):
    if urlconf is None:
        urlconf = settings.ROOT_URLCONF
    return URLResolver(RegexPattern(r'^/'), urlconf)

# ……
```

对于 first_django 项目来说，settings.ROOT_URLCONF 的值为 first_django.urls，在匹配路径时将调用 URLResolver 对象的 resolve()方法：

```
# 源码位置：django/urls/resolvers.py
# ……

class URLResolver:

    # ……

    def resolve(self, path):
        path = str(path)  # path may be a reverse_lazy object
        tried = []
        match = self.pattern.match(path)  # 第一次匹配，输入的 pattern 是 RegexPattern(r'^/')
        if match:
            new_path, args, kwargs = match   # new_path 是匹配的路径
            for pattern in self.url_patterns:
                try:
                    sub_match = pattern.resolve(new_path)
                except Resolver404 as e:
                    # 处理异常
                    # ……
                else:
                    if sub_match:
                        sub_match_dict = {**kwargs, **self.default_kwargs}
                        sub_match_dict.update(sub_match.kwargs)
                        sub_match_args = sub_match.args
                        if not sub_match_dict:
                            sub_match_args = args + sub_match.args
                        # 当前路由
                        current_route = '' if isinstance(pattern, URLPattern) else
str(pattern.pattern)
                        return ResolverMatch(   # 返回一个 ResolverMatch 对象
```

```
                    sub_match.func,
                    sub_match_args,
                    sub_match_dict,
                    sub_match.url_name,
                    [self.app_name] + sub_match.app_names,
                    [self.namespace] + sub_match.namespaces,
                    self._join_route(current_route, sub_match.route),
                )
            # 没有匹配成功，记录尝试匹配的 URLConf 配置
            tried.append([pattern])
        # 没有任何匹配，直接抛出 404 异常
        raise Resolver404({'tried': tried, 'path': new_path})
    raise Resolver404({'path': path})  # 对于第 1 个^/，几乎不会出现不匹配的情况

# ……
```

resolve()方法的实现逻辑非常明确，当在 get_resolver()函数中实例化 URLResolver 类时，传入的 pattern 参数就是 RegexPattern(r'^/')，即得到的 URLResolver 对象中的 pattern 属性值为一个 RegexPattern 对象，其正则属性值为“^/”。因此在第一次调用 self.pattern.match()时，都能匹配成功。以请求 http://192.168.26.110:8000/test_view/hello/为例，下面演示在源码中第一步匹配的结果：

```
(django2-core-test) [root@master first_django]# python manage.py shell
Python 3.8.6 (default, Oct 18 2020, 15:33:08)
[GCC 4.8.5 20150623 (Red Hat 4.8.5-39)] on linux
Type "help", "copyright", "credits" or "license" for more information.
(InteractiveConsole)
>>> from django.urls.resolvers import RegexPattern
>>> path = "/test_view/hello/"
>>> pattern = RegexPattern(r'^/')
>>> pattern.match(path)
('test_view/hello/', (), {})
```

从上面的测试结果可以看出，对于请求路径/test_view/hello/，第一次匹配去掉了前面的“/”，这一点从 match()方法的源码可以看出：

```
# 源码位置：django/urls/resolvers.py
# ……

class RegexPattern(CheckURLMixin):
    regex = LocaleRegexDescriptor('_regex')

    def __init__(self, regex, name=None, is_endpoint=False):
        self._regex = regex
        # ……

    def match(self, path):
        match = self.regex.search(path)  # 使用 search
```

```
        if match:
            kwargs = match.groupdict()
            args = () if kwargs else match.groups()
            # 看返回的结果，后续的匹配会从上一个匹配位置的后续字符串开始
            return path[match.end():], args, kwargs
        return None

    # ……
```

注意，self.regex 是正则表达式的编译结果，它的值为 re.compile(self._regex)，相关的源码追踪可以参见 5.4 节中的介绍。self.regex 值为_compile()方法的返回结果：

```
# 源码位置：django/urls/resolvers.py
# ……

class URLResolver:
    # ……

    def _compile(self, regex):
        """Compile and return the given regular expression."""
        try:
            return re.compile(regex)
        except re.error as e:
            raise ImproperlyConfigured(
                '"%s" is not a valid regular expression: %s' % (regex, e)
            )

    # ……
```

此外，可以通过下面的语句得到 regex 的属性值：

```
>>> pattern = RegexPattern(r'^/')
>>> pattern.regex
re.compile('^/')
```

回到 resolve()方法，在第一次匹配后，new_path 变量其实去掉了“/”的值。而在接下来的循环语句中，最核心的就是 self.url_patterns，在 5.4 节中曾介绍过，它将获取入口的 URLConf 配置：

```
>>> from django.urls.resolvers import URLResolver
>>> from django.urls.resolvers import RegexPattern
>>> resolve = URLResolver(RegexPattern(r'^/'), 'first_django.urls')
>>> resolve.urlconf_module
<module 'first_django.urls' from
'/root/django-core-test/first_django/first_django/urls.py'>
>>> resolve.url_patterns
[<URLResolver <URLPattern list> (admin:admin) 'admin/'>, <URLResolver <module
'test_view.urls' from '/root/django-core-test/first_django/test_view/urls.py'>
(None:None) 'test_view/'>]
```

从上面的输出可以看到，first_django.urls 中定义的 URLConf 共有两个映射关系：“admin/”和“test_view/”。前者是一个 URLPattern 列表，后者是一个模块路径（test_view.urls），它们的真实定义如下：

```
# 源码位置：first_django/urls.py
# ……

urlpatterns = [
    path('admin/', admin.site.urls),
    url('test_view/', include('test_view.urls'))
]
```

结合 url_patterns 属性方法可知（在 5.4 节中曾介绍过，不再赘述），path()方法和 urls()方法最后返回的都是 URLResolver 对象。先追踪 path()方法的源码，内容如下：

```
# 源码位置：django/urls/conf.py
# ……

def _path(route, view, kwargs=None, name=None, Pattern=None):
    if isinstance(view, (list, tuple)):
        # 对include()的处理
        pattern = Pattern(route, is_endpoint=False)
        urlconf_module, app_name, namespace = view
        return URLResolver(
            pattern,
            urlconf_module,
            kwargs,
            app_name=app_name,
            namespace=namespace,
        )
    elif callable(view):
        # view 为视图函数
        pattern = Pattern(route, name=name, is_endpoint=True)
        return URLPattern(pattern, view, kwargs, name)
    else:
        raise TypeError('view must be a callable or a list/tuple in the case of include().')


path = partial(_path, Pattern=RoutePattern)
```

functools 下面的 partial(func)方法主要用来固定 func 中的某个参数，在后续调用时无须指定该参数。例如，下面的操作：

```
>>> from functools import partial
>>> def add(a, b):
...     return a + b
...
>>> add_ = partial(add, 3)  # 固定 add()方法的第 1 个参数为 3，所以后续只需传入一个参数即可
```

```
>>> add_(10)
13
>>> add_(12)
15
```

因此 path()函数是_path()函数传入的一个固定的 Pattern 参数。从_path()函数的源码可知，当传入的 view 参数可被调用时，将直接返回 URLPattern 对象。而对于这里要追踪的 HTTP 请求(/test_view/hello/) 来说，view=admin.site.urls 定义了 Django 内置的一些 URLConf 配置，演示如下：

```
(django2-core-test) [root@master first_django]# python manage.py shell
Python 3.8.6 (default, Oct 18 2020, 15:33:08)
[GCC 4.8.5 20150623 (Red Hat 4.8.5-39)] on linux
Type "help", "copyright", "credits" or "license" for more information.
(InteractiveConsole)
>>> from django.contrib import admin
>>> view = admin.site.urls
>>> view
([<URLPattern '' [name='index']>, <URLPattern 'login/' [name='login']>, <URLPattern
'logout/' [name='logout']>, <URLPattern 'password_change/' [name='password_change']>,
<URLPattern 'password_change/done/' [name='password_change_done']>, <URLPattern
'jsi18n/' [name='jsi18n']>, <URLPattern 'r/<int:content_type_id>/<path:object_id>/'
[name='view_on_site']>, <URLResolver <URLPattern list> (None:None) 'auth/group/'>,
<URLResolver <URLPattern list> (None:None) 'auth/user/'>, <URLPattern
'^(?P<app_label>auth)/$' [name='app_list']>], 'admin', 'admin')
>>> isinstance(view, (list, tuple))
True
```

再来看 url()方法，其源码如下：

```
# 源码位置：django/conf/urls/__init__.py
# ……

def url(regex, view, kwargs=None, name=None):
   return re_path(regex, view, kwargs, name)
```

include()函数、re_path()函数和 path()函数的源码实现均在同一个文件中：

```
# 源码位置：django/urls/conf.py
# ……

def include(arg, namespace=None):
   app_name = None
   if isinstance(arg, tuple):
      try:
         urlconf_module, app_name = arg
      except ValueError:
         # 异常处理，继续抛出异常
         # ……
   else:
```

```
        # 没有 namespace 信息，手工指定
        urlconf_module = arg

    if isinstance(urlconf_module, str):
        urlconf_module = import_module(urlconf_module)  # 注意最终将导入该 urls 模块
    # 得到导入的在 urls 模块中定义的 urlpatterns 变量，这里是定义 URLConf 配置的地方
    patterns = getattr(urlconf_module, 'urlpatterns', urlconf_module)
    app_name = getattr(urlconf_module, 'app_name', app_name) # 获取 app_name 名称
    if namespace and not app_name:
        # 如果有 namespace，但是没有 app_name，直接抛出异常
        # ……
    namespace = namespace or app_name
    if isinstance(patterns, (list, tuple)):
        for url_pattern in patterns:
            pattern = getattr(url_pattern, 'pattern', None)
            if isinstance(pattern, LocalePrefixPattern):
                raise ImproperlyConfigured(
                    'Using i18n_patterns in an included URLconf is not allowed.'
                )
    return (urlconf_module, app_name, namespace)

# ……
re_path = partial(_path, Pattern=RegexPattern)
```

re_path()函数和 path()函数的功能基本一致，只是固定的 Pattern 参数值不同，前者是 RegexPattern 对象，后者是普通的路径表达式 RoutePattern 对象。此外，include()函数也比较简单，它会导入指定的 urls 模块，然后加载其中定义的 urlpatterns 变量，并循环检查在该变量中指定的每个 URLConf 配置，最后返回一个三元组(urlconf_module, app_name, namespace)。其中，第 1 个元素为导入的 urls 模块。下面手工测试一下 include()函数的结果：

```
>>> from django.urls.conf import include
>>> include('test_view.urls')
(<module 'test_view.urls' from
'/root/django-core-test/first_django/test_view/urls.py'>, None, None)
```

继续回到 URLResolver 类的 resolve()方法中，从前面的测试可知，在 first_django 项目中，self.url_patterns 只有两个元素，均为 URLResolver 对象，因此循环中的 pattern 属性值都是 URLResolver 对象。循环中的 sub_match = pattern.resolve(new_path)语句非常关键，它再次调用了该类中的 resolve()方法，不过这次调用的参数及匹配的正则表达式有所不同：

```
>>> from django.urls import get_resolver
>>> resolver = get_resolver()
>>> patterns = resolver.url_patterns
>>> patterns[1].pattern
<django.urls.resolvers.RegexPattern object at 0x7ff40b0d6700>
>>> patterns[1].pattern._regex
```

```
'test_view/'
```

由于第 1 个 URLConf 规则（path('admin/', admin.site.urls)）不匹配测试的 HTTP 请求（前面的测试请求路径为/test_view/hello/），故直接考虑第 2 个配置。

**注意**，当再次调用 URLResolver 对象的 resolve()方法进行匹配时，该对象的 pattern 属性值为 RegexPattern 对象，而该对象的_regex 属性值正好是在 URLConf 中定义的“test_view/”，传入 resolve()方法的 new_path 参数为“test_view/hello/”。再次进入语句 match = self.pattern.match(path)，结果就不一样了：

```
# 接着上面的继续输入
>>> new_path = 'test_view/hello/'
>>> resolve = patterns[1]
>>> resolve.pattern.match(new_path)  # 匹配结果
('hello/', (), {})
>>> resolve.url_patterns
[<URLPattern 'hello/' [name='hello_view']>, <URLPattern 'hello_class/'
[name='hello_view_class']>, <URLPattern 'articles/<int:year>/<int:month>/'>,
<URLPattern
'articles_regex/(?P<year>[0-9]{4})/(?P<month>0[1-9]|1[0-2])/(?P<title>[a-zA-Z0-9-_]+
)/'>, <URLPattern 'articles_regex_extra/(?P<year>[0-9]{4})/'>, <URLPattern
'check_params/'>, <URLPattern 'file_upload/'>, <URLPattern 'file_upload2/'>,
<URLPattern 'login_cookie/'>, <URLPattern 'login_session/'>]
```

这次 resolve.url_patterns 的结果不再是 URLResolver 对象列表，而是 URLPattern 对象列表，对应的 urls.py 文件内容如下：

```
# 源码位置：first_django/test_view/urls.py
# ……

urlpatterns = [
    url('hello/', views.hello_view, name='hello_view'),
    url('hello_class/', views.HelloView.as_view(), name='hello_view_class'),
    path('articles/<int:year>/<int:month>/', views.article_dynamic),
    re_path('articles_regex/(?P<year>[0-9]{4})/(?P<month>0[1-9]|1[0-2])/(?P<title>[a
-zA-Z0-9-_]+)/', views.article_dynamic_regex),
    re_path('articles_regex_extra/(?P<year>[0-9]{4})/', views.year_archive, {'hello':
'world'}),
    url('check_params/', views.check_params),
    url('file_upload/', views.file_upload),
    url('file_upload2/', views.file_upload2),
    url('login_cookie/', views.TestLoginCookieView.as_view()),
    url('login_session/', views.TestLoginSessionView.as_view())
]
```

此时，在 test_view/urls.py 文件中没有 include()函数或者是其他模块，都是 URL 路径和视图函数，偶尔加上 name 参数。继续执行循环下面的 sub_match = pattern.resolve(new_path)语句，请看下面的模拟操作：

```
# 接着上面的继续输入，此时 new_path 变成了'hello/'
>>> new_path = 'hello/'
>>> patterns = resolve.url_patterns
>>> patterns[0]
<URLPattern 'hello/' [name='hello_view']>
>>> patterns[0].resolve(new_path)  # 只有第 1 个能匹配到 new_path 路径
ResolverMatch(func=test_view.views.hello_view, args=(), kwargs={},
url_name=hello_view,
app_names=[], namespaces=[], route=hello/)
>>> patterns[1].resolve(new_path)  # 后面的都匹配不到
>>> patterns[2].resolve(new_path)
```

从上面的结果可知，最终 test_view/urls.py 文件中的 url('hello/', views.hello_view)匹配到了测试的 HTTP 请求（/test_view/hello/）。在匹配成功后，将返回一个 ResolverMatch 对象，该对象中包含了对应的视图函数及相关的额外参数：

```
>>> callback, callback_args, callback_kwargs = patterns[0].resolve(new_path)
>>> callback
<function hello_view at 0x7f125d274af0>
>>> callback_args
()
>>> callback_kwargs
{}
```

为了能演示匹配额外参数，下面以匹配路径“articles_regex_extra/”为例继续演示：

```
# 继续在上面的结果中演示
>>> patterns[4]
<URLPattern 'articles_regex_extra/(?P<year>[0-9]{4})/'>
>>> new_path = 'articles_regex_extra/2019/'
>>> patterns[0].resolve(new_path)  # 第 1 个和第 2 个 URLConf 规则匹配不到
>>> patterns[1].resolve(new_path)
>>> patterns[4].resolve(new_path)
>>> callback, callback_args, callback_kwargs = patterns[4].resolve(new_path)
>>> callback   # 得到视图函数
<function year_archive at 0x7f125d274f70>
>>> callback_args
()
>>> callback_kwargs   # 得到相关参数
{'year': '2019', 'hello': 'world'}
```

至此，对 HTTP 请求的匹配过程就介绍完毕了。总结上述过程，即指定 urls 模块位置（用 include() 方法传入对应的模块路径），读取其中定义的 urlpatterns 变量，依次匹配 URLConf 配置，匹配成功即返回。

## 6.3.2 答疑解惑

最后还有两处细节需要深入源码学习：

◎ 对于带有正则表达式的 URLConf 配置，它是如何匹配请求路径并提取其中参数的？

◎ 在 url()方法中，view 参数为视图函数比较容易理解，但是 views.HelloView.as_view()有何含义，它也是视图函数吗？

第 1 个问题比较明确，就是通过_path()函数中的 Pattern 参数值匹配并提取的。对于 path()函数，其 Pattern 参数值为 RoutePattern 对象；而对于 re_path()函数，其 Pattern 参数值固定为 RegexPattern 对象。从源码的角度来看，RoutePattern 类和 RegexPattern 类中定义的 match()方法在正则表达式上的区别其实并不大：

```
# 源码位置：django/urls/resolvers.py
# ……

class RegexPattern(CheckURLMixin):
    regex = LocaleRegexDescriptor('_regex')

    def __init__(self, regex, name=None, is_endpoint=False):
        self._regex = regex
        self._regex_dict = {}
        self._is_endpoint = is_endpoint
        self.name = name
        self.converters = {}  # 注意这里的 converters 属性值为空字典


    def match(self, path):
        match = self.regex.search(path)   # 正则匹配
        if match:
            kwargs = match.groupdict()  # 获取匹配结果
            # 如果子匹配都没有命名，则把匹配结果赋给 args，否则为空
            args = () if kwargs else match.groups()
            return path[match.end():], args, kwargs
        return None

# ……

class RoutePattern(CheckURLMixin):
    regex = LocaleRegexDescriptor('_route')

    def __init__(self, route, name=None, is_endpoint=False):
        self._route = route
        self._regex_dict = {}
        self._is_endpoint = is_endpoint
        self.name = name
```

```
        # _route_to_regex()函数使得 path()函数支持 foo/<int:pk>这样的 URL 匹配表达式
        self.converters = _route_to_regex(str(route), is_endpoint)[1]

    def match(self, path):
        match = self.regex.search(path)
        if match:
            # RoutePattern 对象不允许出现未命名的子匹配，因此 args 将被忽略
            kwargs = match.groupdict()
            for key, value in kwargs.items():
                converter = self.converters[key]  # 注意，这里的 path()函数支持内置的匹配标签
                try:
                    kwargs[key] = converter.to_python(value)  # 转换
                except ValueError:
                    return None
            return path[match.end():], (), kwargs
        return None

    def _compile(self, route):
        # 从这里可以得到 regex 属性值，注意路径表达式将经过_route_to_regex()函数进行转换
        return re.compile(_route_to_regex(route, self._is_endpoint)[0])

    def __str__(self):
        return str(self._route)

# ……
```

总结一下这两个 Pattern 类的 match()方法。RoutePattern 对象不允许出现未命名的子匹配，此外，使用 path()函数包裹的路径表达式支持<int:pk>、<str:age>等这样的写法。以 6.1.3 节中 test_view 应用下 urls.py 文件中的第 3 个 URLConf 配置为例：

```
# 接着前面的结果继续执行
>>> patterns[2]
<URLPattern 'articles/<int:year>/<int:month>/'>
>>> patterns[2].pattern       # pattern 属性为一个 RoutePattern 对象
<django.urls.resolvers.RoutePattern object at 0x7f125d2791c0>
>>> patterns[2].pattern._route  # 查看_route 属性值
'articles/<int:year>/<int:month>/'
>>> patterns[2].pattern.regex   # 查看 regex 属性值
re.compile('^articles/(?P<year>[0-9]+)/(?P<month>[0-9]+)/$')
>>>
```

**注意**，路径表达式'articles/<int:year>/<int:month>/'实例化的 RoutePattern 对象在生成 regex 属性值时将经过_route_to_regex()函数进行转换。该函数的细节这里不再深究，有兴趣的读者可自行学习。从上面的结果可以看到，_route_to_regex()函数会将<int:year>这样的字符串转换成(?P<year>[0-9]+)，示例如下：

```
>>> from django.urls.resolvers import _route_to_regex
>>> patterns[2].pattern._route
'articles/<int:year>/<int:month>/'
>>> _route_to_regex(patterns[2].pattern._route)
('^articles/(?P<year>[0-9]+)/(?P<month>[0-9]+)/', {'year':
<django.urls.converters.IntConverter object at 0x7f125f222c70>, 'month':
<django.urls.converters.IntConverter object at 0x7f125f222c70>})
```

继续看 RegexPattern 类中的 match()方法，该方法直接针对带正则表达式的 URL 路径，它的实现非常简单，下面以 6.1.3 节中演示的 URLConf 配置中的第 5 个配置为例，演示 match()方法中的代码，操作如下：

```
>>> patterns[4]
<URLPattern 'articles_regex_extra/(?P<year>[0-9]{4})/'>
>>> patterns[4].pattern
<django.urls.resolvers.RegexPattern object at 0x7f125d279880>
>>> regex = patterns[4].pattern.regex
>>> regex
re.compile('articles_regex_extra/(?P<year>[0-9]{4})/')
>>> path = "articles_regex_extra/2021/"
>>> match = regex.search(path)  # 使用正则表达式匹配语句
>>> match
<re.Match object; span=(0, 26), match='articles_regex_extra/2021/'>
>>> match.groupdict()
{'year': '2021'}
```

如果在这里调整 URL 路径表达式，将使得子匹配组没有命名，如下：

```
# 接着上面的代码继续操作
>>> import re
>>> .regex = re.compile('articles_regex_extra/([0-9]{4})/')
>>> path
'articles_regex_extra/2021/'
>>> match = regex.search(path)
>>> match
<re.Match object; span=(0, 26), match='articles_regex_extra/2021/'>
>>> match.groupdict()  # 此时 kwargs 为{}
{}
>>> match.groups()   # 因此 args= ('2021',)
('2021',)
```

这里演示了未命名子匹配组的情况，最终 kwargs 为空字典，而 args 匹配到“2021”字符串。

继续看第 2 个细节，对于 url('hello_class/', views.HelloView.as_view())这样的 URLConf 配置，其中的 as_view()方法是什么？为什么它能将不同的 HTTP 请求（如 GET、POST、PUT、DELETE 请求等）匹配到对应名称的方法中呢？首先看视图类源码，内容如下：

```
# 源码位置：django/views/generic/base.py
# ……

class View:

    # 支持的请求方式
    http_method_names = ['get', 'post', 'put', 'patch', 'delete', 'head', 'options',
'trace']

    def __init__(self, **kwargs):
        for key, value in kwargs.items():
            setattr(self, key, value)

    @classonlymethod
    def as_view(cls, **initkwargs):  # 核心方法
        for key in initkwargs:
            # 处理一些参数异常情况
            # ……

        def view(request, *args, **kwargs):
            # 实例化 View 对象
            self = cls(**initkwargs)
            if hasattr(self, 'get') and not hasattr(self, 'head'):
                self.head = self.get
            # 初始化一些公共属性
            self.setup(request, *args, **kwargs)
            if not hasattr(self, 'request'):
                raise AttributeError(
                    "%s instance has no 'request' attribute. Did you override "
                    "setup() and forget to call super()?" % cls.__name__
                )
            # 核心处理请求语句，调用 self.dispatch()方法处理请求
            return self.dispatch(request, *args, **kwargs)

        # 给函数属性赋值
        view.view_class = cls
        view.view_initkwargs = initkwargs

        # take name and docstring from class
        update_wrapper(view, cls, updated=())
        update_wrapper(view, cls.dispatch, assigned=())

        # 关键语句，as_view()方法返回的是 view，它是函数名
        return view

    # ……
```

重点看 as_view()方法的源码，在该方法内部定义了一个 view()方法，而 as_view()方法最终返回

的是内部 view()方法的方法名。也就是说，as_view()方法返回的其实是一个函数，这一点只需简单测试一下就能验证：

```
>>> from test_view.views import HelloView
>>> HelloView.as_view()
<function HelloView at 0x7f9dfdcd6700>
```

这就清楚了为什么 HelloView.as_view()可以传给 path()函数中的第 2 个参数了。接下来要解决的就是为何 GET 请求能正确映射到 HelloView 中的 get()方法呢？仔细看 view()方法，它的核心处理语句是 self.dispatch(request, *args, **kwargs)。再来看 dispatch()方法的实现源码，答案就十分清楚了：

```
# 源码位置：django/views/generic/base.py
# ……

class View:

   http_method_names = ['get', 'post', 'put', 'patch', 'delete', 'head', 'options',
'trace']

   # ……

   def dispatch(self, request, *args, **kwargs):
      if request.method.lower() in self.http_method_names:
         handler = getattr(self, request.method.lower(), self.http_method_not_allowed)
      else:
         handler = self.http_method_not_allowed
      return handler(request, *args, **kwargs)

   # ……
```

很明显，request.method 就是获取 HTTP 的请求方式，比如 GET 请求、POST 请求等。首先要判断 HTTP 请求方式是否在支持范围内（即是否在 http_method_names 列表中）。如果满足条件，则使用 getattr()方法获取当前对象中 request.method.lower()的属性值，得到 handler 变量，最后调用 handler()方法处理 HTTP 请求。下面简单演示一下针对 HelloView 的各种请求方式，如下：

```
>>> hv = HelloView()
>>> 'get' in hv.http_method_names
True
>>> handler = getattr(hv, 'get')
>>> handler
<bound method HelloView.get of <test_view.views.HelloView object at 0x7f9dfdcddd30>>
>>> 'post' in hv.http_method_names
True
>>> handler = getattr(hv, 'post')
>>> handler
<bound method HelloView.post of <test_view.views.HelloView object at 0x7f9dfdcddd30>>
# 模拟不支持该请求的操作
```

```
>>> from django.http.request import HttpRequest
>>> request = HttpRequest()
>>> request.method = 'xyz'
>>> request.path = '/test_view/hello_class/'
>>> handler(request)
<HttpResponseNotAllowed [GET, POST, PUT, OPTIONS] status_code=405, "text/html;
charset=utf-8">
```

现在应该清楚为何 GET 请求会被 Django 映射到 HelloView 中的 get()方法了吧？核心就是调用 getattr()方法。

## 6.4 视图类与 Mixin 类

在理解了 HTTP 请求路径和视图函数的匹配过程及 Django 定义的 View 类后，就可以继续学习 Django 框架为开发者准备好的其他视图类了。在学习这些视图类的源码之前，应先了解 Mixin 类的概念。在 Django 官方文档中，关于 Mixin 类的介绍可大致翻译如下：

Django 内置的视图类提供了许多功能，但是我们很可能只需要其中一部分功能。例如，你想写一个视图，该视图内容是由模板文件渲染后得到的，但是你又不能使用 TemplateView 类来实现，因为你只想在 POST 请求上使用这个功能，而在 GET 请求上做其他事情。当然，我们可以直接使用 TemplateResponse 类来完成，但是这会导致代码重复。基于这个原因，Django 内部提供了许多离散功能的 Mixin 类。也就是说，Mixin 类就是一些单独功能的类，需配合视图类一起使用，用于组合出各种功能的视图。

### 6.4.1 Mixin 类的源码解析

下面先来看在 Django 框架中定义的几个简单的 Mixin 类，如下：

```
# 源码位置：django/views/generic/base.py
# ……

class ContextMixin:

   extra_context = None

   def get_context_data(self, **kwargs):
      kwargs.setdefault('view', self)       # 设置固定值
      if self.extra_context is not None:
         kwargs.update(self.extra_context) # 更新 kwargs
      return kwargs

# ……
```

```
class TemplateResponseMixin:
    """一个用于辅助渲染模板文件的 Mixin 类"""
    template_name = None            # 模板文件
    template_engine = None          # 模板引擎
    response_class = TemplateResponse  # 模板响应类
    content_type = None             # 响应的 content_type

    def render_to_response(self, context, **response_kwargs):
        response_kwargs.setdefault('content_type', self.content_type)
        # 返回模板响应对象
        return self.response_class(
            request=self.request,
            template=self.get_template_names(),
            context=context,
            using=self.template_engine,
            **response_kwargs
        )

    def get_template_names(self):
        if self.template_name is None:
            raise ImproperlyConfigured(
                "TemplateResponseMixin requires either a definition of "
                "'template_name' or an implementation of 'get_template_names()'")
        else:
            # 正常情况下，返回带有一个元素的列表
            return [self.template_name]

# ……
```

上面定义的这两个 Mixin 类都非常简单，只有几个基础函数。这是 Mixin 类的一大特点，它只实现几个特有的功能，无法单独使用，只能在继承类中使用它定义的一些特殊方法。

## 6.4.2　TemplateView 类的源码解析

在 Django 中，通过继承上面两个 Mixin 类及 View 类可以得到一个模板视图类 TemplateView，它的源码如下：

```
# 源码位置：django/views/generic/base.py
# ……

class TemplateView(TemplateResponseMixin, ContextMixin, View):

    def get(self, request, *args, **kwargs):
        # get_context_data()方法出自 ContextMixin
        context = self.get_context_data(**kwargs)
        # render_to_resopnse()方法出自 TemplateResponseMixin 类
        return self.render_to_response(context)
```

```
# ……
```

TemplateView 类非常典型，它继承了 3 个类：TemplateResponseMixin、ContextMixin 和 View。通过继承 View 类使得 TemplateView 类具备 Django 中的视图功能，因此它定义的 get()方法可以处理对应 URL 路径的 GET 请求。而在 get()方法中调用的 get_context_data()方法来自 ContextMixin 类，该方法可以将从 get()方法中传入的额外参数更新到 context 变量中。render_to_response()方法来自 TemplateResponseMixin 类，该方法可以返回一个 TemplateResponse 对象，该对象会传入模板文件和上下文变量 context。继续追踪 TemplateResponse 对象的实现源码，如下：

```
# 源码位置：django/template/response.py
# ……

class SimpleTemplateResponse(HttpResponse):
   rendering_attrs = ['template_name', 'context_data', '_post_render_callbacks']

   def __init__(self, template, context=None, content_type=None, status=None,
            charset=None, using=None):
      self.template_name = template  # 模板文件名称
      self.context_data = context    # 上下文环境变量，用于转换模板变量

      self.using = using             # 模板引擎，在第 4 章中介绍过

      self._post_render_callbacks = []

      # ……

      # 是否已渲染标识
      self._is_rendered = False

   # ……

   def add_post_render_callback(self, callback):
      if self._is_rendered:
         callback(self)
      else:
         self._post_render_callbacks.append(callback)

   def render(self):
      # 模板响应的核心
      retval = self
      if not self._is_rendered:  # 如果模板内容没有渲染
         # 调用属性方法 self.rendered_content 获得渲染后的内容，此时设置_is_rendered 为 True
         self.content = self.rendered_content
         # 如果有后续的回调处理，则继续执行下面的代码
         for post_callback in self._post_render_callbacks:
```

```
            newretval = post_callback(retval)
            if newretval is not None:
                retval = newretval
        return retval

    # ……

class TemplateResponse(SimpleTemplateResponse):
    rendering_attrs = SimpleTemplateResponse.rendering_attrs + ['_request']

    def __init__(self, request, template, context=None, content_type=None,
            status=None, charset=None, using=None):
        super().__init__(template, context, content_type, status, charset, using)
        self._request = request
```

TemplateResponse 对象的源码内容非常简单，它的核心源码在父类 SimpleTemplateResponse 中。而对于 SimpleTemplateResponse 类中的源码，我们主要对 render()方法进行说明。首先，在 SimpleTemplateResponse 类中有一个_is_rendered 属性，它表示该响应类中的模板文件是否已被渲染。刚开始时（初始化方法__init__()），该属性值被设置为 False。因此当第一次调用该响应类的 render()方法时，not self._is_rendered 为 True，进入该分支下执行。该分支下的 self.content = self.rendered_content 语句是整个 render()方法的核心。首先看该语句的后半部分 self.rendered_content，与它相关的源码如下：

```
# 源码位置：django/template/response.py
# ……

class SimpleTemplateResponse(HttpResponse):
    # ……

    def resolve_template(self, template):
        """得到一个 Template 对象"""
        if isinstance(template, (list, tuple)):
            return select_template(template, using=self.using)
        elif isinstance(template, str):
            # 单个 template 文件，直接使用 get_template()方法获取 Template 对象
            return get_template(template, using=self.using)
        else:
            return template

    def resolve_context(self, context):
        return context

    @property
    def rendered_content(self):
        template = self.resolve_template(self.template_name)  # 得到一个 Template 对象
        context = self.resolve_context(self.context_data)     # 得到上下文变量
```

```
        # 调用 Template 对象的 render()方法
        content = template.render(context, self._request)
        return content    # 返回渲染后的内容

    # ……
```

在学完第 4 章内容后就不难理解 rendered_conten()方法了，该方法通过 resolve_template()方法可以获得一个 Template 对象，然后得到上下文变量 context，接着调用 Template 对象中的方法对模板文件进行渲染，得到渲染后的内容 content。

TemplateResponse 对象中的 content 属性的赋值过程参见如下源码：

```
# 源码位置：django/template/response.py
# ……

class SimpleTemplateResponse(HttpResponse):
    # ……

    @property
    def content(self):
        if not self._is_rendered:
            raise ContentNotRenderedError(
                'The response content must be rendered before it can be accessed.'
            )
        return super().content

    @content.setter
    def content(self, value):
        """设置相应的具体内容"""
        HttpResponse.content.fset(self, value)  # fset()是一个用来设置属性值的函数
        self._is_rendered = True

    # ……
```

当给 content 属性赋值时，@content.setter 装饰的 content()函数将生效，此时将设置响应对象的 content 属性值，同时设置_is_rendered 的属性值为 True，表示已经渲染完成。对于响应类 HttpResponse 的使用，还记得 6.3 节中关于_get_response()方法的源码吗？

```
# 源码位置：django/core/handlers/base.py
# ……

class BaseHandler:
    # ……

    def _get_response(self, request):
        response = None

        # ……
```

```
        if response is None:
            # ……

        elif hasattr(response, 'render') and callable(response.render):
            # ……

            try:
                # 进行渲染，得到最终的响应结果
                response = response.render()
            except Exception as e:
                response = self.process_exception_by_middleware(e, request)

        # 得到初步的请求响应并返回
        return response
# ……
```

当视图函数返回的结果为 TemplateResponse 对象时，由于 TemplateResponse 对象有 render()方法，因此 hasattr(response, 'render') and callable(response.render)分支判断为 True。在进入该分支后，通过调用 TemplateResponse 对象的 render()方法可重新得到新的响应结果。

**注意**，该 render()方法的返回值其实是当前对象（render()方法中的第一句为 retval = self），渲染后的内容被保存在当前 TemplateResponse 对象的 content 属性中，此时的 TemplateResponse 对象就可以被看作一个完整的响应了（当没有渲染时，缺少了响应内容）。

### 6.4.3　RedirectView 类的源码解析

接下来学习在 django/views/generic/base.py 文件中定义的最后一个视图类：RedirectView。该类比较简单，主要用于实现重定向功能，源码如下：

```
# 源码位置：django/views/generic/base.py
# ……

class RedirectView(View):
    permanent = False
    url = None
    pattern_name = None
    query_string = False

    def get_redirect_url(self, *args, **kwargs):
        """
        返回重定向的 URL 地址
        """
        if self.url:
            url = self.url % kwargs
```

```
        elif self.pattern_name:
            url = reverse(self.pattern_name, args=args, kwargs=kwargs)
        else:
            return None

        args = self.request.META.get('QUERY_STRING', '')  # 获取请求参数
        if args and self.query_string:
            url = "%s?%s" % (url, args)  # 拼接成完整的URL地址
        return url

    def get(self, request, *args, **kwargs):
        url = self.get_redirect_url(*args, **kwargs)
        if url:
            if self.permanent:  # 永久重定向
                return HttpResponsePermanentRedirect(url)
            else:
                return HttpResponseRedirect(url)  # 临时重定向
        else:
            # 打印重定向的告警信息
            # ……
            return HttpResponseGone()

    #无论什么方法，均使用get()方法返回响应对象
    def head(self, request, *args, **kwargs):
        return self.get(request, *args, **kwargs)

    def post(self, request, *args, **kwargs):
        return self.get(request, *args, **kwargs)

    def options(self, request, *args, **kwargs):
        return self.get(request, *args, **kwargs)

    def delete(self, request, *args, **kwargs):
        return self.get(request, *args, **kwargs)

    def put(self, request, *args, **kwargs):
        return self.get(request, *args, **kwargs)

    def patch(self, request, *args, **kwargs):
        return self.get(request, *args, **kwargs)
```

RedirectView 类的实现非常简单，它只是单纯继承了 View 类，没有继承任何其他的 Mixin 类。在该类中重点实现了 get()方法，首先通过 get_redirect_url()方法得到重定向的 URL 地址。如果有重定向地址，就根据临时重定向还是永久重定向返回重定向响应对象。如果没有重定向地址，则直接返回 HttpResponseGone 对象。该视图的其他方法均在内部调用 get()方法进行请求处理。

## 6.4.4　DetailView 类和 ListView 类的源码解析

接下来重点介绍 Django 提供的两个典型视图类：DetailView 和 ListView。首先给出这两个类的使用示例，然后再对相关视图类的源码进行解读。

### DetailView 类的使用示例

这里使用在第 3 章中测试的 DjangoBooks 模型（位于 shell_test 应用中）完成本次实验。在之前测试的数据库中，曾向该表中插入了 10 条测试数据，内容如图 6-1 所示。

```
mysql> use django_book
Reading table information for completion of table and column names
You can turn off this feature to get a quicker startup with -A

Database changed
mysql> select id, book_name, author, price from django_books;
+----+------------------------------------------+--------+-------+
| id | book_name                                | author | price |
+----+------------------------------------------+--------+-------+
|  1 | Django源码笔记                           | spyinx |   109 |
|  2 | Ansible核心源码剖析与项目实战            | spyinx |   119 |
|  3 | Python自动化运维                         | test   |    79 |
|  4 | 中国电信媒体存储自主研发之路             | group  |    89 |
|  5 | 大国崛起                                 | xyz    |    59 |
|  6 | 小郭历险记                               | mmmm   |    39 |
|  7 | Django项目实战                           | hhhhh  |    79 |
|  8 | 精通Django框架                           | zzzzz  |   129 |
|  9 | Django实战                               | test   |    89 |
| 10 | Django3.0从入门到精通                    | group  |    40 |
+----+------------------------------------------+--------+-------+
10 rows in set (0.00 sec)

mysql> 
```

图 6-1

第 1 步，准备一个模板文件 detail_view_test.html（放到 first_django 项目的 template 目录中，其中与模板层相关的配置参见 4.1 节），内容如下：

```
(django2-core-test) [root@master first_django]# cat templates/detail_view_test.html
<p>测试 DetailView</p>
<h1>图书名称:{{ books.book_name }}</h1>
<h1>作者名称:{{ books.author }}</h1>
<h1>图书价格:{{ books.price }}</h1>
```

第 2 步，准备视图类 TestDetailView，该类会继承 Django 内置的 DetailView 类。这里不用添加任何方法，只需设置相关类属性即可：

```
# 源码位置：test_view/views.py
# ……
```

```
class TestDetailView(DetailView):
    model = DjangoBooks
    pk_url_kwarg = 'id'
    template_name = "detail_view_test.html"
    context_object_name = 'books'
    # query_pk_and_slug = True
    # slug_field = 'author'
    # slug_url_kwarg = 'author'
```

第 3 步，完成对应的 URLConf 配置，如下：

```
(django2-core-test) [root@master first_django]# cat test_view/urls.py
# ……

urlpatterns = [
    # ……
    path('detail_view/<int:id>/', views.TestDetailView.as_view()), # 测试 DetailView 类
]
```

第 4 步，使用 runserver 命令启动 first_django 项目，并监听 8000 端口：

```
(django2-core-test) [root@master first_django]# python manage.py runserver
192.168.26.110:8000
Watching for file changes with StatReloader
Performing system checks...

# ……
April 30, 2021 - 15:47:46
Django version 2.2.16, using settings 'first_django.settings'
Starting development server at http://192.168.26.110:8000/
Quit the server with CONTROL-C.
```

第 5 步，使用 curl 命令模拟 HTTP 请求，测试该视图，具体操作如下：

```
[store@ceph-1 ~]$ curl http://192.168.26.110:8000/test_view/detail_view/9/
<p>测试 DetailView</p>
<h1>图书名称:Django 实战</h1>
<h1>作者名称:test</h1>
<h1>图书价格:89.0</h1>

[store@ceph-1 ~]$ curl http://192.168.26.110:8000/test_view/detail_view/1/
<p>测试 DetailView</p>
<h1>图书名称:Django 源码笔记</h1>
<h1>作者名称:spyinx</h1>
<h1>图书价格:109.0</h1>

[store@ceph-1 ~]$ curl http://192.168.26.110:8000/test_view/detail_view/4/
<p>测试 DetailView</p>
<h1>图书名称:中国电信媒体存储自主研发之路</h1>
<h1>作者名称:group</h1>
```

```
<h1>图书价格:89.0</h1>

[store@ceph-1 ~]$ curl -XPOST http://192.168.26.110:8000/test_view/detail_view/4/ # 不
支持 POST 请求
[store@ceph-1 ~]$
```

前面定义的 TestDetailView 类似乎什么都没做，只是指定了要操作的模型表（model = DjangoBooks）和对应的模板文件（template_name = "detail_view_test.html"）等属性，就可以实现上面的请求处理。该视图类的内部是如何工作的？为什么一定要用/test_view/detail_view/4/这样的写法呢？在后续分析完 DetailView 类的源码后，一切就会豁然开朗。下面继续完成一个关于 ListView 类的使用示例。

### ListView 类的使用示例

与 DetailView 类的使用示例一样，仍然使用 DjangoBooks 模型进行测试。

第 1 步，在 template 目录下准备一个模板文件 list_view_test.html，内容如下：

```
(django2-core-test) [root@master first_django]# cat templates/list_view_test.html
<html>
<head>
<style type="text/css">
   .page{
     margin-top: 10px;
     font-size: 14px;
   }

   .member-table {
     width: 50%;
     text-align: center;
   }
</style>
</head>
<body>

<p>图书信息-第{{ page_obj.number }}页/共{{ paginator.num_pages }}页, 每页
{{ paginator.per_page }}条, 总共{{ paginator.count }}条</p>
<div>
<table border="1" class="member-table">
  <thead>
  <tr>
    <th>图书名称</th>
    <th>作者</th>
    <th>性别</th>
    <th>价格</th>
    <th>出版日期</th>
  </tr>
```

```
 </thead>
 <tbody>
 {% for book in books %}
 <tr>
   <td>{{ book.book_name }}</td>
   <td>{{ book.author }}</td>
   {% if book.sex == 0 %}
   <td>男</td>
   {% else %}
   <td>女</td>
   {% endif %}
   <td>{{ book.price }}</td>
   <td>{{ book.publish_date|date:'Y' }}</td>
 </tr>
 {% endfor %}
  </tbody>
</table>
<div >
<div class="page">
</div>
</div>
</div>

</body>

</html>
```

第 2 步，准备一个视图类 TestListView，该类继承自 Django 内部的 ListView 类。此外，和 DetailView 类的使用示例一样，给 TestListView 类赋予相关的属性值，如下：

```
# 源码位置：test_view/views.py
# ……

class TestListView(ListView):
    template_name = 'list_view_test.html'  # 关联渲染的模板文件
    model = DjangoBooks  # 关联操作的模型类
    paginate_by = 4    # 指定分页大小
    ordering = ["-price"]  # 指定展示列表的排序字段
    context_object_name = "books"  # 在模板文件中会用到，给表中查询的列表数据关联一个变量
```

第 3 步，准备对应视图的 URLConf 配置，如下：

```
(django2-core-test) [root@master first_django]# cat test_view/urls.py
# ……

urlpatterns = [
     # ……
     url('list_view/', views.TestListView.as_view()),   # 测试 ListView 类
]
```

第 4 步，启动 first_django 项目，在浏览器上访问该路径地址。注意，在请求的 URL 路径上可以带有 query 参数，用于指定请求第几页数据。测试结果如图 6-2 所示。

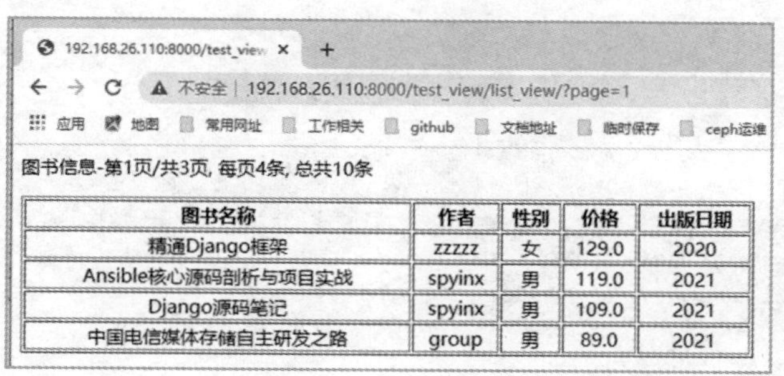



图书信息-第1页/共3页, 每页4条, 总共10条

| 图书名称 | 作者 | 性别 | 价格 | 出版日期 |
|---|---|---|---|---|
| 精通Django框架 | zzzzz | 女 | 129.0 | 2020 |
| Ansible核心源码剖析与项目实战 | spyinx | 男 | 119.0 | 2021 |
| Django源码笔记 | spyinx | 男 | 109.0 | 2021 |
| 中国电信媒体存储自主研发之路 | group | 男 | 89.0 | 2021 |

（a）



图书信息-第3页/共3页, 每页4条, 总共10条

| 图书名称 | 作者 | 性别 | 价格 | 出版日期 |
|---|---|---|---|---|
| Django3.0从入门到精通 | group | 男 | 40.0 | 2021 |
| 小郭历险记 | mmmm | 女 | 39.0 | 2021 |

（b）

图 6-2

从图 6-2 中可以看到，只需简单继承 ListView 类并指定几个属性，就能实现基本的分页展示数据，并能显示分页相关的信息。

### DetailView 类的源码解析

在看完上面的两个示例后，是不是觉得有些迷惑？Django 究竟提供了哪些有用的视图类，这些视图类又有哪些属性，它们分别表示什么含义？下面我们一起查看 Django 的源码，找到这些问题的答案。先看 DetailView 类，其源码实现如下：

```
# 源码位置：django/views/generic/detail.py
# ……

class DetailView(SingleObjectTemplateResponseMixin, BaseDetailView):
    """
    渲染一个对象的详细视图
    """
```

可以看到，DetailView 类除了继承 SingleObjectTemplateResponseMixin 类和 BaseDetailView 类，没有任何多余的代码。继续看 SingleObjectTemplateResponseMixin 类和 BaseDetailView 类的实现源码，

如下：

```
# 源码位置：django/views/generic/detail.py
# ……

class BaseDetailView(SingleObjectMixin, View):

    # 只实现了 get()方法，当父类没有实现其他方法时，父类将只支持 GET 请求
    def get(self, request, *args, **kwargs):
        self.object = self.get_object()  # 获取模型对象，相当于数据库中的一条记录
        context = self.get_context_data(object=self.object) # 获取上下文变量，用于渲染模板
        return self.render_to_response(context)  # 返回 TemplateResponse 对象


class SingleObjectTemplateResponseMixin(TemplateResponseMixin):
    template_name_field = None
    template_name_suffix = '_detail'  # 这里的变量将用于处理异常情况

    def get_template_names(self):

        try:
            names = super().get_template_names()  # 获取模板文件名称
        except ImproperlyConfigured:
            # 处理异常，此时会根据_detail 后缀在模板目录下搜索可能的模板文件
            # ……

        return names
```

在 BaseDetailView 类中只定义了一个 get()方法，当父类没有实现其他方法时，父类将只支持 GET 请求。get()方法的执行逻辑非常明确，即首先通过 self.get_object()方法（来自继承的 SingleObjectMixin 类）获取模型对象，对应着数据表中的一条记录。其次通过继承的 get_context_data()方法将得到的模型对象保存在上下文变量 context 中。最后调用继承的 render_to_response()方法返回 TemplateResponse 对象。Django 将返回渲染后的模板文件内容，也就是前面看到的 curl 命令的请求结果。

SingleObjectTemplateResponseMixin 类继承自 TemplateResponseMixin 类，该类重新实现了父类中的 get_template_names()方法，并在原方法的基础上增加了异常处理，即当没有指定 template_name 属性时，会按照一定的规则搜索可能的模板文件，最终返回待渲染的模板文件列表。

继续看 BaseDetailView 的父类 SingleObjectMixin 的实现源码，这里有几个非常重要的方法需要重点分析，具体如下：

```
# 源码位置：django/views/generic/detail.py
# ……

class SingleObjectMixin(ContextMixin):  # 单个对象的 Mixin 类
```

```
    model = None            # 继承类必填，用于关联操作的模型对象
    queryset = None         # 指定 QuerySet 对象
    slug_field = 'slug'     # 指定相关字段名
    context_object_name = None
    slug_url_kwarg = 'slug' # 在请求中传入该值的参数名
    pk_url_kwarg = 'pk'       # 在请求中传入 pk 字段的参数名
    query_pk_and_slug = False

    def get_object(self, queryset=None):

        if queryset is None:
            queryset = self.get_queryset()  # 通过 get_queryset()方法得到 QuerySet 对象

        pk = self.kwargs.get(self.pk_url_kwarg)
        slug = self.kwargs.get(self.slug_url_kwarg)
        if pk is not None:
            # 如果传入了 pk 参数，就调用 filter()方法得到新的 QuerySet 对象
            queryset = queryset.filter(pk=pk)

        if slug is not None and (pk is None or self.query_pk_and_slug):
            slug_field = self.get_slug_field()
            queryset = queryset.filter(**{slug_field: slug})  # 继续得到新的 QuerySet 对象

        # 处理异常情况
        # ……

        try:
            # 根据 QuerySet 对象获取单条记录
            obj = queryset.get()
        except queryset.model.DoesNotExist:
            # 如果没有单条记录，抛出 404 异常
            raise Http404(_("No %(verbose_name)s found matching the query") %
                          {'verbose_name': queryset.model._meta.verbose_name})

        # 返回查到的单条记录
        return obj

    def get_queryset(self):

        if self.queryset is None:
            if self.model:
                # 通过 Model 对象可以得到包含对应数据表中所有记录的 QuerySet 对象
                return self.model._default_manager.all()
            else:
                # 如果没既有设置 self.queryset，也没有设置 self.model，直接抛出异常
                raise ImproperlyConfigured(
                    "%(cls)s is missing a QuerySet. Define "
                    "%(cls)s.model, %(cls)s.queryset, or override "
```

```
                "%(cls)s.get_queryset()." % {
                    'cls': self.__class__.__name__
                }
            )
        return self.queryset.all()

    def get_slug_field(self):
        return self.slug_field

    def get_context_object_name(self, obj):
        # 这里返回的变量将影响模板文件中变量的使用
        if self.context_object_name:
            return self.context_object_name
        elif isinstance(obj, models.Model):
            return obj._meta.model_name
        else:
            return None

    def get_context_data(self, **kwargs):
        context = {}
        if self.object:
            context['object'] = self.object
            context_object_name = self.get_context_object_name(self.object)
            if context_object_name:
                # 从这里能看到 context_object_name 和 self.context_object_name 的作用
                context[context_object_name] = self.object
        # 更新 context
        context.update(kwargs)
        # 再次调用父类的 get_context_data()方法
        return super().get_context_data(**context)

# ……
```

首先分析核心方法 get_object()，它的功能是获取对应模型表中的单条数据。在 get_object()方法中，将使用 self.model 或者 self.queryset 属性值去关联操作的模型表，其中，self.queryset 可以由模型对象得到，因此在继承该 Mixin 类时必须要有模型对象。注意语句 pk=self.kwargs.get(self.pk_url_kwarg)，在 DetailView 类的示例中，我们设置了视图类 TestDetailView 的 pk_url_kwarg 属性值为 id。因此，这里的 pk 变量即获得传入参数中“id”对应的参数值，而这个值是从 kwargs 属性值中获取的。那这个属性值又是在哪里进行赋值的呢？在 VSCode 中全局搜索 self.kwargs 关键字，结果如下：

```
# 源码位置：django/views/generic/base.py
# ……

class View:
```

```
    @classonlymethod
    def as_view(cls, **initkwargs):
        # ……

        def view(request, *args, **kwargs):
            # ……
            self.setup(request, *args, **kwargs) # 这里的 kwargs 来自视图函数
            # ……
            return self.dispatch(request, *args, **kwargs)

        # ……

        return view

    def setup(self, request, *args, **kwargs):
        self.request = request
        self.args = args
        self.kwargs = kwargs  # 视图类中的 kwargs 属性值就是在这里进行赋值的

    # ……
```

从上面的代码可知，视图类中的 kwargs 属性值来自 setup()方法的赋值，而该方法中的参数来自最终处理 URL 请求的视图函数 view()中的 kwargs 参数。根据 6.3 节的源码解读经验可知，在 URLConf 配置中的 detail_view/<int:id>/会将 HTTP 请求中匹配的 ID 参数形成字典放到视图函数的 kwargs 参数中。因此，对于请求路径/test_view/detail_view/10/，最终处理请求的视图函数 view()中的 kwargs 值为{'id':10}。接着调用 setup()方法设置 self.kwargs 值为{'id':10}。再回到前面的 get_object()方法中，pk 变量得到的就是 10。当 pk 变量不为 None 时，将调用 QuerySet 对象的 filter(pk=pk)方法得到一个新的 QuerySet 对象。继续看后续的代码，可以看到 slug 也是一个用于查询的字段，slug_field 属性值就是用于指定的查询字段。关于 slug_field 字段的使用会有一个限制，对应的 if 语句如下：

```
if slug is not None and (pk is None or self.query_pk_and_slug)
```

从上面的语句可以了解 query_pk_and_slug 属性的作用。在默认情况下，该属性值为 False。如果同时有 pk 值和 slug_field 字段的值，则上述判断为 False，slug_field 字段的值无法起到过滤作用。只有当设置 query_pk_and_slug 属性为 True，同时设置 pk 值和 slug_field 字段的值时，后者才能作为过滤参数进行过滤，即执行 queryset.filter(**{slug_field: slug})语句。最后执行 QuerySet 对象的 get()方法从对应的数据表中获取单条记录（obj = queryset.get()）。此时，SingleObjectMixin 类中的 get_context_data()方法生成用于渲染模板文件的上下文变量 context，从该处源码中能看到 context_object_name 属性值的作用。最终，通过 get_object()方法查询得到的模型对象将保存在上下文变量 context 中，对应的键值正是 context_object_name 属性值。如果没有设置该属性值，则使用 obj._meta.model_name 值代替（参见 get_context_object_name()方法）。

### ListView 类的源码解析

继续分析在 django/views/generic/list.py 文件中定义的 ListView 类，其源码如下：

```
# 源码位置：django/views/generic/list.py
# ……

class ListView(MultipleObjectTemplateResponseMixin, BaseListView):
    """
    渲染一个对象列表
    """
```

与 DetailView 类一样，ListView 类通过继承 Mixin 类和 View 类实现新用途的视图类。先看其父类的源码，如下：

```
# 源码位置：django/views/generic/list.py
# ……

class BaseListView(MultipleObjectMixin, View):
    """ 展示对象列表的基本视图类 """
    def get(self, request, *args, **kwargs):
        self.object_list = self.get_queryset()  # 获取查询的对象列表，对应数据库表中的记录列表
        allow_empty = self.get_allow_empty()    # 是否允许空结果

        if not allow_empty:
            # get_paginate_by()方法继承自父类 MultipleObjectMixin，用于获取分页信息
            if self.get_paginate_by(self.object_list) is not None and 
hasattr(self.object_list, 'exists'):
                # exists()方法用于判断是否有查询结果
                is_empty = not self.object_list.exists()
            else:
                # 否则直接通过 not 判断是否为 None
                is_empty = not self.object_list
            if is_empty:  # 如果没有模型对象，且不允许为空，抛出 404 异常
                raise Http404(_("Empty list and '%(class_name)s.allow_empty' is False.")
                  % {'class_name': self.__class__.__name__,})
        # 获取渲染模板为上下文变量 context
        context = self.get_context_data()
        # 返回 TemplateResponse 对象，最终返回渲染后的模板文件内容
        return self.render_to_response(context)


class MultipleObjectTemplateResponseMixin(TemplateResponseMixin):
    template_name_suffix = '_list'

    def get_template_names(self):
        try:
            names = super().get_template_names()  # 获取模板文件名
```

```
        except ImproperlyConfigured:
            # 处理未设置模板文件的情况，有默认的文件路径
            # ……
        return names
```

在 BaseListView 类中定义了一个 get()方法，对应处理 HTTP 中的 GET 请求。该 get()方法的逻辑比较简单，即通过 get_queryset()方法获取对象列表，对应数据库中的多条记录。在后续的源码分析中可以看到，这里返回的 QuerySet 对象中只有排序信息。此外，如果设置了 allow_empty 属性值为 False（默认为 False），还需要考虑是否有的查询结果为空。如果为空，则抛出 404 异常。接下来的 get_context_data()方法则是真正获取对应的模型列表的地方，这里将得到分页数据及分页信息，返回包含这些数据的上下文变量 context。最后，调用 render_to_response()方法，返回 TemplateResponse 对象。

MultipleObjectTemplateResponseMixin 类同样继承自 TemplateResponseMixin 类，同样重新实现了 get_template_names()方法，该方法会返回待渲染的模板文件列表。如果没有待渲染的模板文件列表，则会按照一定规则返回默认的模板文件列表（具体可参见该方法中异常处理部分的代码）。

## 6.4.5　MultipleObjectMixin 类的源码解析

对于 View 类和 TemplateResponseMixin 类，此处不再赘述。继续追踪 MultipleObjectMixin 类的源码，内容如下：

```
# 源码位置：django/views/generic/list.py
# ……

class MultipleObjectMixin(ContextMixin):
    allow_empty = True
    queryset = None
    model = None   # 关联查询模型对象，对应着数据表
    paginate_by = None           # 设置每页记录数
    paginate_orphans = 0
    context_object_name = None   # 指定查询对象列表在上下文变量中的 key
    paginator_class = Paginator  # 指定分页处理类，默认在 Django 内部实现
    page_kwarg = 'page'      # 在请求中传递查询第几页的参数名，
    # 例如，?page=1 表示查询第 1 页数据
    ordering = None              # 查询结果按 ordering 进行排序

    def get_queryset(self):
        if self.queryset is not None:
            queryset = self.queryset
            if isinstance(queryset, QuerySet):
                # 默认查询所有结果的 QuerySet 对象，用于显示后续的分页信息
                queryset = queryset.all()
        elif self.model is not None:
```

```
        queryset = self.model._default_manager.all()
    else:
        # 抛出无 QuerySet 对象异常
        # ……

    # 获取排序字段
    ordering = self.get_ordering()
    if ordering:
        if isinstance(ordering, str):
            # 如果是 str，直接转成单个元素的元组
            ordering = (ordering,)
        # 直接在 QuerySet 对象中加入排序部分并得到新的 QuerySet 对象
        queryset = queryset.order_by(*ordering)

    return queryset

def get_ordering(self):
    return self.ordering

def paginate_queryset(self, queryset, page_size):
    # 得到一个分页对象
    paginator = self.get_paginator(
        queryset, page_size, orphans=self.get_paginate_orphans(),
        allow_empty_first_page=self.get_allow_empty())
    page_kwarg = self.page_kwarg
    # 从 URL 的请求参数中获取对应 page_kwarg 参数的值，即提取第几页的数据，默认为 1
    page = self.kwargs.get(page_kwarg) or self.request.GET.get(page_kwarg) or 1
    try:
        page_number = int(page)     # 转成 int 类型
    except ValueError:
        if page == 'last':          # 支持字符串 "last" 写法，表示最后一页
            page_number = paginator.num_pages
        else:
            raise Http404(_("Page is not 'last', nor can it be converted to an int."))
    try:
        # 从返回的结果可以看到，分页的核心在于分页器类，即 Paginator 类
        page = paginator.page(page_number)
        return (paginator, page, page.object_list, page.has_other_pages())
    except InvalidPage as e:
        raise Http404(_('Invalid page (%(page_number)s): %(message)s') % {
            'page_number': page_number,
            'message': str(e)
        })

def get_paginate_by(self, queryset):
    return self.paginate_by

def get_paginator(self, queryset, per_page, orphans=0,
```

```
                allow_empty_first_page=True, **kwargs):
        # 返回分页器类的一个对象
        return self.paginator_class(
            queryset, per_page, orphans=orphans,
            allow_empty_first_page=allow_empty_first_page, **kwargs)

    def get_paginate_orphans(self):
        return self.paginate_orphans

    def get_allow_empty(self):
        return self.allow_empty

    def get_context_object_name(self, object_list):
        if self.context_object_name:
            return self.context_object_name
        elif hasattr(object_list, 'model'):
            return '%s_list' % object_list.model._meta.model_name
        else:
            return None

    def get_context_data(self, *, object_list=None, **kwargs):

        queryset = object_list if object_list is not None else self.object_list
        # 直接返回 paginate_by 属性值，可以看到其含义为每页的大小
        page_size = self.get_paginate_by(queryset)
        # 如果设置了 context_object_name 属性值，直接返回该值
        context_object_name = self.get_context_object_name(queryset)
        if page_size:
            # 核心调用，将得到分页信息和当前页的数据
            paginator, page, queryset, is_paginated = self.paginate_queryset(queryset,
page_size)
            context = {
                'paginator': paginator,
                'page_obj': page,
                'is_paginated': is_paginated,
                'object_list': queryset
            }
        else:
            # 如果没有设置 paginate_by 属性值，则默认不分页，在上下文变量 context 中不会有分页信息
            context = {
                'paginator': None,
                'page_obj': None,
                'is_paginated': False,
                'object_list': queryset
            }
        # 设置得到的某页数据，将得到的分页数据设置到上下文变量 context 的对应键中
        if context_object_name is not None:
            context[context_object_name] = queryset
```

```
        context.update(kwargs)
        # 返回用于渲染模板文件的上下文变量 context
        return super().get_context_data(**context)

# ……
```

上述源码内容比较基础，并且有详细的注释，这里只需重点关注两个方法：get_queryset()方法和 get_context_data()方法。通过这两个方法能迅速了解 MultipleObjectMixin 类中各种属性的含义。

（1）get_queryset()方法，该方法是根据继承类中指定的 model 或者 queryset 属性值来得到模型对象的。当然，与 DetailView 类一样，只设置模型类也是可以的，根据该模型类能得到一个默认查询关联表中全部数据的 QuerySet 对象。在 get_queryset()方法中只用到了 ordering，它会作为 order_by()的参数，最终得到包含排序信息的 QuerySet 对象。

（2）get_context_data()方法，从该方法中的第二行语句 page_size = self.get_paginate_by(queryset)可知，MultipleObjectMixin 类中的 paginate_by 属性值即为指定每页展示的对象数。当设置了 paginate_by 属性值时，get_context_data()方法会调用 self.paginate_queryset()方法获取分页数据，同时返回分页信息。最终这些分页信息（包括分页数据）将保存在上下文变量 context 中。如果没有设置 paginate_by 属性值，就不需要做分页处理。当分页变量设置为 None 或者 False 时，将获取对应数据表中的全部数据。

分页的核心操作是调用 self.paginate_queryset()方法，该方法会从请求参数中提取出 page 参数值，它表示要查询第几页数据。self.paginate_queryset()方法的核心是得到一个分页器对象 paginator，然后执行 page = paginator.page(page_number)语句，最终在得到的 page 中就包含了如第几页、页大小、总页数及该页的数据列表等信息。

## 6.4.6 Paginator 类的源码解析

继续追踪 Paginator 类（分页器类）的源码，内容如下：

```
# 源码位置：django/core/paginator.py
# ……

class Paginator:

    def __init__(self, object_list, per_page, orphans=0,
                 allow_empty_first_page=True):
        self.object_list = object_list      #包含全部查询数据的 QuerySet 对象
        self._check_object_list_is_ordered()  # 检查在 QuerySet 对象中是否有排序信息
        self.per_page = int(per_page)       # 每页大小
        self.orphans = int(orphans)         # 最后一页显示的数据数目
        self.allow_empty_first_page = allow_empty_first_page
```

```
def validate_number(self, number):
    try:
        if isinstance(number, float) and not number.is_integer():
            raise ValueError
        number = int(number)
    except (TypeError, ValueError):
        raise PageNotAnInteger(_('That page number is not an integer'))
    if number < 1:
        raise EmptyPage(_('That page number is less than 1'))
    if number > self.num_pages:
        if number == 1 and self.allow_empty_first_page:
            pass
        else:
            raise EmptyPage(_('That page contains no results'))
    return number

def get_page(self, number):
    try:
        number = self.validate_number(number)  # 校验输入的页号 number
    except PageNotAnInteger:
        number = 1
    except EmptyPage:
        number = self.num_pages
    return self.page(number)

def page(self, number):
    number = self.validate_number(number)
    bottom = (number - 1) * self.per_page
    top = bottom + self.per_page
    if top + self.orphans >= self.count:
        top = self.count
    # 获取第 number 页的内容，返回 Page 对象
    return self._get_page(self.object_list[bottom:top], number, self)

def _get_page(self, *args, **kwargs):
    return Page(*args, **kwargs)

@cached_property
def count(self):
    c = getattr(self.object_list, 'count', None)
    if callable(c) and not inspect.isbuiltin(c) and method_has_no_args(c):
        # 如果 c 是可调用的，且不是内置函数，也没有参数，则直接调用返回
        return c()  # 如果是 QuerySet 对象，则返回 c()
    return len(self.object_list)

@cached_property
def num_pages(self):
    """返回总页数"""
```

```
        if self.count == 0 and not self.allow_empty_first_page:
            return 0
        # orphans 表示最后一页的数据大小
        hits = max(1, self.count - self.orphans)
        # 减去最后一页
        return ceil(hits / self.per_page)

    @property
    def page_range(self):
        return range(1, self.num_pages + 1)  # 页范围

    def _check_object_list_is_ordered(self):
        ordered = getattr(self.object_list, 'ordered', None)
        if ordered is not None and not ordered:
            # 如果在 QuerySet 对象中没有排序信息，打印告警信息
            # ……
```

Paginator 类的源码并不复杂，下面通过实战演示 Paginator 类中各方法的含义，操作如下：

```
>>> from django.core.paginator import Paginator
>>> from shell_test.models import DjangoBooks
>>> object_list = DjangoBooks.objects.all().order_by(*['-price'])
 # 如果没有排序，会产生告警信息
>>> c = getattr(object_list, 'count')  # 其实就是 object_list.count()，获取记录总数
>>> c()
10
>>> paginator = Paginator(object_list, 4)      # 页大小为 4
>>> paginator.validate_number('xxx')           # 输入错误页码
Traceback (most recent call last):
  File
"/root/.pyenv/versions/django2-core-test/lib/python3.8/site-packages/django/core/pag
inator.py", line 43, in validate_number
    number = int(number)
ValueError: invalid literal for int() with base 10: 'xxx'

During handling of the above exception, another exception occurred:

Traceback (most recent call last):
  File "<console>", line 1, in <module>
  File
"/root/.pyenv/versions/django2-core-test/lib/python3.8/site-packages/django/core/pag
inator.py", line 45, in validate_number
    raise PageNotAnInteger(_('That page number is not an integer'))
django.core.paginator.PageNotAnInteger: That page number is not an integer
>>> paginator.validate_number(5)   # 总共 3 页，5>3，所以抛出异常
Traceback (most recent call last):
  File "<console>", line 1, in <module>
```

```
 File
"/root/.pyenv/versions/django2-core-test/lib/python3.8/site-packages/django/core/paginator.py", line 52, in validate_number
   raise EmptyPage(_('That page contains no results'))
django.core.paginator.EmptyPage: That page contains no results
>>> paginator.validate_number(2)
2
>>> paginator.validate_number('3')
3
>>> page = paginator.page(2)     # 获取第 2 页数据，作用与 paginator.get_page(2)相同
>>> page                         # 得到的是一个 Page 对象
<Page 2 of 3>
>>> page.object_list             # Page 对象的 object_list 是当前页的 QuerySet 对象
<QuerySet [<DjangoBooks: <Django 实战, test>>, <DjangoBooks: <Python 自动化运维, test>>,
<DjangoBooks: <Django 项目实战, hhhhh>>, <DjangoBooks: <大国崛起, xyz>>]>
>>> page.number                  # Page 对象的 number 是当前页号
2
>>> page.has_other_pages()       # 是否有上一页或者下一页
True
```

在上面的代码中，最后演示的是 MultipleObjectMixin 中 paginate_queryset()方法的核心语句。在正常情况下，该方法将返回一个四元组，其中的元素分别表示分页器（Paginator）对象、页（Page）对象、当前页号和是否有其他页（True 或 False）。Page 类的源码十分简单，如下：

```
# 接着上面得到的 Page 对象继续操作
>>> len(page)  # 对应魔法函数__len__()
4
>>> page[2]   # 对应魔法函数__getitem__()
<DjangoBooks: <Django 项目实战, hhhhh>>
>>> page.has_next()      # 是否有下一页
True
>>> page.has_previous()  # 是否有上一页
True
>>> page.next_page_number()  # 下一页的页号
3
>>> page.previous_page_number()  # 上一页的页号
1
>>> page.start_index()  # 全局列表数据中的起始索引
5
>>> page.end_index()  # 全局列表数据中的结束索引
8
```

对于在 date.py 文件和在 edit.py 文件中定义的各种 Mixin 类和视图类就不再一一介绍了，有兴趣的读者可以自行分析。下面分析视图层中的最后一部分内容，即 Session 相关的源码。

# 6.5 追踪 Session 相关的源码

Session 相关的源码主要位于 django/contrib/sessions 目录中，本书会介绍该目录下的大部分源码。

## 6.5.1 Session 相关的配置

Session 相关的配置默认位于 django/conf/global_settings.py 文件中，具体内容如下：

```
# 源码位置：django/conf/global_settings.py
# ……

############
# SESSIONS    #
############
# 指定后端缓存 Session 数据的引擎
SESSION_CACHE_ALIAS = 'default'
# Cookie 中保存 Session 数据的 key，可随意指定
SESSION_COOKIE_NAME = 'sessionid'
# cookie 的过期时间，以秒为单位，默认是 2 周
SESSION_COOKIE_AGE = 60 * 60 * 24 * 7 * 2
# 类似"example.com"这样的字符串，或者设置为 None，即使用标准的域名 cookie
SESSION_COOKIE_DOMAIN = None
# 是否使用安全的 session cookie，即只能是 https 请求
SESSION_COOKIE_SECURE = False
# session cookie 的路径
SESSION_COOKIE_PATH = '/'
# 是否使用 HttpOnly 标记
SESSION_COOKIE_HTTPONLY = True
# 只能是'Lax', 'Strict', 或者 None.
SESSION_COOKIE_SAMESITE = 'Lax'
# 该参数决定是否在每次请求时都保存 Session 数据
SESSION_SAVE_EVERY_REQUEST = False
# 当浏览器关闭时，该参数将决定用户的 session cookie 数据是否过期
SESSION_EXPIRE_AT_BROWSER_CLOSE = False
# 指定保存 Session 数据的方式
SESSION_ENGINE = 'django.contrib.sessions.backends.db'
# 如果使用文件保存 Session 数据，则这里表示这些数据文件所在的目录，否则为 None
SESSION_FILE_PATH = None
# 指定用于序列化 Session 数据的类
SESSION_SERIALIZER = 'django.contrib.sessions.serializers.JSONSerializer'

# ……
```

最核心的语句是指定保存 Session 数据的方式，即 SESSION_ENGINE 变量设置的值。该值代表的是 django/contrib/sessions/backends/db.py 文件中的代码。在 django/contrib/sessions/backends 目录下，

一共有 7 个代码文件，其中，在__init__.py 文件中没有内容，在 base.py 文件中定义了 SessionBase 类，该基类实现了 4 个魔法函数，如下：

```
# 源码位置：django/contrib/sessions/backends/base.py
# ……

class SessionBase:

    TEST_COOKIE_NAME = 'testcookie'
    TEST_COOKIE_VALUE = 'worked'

    __not_given = object()

    def __init__(self, session_key=None):
        self._session_key = session_key  # Session 数据对应的 key
        self.accessed = False
        self.modified = False
        # 根据配置中设置的 SESSION_SERIALIZER 值，导入 Session 数据的序列化模块
        self.serializer = import_string(settings.SESSION_SERIALIZER)

    def __contains__(self, key):
        return key in self._session    # 判断 key 是否包含在_session 属性中

    def __getitem__(self, key):
        return self._session[key]      # 获取_session 属性值中 key 对应的 value 值

    def __setitem__(self, key, value):
        self._session[key] = value     #操作_session 属性值，设置 modified 值为 True
        self.modified = True

    def __delitem__(self, key):
        del self._session[key]         # 删除_session属性值中对应的key,设置modified值为True
        self.modified = True

    # ……
```

上面的魔法函数在 3.2 节中有详细的介绍，此处不再赘述。从上面的代码可以看到，在魔法函数中操作的都是_session 属性值。那么_session 属性值一开始是在哪里赋值的呢？继续看后面的源码：

```
# 源码位置：django/contrib/sessions/backends/base.py
# ……

class SessionBase:

    # ……

    def _get_session(self, no_load=False):
        self.accessed = True
```

```
        try:
            return self._session_cache  # 因为此时没有_session_cache属性值，所以会抛出异常
        except AttributeError:
            if self.session_key is None or no_load:
                self._session_cache = {}
            else:
                self._session_cache = self.load()  # 在SessionBase类中，load()方法并未实现
        return self._session_cache

    _session = property(_get_session) # _session其实是一个类属性函数

    # ……
```

在 Django 的 shell 模式下进行简单测试，操作如下：

```
(django2-core-test) [root@master first_django]# python manage.py shell
Python 3.8.6 (default, Oct 18 2020, 15:33:08)
[GCC 4.8.5 20150623 (Red Hat 4.8.5-39)] on linux
Type "help", "copyright", "credits" or "license" for more information.
(InteractiveConsole)
>>> from django.contrib.sessions.backends.base import SessionBase
>>> SessionBase._session
<property object at 0x7ff2b0e6bd10>
>>> sb = SessionBase()
>>> sb._session
{}
>>> sb = SessionBase()
>>> sb._session
{}
>>> sb.modified
False
>>> sb['hello'] = 'world'
>>> sb._session
{'hello': 'world'}
>>> sb.modified
True
>>> 'hello' in sb
True
>>> 'hellox' in sb
False
>>> del sb['hello']
>>> sb._session
{}
```

此外，SessionBase 类还实现了许多简单的方法，如 get()、pop()、setdefault()、update()、has_key()、keys()、value()和 items()等，在这些方法的内部同样是操作_session 属性值，它们的实现源码如下：

```
# 源码位置：django/contrib/sessions/backends/base.py
# ……
```

```
class SessionBase:

    TEST_COOKIE_NAME = 'testcookie'
    TEST_COOKIE_VALUE = 'worked'

    def get(self, key, default=None):
        return self._session.get(key, default)  # 从_session 属性中获取 key 对应的 value 值

    def pop(self, key, default=__not_given):
        self.modified = self.modified or key in self._session
        args = () if default is self.__not_given else (default,)
        return self._session.pop(key, *args)  # 从_session 属性值中弹出元素

    def setdefault(self, key, value):
        if key in self._session:
            return self._session[key]  # 如果 key 已经存在，直接返回对应的 value 值
        else:
            self.modified = True  # 如果 key 不存在，则设置 key，并设置 modified 属性值为 True
            self._session[key] = value
            return value

    def set_test_cookie(self):
        # 调用魔法函数__setitem__()，同样是操作_session 属性值
        self[self.TEST_COOKIE_NAME] = self.TEST_COOKIE_VALUE

    def test_cookie_worked(self):
        # 检查对应的键值对是否设置成功
        return self.get(self.TEST_COOKIE_NAME) == self.TEST_COOKIE_VALUE

    def delete_test_cookie(self):
        # 删除_session 属性值中对应的 key
        del self[self.TEST_COOKIE_NAME]

    # ……

    def update(self, dict_):
        self._session.update(dict_)    # 更新_session 属性值
        self.modified = True

    def has_key(self, key):
        return key in self._session    # 判断某个 key 是否存在

    def keys(self):
        return self._session.keys()    # 获取_session 属性值中所有的 key

    def values(self):
        return self._session.values()  # 获取_session 属性值中所有的 value 值
```

```
    def items(self):
        return self._session.items()   # 获取_session 属性值中所有的键值对

    def clear(self):
        self._session_cache = {}       # 清除相关数据
        self.accessed = True
        self.modified = True

    def is_empty(self):
        try:
            return not self._session_key and not self._session_cache
        except AttributeError:
            return True

# ……
```

上面的方法都非常简单，下面演示对这些方法的调用，操作如下：

```
>>> from django.contrib.sessions.backends.base import SessionBase
>>> sb = SessionBase()
>>> sb.setdefault('hello', 'world')
'world'
>>> sb._session
{'hello': 'world'}
>>> sb.get('hello')
'world'
>>> sb.pop('hello')
'world'
>>> sb._session
{}
>>> sb.set_test_cookie()
>>> sb._session
{'testcookie': 'worked'}
>>> sb.test_cookie_worked()
True
>>> sb.delete_test_cookie()
>>> sb.test_cookie_worked()
False
>>> sb._session
{}
>>> sb.setdefault('k1', 'v1')
'v1'
>>> sb.setdefault('k2', 'v2')
'v2'
>>> sb.setdefault('k3', 'v3')
'v3'
>>> sb._session
```

```
{'k1': 'v1', 'k2': 'v2', 'k3': 'v3'}
>>> sb.update({'hello': 'world', 'k3': 'new3'})  # 测试 update()方法
>>> sb._session
{'k1': 'v1', 'k2': 'v2', 'k3': 'new3', 'hello': 'world'}
>>> sb.has_key('k3')
True
>>> sb.has_key('k4')
False
>>> sb.keys()
dict_keys(['k1', 'k2', 'k3', 'hello'])
>>> sb.values()
dict_values(['v1', 'v2', 'new3', 'world'])
>>> sb.items()
dict_items([('k1', 'v1'), ('k2', 'v2'), ('k3', 'new3'), ('hello', 'world')])
>>>
```

继续看 BaseSession 类中的 encode()方法和 decode()方法，源码如下：

```
# 源码位置：django/contrib/sessions/backends/base.py
# ……

class SessionBase:
   # ……

   def _hash(self, value):
      key_salt = "django.contrib.sessions" + self.__class__.__name__
      return salted_hmac(key_salt, value).hexdigest()

   def encode(self, session_dict):
      serialized = self.serializer().dumps(session_dict) # 对 session_dict 数据进行序列化
      hash = self._hash(serialized)  # 计算 hash 值
      # 对结果再次编码
      return base64.b64encode(hash.encode() + b":" + serialized).decode('ascii')

   def decode(self, session_data):
      # 按照上述过程解码
      encoded_data = base64.b64decode(session_data.encode('ascii'))
      try:
         # could produce ValueError if there is no ':'
         hash, serialized = encoded_data.split(b':', 1)
         expected_hash = self._hash(serialized)
         if not constant_time_compare(hash.decode(), expected_hash):
            raise SuspiciousSession("Session data corrupted")
         else:
            return self.serializer().loads(serialized)
      except Exception as e:
         # 打印异常信息
         # ……
```

```
        return {}
```

上面的源码涉及一些加密的内容，但并不复杂，而且也不用完全掌握其编码和解码的细节，只需掌握对应方法的输入和输出，理解其功能即可。下面继续操作 encode()方法和 decode()方法，同时一步步演示编码和解码的过程，具体操作如下：

```
# 接着前面的操作继续执行
>>> from django.utils.crypto import salted_hmac
>>> text = '待计算 hash 的文本'
>>> salted_hmac('xxx', text).hexdigest()
'e40d31ebb4258639baa73ef7d61b7a532f4fe907'
>>> key_salt = "django.contrib.sessions" + sb.__class__.__name__
>>> key_salt
'django.contrib.sessionsSessionBase'
>>> salted_hmac(key_salt, text).hexdigest()
'a743f637d27bfccff006835c41e223d481ed9a44'
>>> sb._hash(text)
'a743f637d27bfccff006835c41e223d481ed9a44'
>>> sb.serializer
<class 'django.core.signing.JSONSerializer'>
>>> sb._session
{'k1': 'v1', 'k2': 'v2', 'k3': 'new3', 'hello': 'world'}
>>> serialized = sb.serializer().dumps(sb._session)   # encode()方法中的第一行代码
>>> serialized
b'{"k1":"v1","k2":"v2","k3":"new3","hello":"world"}'
>>> hash = sb._hash(serialized)                        # encode()方法中的第二行代码
>>> hash
'b70c0b7e72d7f2d3dd7ad5d8de8cd4ca024c8f11'
>>> import base64
>>> base64.b64encode(hash.encode() + b":" + serialized).decode('ascii')
# encode()方法中的第三行代码
'YjcwYzBiN2U3MmQ3ZjJkM2RkN2FkNWQ4ZGU4Y2Q0Y2EwMjRjOGYxMTp7ImsxIjoidjEiLCJrMiI6InYyIiw
iazMiOiJuZXczIiwiaGVsbG8iOiJ3b3JsZCJ9'
>>> session_data = sb.encode(sb._session)              # 直接使用 encode()方法得到的结果
>>> session_data
'YjcwYzBiN2U3MmQ3ZjJkM2RkN2FkNWQ4ZGU4Y2Q0Y2EwMjRjOGYxMTp7ImsxIjoidjEiLCJrMiI6InYyIiw
iazMiOiJuZXczIiwiaGVsbG8iOiJ3b3JsZCJ9'
>>> encoded_data = base64.b64decode(session_data.encode('ascii'))
# decode()方法中的第一行代码
>>> encoded_data # 该数据有两部分，分号前的为 hash.encode()的结果，分号后的为 serialized
b'b70c0b7e72d7f2d3dd7ad5d8de8cd4ca024c8f11:{"k1":"v1","k2":"v2","k3":"new3","hello":
"world"}'
>>> hash, serialized = encoded_data.split(b':', 1)
>>> hash
b'b70c0b7e72d7f2d3dd7ad5d8de8cd4ca024c8f11'
>>> serialized
b'{"k1":"v1","k2":"v2","k3":"new3","hello":"world"}'
>>> expected_hash = sb._hash(serialized)  # 直接使用_hash()方法计算 serialized 的 hash 值
```

```
>>> expected_hash
'b70c0b7e72d7f2d3dd7ad5d8de8cd4ca024c8f11'
>>> hash.decode()
'b70c0b7e72d7f2d3dd7ad5d8de8cd4ca024c8f11'
# 如果上面计算的 hash 值与编码中得到的 hash 变量一致，直接反序列化结果后返回
>>> sb.serializer().loads(serialized)
{'k1': 'v1', 'k2': 'v2', 'k3': 'new3', 'hello': 'world'}
>>> sb.decode(session_data)              # 直接使用 decode()方法得到解码后的结果
{'k1': 'v1', 'k2': 'v2', 'k3': 'new3', 'hello': 'world'}
```

通过上面的操作，相信读者已经理解了 encode()方法和 decode()方法的实现细节。接下来是几个关于过期时间的方法，其源码如下：

```
# 源码位置：django/contrib/sessions/backends/base.py
# ……

class SessionBase:
    # ……

    def get_expiry_age(self, **kwargs):
        # 获取过期时间，以秒为单位，返回一个数值
        try:
            modification = kwargs['modification']  # 从传入参数中获取修改时间
        except KeyError:
            modification = timezone.now()  # 如果没有，则默认修改时间为当前时间
        try:
            expiry = kwargs['expiry']  # 获取过期时间
        except KeyError:
            # 从_session 属性值中获取_session_expiry 对应的值
            expiry = self.get('_session_expiry')

        if not expiry:   #同时检查 expiry 值为 None 和 0 的情况
            return settings.SESSION_COOKIE_AGE  # 返回 settings 配置中设置的值
        if not isinstance(expiry, datetime):
            return expiry  # 如果不是 datetime 对象，则直接返回
        delta = expiry - modification  # 否则计算过期时间和修改时间差的秒数值并返回
        return delta.days * 86400 + delta.seconds

    def get_expiry_date(self, **kwargs):
        # 获取过期日期
        try:
            modification = kwargs['modification']
        except KeyError:
            modification = timezone.now()
        # Same comment as in get_expiry_age
        try:
            expiry = kwargs['expiry']
        except KeyError:
```

```
        expiry = self.get('_session_expiry')

    if isinstance(expiry, datetime):
        return expiry
    expiry = expiry or settings.SESSION_COOKIE_AGE  # Checks both None and 0 cases
    # 计算过期时间
    return modification + timedelta(seconds=expiry)

def set_expiry(self, value):
    # 设置过期时间
    if value is None:
        # 如果 value 值为空，则删除_session 属性值中的_session_expiry 并返回
        try:
            del self['_session_expiry']
        except KeyError:
            pass
        return
    if isinstance(value, timedelta):
        # 如果是 timedelat，则和当前时间相加得到过期时间
        value = timezone.now() + value
    # 设置_session 属性值中_session_expiry 的键值
    self['_session_expiry'] = value

def get_expire_at_browser_close(self):
    if self.get('_session_expiry') is None:
        return settings.SESSION_EXPIRE_AT_BROWSER_CLOSE
    return self.get('_session_expiry') == 0

# ……
```

这些方法并不复杂，通过注释很容易就能看出其执行逻辑。下面演示如何调用这 4 个方法 get_expiry_age()、get_expiry_date()、set_expiry()和 get_expire_at_browser_close()，具体操作如下：

```
# 接着前面的操作继续执行
>>> '_session_expiry' in sb
False
>>> sb.get_expire_at_browser_close()  # 在 settings 中默认值为 False
False
>>> sb['_session_expiry'] = 0
>>> sb.get_expire_at_browser_close()
True
>>> sb.set_expiry(None) # 清除_session_expiry
>>> '_session_expiry' in sb
False
>>> sb['_session_expiry'] = 10   # 设置 10s 后过期
>>> sb.get_expiry_age()       # 如果设置的是数值，则直接返回
10
>>> import datetime
```

```
>>>
datetime.datetime.now().strftime('%Y-%m-%d %H:%M:%S');sb.get_expiry_date().strftime(
'%Y-%m-%d %H:%M:%S')  # 正好 10s，放在一起执行，时间误差非常小
'2021-05-09 22:37:58'
'2021-05-09 22:38:08'
>>> sb['_session_expiry'] = datetime.datetime(2021, 5, 10, 8, 30)
>>> modification = datetime.datetime(2021, 5, 9, 22, 30)
# 过期时间减去修改时间正好等于 10 小时，即 10 ×3600 = 36000
>>> sb.get_expiry_age(**{'modification': modification})
36000
>>> sb.get_expiry_date()
datetime.datetime(2021, 5, 10, 8, 30)
```

**注意**，如果在上面调用 get_expiry_date()方法时出现 8 小时误差，则需要在 first_django/settings.py 文件中设置 USE_TZ = False。

最后，在 BaseSession 类中还定义了数个需要在子类中实现的方法，如 exists()、create()、save()、delete()、load()和 clear_expired()等，其中，clear_expired()方法为类方法。这些方法会在后续研究子类实现时再去分析。

## 6.5.2　Session 的存储引擎

接下来分析 Session 的两种典型存储引擎：内存存储引擎（cache.py）和数据库存储引擎（db.py）。文件存储（file.py）等学习思路和它们一致，不再赘述。

先来看内存存储引擎中的源码（地址为 django/contrib/sessions/backends/cache.py），这里的源码和 5.3 节中介绍的缓存模块的源码密切相关，请在掌握前面的知识后再来学习这里的源码：

```
# 源码位置：django/contrib/sessions/backends/cache.py
# ……

# 导入核心模块下的 caches，这个模块在 5.3 节中曾详细介绍过
from django.core.cache import caches

KEY_PREFIX = "django.contrib.sessions.cache"  # 保存在缓存中的 key 的前缀


class SessionStore(SessionBase):
    """
    基于缓存的 Session 存储
    """
    cache_key_prefix = KEY_PREFIX

    def __init__(self, session_key=None):
        # 直接使用 core 目录下的缓存模块作为这里的缓存核心
        self._cache = caches[settings.SESSION_CACHE_ALIAS]
```

```
        super().__init__(session_key)

    @property
    def cache_key(self):
        # 返回缓存数据的 key：前缀+调用父类方法创建的字符串
        return self.cache_key_prefix + self._get_or_create_session_key()

    def load(self):
        try:
            session_data = self._cache.get(self.cache_key)  # 从内存中加载对应 key 的数据
        except Exception:
            session_data = None
        if session_data is not None:
            return session_data
        self._session_key = None
        return {}

    def create(self):
        for i in range(10000):
            self._session_key = self._get_new_session_key()  # 获取 session_key
            try:
                self.save(must_create=True)
            except CreateError:
                continue
            self.modified = True
            return
        raise RuntimeError(
            "Unable to create a new session key. "
            "It is likely that the cache is unavailable.")

    def save(self, must_create=False):
        if self.session_key is None:
            return self.create()
        if must_create:
            func = self._cache.add      # add()方法
        elif self._cache.get(self.cache_key) is not None:
            func = self._cache.set      # set()方法
        else:
            raise UpdateError
        result = func(self.cache_key,
                      self._get_session(no_load=must_create),
                      self.get_expiry_age())
        if must_create and not result:
            raise CreateError

    def exists(self, session_key):
        # 判断对应的 key 是存在，在内存字典中进行判断
        return bool(session_key) and (self.cache_key_prefix + session_key) in self._cache
```

```
    def delete(self, session_key=None):
        if session_key is None:
            if self.session_key is None:
                return
            session_key = self.session_key
        # 删除内存字典中对应的 key
        self._cache.delete(self.cache_key_prefix + session_key)

    @classmethod
    def clear_expired(cls):
        pass
```

cache.py 文件中的 SessionStore 类继承自 SessionBase 类，因此它必须实现 Session 存储的几个核心方法，比如 save()、load()、exists()和 delete()等。这部分代码依赖 core 目录下的缓存模块，在 5.3 节中曾详细分析过，不再赘述。下面直接演示这些方法的使用，具体操作如下：

```
>>> from django.contrib.sessions.backends.cache import SessionStore
# 当没有设置 session_key 属性值时，除前缀部分外，后缀由随机字符串生成
>>> ss = SessionStore()
>>> ss.cache_key
'django.contrib.sessions.cache0br3tfhfkmzmf6jtwnry6subuxmjuphf'
sss>>> ss._cache  # 注意这里的_cache 属性值为 LocMemCache 对象
<django.core.cache.backends.locmem.LocMemCache object at 0x7f60540c1310>
>>> ss._cache._cache   # LocMemCache 对象的_cache 属性值是最终保存数据的地方
OrderedDict()
>>> ss.setdefault('hello', 'world')
'world'
>>> ss.setdefault('k1', 'v1')
'v1'
>>> ss._session  # 模拟 Session 数据
{'hello': 'world', 'k1': 'v1'}
>>> ss.save()    # 使用 save()方法保存，实际上是保存到 LocMemCache 对象的_cache 属性值中
>>> ss._cache._cache  # 查看保存到缓存中的数据
OrderedDict([(':1:django.contrib.sessions.cache5zl52kjubrbab1b31wcfcf2x06accqgk',
b'\x80\x05\x95\x0e\x00\x00\x00\x00\x00\x00\x00}\x94\x8c\x02k1\x94\x8c\x02v1\x94s.'),
(':1:django.contrib.sessions.cache77vs5yfj7y93x2huqll7z7esi6x68ife', b'\x80\x05\x95\
x1f\x00\x00\x00\x00\x00\x00\x00}\x94(\x8c\x05hello\x94\x8c\x05world\x94\x8c\x02k1\x9
4\x8c\x02v1\x94u.')])
>>> ss.cache_key
'django.contrib.sessions.cache5zl52kjubrbab1b31wcfcf2x06accqgk'
>>> ss.session_key
'0bdige2hqay1kw2iiziqkd16t3urgf0b'
>>> ss.load()  # 加载缓存的数据
{'hello': 'world', 'k1': 'v1'}
>>> ss.exists(ss.session_key)
True
>>> ss.exists('xx')
```

```
False
>>> ss.delete()         # 清除 Session 数据
>>> ss._cache._cache    # 最后查看，发现数据已被清除
OrderedDict()
```

数据库存储引擎的源码如下：

```
# 源码位置：django/contrib/sessions/backends/db.py
# ……

class SessionStore(SessionBase):
    """
    实现用数据库保存 Session 数据
    """
    def __init__(self, session_key=None):
        super().__init__(session_key)

    @classmethod
    def get_model_class(cls):
        from django.contrib.sessions.models import Session  # 导入 Session 模型类
        return Session

    @cached_property
    def model(self):
        return self.get_model_class()  # 通过 SessionStore.model 就能获取 Session 模型类

    def _get_session_from_db(self):
        try:
            return self.model.objects.get(        # 这是 Django 中最常见的 ORM 操作
                session_key=self.session_key,
                expire_date__gt=timezone.now()    # 获取未过期的数据
            )
        except (self.model.DoesNotExist, SuspiciousOperation) as e:
            if isinstance(e, SuspiciousOperation):
                logger = logging.getLogger('django.security.%s' % e.__class__.__name__)
                logger.warning(str(e))
            self._session_key = None

    def load(self):
        s = self._get_session_from_db()  # 获取单条记录
        # 对记录中的 session_data 进行解码，得到相应的字典结果
        return self.decode(s.session_data) if s else {}

    def exists(self, session_key):
        return self.model.objects.filter(session_key=session_key).exists()

    def create(self):
        while True:
```

```
            self._session_key = self._get_new_session_key()
            try:
                self.save(must_create=True)
            except CreateError:
                # Key wasn't unique. Try again.
                continue
            self.modified = True
            return

    def create_model_instance(self, data):
        return self.model(   # 创建模型对象
            session_key=self._get_or_create_session_key(), # 生成 session_key
            session_data=self.encode(data),       # 对 Session 数据进行编码后保存
            expire_date=self.get_expiry_date(), # 获取过期日期，父类方法
        )

    def save(self, must_create=False):
        if self.session_key is None:
            return self.create()
        data = self._get_session(no_load=must_create)    # 获取 Session 数据
        obj = self.create_model_instance(data)           # 获取模型对象
        using = router.db_for_write(self.model, instance=obj)  # using 为指定使用的数据库
        try:
            with transaction.atomic(using=using):
                # 保存 Session 数据到数据库中
                obj.save(force_insert=must_create, force_update=not must_create,
using=using)
        except IntegrityError:
            if must_create:
                raise CreateError
            raise
        except DatabaseError:
            if not must_create:
                raise UpdateError
            raise

    def delete(self, session_key=None):
        if session_key is None:
            if self.session_key is None:
                return
            session_key = self.session_key
        try:
            # 删除对应的 Session 记录
            self.model.objects.get(session_key=session_key).delete()
        except self.model.DoesNotExist:
            pass

    @classmethod
```

```
    def clear_expired(cls):
        # 删除过期的 Session 记录
        cls.get_model_class().objects.filter(expire_date__lt=timezone.now()).delete()
```

在 db.py 中定义的 SessionStore 类同样继承自 SessionBase 类。很明显，在该 SessionStore 类中定义的 save()、exists()、create()、delete()方法都与数据库操作有关，用的正是 Django 内置的 ORM 操作语句。继续在 Django 的 shell 模式下演示对上述类和相关方法的调用，以便读者理解其含义，具体操作如下：

```
# 通过迁移命令生成 django_session 表，SessionStore 类中的方法主要是操作该表
(django2-core-test) [root@master first_django]# python manage.py migrate sessions
System check identified some issues:

# 忽略告警信息
# ……
Operations to perform:
  Apply all migrations: sessions
Running migrations:
  Applying sessions.0001_initial... OK

(django2-core-test) [root@master first_django]# python manage.py shell
Python 3.8.6 (default, Oct 18 2020, 15:33:08)
[GCC 4.8.5 20150623 (Red Hat 4.8.5-39)] on linux
Type "help", "copyright", "credits" or "license" for more information.
(InteractiveConsole)
>>> from django.contrib.sessions.backends.db import SessionStore
>>> ss = SessionStore()
>>> ss.model
<class 'django.contrib.sessions.models.Session'>
>>> ss.load()      # 从数据库中获取数据，一开始为空
{}
>>> ss.setdefault('session_id', '1111111')
'1111111'
>>> ss.setdefault('hello', 'world')
'world'
>>> ss._session   # 前面设置的键值对全部保存在_session 属性值中
{'session_id': '1111111', 'hello': 'world'}
>>> ss._get_session()
{'session_id': '1111111', 'hello': 'world'}
# 通过 encode()方法编码_session 属性值，这是后面保存到数据库中的 session_data 字段的值
>>> ss.encode(ss._session)
'ZGQzNTU3NTYyNGExMDkyMzVmZmMxZDliMmNkMThmOTkwZGZlNTE2Yzp7InNlc3Npb25faWQiOiIxMTExMTE
xIiwiaGVsbG8iOiJ3b3JsZCJ9'
>>> ss.save()  # 保存到数据库中，此时_session 属性值被编码保存
>>> ss.model.objects.all()
<QuerySet [<Session: ugzh49cd7zg7cmnc44e9i55xfhtkkatv>]>
>>> ss.session_key
```

```
'ugzh49cd7zg7cmnc44e9i55xfhtkkatv'
# 在保存的记录中，把 session_data 字段的值记为编码后的 Session 数据
>>> ss.model.objects.all()[0].session_data
'ZGQzNTU3NTYyNGExMDkyMzVmZmMxZDliMmNkMThmOTkwZGZlNTE2Yzp7InNlc3Npb25faWQiOiIxMTExMTE
xIiwiaGVsbG8iOiJ3b3JsZCJ9'
>>> ss.load()  # 用 load()方法加载 Session 数据，并对 session_data 字段进行解码
{'session_id': '1111111', 'hello': 'world'}
>>> ss.model.objects.all()[0].expire_date  # 默认过期时间是 2 周，即默认配置的值
datetime.datetime(2021, 5, 24, 11, 50, 41, 213574)
>>> import datetime
>>> datetime.datetime.now()      # 过了几分钟之后才打印当前时间
datetime.datetime(2021, 5, 10, 11, 53, 37, 432469)
>>> ss.delete()                  # 删除数据库中的记录，根据 session_key 字段进行删除
>>> ss.model.objects.all()
<QuerySet []>
>>> ss.clear_expired()           #用 clear_expired()方法直接从数据库中删除已过期的记录
```

至此，关于 Session 数据的存储核心就介绍完毕了，后续的文件存储等方式留给读者自行学习，同样是定义 SessionStore 类（继承 SessionBase 类），实现核心的 save()、exists()、load()和 delete()等方法。

通过对上述代码的解读，现在我们可以解答 6.1 节中实验 5 的相关问题了，其中，关于 Session 数据是如何存储的问题已经在上面的代码分析和实战中进行了解答。现在思考 Django 在操作 Session 数据时是如何判断数据是否过期的呢？首先看在 db.py 文件中用来获取 Session 数据的_get_session_from_db()方法的源码：

```
# 源码位置：django/contrib/sessions/backends/db.py
# ……

class SessionStore(SessionBase):
    # ……

    def _get_session_from_db(self):
        try:
            return self.model.objects.get(
                session_key=self.session_key,
                expire_date__gt=timezone.now()  # 获取未过期的数据
            )
        except (self.model.DoesNotExist, SuspiciousOperation) as e:
            # 打印告警信息
            # ……
            self._session_key = None

    # ……
```

在生成 Session 数据时，如果没有设置过期时间，则在 Django 内部有默认的 Session 数据过期时

间（可能设置过期秒数，也可能直接设置过期时间）。因此，当把 Session 数据保存在数据库中时会有一个过期时间，而在查询对应的 Session 数据时，会带上过期条件，即过期时间要大于当前时间。如果 Session 数据已经过期，就无法查询该 Session 数据，实现了 Session 数据已过期功能。而对于 cache.py 文件中的代码而言，其依赖 django/core 目录下的缓存模块。在 5.3 节中，我们曾解读和测试过该缓存模块的过期功能，get()方法在获取缓存数据前会先判断该数据是否已过期（同样会对保存的数据设置过期时间）。如果没有过期，返回相应值，否则返回 None。

## 6.5.3 SessionBase 类中的代码文件

接下来看一些比较简单的代码文件，首先是在 SessionBase 类中指定默认的序列化器。默认的序列化器的定义如下：

```
# 源码位置：django/conf/global_settings.py
# ……

SESSION_SERIALIZER = 'django.contrib.sessions.serializers.JSONSerializer'

# ……
```

而 SessionBase 类中的默认的序列化器的定义如下：

```
# 源码位置：django/contrib/sessions/serializers.py
# ……

import pickle

from django.core.signing import JSONSerializer as BaseJSONSerializer

# ……

JSONSerializer = BaseJSONSerializer
```

继续追踪 BaseJSONSerializer 序列化器的实现源码，如下：

```
# 源码位置：django/core/signing.py
# ……

class JSONSerializer:
    def dumps(self, obj):
        return json.dumps(obj, separators=(',', ':')).encode('latin-1')

    def loads(self, data):
        return json.loads(data.decode('latin-1'))

# ……
```

可以看到，默认的序列化器实际上就是简单封装了 JSON 模块，而 dumps()方法和 loads()方法的核心是调用 json.dumps()方法和 json.load()方法：

```
>>> from django.core.signing import JSONSerializer
>>> serializer = JSONSerializer()
>>> serializer.dumps({'hello': 'world', 'session_id': 'xxxxxx1'})
b'{"hello":"world","session_id":"xxxxxx1"}'
>>> data = serializer.dumps({'hello': 'world', 'session_id': 'xxxxxx1'})
>>> data
b'{"hello":"world","session_id":"xxxxxx1"}'
>>> serializer.loads(data)
{'hello': 'world', 'session_id': 'xxxxxx1'}
```

### Session 表的抽象类和抽象管理类

继续看在 django/contrib/sessions/base_session.py 文件中定义的相关类，源码如下：

```
# 源码位置：django/contrib/sessions/base_session.py
# ……

class BaseSessionManager(models.Manager):
    def encode(self, session_dict):
        session_store_class = self.model.get_session_store_class()
        # 最终调用 SessionStore 对象的 encode()方法
        return session_store_class().encode(session_dict)

    def save(self, session_key, session_dict, expire_date):
        # 对 Session 数据进行编码，即调用 encode()方法
        s = self.model(session_key, self.encode(session_dict), expire_date)
        if session_dict:
            s.save()        # 保存该记录
        else:
            s.delete()      # 如果没有 Session 数据，清除相关记录
        return s


class AbstractBaseSession(models.Model):
    # django_session 表中的三个字段
    session_key = models.CharField(_('session key'), max_length=40, primary_key=True)
    session_data = models.TextField(_('session data'))
    expire_date = models.DateTimeField(_('expire date'), db_index=True)

    objects = BaseSessionManager()  # 定义 Manager 对象

    class Meta:
        abstract = True
        verbose_name = _('session')
        verbose_name_plural = _('sessions')
```

```
    def __str__(self):
        return self.session_key

    @classmethod
    def get_session_store_class(cls):
        raise NotImplementedError

    def get_decoded(self):
        session_store_class = self.get_session_store_class()
        # 对编码的 Session 数据进行解码
        return session_store_class().decode(self.session_data)
```

上面定义了关于 Session 表的抽象类和抽象管理类，从中可以看到在 Session 表中定义的基本字段，以及当把 Session 数据保存到数据库中时对 session_data 字段的编码处理。

### Session 的模型类及管理类

继续看在 django/contrib/sessions/models.py 文件中定义的 Session 的模型类及管理类，源码如下：

```
# 源码位置：django/contrib/sessions/models.py

from django.contrib.sessions.base_session import (
    AbstractBaseSession, BaseSessionManager,   # 导入前面定义的抽象类
)


class SessionManager(BaseSessionManager):
    use_in_migrations = True                    # 在简单继承父类后，定义一个属性值


class Session(AbstractBaseSession):

    objects = SessionManager()                  # 定义 Session 表的管理器

    @classmethod
    def get_session_store_class(cls):
        from django.contrib.sessions.backends.db import SessionStore
        # 获取 SessionStore 类，主要使用该类中的 encode()方法和 decode()方法
        return SessionStore

    class Meta(AbstractBaseSession.Meta):
        db_table = 'django_session'  # 定义该模型类关联的表
```

看完该 Session 模型类的定义再结合其父类的实现，就能理解该模型表所定义的字段，以及对一些字段的特殊处理（如编码和解码）了。

### 中间件模块

Django 在内部是如何处理每次请求的 Session 数据的呢？答案是使用 Django 内部的中间件模块。每个基于 Django 框架创建的项目在 setting.py 文件中都会定义如下变量：

```
# 源码位置：django/contrib/sessions/middleware.py
# ……

MIDDLEWARE = [
    'django.middleware.security.SecurityMiddleware',
    'django.contrib.sessions.middleware.SessionMiddleware',
    'django.middleware.common.CommonMiddleware',
    'django.middleware.csrf.CsrfViewMiddleware',
    'django.contrib.auth.middleware.AuthenticationMiddleware',
    'django.contrib.messages.middleware.MessageMiddleware',
    'django.middleware.clickjacking.XFrameOptionsMiddleware',
]
```

在 MIDDLEWARE 变量中指定的各模块路径是 Django 内部的中间件类路径，这部分内容将在第 7 章详细介绍。这里只需知道，外部的 HTTP 请求将依次被这些中间件类的 process_request()方法进行处理，最后才交给视图函数处理。而视图函数返回的响应对象同样会依次经过这些中间件类的 process_response()方法进行，处理之后再返回给客户端。在 MIDDLEWARE 变量中定义的第 2 个中间件类就是用来处理请求的 Session 数据的，该中间件类的源码如下：

```
# 源码位置：django/contrib/sessions/middleware.py
# ……

class SessionMiddleware(MiddlewareMixin):
    def __init__(self, get_response=None):
        self.get_response = get_response
        engine = import_module(settings.SESSION_ENGINE) # 获取 Session 数据的存储引擎
        self.SessionStore = engine.SessionStore  # 获取存储引擎中的 SessionStore 类

    def process_request(self, request):
        # 从 cookie 数据中获取对应 key 保存的 Session 数据
        session_key = request.COOKIES.get(settings.SESSION_COOKIE_NAME)
        # 设置请求的 session 属性值为一个 SessionStore 对象
        request.session = self.SessionStore(session_key)

    def process_response(self, request, response):
        try:
            accessed = request.session.accessed  # 当获取 Session 数据时会设置为 True
            modified = request.session.modified  # 当修改 Session 数据时会设置为 True
            empty = request.session.is_empty()   # 判断 Session 数据是否为空
        except AttributeError:
            pass
        else:
```

```
        if settings.SESSION_COOKIE_NAME in request.COOKIES and empty:
            # 调用 Resopnse 对象的 delete_cookie()方法，删除相关的 cookie 信息
            response.delete_cookie(
                settings.SESSION_COOKIE_NAME,
                path=settings.SESSION_COOKIE_PATH,
                domain=settings.SESSION_COOKIE_DOMAIN,
                samesite=settings.SESSION_COOKIE_SAMESITE,
            )
        else:
            if accessed:
                patch_vary_headers(response, ('Cookie',))
            if (modified or settings.SESSION_SAVE_EVERY_REQUEST) and not empty:
                if request.session.get_expire_at_browser_close():
                    max_age = None
                    expires = None
                else:
                    max_age = request.session.get_expiry_age()    # 获取过期秒数
                    expires_time = time.time() + max_age          # 计算过期时间
                    expires = http_date(expires_time)             # 格式转换
                if response.status_code != 500:
                    try:
                        request.session.save()
                    except UpdateError:
                        # 更新 Session 数据异常
                        # ……

                    # 设置 cookie 信息，将关键的 session_key 保存到 cookie 中
                    response.set_cookie(
                        settings.SESSION_COOKIE_NAME,
                        request.session.session_key, max_age=max_age,
                        expires=expires, domain=settings.SESSION_COOKIE_DOMAIN,
                        path=settings.SESSION_COOKIE_PATH,
                        secure=settings.SESSION_COOKIE_SECURE or None,
                        httponly=settings.SESSION_COOKIE_HTTPONLY or None,
                        samesite=settings.SESSION_COOKIE_SAMESITE,
                    )
    # 最后返回响应对象
    return response
```

看完上述代码后，就能理解实验 5 中的相关操作语句了。例如，在该示例中，是通过 request.session 语句设置和获取 Session 数据的。由 process_request()方法中的代码可知，request.session 其实就是 Session 存储引擎中的 SessionStore 对象。对该对象的操作，以及查询 key 是否过期的功能均在前面演示过，不再赘述。

# 6.6　答疑解惑

在学完 Django 的视图层的源码后，就可以解答 6.1 节中提出的问题了。下面先全面汇总待解答的问题：

问题 1：在 Django 中如何实现 URLConf 配置与路径的匹配？

问题 2：url()函数和 path()函数有何区别？include()函数呢？

问题 3：视图类的功能是什么？为什么它能自动映射不同的请求到不同的方法中？为什么在 URLConf 配置中要用 HelloView.as_view()语句，而不能用 HelloView 语句本身呢？

问题 4：在视图类实验中出现的@csrf_exempt 有何作用，可以不用吗？

问题 5：在 Django 框架中，是在哪里将 HTTP 的请求参数传到 request.GET、request.POST 及 request.FILES 等属性上的？

问题 6：在文件上传实验中，在使用 request.body 语句时会报错？错误原因是什么？如何解决？

问题 7：在上传文件的第 2 个实验中，保存文件到本地的代码在哪里？upload_to 参数究竟有何作用，是否有其他写法呢？

问题 8：在 Django 项目中，Session 数据的存储过程和过期时间判断的逻辑是怎样的？

接下来我们逐一梳理上述问题的答案。对于已经在前面解答过的问题将直接跳过，对于没有在前面明确解答过的，将根据前面的源码进行解答。

(1) 先看问题 1，这个问题的答案在 6.3 节的源码中曾分析过。我们追踪了在 Django 中处理 HTTP 请求的最核心的方法：_get_response()。通过对该方法的源码追踪，可以看到在 Django 中是如何实现 URLConf 配置与路径匹配的过程的。在该方法中，实现匹配 URL 路径的核心语句如下：

```
resolver_match = resolver.resolve(request.path_info)
```

顺着该语句追踪下去，就能看到在 Django 源码中匹配请求路径的逻辑。resolve()方法的匹配逻辑可简单总结如下：

- ◎ 以项目配置的 ROOT_URLCONF 变量作为入口点，导入该变量指定的模块路径，并获取该模块中 urlpatterns 变量指定的 URLConf 配置。
- ◎ 将 urlpatterns 变量中的 URLConf 配置与请求的真实路径进行匹配。
- ◎ 如果是<URL 路径，视图函数>这样的 URLConf 配置，则匹配成功，直接返回 ResolverMatch()对象，匹配过程结束。
- ◎ 如果是<URL 路径，include()、url()或者 admin.site.urls>这样的 URLConf 配置，则匹配成功后会继续返回 URLResolver 对象，递归调用 resolve()函数，匹配对应模块中的 urlpatterns

变量指定的 URLConf 配置，最后找到匹配结果。

◎ 如果匹配不到，则抛出 Resolver404 异常，对请求客户端来说就是 404 Not Found。

比如前面的/test_view/hello/请求路径，第一次传入的匹配是“^/”，接着去掉开始的“/”，再去匹配路径 test_view/hello/。在成功匹配到 url('test_view/', include('test_view.urls'))后，去掉匹配的“test_view/”，然后导入 test_view.urls 模块，继续递归调用 URLResolver 对象中的 resovle()函数。此时的匹配路径变成了“hello/”。当再次扫描 test_view.urls 模块中的 URLConf 配置时，便能成功匹配到 url('hello/', views.hello_view, name='hello_view')配置了，返回最终的 ResolverMatch()对象。

(2)问题2比较简单，前面曾追踪过 url()函数和path()函数的源码。url()函数最终调用的是 re_path()函数，它就定义在 path()函数下：

```
# 源码位置：django/urls/conf.py
# ……

path = partial(_path, Pattern=RoutePattern)
re_path = partial(_path, Pattern=RegexPattern)
```

从上面的源码可以看到，这两个函数最终都是调用_path()函数，但是传给_path()函数的 Pattern 参数却有所不同。path()函数的 Pattern 参数为 RoutePattern 类，而 re_path 函数的 Pattern 参数为 RegexPattern 类。根据 6.3 节的分析可知，在 path()函数中设置的 URL 路径除支持正则表达式的写法外，还支持<str:name>/<int:age>这样的写法，而后者只支持正则表达式的写法。include()函数源码在 6.3 节中已经分析过，它将返回一个三元组(urlconf_module, app_name, namespace)，其中，urlconf_module 正对应导入的包含 URLConf 配置的模块。

(3) 对于问题 3，在 6.3 节的最后已经讲解得非常清楚了。视图类实现了具有 RESTful 风格的 API 接口，只需继承视图类并实现 get()、post()、put()等方法，就可以实现同一个 URL 路径的 GET、POST 和 PUT 请求。至于为什么相关方式的请求能映射到视图类的对应方法中，在 6.3 节中也已经分析得很清楚了，不再赘述。

在 URLConf 中要写成 HelloView.as_view()方式的原因是，它的结果是一个可调用的方法。因此，在前面的 URL 匹配成功后，将由该方法处理对应的 HTTP 请求。HelloView 本身是一个类，并没有实现魔法函数__call__()，因此它不能被调用，也就不能被放到 URLConf 配置的第 2 个位置上。

(4) 问题 4 中的@csrf_exempt 装饰器可用于避免 HTTP 请求在 Django 内部进行 CSRF 校验，而相关的校验过程及 csrf_token 生成原理将在第 7 章中进行详细分析。为了避免在测试中对 POST、PUT 或 DELETE 请求进行安全校验，需要在 dispatch()方法上加上@csrf_exempt 装饰器。实际上，不用该装饰器的方法也很简单，在 first_django/settings.py 的中间件变量中注释掉和 csrf_token 校验相关的字符串模块即可：

```
[root@master first_django]# cat first_django/settings.py | grep MIDDLEWARE -A8
MIDDLEWARE = [
    'django.middleware.security.SecurityMiddleware',
    'django.contrib.sessions.middleware.SessionMiddleware',
    'django.middleware.common.CommonMiddleware',
    # 'django.middleware.csrf.CsrfViewMiddleware',  # 注释掉这里即可
    'django.contrib.auth.middleware.AuthenticationMiddleware',
    'django.contrib.messages.middleware.MessageMiddleware',
    'django.middleware.clickjacking.XFrameOptionsMiddleware',
]
```

（5）由于问题 5 在前面并没有涉及，所以需要进一步追踪源码才能解锁答案。回到处理 HTTP 请求的最早的源码，如下：

```
# 源码位置：django/core/handlers/wsgi.py
# ……

class WSGIHandler(base.BaseHandler):
    request_class = WSGIRequest

    # ……

    def __call__(self, environ, start_response):
        set_script_prefix(get_script_name(environ))
        signals.request_started.send(sender=self.__class__, environ=environ)
        # 请求入口，在这里得到了 request 变量
        request = self.request_class(environ)
        response = self.get_response(request)

        # ……
        return response

# ……
```

看上面的源码是不是非常熟悉？在 6.3 节中介绍的 Django 内部处理 HTTP 请求的入口源码用的正是这里的魔法函数__call__()。从上面的源码可以看到出，Django 内部的 request 变量其实是一个 WSGIRequest 对象。继续追踪 WSGIRequest 类的实现源码便可知 request.GET、request.POST 及 request.FILES 的来源了：

```
# 源码位置：django/core/handlers/wsgi.py
# ……

class WSGIRequest(HttpRequest):
    def __init__(self, environ):
        script_name = get_script_name(environ)
        path_info = get_path_info(environ) or '/'
        self.environ = environ
```

```
        self.path_info = path_info
        self.path = '%s/%s' % (script_name.rstrip('/'), path_info.replace('/', '', 1))
        self.META = environ
        self.META['PATH_INFO'] = path_info
        self.META['SCRIPT_NAME'] = script_name
        self.method = environ['REQUEST_METHOD'].upper() # 获取请求方式
        self.content_type, self.content_params =
cgi.parse_header(environ.get('CONTENT_TYPE', ''))
        if 'charset' in self.content_params:
            try:
                codecs.lookup(self.content_params['charset'])
            except LookupError:
                pass
            else:
                self.encoding = self.content_params['charset']
        try:
            content_length = int(environ.get('CONTENT_LENGTH'))
        except (ValueError, TypeError):
            content_length = 0
        self._stream = LimitedStream(self.environ['wsgi.input'], content_length)
        self._read_started = False
        self.resolver_match = None

    def _get_scheme(self):
        return self.environ.get('wsgi.url_scheme')

    @cached_property
    def GET(self):
        # 按 WSGI 标准来说，QUERY_STRING 可能会被移除
        raw_query_string = get_bytes_from_wsgi(self.environ, 'QUERY_STRING', '')
        # 从 self.environ 中获取 QUERY_STRING 对应的值，对应 URL 路径中“?”后的字符串
        return QueryDict(raw_query_string, encoding=self._encoding)

    def _get_post(self):
        if not hasattr(self, '_post'):
            self._load_post_and_files() # 载入 post 数据和文件数据
        return self._post

    def _set_post(self, post):
        self._post = post  # 设置_post 属性值

    @cached_property
    def COOKIES(self):
        # 获取 cookies 数据
        raw_cookie = get_str_from_wsgi(self.environ, 'HTTP_COOKIE', '')
        return parse_cookie(raw_cookie) # 将 cookies 数据解析成字典形式

    @property
```

```
    def FILES(self):
        if not hasattr(self, '_files'):
            self._load_post_and_files()
        return self._files  # 获取文件数据

POST = property(_get_post, _set_post)  # 获取_post 属性值

# ……
```

由上面的源码可知，request.GET 的属性值来源于 environ 变量中 QUERY_STRING 对应的值，raw_query_string 其实就是请求的 URL 路径“？”后面的字符串。最后返回该变量初始化后的 QueryDict 对象，该对象会自动解析“a=x&b=y”这样的字符串，得到一个类字典的结果。WSGIRequest 对象中的 request.POST 和 request.FILES 的属性值来源也很明确，即分别来自该对象的_post 和_files 属性值。这两个属性值的核心获取方法为_load_post_and_files()，对于该方法的细节不再赘述，有兴趣的读者可以自行学习。

（6）问题 6 比较容易处理，直接看 body()方法的源码即可（该方法位于 HttpRequest 类中）：

```
# 源码位置：django/http/request.py
# ……

class HttpRequest:

    # ……

    @property
    def body(self):
        if not hasattr(self, '_body'):
            # 调试时可在源码中添加该 print()语句
            # print('_read_started={}'.format(self._read_started))
            if self._read_started:
                # 这就是 6.1 节的实验 3 中上传文件抛出的异常
                raise RawPostDataException(
"You cannot access body after reading from request's data stream"
)

            # 检查请求数据大小是否超出项目设置值，如果超出，抛出异常
            # ……

            try:
                self._body = self.read()
            except IOError as e:
                raise UnreadablePostError(*e.args) from e
            # self._stream 属性在这里赋值
            self._stream = BytesIO(self._body)
        return self._body
```

```
# ……
```

从上面的源码可以看到，6.1 节的实验 3 中抛出的 RawPostDataException 异常，其发生需要满足两个条件：

（1）Request 对象中没有_body 属性。

（2）Request 对象中的_read_started 属性值为 True。

在调试时，可以在 if self._read_started 语句的上方加一条 print()语句，即打印 self._read_started 的值。然后分别测试不带文件的 HTTP 请求和带文件的 HTTP 请求。在测试后可以发现，在获取 WSGIRequest 对象的_body 属性值时，对于不带文件的 HTTP 请求，WSGIRequest 对象的_read_started 属性值为 False；而对于带文件的 HTTP 请求，WSGIRequest 对象的_read_started 属性值为 True。因此，后者将抛出 RawPostDataException 异常，并提示“如果已经从请求的数据流中读取数据，就不能再访问 body 属性值”。_read_started 属性值的含义是什么呢？继续看下面的源码：

```
# 源码位置：django/http/request.py
# ……

class HttpRequest:

    # ……

    def read(self, *args, **kwargs):
        # 只有在调用 read()方法和 readline()方法时，才会设置_read_started 属性值为 True
        self._read_started = True
        try:
            return self._stream.read(*args, **kwargs)  # 读取请求流中的数据
        except IOError as e:
            raise UnreadablePostError(*e.args) from e

    def readline(self, *args, **kwargs):
        self._read_started = True
        try:
            return self._stream.readline(*args, **kwargs)
        except IOError as e:
            raise UnreadablePostError(*e.args) from e

    # ……
```

在 HttpRequest 请求中，read()方法和 readline()方法会设置_read_started 属性值。_read_started 属性值的含义是，Django 已经从请求的数据流中读取了相关数据。如果再次调用 read()方法，读取数据流中的数据就是空。可以看到，在前面的 body()中确实会再次调用请求对象的 read()方法。如果请求流中的数据已经被读取完毕，则再次调用 read()方法并不会报错，返回的是空字符串。为了避免重复读取数据流，对于这种情况 Django 将主动抛出一个异常，即 RawPostDataException 异常。继续看

read()方法中出现的_stream 属性值的来源，可以从 WSGIRequest 的初始化方法中找到，其源码如下：

```
# 源码位置：django/core/handlers/wsgi.py
# ……

class WSGIRequest(HttpRequest):
    def __init__(self, environ):
        # ……

        # 最开始在这里赋值
        self._stream = LimitedStream(self.environ['wsgi.input'], content_length)
        self._read_started = False
        self.resolver_match = None

    # ……

# ……
```

可以看到，WSGIRequest 对象的_stream 属性值和 GET、POST 属性一样，均来源于请求的环境变量 environ。

（7）问题 7 比较简单，在上传文件的第 2 个实验中，对于上传的文件使用了 FileField 字段进行映射，因此只需找到该 FileField 类对应的源码即可，其内容如下：

```
# 源码位置：django/db/models/fields/files.py
# ……

class FileField(Field):

    attr_class = FieldFile   # 非常关键

    descriptor_class = FileDescriptor

    description = _("File")

    def __init__(self, verbose_name=None, name=None, upload_to='',
                 storage=None, **kwargs):
        self._primary_key_set_explicitly = 'primary_key' in kwargs

        # 用于操作文件的文件模块 default_storage，可参考 5.6 节的内容
        self.storage = storage or default_storage
        self.upload_to = upload_to

        kwargs.setdefault('max_length', 100)
        super().__init__(verbose_name, name, **kwargs)

    # ……
```

```
    def pre_save(self, model_instance, add):
        file = super().pre_save(model_instance, add)
        if file and not file._committed:
            # 在保存模型对象的数据到数据库之前，先保存指定的文件到本地
            file.save(file.name, file.file, save=False)
        return file

    # ……

    def generate_filename(self, instance, filename):

        if callable(self.upload_to):
            filename = self.upload_to(instance, filename)
        else:
            dirname = datetime.datetime.now().strftime(self.upload_to)
            filename = posixpath.join(dirname, filename)
        # 返回文件路径，在该方法中会对文件名进行校验
        return self.storage.generate_filename(filename)

    # ……
```

从 FileField 类的 generate_filename()方法中能看出 upload_to 属性的作用。如果其值是可调用的（会传入文件名参数），直接返回其调用结果作为生成文件的路径（参考 6.1 节中的文件上传实验）。对于其他情况，则会调用 datetime.datetime.now().strftime()得到保存文件的目录。因此 upload_to 属性值必须是时间格式，例如“%Y/%d/%m”，这样会得到一个保存路径，而该值将被记录到数据库中。下面看一个简单的示例：

```
>>> import posixpath
>>> import datetime
>>> from django.core.files.storage import default_storage
>>> dirname = datetime.datetime.now().strftime("%Y/%d/%m")
>>> dirname
'2021/12/05'
>>> filename = posixpath.join(dirname, "test.txt")
>>> filename
'2021/12/05/test.txt'
>>> default_storage.exists(filename)
False
>>> default_storage._wrapped
<django.core.files.storage.FileSystemStorage object at 0x7f7048103e80>
>>> default_storage.generate_filename(filename)
'2021/12/05/test.txt'
```

接下来追踪保存文件的逻辑。在 FileField 类中有一个 pre_save()方法，通过该方法的源码可以确认此处正是 Django 保存上传文件到本地的地方。如果不理解上面的源码，可以在源码中打印 pre_save()方法中的 file 变量的结果，它是一个 FieldFile 对象：

```
# 源码位置：django/db/models/fields/files.py
# ……

class FieldFile(File):

    def __init__(self, instance, field, name):
        super().__init__(None, name)  # Field 对象名
        self.instance = instance      # Field 所属的模型对象
        self.field = field
        self.storage = field.storage
        self._committed = True

    # ……

    def _get_file(self):
        self._require_file()
        if getattr(self, '_file', None) is None:
            # 通过 Storage 对象（默认是 FileSystemStorage 对象）打开文件
            self._file = self.storage.open(self.name, 'rb')
        return self._file

    def _set_file(self, file):
        self._file = file

    def _del_file(self):
        del self._file

    # 关于 file 属性值
    file = property(_get_file, _set_file, _del_file)

    # ……

    def save(self, name, content, save=True):
        name = self.field.generate_filename(self.instance, name)
        self.name = self.storage.save(name, content, max_length=self.field.max_length)
        setattr(self.instance, self.field.name, self.name)
        self._committed = True

        # Save the object because it has changed, unless save is False
        if save:
            self.instance.save()

    # ……
```

FieldFile 类中的 save()方法的作用也非常明确，首先通过 self.storage 属性的 save()方法保存文件到本地，如果传入的 save 参数为 True，则调用该模型对象的 save()方法将记录保存到数据库中。这里使用模型对象保存上传文件到本地的核心语句就是 self.storage.save()。如果继续追踪其 save()方法

的核心逻辑，就回到 5.6 节中的内容了，不再赘述。最后还有一个疑问，Django 会在哪里调用 Field 对象的 pre_save()方法呢？其实只需简单全局搜索一下就能找到，下面直接给出调用位置，如下：

```
# 源码位置：django/db/models/base.py
# ……

class Model(metaclass=ModelBase):

    # ……

    def _save_table(self, raw=False, cls=None, force_insert=False,
                    force_update=False, using=None, update_fields=None):

        # ……

        if pk_set and not force_insert:
            base_qs = cls._base_manager.using(using)
            # f 为每个 Field 对象，这里指定 save 参数为 False
            values = [(f, None, (getattr(self, f.attname) if raw else f.pre_save(self,
False))) for f in non_pks]
            forced_update = update_fields or force_update
            updated = self._do_update(base_qs, using, pk_val, values, update_fields,
                                      forced_update)
            if force_update and not updated:
                raise DatabaseError("Forced update did not affect any rows.")
            if update_fields and not updated:
                raise DatabaseError("Save with update_fields did not affect any rows.")

        # ……
```

通过 3.4 节的内容可知，在 6.1 节的上传文件实验中出现的 FileModel.objects.create()的执行顺序如图 6-3 所示。

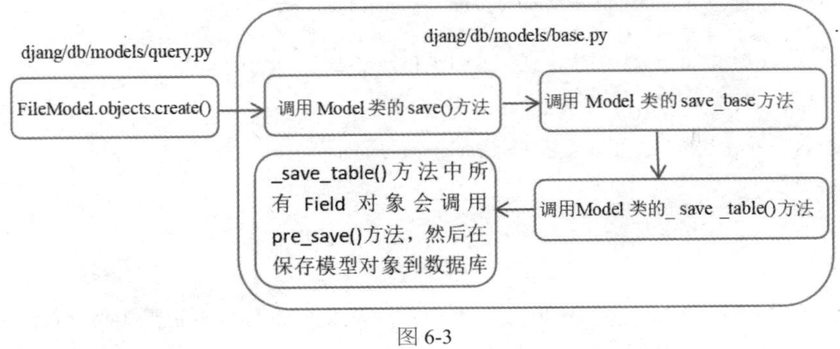


图 6-3

（8）问题 8 已经在 6.5 节的源码分析中完整解答了，不再赘述。

## 6.7 小结

本章重点解读了在 Django 中处理 HTTP 请求的核心源码，帮助读者梳理 Django 框架中 HTTP 请求被映射到对应的视图函数的过程。此外，本章详细介绍了视图层中定义的各种 View 类，同时使用丰富的实战案例帮助读者掌握这些视图类的使用。最后，简单分析了在 Django 中与 Session 相关的源码，帮助读者理解 Session 数据底层存储的过程和过期判断的逻辑。

# 第 7 章 Django的中间件原理

本章分析中间件（Middleware）相关的源码，厘清中间件组件的执行逻辑。最后以 CsrfViewMiddleware 中间件为例，剖析 csrf_token 的产生与校验过程。

## 7.1 配置中间件

在 Django 项目中，在编写处理 HTTP 请求的视图函数后，当 HTTP 请求到来时，并不会把 HTTP 请求直接交给视图函数进行处理，而是要先经过中间件进行预处理。HTTP 请求需要经过哪些中间件是在 Django 项目的 settings.py 文件中配置的。以 first_django 项目为例，在创建 first_django 项目后，在 settings.py 文件中关于中间件的配置如下：

```
[root@master first_django]# cat first_django/settings.py | grep MIDDLEWARE -A 8
MIDDLEWARE = [
    'django.middleware.security.SecurityMiddleware',
    'django.contrib.sessions.middleware.SessionMiddleware',
    'django.middleware.common.CommonMiddleware',
    'django.middleware.csrf.CsrfViewMiddleware',
    'django.contrib.auth.middleware.AuthenticationMiddleware',
    'django.contrib.messages.middleware.MessageMiddleware',
    'django.middleware.clickjacking.XFrameOptionsMiddleware',
]
```

当然，开发人员也可以按照自己的需要，自定义中间件，然后配置在 settings.py 文件的 MIDDLEWARE 变量中。当有 HTTP 请求到来时，这个 HTTP 请求就会按照顺序一层一层地经过列表中配置的中间件，只有符合所有中间件处理要求的 HTTP 请求才能到达相应的视图函数。比如，我们编写一个中间件，用于对 HTTP 请求的权限进行校验，实现 Web 项目的登录和权限管理功能。首先思考以下几个问题：

- ◎ 中间件是在什么时候加载的？
- ◎ 中间件处理请求的流程是怎样的？
- ◎ 常用的中间件有哪些？

◎ 如何自定义中间件？

这些问题将在学习中间件源码后一一得到解答，以帮助读者理解 HTTP 请求在 Django 框架中的处理过程。

## 7.2 加载中间件

在启动 Django 项目的过程中，会创建一个 django.core.handlers.wsgi.WSGIHandler 对象，用于处理接收的请求。通过查看 WSGIHandler 类的源码可知，中间件的加载就是在初始化过程中完成的（这些分别在第 5 章和第 6 章中介绍过）：

```
# 源码位置：django/core/handler/wsgi.py
# ……

class WSGIHandler(base.BaseHandler):
    request_class = WSGIRequest

    def __init__(self, *args, **kwargs):
        super().__init__(*args, **kwargs)
        self.load_middleware()
# ……
```

通过上面的源码，可以看到加载中间件是调用 load_middleware()方法完成的，所以具体的加载过程就在该方法的源码中。继续追踪源码，可以看到 load_middleware()方法是 WSGIHandler 类通过继承 django.core.handlers.base.BaseHandler 类得到的，如下：

```
# 源码位置：django/core/handler/base.py
# ……

class BaseHandler:
    _view_middleware = None
    _template_response_middleware = None
    _exception_middleware = None

    # 定义变量指向中间件的调用链
    _middleware_chain = None

    def load_middleware(self):
        self._view_middleware = []
        self._template_response_middleware = []
        self._exception_middleware = []

        # 装饰处理请求的函数，使其在捕捉到异常时返回响应
        handler = convert_exception_to_response(self._get_response)
        # 反向遍历 settings.py 文件中定义的中间件列表
```

```
        for middleware_path in reversed(settings.MIDDLEWARE):
            # 引入中间件类
            middleware = import_string(middleware_path)
            try:
                # 把请求处理函数作为初始化参数，创建中间件类的实例对象
                mw_instance = middleware(handler)
            except MiddlewareNotUsed as exc:
                # 处理异常
                # ……

            if mw_instance is None:
                # 处理异常，没有对应位置的中间件类
                raise ImproperlyConfigured(
                    'Middleware factory %s returned None.' % middleware_path
                )

            # 判断中间件实例对象是否具有下列属性，并放入相应列表中
            if hasattr(mw_instance, 'process_view'):
                self._view_middleware.insert(0, mw_instance.process_view)
            if hasattr(mw_instance, 'process_template_response'):
                self._template_response_middleware.append(
                     mw_instance.process_template_response
                )
            if hasattr(mw_instance, 'process_exception'):
                self._exception_middleware.append(mw_instance.process_exception)

            # 装饰函数，注意，此时的请求处理函数已经是可调用的中间件对象
            handler = convert_exception_to_response(mw_instance)

        # _middleware_chain 指向了在 settings.py 文件中定义的中间件列表的第 1 个中间件实例对象
        self._middleware_chain = handler

# ……
```

为了更好地理解中间件的加载过程，我们在 first_django 工程目录中进入 Python 命令行模式，模拟 load_middleware()方法的执行过程。具体操作如下：

```
>>> import os
>>> import django
>>> from django.core.handlers.wsgi import WSGIHandler
>>>
>>> os.environ.setdefault('DJANGO_SETTINGS_MODULE', 'django_core_test.settings')
'django_core_test.settings'
>>> django.setup(set_prefix=False)
>>> def _get_response():
...     print("hello")
...
>>> handler = _get_response
```

```
>>> print(handler)
<function _get_response at 0x0000025E47955488>
>>>
>>> from django.conf import settings
>>> from django.utils.module_loading import import_string
>>> for middleware_path in reversed(settings.MIDDLEWARE):
...     middleware = import_string(middleware_path)
...     mw_instance = middleware(handler)
...     handler = mw_instance
...     print(handler)
...
<django.middleware.clickjacking.XFrameOptionsMiddleware object at 0x0000025E4794EB70>
<django.contrib.messages.middleware.MessageMiddleware object at 0x0000025E47986390>
<django.contrib.auth.middleware.AuthenticationMiddleware object at 0x0000025E479863C8>
<django.middleware.csrf.CsrfViewMiddleware object at 0x0000025E4799A4E0>
<django.middleware.common.CommonMiddleware object at 0x0000025E4799A668>
<django.contrib.sessions.middleware.SessionMiddleware object at 0x0000025E4799A9E8>
<django.middleware.security.SecurityMiddleware object at 0x0000025E4799A9B0>
>>>
```

在上面的演示过程中，为了便于理解，这里定义了一个简单的函数_get_response()代替 BaseHandler 类的成员函数_get_response()，并跳过了包装这一步，直接用变量 handler 指向它，然后反向遍历在 settings.py 文件中定义的中间件列表，引入相应的中间件类，并实例化。可以看到，变量 handler 从最开始指向的_get_response()函数，变成了依次指向 XFrameOptionsMiddleware 类的实例对象、MessageMiddleware 类的实例对象等。在遍历结束后，变量 handler 指向的是 SecurityMiddleware 类的实例对象。而 SecurityMiddleware 类是中间件列表中的第 1 个类。所以，当 HTTP 请求到来时，最先调用的是 SecurityMiddleware 类的实例对象。

## 7.3　中间件的处理流程

### 7.3.1　中间件的请求处理流程

Django 项目在启动过程中会创建一个 WSGIHandler 类的对象，用来处理接收的请求。而在 WSGIHandler 类中又定义了一个魔法函数__call__ ()，所以它的实例对象是可以直接调用的。整个过程的概述就是，在 Django 运行过程中，每接收一个请求，就会创建一个线程，该线程调用 server 实例的处理函数对请求进行处理。而在调用这个处理函数时，会实例化 WSGIRequestHandler 类。WSGIRequestHandler 类在初始化时调用的函数又使用了 Django 启动时创建的 WSGIHandler 对象作为参数，并且直接调用该对象。因此，整个中间件的处理流程入口就在魔法函数__call__()中，源码如下：

```
# 源码位置：django/core/handler/wsgi.py
# ……

class WSGIHandler(base.BaseHandler):
    request_class = WSGIRequest
    # ……

    def __call__(self, environ, start_response):
        # ……

        response = self.get_response(request)

        # ……

        return response
```

由 6.3 节的分析可知，这里的 get_response()方法是处理 HTTP 请求的入口。继续追踪该方法的源码，如下：

```
# 源码位置：django/core/handler/base.py
# ……

class BaseHandler:

    # ……

    def get_response(self, request):
        set_urlconf(settings.ROOT_URLCONF)
        # 处理中间件链
        response = self._middleware_chain(request)
        response._closable_objects.append(request)
        # ……

        return response

    # ……
```

BaseHandler 类中的 get_response()方法的主要功能是：对输入的特定的 HttpRequest 请求返回一个 HttpResponse 对象。本章主要讲解的是如何调用中间件，因此在上述代码中，response = self._middleware_chain(request)语句是重点分析对象。在前面分析中间件的加载过程中，已知变量_middleware_chain会指向中间件列表的第 1 个中间件类实例对象(其实是指向包装后的可调用对象)。下面以创建 django 工程时默认配置的中间件列表为例进行分析，_middleware_chain 指向的是 SecurityMiddleware 类的实例对象，而 SecurityMiddleware 类的源码如下：

```
# 源码位置：django/middleware/security.py
# ……
```

```
class SecurityMiddleware(MiddlewareMixin):
    def __init__(self, get_response=None):
        # ……
        self.get_response = get_response

    def process_request(self, request):
        path = request.path.lstrip("/")
        if (self.redirect and not request.is_secure() and
                not any(pattern.search(path)
                        for pattern in self.redirect_exempt)):
            host = self.redirect_host or request.get_host()
            return HttpResponsePermanentRedirect(
                "https://%s%s" % (host, request.get_full_path())
            )

    def process_response(self, request, response):
        # ……
```

SecurityMiddleware 类继承了 MiddlewareMixin 类，而且在它的初始化方法中有一个参数 get_response，这个参数的作用是获取请求的响应。如果查看列表的中间件源码，就会发现它们的初始化方法都有一个名为 get_response 的参数。结合前面的分析可知，SecurityMiddleware 类在初始化时传入的参数是 SessionMiddleware 类的实例对象，即成员变量 get_response 指向的是 SessionMiddleware 类的实例对象。依次类推，最后，XFrameOptionsMiddleware 类在初始化时传入的参数是 BaseHandler 类的成员方法_get_response()。前面介绍过，SecurityMiddleware 类的实例是调用对象，那么它的魔法函数__call__()是在哪里实现的？答案是在其父类 MiddlewareMixin 中实现的，这也是理解中间件链式处理 HttpRequest 的关键。MiddlewareMixin 类的源码如下：

```
# 源码位置：django/utils/deprecation.py
# ……

class MiddlewareMixin:
    def __init__(self, get_response=None):
        self.get_response = get_response
        super().__init__()

    def __call__(self, request):
        response = None
        if hasattr(self, 'process_request'):
            # 调用子类的 process_request()方法
            response = self.process_request(request)
        # 如果 response 为 None，则调用实例的 get_response()方法
        response = response or self.get_response(request)
        if hasattr(self, 'process_response'):
            response = self.process_response(request, response)
        return response
```

MiddlewareMixin 类中的源码并不多，只有初始化方法__init__()和魔法函数__call__()。这里重点介绍魔法函数__call__()。在该函数中，首先判断实例是否有 process_request 属性，如果有，就调用它来处理 request，很显然这个属性只能在子类中定义。再来看 SecurityMiddleware 类的源码，它实现了一个名为 process_request 的成员方法，用来判断是否需要重定向。如果需要重定向，就返回一个 HttpResponsePermanentRedirect 对象，否则什么都不返回。如果返回了 HttpResponsePermanentRedirect 对象，则 response 变量的值为非 None，即 self.get_response(request) 这一语句就不会被执行了。如果什么都不返回，则 response 变量的值仍然为 None，self.get_response(request)这一语句会被执行。

如果执行 get_response()方法来处理 request 请求，那么实际上运行的代码是怎样的呢？当前的中间件是 SecurityMiddleware，它的实例对象在初始化时经 get_response 属性指向了中间件列表中下一个中间件类的对象，即 SessionMiddleware 类的对象。下面查看 SessionMiddleware 类的实现源码：

```
# 源码位置：django/contrib/sessions/middleware.py
# ……

class SessionMiddleware(MiddlewareMixin):
    def __init__(self, get_response=None):
        self.get_response = get_response
        engine = import_module(settings.SESSION_ENGINE)
        self.SessionStore = engine.SessionStore

    def process_request(self, request):
        session_key = request.COOKIES.get(settings.SESSION_COOKIE_NAME)
        request.session = self.SessionStore(session_key)

    def process_response(self, request, response):
        # ……
```

从上面的源码可以看到，SessionMiddleware 类也继承了 MiddlewareMixin 类，在初始化时会为自己的 get_response 属性赋值，并且实现了 process_request()和 process_response()两个方法。这与 SecurityMiddleware 类十分相似。所以，当调用 SecurityMiddleware 类的 get_response(request)方法时，实际上会进入 MiddlewareMixin 类定义的魔法函数__call__()中，只不过 get_response 指向了下一个中间件对象，而 process_request 是当前 SessionMiddleware 类的成员方法。这样就实现了用中间件链对 request 请求进行处理。

继续看中间件类 CommonMiddleware 的源码，会发现它同样继承了 MiddlewareMixin 类，也实现了 process_request()和 process_response()两个方法。所以，它处理 request 请求的原理和过程与 SessionMiddleware 类十分相似。

如果某一中间件的 process_request()方法返回了非 None 的结果，那么中间件的链式处理就中止了，此时会进入中间件的响应处理流程。

## 7.3.2　中间件的响应处理流程

当某个中间件的 process_request()方法返回了非 None 的结果，或者请求在经过了所有的中间件之后被 BaseHandler 类中的_get_response()方法处理并得到了返回值时，则开始了中间件的响应处理流程。这个流程的顺序与中间件请求处理流程的顺序完全相反，是一个自下而上的过程。

以 first_django 工程的默认中间件列表为例，假设请求的处理链在第 3 个中间件中止了，即 CommonMiddleware 类的 process_request()方法返回了非 None 的结果。通过 7.3.1 节的分析可知，CommonMiddleware 类的 process_request()方法是在其父类 MiddlewareMixin 的魔法函数__call__()中被调用的，为了方便追踪接下来的流程，再看一下 MiddlewareMixin 类中的源码，如下：

```
# 源码位置：django/utils/deprecation.py
# ……

class MiddlewareMixin:
    # ……

    def __call__(self, request):
        # ……

        response = response or self.get_response(request)
        if hasattr(self, 'process_response'):
            # 调用子类的 process_response 成员函数
            response = self.process_response(request, response)
        return response
```

从上面的源码可以看到，在调用了 CommonMiddleware 类的 get_response()方法之后，程序会先判断实例对象是否有 process_response 属性。如果有，就调用相应的方法。而在 CommonMiddleware 类中恰好有名为 process_response()的方法，源码如下：

```
# 源码位置：django/middleware/common.py
# ……

class CommonMiddleware(MiddlewareMixin):
    # ……

    def process_response(self, request, response):

        if response.status_code == 404:
            if self.should_redirect_with_slash(request):
                return
 self.response_redirect_class(self.get_full_path_with_slash(request))

        if not response.streaming and not response.has_header('Content-Length'):
            response['Content-Length'] = str(len(response.content))
```

```
        return response
```

process_response()方法的作用是：当 response 状态码为 404 时，如果给请求的原始路径加上斜杠作为结尾能变成有效路径，那么就做重定向，否则返回响应 response。这样做的目的是分析中间件对响应 response 的处理流程，因此继续假设 CommonMiddleware 类中的 process_response()方法可以直接返回 response。由于上一个 SecurityMiddleware 类的 get_response 属性指向的是 CommonMiddleware 对象，因此在返回 response 之后又回到了 SecurityMiddleware 类的父类 MiddlewareMixin 的魔法函数__call__()中。接下来判断 SecurityMiddleware 类的实例对象是否有 process_response 属性，如果有，就调用该属性代表的方法。这样，Django 的中间件模块就实现了对响应 response 的链式处理。

在顶层的中间件 django.middleware.security.SecurityMiddleware 的 process_response()方法返回 response 之后，回到 wsgiref.handlers.BaseHandler 类的 get_response()方法。get_response()方法是在其子类 WSGIHandler 的魔法函数__call__()中调用的。至此，HTTP 请求在中间件层的整个处理流程就结束了。

### 7.3.3 中间件的其他钩子方法

在前面两个小节中，我们分析了 process_request()和 process_response()两个方法的调用链。接下来看 process_view()、process_template_response()和 process_exception()这三个钩子方法是如何被调用的。首先，把这三个方法分别放入 BaseHandler 类的三个不同的列表型成员变量中，源码如下：

```
# 源码位置：django/core/handler/base.py
# ……

class BaseHandler:
    # ……

    def load_middleware(self):
        self._view_middleware = []
        self._template_response_middleware = []
        self._exception_middleware = []
        # ……

        # 反向遍历中间件列表
        for middleware_path in reversed(settings.MIDDLEWARE):
            # ……

            if hasattr(mw_instance, 'process_view'):
                # 作为第 1 个插入列表，最后得到的列表顺序与中间件列表顺序一致
                self._view_middleware.insert(0, mw_instance.process_view)
            if hasattr(mw_instance, 'process_template_response'):
                # 插入响应中间件列表，最后得到的列表顺序与中间件列表顺序相反
```

```
            self._template_response_middleware.append(
                mw_instance.process_template_response
            )
        if hasattr(mw_instance, 'process_exception'):
            # 插入异常中间件列表，最后得到的列表顺序与中间件列表顺序相反
            self._exception_middleware.append(mw_instance.process_exception)

        # ......

    # ......
```

BaseHandler 类中的三个类属性 _view_middleware、_template_response_middleware 和 _exception_middleware，分别保存了所有待执行中间件类中的相关方法。这些方法将在_get_response()方法中被循环调用。因此 process_view()、process_template_response()和 process_exception()这三个钩子方法也将在此处被调用。此外，通过前面的分析可知，中间件列表的最后一个类的 get_response 属性指向的是 django.core.handlers.base.BaseHandler 类的_get_response()方法，对应的视图函数也将在其中被调用。_get_response()方法的源码如下：

```
# 源码位置：django/core/handler/base.py
# ......

class BaseHandler:
    # ......

    def _get_response(self, request):
        response = None

        # 前面介绍过相关源码
        # ......

        # 调用 process_view()方法，顺序与中间件列表相同
        for middleware_method in self._view_middleware:
            response = middleware_method(request, callback, callback_args,
callback_kwargs)
            if response:
                break

        if response is None:
            wrapped_callback = self.make_view_atomic(callback)
            try:
                # view 视图函数会对请求进行处理，得到响应 response
                response = wrapped_callback(request, *callback_args, **callback_kwargs)
            except Exception as e:
                # 如果 view 视图函数在处理请求时发生异常，
                # 则 process_exception_by_middleware()方法会调用 process_exception()方法
                response = self.process_exception_by_middleware(e, request)
```

```
        if response is None:
            # 处理异常
            # ……
        elif hasattr(response, 'render') and callable(response.render):
            # 调用process_template_response()方法，顺序与中间件列表相反
            for middleware_method in self._template_response_middleware:
                response = middleware_method(request, response)
                # 处理异常
                # ……
            try:
                # 对响应 response 进行渲染
                response = response.render()
            except Exception as e:
                # 抛出异常，process_exception_by_middleware()方法会调用process_exception()方法
                response = self.process_exception_by_middleware(e, request)

        return response

    def process_exception_by_middleware(self, exception, request):

        # 调用process_exception()方法，顺序与中间件列表相反
        for middleware_method in self._exception_middleware:
            response = middleware_method(request, exception)
            if response:
                return response
        raise
```

通过上面的源码可知，process_view()方法是在所有中间件的 process_request()方法之后、view 视图函数之前被调用的；process_template_response()方法是在 view 视图函数之后被调用的；process_exception()方法是在发生异常时被调用的。同时，这三个钩子方法都在所有中间件的 process_response()方法之前被调用。

# 7.4 常用的中间件

本节介绍 Django 中最常用的 CsrfViewMiddleware 中间件，重点介绍该中间件 csrf_token 的产生与校验过程。

## 7.4.1 Django 内置的中间件类

Django 内置的中间件类主要分布在 django/middleware 目录和 django/contrib 目录中，这里可以根据需要直接在 settings.py 文件中配置使用 Django 内置的中间件类。

- ◎ django/middleware/cache.py：UpdateCacheMiddleware 类、FetchFromCacheMiddleware 类和 CacheMiddleware 类。
- ◎ django/middleware/clickjacking.py：XFrameOptionsMiddleware 类。
- ◎ django/middleware/common.py：CommonMiddleware 类和 BrokenLinkEmailsMiddleware 类。
- ◎ django/middleware/csrf.py：CsrfViewMiddleware 类。
- ◎ django/middleware/gzip.py：GZipMiddleware 类。
- ◎ django/middleware/http.py：ConditionalGetMiddleware 类。
- ◎ django/middleware/locale.py：LocaleMiddleware 类。
- ◎ django/middleware/security.py：SecurityMiddleware 类。
- ◎ django/contrib/admin/tests.py：CSPMiddleware 类。
- ◎ django/contrib/admindocs/middleware.py：XViewMiddleware 类。
- ◎ django/contrib/auth/middleware.py：AuthenticationMiddleware 类和 RemoteUserMiddleware 类。
- ◎ django/contrib/flatpages/middleware.py：FlatpageFallbackMiddleware 类。
- ◎ django/contrib/messages/middleware.py：MessageMiddleware 类。
- ◎ django/contrib/redirects/middleware.py：RedirectFallbackMiddleware 类。
- ◎ django/contrib/sessions/middleware.py：SessionMiddleware 类。
- ◎ django/contrib/sites/middleware.py：CurrentSiteMiddleware 类。

### 7.4.2　CsrfViewMiddleware 中间件

在前面的内容中，我们已经对 CommonMiddleware、SecurityMiddleware 和 SessionMiddleware 这三个类进行了简单分析。接下来重点分析 CsrfViewMiddleware 中间件。在 Django 中，CsrfViewMiddleware 中间件通过校验 csrf token 来防范跨站请求伪造（Cross-site request forgery，CSRF）攻击，根据前面的分析经验，这里主要通过下面三个钩子方法了解该中间件的功能与作用。

（1）process_request()方法。在 CsrfViewMiddleware 中间件中，该方法的源码如下：

```
# 源码位置：django/middleware/csrf.py
# ……

def _get_new_csrf_string():
   return get_random_string(CSRF_SECRET_LENGTH, allowed_chars=CSRF_ALLOWED_CHARS)

def _salt_cipher_secret(secret):
   salt = _get_new_csrf_string()
   chars = CSRF_ALLOWED_CHARS
   pairs = zip((chars.index(x) for x in secret), (chars.index(x) for x in salt))
   cipher = ''.join(chars[(x + y) % len(chars)] for x, y in pairs)
```

```
    return salt + cipher

# ……

def _get_new_csrf_token():
    return _salt_cipher_secret(_get_new_csrf_string())

# ……

def _sanitize_token(token):
    # Allow only ASCII alphanumerics
    # 如果 csrf_token 不是由字母或数字组成的，就生成一个新的
    if re.search('[^a-zA-Z0-9]', token):
        return _get_new_csrf_token()
    #如果 csrf_token 是由字母或数字组成的，且长度为 64，直接返回原值
    elif len(token) == CSRF_TOKEN_LENGTH:
        return token
    #如果 csrf_token 是由字母或数字组成的，且长度为 32，则以 csrf_token 为参数生成新的值
    elif len(token) == CSRF_SECRET_LENGTH:
        return _salt_cipher_secret(token)
    return _get_new_csrf_token()

# ……

class CsrfViewMiddleware(MiddlewareMixin):
    # ……

    def _get_token(self, request):
        # 如果在 settings.py 文件中没有设置 CSRF_USE_SESSIONS 值，
        # 则使用 django/conf/global_settings.py 文件中的值，默认为 False
        if settings.CSRF_USE_SESSIONS:
            try:
                # 如果在 Session 数据中有 csrf_token 值，则直接返回 csrf_token 值
                return request.session.get(CSRF_SESSION_KEY)
            except AttributeError:
                # 异常处理
                # ……
        else:
            try:
                # 从 cookie 中获取 csrf_token 值
                cookie_token = request.COOKIES[settings.CSRF_COOKIE_NAME]
            except KeyError:
                # 如果在 cookie 中没有 csrf_token 值，则触发异常，返回 None
                return None

            # 得到符合格式要求的 csrf_token 值
            csrf_token = _sanitize_token(cookie_token)
            if csrf_token != cookie_token:
```

```
            request.csrf_cookie_needs_reset = True
        return csrf_token

    # ……

    def process_request(self, request):
        csrf_token = self._get_token(request)
        if csrf_token is not None:
            request.META['CSRF_COOKIE'] = csrf_token

    # ……
```

从上面的源码可以看到，process_request()方法的主要作用是获取 csrf_token 值，并保存到 request.META 属性中。获取 csrf_token 值的方式有两种，分别是从 Session 数据中获取和从 cookie 中获取。如果在 Session 数据中存在 csrf_token 值，则直接返回该值。当从 cookie 中获取 csrf_token 值时，如果在 cookie 中没有相应的值，则得到 None；如果 cookie 中的值不符合规范，则重新生成 csrf_token 值。新的 csrf_token 值主要调用了_salt_cipher_secret()函数。_salt_cipher_secret()函数的参数 secret 是一个长度为 32 的字符串，返回值是一个 64 位的随机字符串。

继续看 process_view()方法的源码实现，如下：

```
# 源码位置：django/middleware/csrf.py
# ……

class CsrfViewMiddleware(MiddlewareMixin):
    def _accept(self, request):
        request.csrf_processing_done = True
        return None

    def _reject(self, request, reason):
        response = _get_failure_view()(request, reason=reason)
        # ……
        return response

    # ……

    def process_view(self, request, callback, callback_args, callback_kwargs):
        # 如果 request 的 csrf_processing_done 的属性为 True，则跳过校验
        if getattr(request, 'csrf_processing_done', False):
            return None

        # 如果视图函数配置了 csrf_exempt 装饰器，则跳过校验
        if getattr(callback, 'csrf_exempt', False):
            return None

        # 除 GET、HEAD、OPTIONS 和 TRACE 这四种请求方法外，其他都需要校验
        if request.method not in ('GET', 'HEAD', 'OPTIONS', 'TRACE'):
```

```
# 如果 request 设置了_dont_enforce_csrf_checks 的属性为 True，则跳过校验
if getattr(request, '_dont_enforce_csrf_checks', False):
    return self._accept(request)

# HTTPS 协议请求
if request.is_secure():
    # 获取请求的跳转来源
    referer = request.META.get('HTTP_REFERER')
    if referer is None:
        return self._reject(request, REASON_NO_REFERER)

    referer = urlparse(referer)

    # 处理一些错误情况
    if '' in (referer.scheme, referer.netloc):
        return self._reject(request, REASON_MALFORMED_REFERER)

    if referer.scheme != 'https':
        return self._reject(request, REASON_INSECURE_REFERER)

    good_referer = (
        settings.SESSION_COOKIE_DOMAIN
        if settings.CSRF_USE_SESSIONS
        else settings.CSRF_COOKIE_DOMAIN
    )
    if good_referer is not None:
        server_port = request.get_port()
        if server_port not in ('443', '80'):
            good_referer = '%s:%s' % (good_referer, server_port)
    else:
        try:
            good_referer = request.get_host()
        except DisallowedHost:
            pass

    good_hosts = list(settings.CSRF_TRUSTED_ORIGINS)
    if good_referer is not None:
        good_hosts.append(good_referer)

    if not any(is_same_domain(referer.netloc, host) for host in good_hosts):
        reason = REASON_BAD_REFERER % referer.geturl()
        return self._reject(request, reason)

# 从 request.META 中获取 csrf_token
csrf_token = request.META.get('CSRF_COOKIE')
if csrf_token is None:
    return self._reject(request, REASON_NO_CSRF_COOKIE)
```

```
            request_csrf_token = ""
            if request.method == "POST":   # 校验 POST 请求
                try:
                    request_csrf_token = request.POST.get('csrfmiddlewaretoken', '')
                except IOError:
                    pass

            if request_csrf_token == "":
                request_csrf_token = request.META.get(settings.CSRF_HEADER_NAME, '')

            # 得到符合格式要求的 request_csrf_token 值
            request_csrf_token = _sanitize_token(request_csrf_token)
            if not _compare_salted_tokens(request_csrf_token, csrf_token):
                return self._reject(request, REASON_BAD_TOKEN)

        return self._accept(request)
# ……
```

下面简单梳理一下 process_view()方法的执行流程：

第 1 步，判断视图方法是否有 csrf_exempt 属性。相当于在视图方法中添加了@csrf_exempt 装饰器，这样就不用检验 csrf_token 值了，直接返回 None，并进入下一个中间件的 process_view()方法继续执行，最后由视图函数去处理 HTTP 请求。

第 2 步，对于 GET、HEAD、OPTIONS 和 TRACE 这四种请求方法，不需要校验 csrf_token 值，直接跳到最后执行 self._accept(request)方法。

第 3 步，对于 POST、PUT 和 DELETE 等请求方法，则需要先从 request.META 中取出 csrf_token 值，这个值是在执行 process_request()方法时放入其中的。对于 POST 请求，需要从请求参数中取得 request_csrf_token 值；对于 PUT、DELETE 等请求，则需要从 request.META 中取得 request_csrf_token 值，然后对 csrf_token 值和 request_csrf_token 值进行比较。

对 csrf_token 值和 request_csrf_token 值进行比较的函数是_compare_salted_tokens()，源码如下：

```
# 源码位置：django/middleware/csrf.py
# ……

def _unsalt_cipher_token(token):

    salt = token[:CSRF_SECRET_LENGTH]
    token = token[CSRF_SECRET_LENGTH:]
    chars = CSRF_ALLOWED_CHARS
    pairs = zip((chars.index(x) for x in token), (chars.index(x) for x in salt))
    secret = ''.join(chars[x - y] for x, y in pairs)  # Note negative values are ok
    return secret
```

```
# ……

def _compare_salted_tokens(request_csrf_token, csrf_token):
    return constant_time_compare(
        _unsalt_cipher_token(request_csrf_token),
        _unsalt_cipher_token(csrf_token),
    )

# ……
```

_unsalt_cipher_token()函数是 token 的解密函数，也是_salt_cipher_secret()函数的逆过程。调用_unsalt_cipher_token()函数可以得到生成 csrf_token 值的 secret 密钥字符串。可以在 Django 的 shell 模式下使用这两个函数：

```
>>> from django.middleware.csrf import _get_new_csrf_token, _unsalt_cipher_token
>>> x1 = _get_new_csrf_token()
>>> x2 = _get_new_csrf_token()
>>> x3 = _get_new_csrf_token()
>>> print('x1={}\nx2={}\nx3={}'.format(x1, x2, x3))
x1=5VQjifCj9cG4159lRX6ep8WhjvTIlMdfwpAAIeq9mkAFQHIo6De4lL5B4DBKuUvW
x2=nWfWTukuivi7GEqDEjWQWNSORbJyqoseCXszkbBK4e25oyRmHleZ3kyy5CDsA89j
x3=4n9P9U7Xcc4vOAy2EWylpyfFH6XwBP3b8IMkmS5DorlVSdArCjm2TPTHGj3U5qSH
>>> _unsalt_cipher_token(x1)
'BEUrA9Y0ni4LZMJdpQi06NjuViScjisR'
>>> _unsalt_cipher_token(x2)
'pbnNBRrqWTU8S4BTdcsjhHQUoB44kURf'
>>> _unsalt_cipher_token(x3)
'evNFn88QmprAeNcz8xYREr0c9ngyELZG'
>>>
```

只有 csrf_token 值和 request_csrf_token 值解密得到的 secret 相同，校验才能通过。也就是说，csrf_token 值和 request_csrf_token 值是通过同一个 secret 密钥字符串生成的。通过上面的分析可知，如果 csrf_token 值或 request_csrf_token 值是通过调用_sanitize_token()函数新生成的，并且如果用到的密钥是随机生成的字符串，则校验不可能通过。那么，这两者是在什么地方产生关联的呢？

如果在开发中使用 Django 框架的 tepmplate 模板进行渲染，那么会在上下文变量 context 中设置一个隐藏的 input 标签来保存 request_csrf_token 值，具体过程涉及 Django 的模板层源码，可参考第 4 章内容，源码如下：

```
# 源码位置：django/template/backends/utils.py
# ……

def csrf_input(request):
    return format_html(
        '<input type="hidden" name="csrfmiddlewaretoken" value="{}">',
        get_token(request))
```

```
csrf_input_lazy = lazy(csrf_input, SafeText, str)
csrf_token_lazy = lazy(get_token, str)
```

从上面的源码可以看到，input 标签的值是通过 get_token()函数获得的，而 get_token()函数又是在 django/middleware/csrf.py 文件中实现的。get_token()函数的源码如下：

```
# 源码位置：django/middleware/csrf.py
# ......

def get_token(request):
    if "CSRF_COOKIE" not in request.META:
        csrf_secret = _get_new_csrf_string()
        request.META["CSRF_COOKIE"] = _salt_cipher_secret(csrf_secret)
    else:
        csrf_secret = _unsalt_cipher_token(request.META["CSRF_COOKIE"])
    request.META["CSRF_COOKIE_USED"] = True
    return _salt_cipher_secret(csrf_secret)

def rotate_token(request):
    request.META.update({
        "CSRF_COOKIE_USED": True,
        "CSRF_COOKIE": _get_new_csrf_token(),
    })
    request.csrf_cookie_needs_reset = True

# ......
```

在 get_token()函数中，csrf_token 值和 request_csrf_token 值产生了关联。在之前的分析中，process_request()方法和 process_view 中的 csrf_token 值是从 request.META 中得到的。如果在 Django 项目中使用 Django 内置的 login()函数进行用户登录认证，就会调用 rotate_token()函数，即把 csrf_token 值放到 request.META 中。当然，如果 csrf_token 值不在 request.META 中，则 get_token()函数会使用随机密钥字符串生成一个新的。如果 csrf_token 值已经在 request.META 中，则将其解密得到 secret 密钥字符串，再用它生成一个新的 request_csrf_token 值。

最后看 process_response()方法的源码，如下：

```
# 源码位置：django/middleware/csrf.py
# ......

class CsrfViewMiddleware(MiddlewareMixin):
    # ......

    def _set_token(self, request, response):
        if settings.CSRF_USE_SESSIONS:
            if request.session.get(CSRF_SESSION_KEY) != request.META['CSRF_COOKIE']:
                request.session[CSRF_SESSION_KEY] = request.META['CSRF_COOKIE']
```

```
        else:
            # 在响应头中设置 cookie 信息
            response.set_cookie(
                settings.CSRF_COOKIE_NAME,
                request.META['CSRF_COOKIE'],
                max_age=settings.CSRF_COOKIE_AGE,
                domain=settings.CSRF_COOKIE_DOMAIN,
                path=settings.CSRF_COOKIE_PATH,
                secure=settings.CSRF_COOKIE_SECURE,
                httponly=settings.CSRF_COOKIE_HTTPONLY,
                samesite=settings.CSRF_COOKIE_SAMESITE,
            )
            patch_vary_headers(response, ('Cookie',))

    # ……

    def process_response(self, request, response):
        if not getattr(request, 'csrf_cookie_needs_reset', False):
            if getattr(response, 'csrf_cookie_set', False):
                return response

        if not request.META.get("CSRF_COOKIE_USED", False):
            return response

        # 更新 response 的 cookie
        self._set_token(request, response)
        response.csrf_cookie_set = True
        return response
```

process_view()方法会检查 cookie 是否需要更新，如果需要更新，就更新 session 或 response 中 cookie 的内容，并将 cookie 已更新的标志设置为 True。

通过前面的分析可以总结出，中间件 CsrfViewMiddleware 防范 CSRF 攻击的主要原理是比较 csrf_token 值和 request_csrf_token 值。其中，csrf_token 值不会变，而 request_csrf_token 值会随着 input 标签的重新渲染而发生变化。

## 7.5 自定义中间件

结合前面对中间件的分析，下面我们自定义一个简单的中间件，步骤如下：

第 1 步，定义一个中间件类，该类继承自 django.utils.deprecation.MiddlewareMixin 类。

第 2 步，在自定义的中间件类的内部，实现 process_request()、process_view()和 process_response() 这三个钩子方法。

第 3 步，将自定义的中间件类加入 settings.py 文件的 MIDDLEWARE 列表中。

下面就按照上面的步骤实现一个自定义的中间件类。

（1）创建一个名为 middleware 的目录，同时在这个目录中创建一个空的文件__init__.py。

（2）在 middleware 目录中创建一个名为 test.py 的文件，用来编写中间件代码。中间件类名为 TestMiddleware，在该类中先实现 process_request()方法。这里主要是对请求的 URL 进行过滤，只有以字符串 test 结尾的 URL 才能通过，否则返回拒绝访问的信息，代码如下所示：

```
import re

from django.http import JsonResponse
from django.utils.deprecation import MiddlewareMixin


class TestMiddleware(MiddlewareMixin):
   pattern = re.compile(r'.*test$')
   def process_request(self, request):
      request_url = request.path_info
      if self.pattern.match(request_url):
         return None
      else:
         return JsonResponse({'status': 'refused', 'message': 'url not match'})
```

（3）将 TestMiddleware 中间件类加到 settings.py 文件中的 MIDDLEWARE 列表的末尾。

（4）接下来实现一个简单的视图函数用于测试，内容如下：

```
from django.http import JsonResponse
from django.views.decorators.http import require_http_methods

@require_http_methods(['GET'])
def get_test(request):
   info = {
      'view': 'get_test',
      'description': 'django middleware test'
   }
   return JsonResponse(info)

@require_http_methods(['GET'])
def get_experiment(request):
   info = {
      'view': 'get_experiment',
      'description': 'django middleware experiment'
   }
   return JsonResponse(info)
```

为了测试自定义的中间件类 TestMiddleware 的效果，继续创建一个 HTML 文件，使用浏览器 Django 项目发送请求，代码如下：

```
<!DOCTYPE html>
<meta charset="utf-8" />
<title>Middleware Test</title>
<body>
<h2>Middleware Test</h2>
<button onclick="doSend()">发送</button>
<script src="./jquery-3.5.1.min.js"></script>
<script language="javascript"type="text/javascript">
var requestUrl = "http://127.0.0.1:9002/backend/get_test";

function sendGet(){
    $.ajax({
        method: "get",
        url: requestUrl,
        data:{},
        dataType: "json",
        headers: {},
        success: function(data){
            console.log(data);
        },
        error: function(e) {

        }
    })
  };

  function doSend(message) {
    sendGet();
  }
</script>
</body>
</html>
```

我们将 Django 工程打包后在本地的服务器上发布启动，用浏览器打开 HTML 文件，并按 F12 键打开调试界面，单击 HTML 页面上的“发送”按钮，向视图函数 get_test()发送 GET 请求，结果如图 7-1 所示。

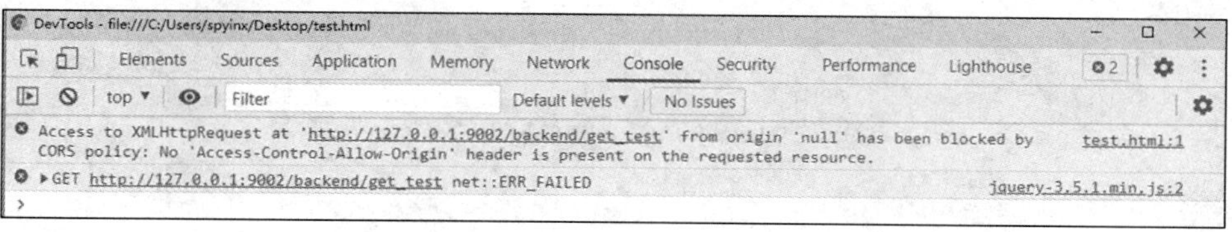


图 7-1

浏览器在发送请求之后，并没有如预期般地接到后台返回的响应值，而是发生了错误。通过错误提示可以看出，该错误是跨域请求错误，即响应的 header 缺少 Access-Control-Allow-Origin。这是由浏览器内部的同源策略机制导致的，而后台的视图函数 get_test()已经正确处理请求并返回了。为了解决该问题，可以在中间件类 TestMiddleware 中实现一个 process_response()函数，将视图函数返回的响应状态和内容打印出来：

```
class TestMiddleware(MiddlewareMixin):
   # ……

   def process_response(self, request, response):
      print(response.status_code)
      print(response.content)
      return response
```

打印出来的视图函数返回的响应状态码和内容如图 7-2 所示。可以看到，请求已经到达了视图函数 get_test()，并且被正确处理了。

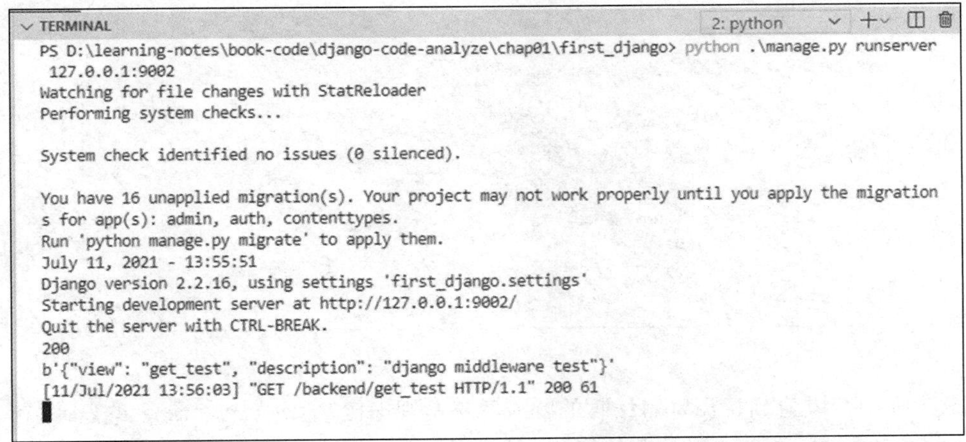

图 7-2

为了解决跨域问题，需要在响应的 header 中加入 Access-Control-Allow-Origin。简单起见，允许所有的请求来源都可以访问资源，并且在中间件类 TestMiddleware 的 process_response()函数中完成这一步。当然，也可以根据需要设置不同的跨域请求策略：

```
class TestMiddleware(MiddlewareMixin):
   # ……

   def process_response(self, request, response):
      response['Access-Control-Allow-Origin'] = '*'
      return response
```

再次向 get_test()视图函数发送请求，前台打印返回结果如图 7-3 所示。此时的浏览器已经能正确地处理响应值了，这说明在 process_response()函数中添加的响应头起作用了。

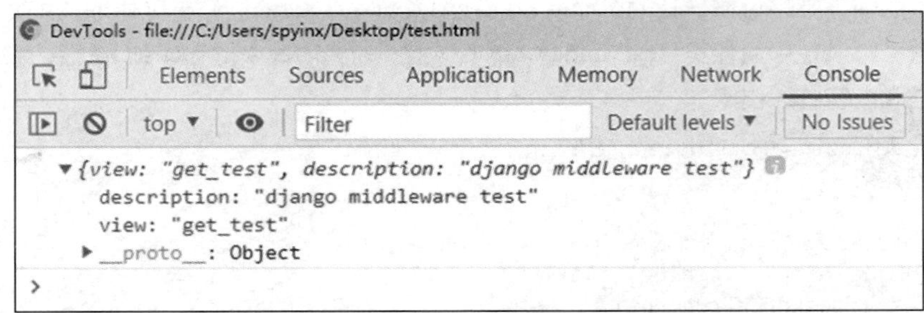


图 7-3

修改 HTML 文件中的变量 requestUrl，设置为 http://127.0.0.1:9002/backend/get_experiment，继续向视图函数 get_experiment()发送 GET 请求。按 F12 键打开浏览器调试界面，查看返回值，结果如图 7-4 所示。

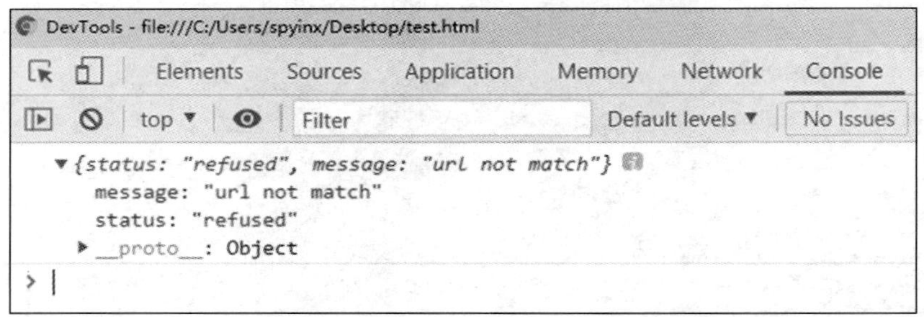


图 7-4

从图 7-4 中可以看到，浏览器接收的响应并不是视图函数 get_experiment()的返回值，而是 process_request()函数返回的拒绝访问的信息。这说明视图函数 get_experiment()的请求 URL 不是以 test 结尾的，请求没有到达视图函数，在中间件类中设置的过滤规则起了作用。

在这个简单的示例中，我们自定义了一个中间件类 TestMiddleware，其中，prcess_request()函数将对请求 URL 进行过滤，并且在 process_response()函数中设置了跨域请求策略。在实际使用中，可以根据需要在中间件的 process_request()、process_view()、process_response()、process_template_response()和 process_exception()这五个钩子方法中实现各种代码逻辑，以达到改变全局输入和输出的目的。

## 7.6　小结

本章主要分析了中间件的原理和使用方法。首先介绍了如何在 Django 工程中配置中间件列表及中间件的五个钩子方法；其次分析了中间件的加载过程和链式处理的原理；然后给出了 Django 框架中内置的全部中间件类；接着重点分析了防范 CSRF 攻击的中间件 CsrfViewMiddleware 的原理；最后自定义了一个中间件类 TestMiddleware，在其中实现了 process_request()函数和 process_response()函数。充分利用中间件，不仅可以减少代码冗余，还可以提升代码的健壮性。

# 第 8 章
# 解读Django中的辅助代码

本章分析 django/utils 目录中的源码文件，从代码角度体会 Django 框架的大和全。同时这部分的代码将扩展读者对 Django 框架的认识，理解其额外功能，对学习 Python 也有所帮助。

## 8.1　自动重载

Django 中的 runserver 命令主要用于调试 Django 项目，默认在调试期间，当有任何文件发生改动时，该调试命令都会被重新加载，以使改动生效。这一功能对开发者而言十分友好。如果不想在调试中出现自动重载功能，可以设置--noreload 选项：

```
(django2-core-test) [root@master first_django]# python manage.py runserver
192.168.26.110:8000 --noreload  # 在设置--noreload 选项后，就不会自动重载了。默认会自动重载
```

那么 Django 框架是如何实现自动重载功能的呢？这就要看 runserver 命令的源码了，基于第 2 章的分析经验，直接定位 runserver 命令的核心语句，如下：

```
# 源码位置：django/core/management/command/runserver.py
# ……

class Command(BaseCommand):
    # ……

    def add_arguments(self, parser):
        # ……

        # 注意，应设置 action 为 store_false，在设置该选项后，
        # options['use_reloader']为 False
        parser.add_argument(
            '--noreload', action='store_false', dest='use_reloader',
            help='Tells Django to NOT use the auto-reloader.',
        )

    # ……
```

```
    def run(self, **options):
        # 在设置--noreload 选项之后，use_reloader=False
        use_reloader = options['use_reloader']

        if use_reloader:
            # 使用自动重载模式运行 Web 服务，启动服务的核心函数仍然是 inner_run()方法
            autoreload.run_with_reloader(self.inner_run, **options)
        else:
            self.inner_run(None, **options)

    # ……
```

从上面的源码可知，当给 runserver 命令加上--noreload 选项后，use_reloader 变量为 False，即不进行自动重载。在默认情况下，use_reloader 变量为 True，会执行 autoreload.run_with_reloader()语句启动自动重载服务，即定时监听该 Django 项目下的所有文件，一旦发现改动，就立即重新加载服务，确保改动生效。下面从该语句入手，继续追踪启动服务的源码：

```
# 源码位置：django/utils/autoreload.py
# ……

DJANGO_AUTORELOAD_ENV = 'RUN_MAIN'

# ……

def run_with_reloader(main_func, *args, **kwargs):
    signal.signal(signal.SIGTERM, lambda *args: sys.exit(0))
    try:
        if os.environ.get(DJANGO_AUTORELOAD_ENV) == 'true':
            reloader = get_reloader()
            logger.info('Watching for file changes with %s', reloader.__class__.__name__)
            start_django(reloader, main_func, *args, **kwargs)
        else:
            # 第一次会进入这里执行
            exit_code = restart_with_reloader()
            sys.exit(exit_code)
    except KeyboardInterrupt:
        pass
```

由于一开始没有设置 DJANGO_AUTORELOAD_ENV 的环境变量，所以上述代码中的 if 判断为 False，即执行 restart_with_reloader()方法：

```
# 源码位置：django/utils/autoreload.py
# ……

def restart_with_reloader():
    new_environ = {**os.environ, DJANGO_AUTORELOAD_ENV: 'true'}
    args = get_child_arguments()
```

```
    while True:
        # 当项目目录下的文件被修改时，线程退出，退出码为 exit_code
        exit_code = subprocess.call(args, env=new_environ, close_fds=False)
        if exit_code != 3:
            return exit_code

# ……
```

从上面的源码可以看到，在 restart_with_reloader() 方法中设置了环境变量 DJANGO_AUTORELOAD_ENV 为 true 的字符串。此外，get_child_arguments()方法的源码如下：

```
# 源码位置：django/utils/autoreload.py
# ……

def get_child_arguments():
    import django.__main__

    # sys.executable 是 Python 命令的路径
    args = [sys.executable] + ['-W%s' % o for o in sys.warnoptions]
    if sys.argv[0] == django.__main__.__file__:
        # The server was started with `python -m django runserver`.
        args += ['-m', 'django']
        args += sys.argv[1:]
    else:
        args += sys.argv
    return args

# ……
```

对于 python manage.py runserver 192.168.26.110:8000 命令来说，最终得到的 args 变量如下：

```
['/root/.pyenv/versions/django2-core-test/bin/python', 'manage.py', 'runserver',
'192.168.26.110:8000']
```

**注意**，对于上述方法中的一些变量，可以在 Python 交互模式下打印，例如下面的操作：

```
>>> import sys
>>> sys.executable
'/root/.pyenv/versions/django2-core-test/bin/python'
>>> ['-W%s' % o for o in sys.warnoptions]
[]
>>> django.__main__.__file__
'/root/.pyenv/versions/django2-core-test/lib/python3.8/site-packages/django/__main__
.py'
```

回到 restart_with_reloader()方法中，可知 restart_with_reloader()方法最终在 while 循环中继续执行前面的 Django 命令（即调用 Python 内置的 subprocess 模块执行得到的 args 变量）。

**注意**，这里将传入设置的环境变量{**os.environ, DJANGO_AUTORELOAD_ENV: 'true'}，而在

执行 subprocess.call()方法后，程序将再次跳转到 run_with_reloader()方法中执行 runserver 命令（这次和上次并不是同一个线程）。此时由于已设置相应的环境变量值，所以程序进入 if 分支继续执行，调用 get_reloader()方法得到一个重载对象：

```
# 源码位置：django/utils/autoreload.py
# ……

def get_reloader():
    """返回重载对象，如果没有相关模块支持，则默认返回 StatReloader 对象"""
    try:
        WatchmanReloader.check_availability()
    except WatchmanUnavailable:
        return StatReloader()
    return WatchmanReloader()

# ……
```

继续追踪 WatchmanReloader 类中的 check_availability()方法，该方法会检查当前系统是否安装了 pywatchman 模块。如果没有安装 pywatchman 模块，直接抛出异常：

```
# 源码位置：django/utils/autoreload.py
# ……

class WatchmanReloader(BaseReloader):
    # ……

    @classmethod
    def check_availability(cls):
        if not pywatchman:   # 如果没有安装 pywatchman 模块，则直接抛出异常
            raise WatchmanUnavailable('pywatchman not installed.')
        client = pywatchman.client(timeout=0.1)
        try:
            result = client.capabilityCheck()
        except Exception:
            # 服务没有启动，抛出异常
            raise WatchmanUnavailable('Cannot connect to the watchman service.')
        version = get_version_tuple(result['version'])
        # 需要满足版本要求
        if version < (4, 9):
            raise WatchmanUnavailable('Watchman 4.9 or later is required.')

# ……
```

一般情况下是没有安装 pywatchman 模块的，所以 get_reloader()方法会返回 StatReloader 对象。继续追踪 StatReloader 类的实现源码，如下：

```
# 源码位置：django/utils/autoreload.py
# ……
```

```
class StatReloader(BaseReloader):
    SLEEP_TIME = 1  # 每秒检查一次变化情况

    def tick(self):
        mtimes = {}
        while True:
            for filepath, mtime in self.snapshot_files():  # 遍历上一次快照中的文件情况
                old_time = mtimes.get(filepath)        # 在 mtimes 中保存了该文件的上次修改时间
                mtimes[filepath] = mtime               # 更新时间
                if old_time is None:
                    # 第一次发现只打印告警信息
                    logger.debug('File %s first seen with mtime %s', filepath, mtime)
                    continue
                elif mtime > old_time:
                    self.notify_file_changed(filepath)      # 通知文件有变化

            time.sleep(self.SLEEP_TIME)                     # 休息 1s
            yield

    def snapshot_files(self):
        # watched_files 可能产生重复路径
        seen_files = set()
        for file in self.watched_files():
            if file in seen_files:
                continue
            try:
                # 通过 stat()方法获取文件相关属性，st_mtime 是文件的最近修改时间
                mtime = file.stat().st_mtime
            except OSError:
                # 当文件不存在时，抛出异常并进入这里
                continue
            seen_files.add(file)
            yield file, mtime   # 生成器返回当前找到的文件及该文件的最新更新时间

    @classmethod
    def check_availability(cls):
        return True

# ……
```

StatReloader 类继承自 BaseReloader 类，从它实现的几个方法来看，对应的功能十分明确：

- ◎ snapshot_files()：该方法配合 yield 返回一个迭代器，每次返回文件列表（由父类方法 self.watched_files()得到）中的一个文件（file）及该文件的最新更新时间（mtime）。
- ◎ tick()：该方法会一直循环 self.snapshot_files()的结果，然后对比上次该文件的最新更新时间。如果有变化，则调用 self.notify_file_changed()方法进行通知。

◎　check_availability()：检查该重载器是否有效。该方法直接返回 True，即无任何使用条件。

从 StatReloader 类的源码可以看出该重载器的一些功能，但有些实现细节需要查看其父类的源码，比如 watched_files()方法的实现、tick()方法如何被调用等。继续追踪 BaseReloader 类的源码，如下：

```
# 源码位置：django/utils/autoreload.py
# ……

class BaseReloader:
    def __init__(self):
        self.extra_files = set()
        self.directory_globs = defaultdict(set)
        self._stop_condition = threading.Event()  # 创建一个事件管理标志

    def watch_dir(self, path, glob):
        path = Path(path)                   # 使用 Python 内置的 Path 模块
        try:
            path = path.absolute()          # 获取绝对路径
        except FileNotFoundError:
            # 打印调试信息
            # ……
            return
        logger.debug('Watching dir %s with glob %s.', path, glob)
        self.directory_globs[path].add(glob)  # 更新 self.directory_globs 值

    def watch_file(self, path):
        path = Path(path)  # Path 模块
        if not path.is_absolute():
            raise ValueError('%s must be absolute.' % path)
        logger.debug('Watching file %s.', path)
        self.extra_files.add(path)  # 更新 self.extra_files 值

    def watched_files(self, include_globs=True):
        """
        yield 中所有需要被监控的文件
        """
        yield from iter_all_python_module_files()   # 所有模块文件，包括项目中的代码文件
        yield from self.extra_files                 # 所有 self.extra_files 中的文件
        if include_globs:
            for directory, patterns in self.directory_globs.items():
                for pattern in patterns:
                    yield from directory.glob(pattern)

    def wait_for_apps_ready(self, app_reg, django_main_thread):
        # 直到 Django 报告 apps 均已被加载后再返回
        while django_main_thread.is_alive():
            if app_reg.ready_event.wait(timeout=0.1):
                return True
```

```
        else:
            logger.debug('Main Django thread has terminated before apps are ready.')
            return False

    def run(self, django_main_thread):
        logger.debug('Waiting for apps ready_event.')
        self.wait_for_apps_ready(apps, django_main_thread)
        from django.urls import get_resolver
        try:
            get_resolver().urlconf_module
        except Exception:
            pass
        logger.debug('Apps ready_event triggered. Sending autoreload_started signal.')
        autoreload_started.send(sender=self)
        # 最核心的调用
        self.run_loop()

    def run_loop(self):
        ticker = self.tick()
        while not self.should_stop:
            try:
                # 迭代，实现每秒监测文件变化。该 tick()方法由父类实现
                next(ticker)
            except StopIteration:
                break
        self.stop()

    def tick(self):
        raise NotImplementedError('subclasses must implement tick().')

    @classmethod
    def check_availability(cls):
        raise NotImplementedError('subclasses must implement check_availability().')

    def notify_file_changed(self, path):
        # 通知文件已发生改变，其实就是发送信号
        results = file_changed.send(sender=self, file_path=path)
        logger.debug('%s notified as changed. Signal results: %s.', path, results)
        if not any(res[1] for res in results):
            # 打印文件变化信息并退出
            trigger_reload(path)

    @property
    def should_stop(self):
        # 返回事件管理标志是否被设置
        return self._stop_condition.is_set()

    def stop(self):
```

```
        # 设置事件管理标志，表示停止
        self._stop_condition.set()

# ……
```

从上面的源码可以看到，调用自动重载器的入口是 run()方法。run()方法的核心是 run_loop()方法。此外，在该基类中定义了获取所有监听文件的 watched_files()方法。在默认情况下，在 watched_files()方法中，只有 iter_all_python_module_file()方法有文件数据。下面是获取所有 Python 模块文件方法的一个调用示例：

```
>>> from django.utils.autoreload import iter_all_python_module_files
>>> files = iter_all_python_module_files()
>>> type(files)
<class 'frozenset'>
>>> len(list(files))
513
```

从上面的代码可以看到，一共监听了 513 个文件。由于文件数量太多，这里先查看 5 个文件的路径，看看 Django 究竟会监测哪些地方的代码文件：

```
>>> list(files)[:5]
[PosixPath('/root/django-core-test/first_django/manage.py'),
PosixPath('/root/.pyenv/versions/3.8.6/lib/python3.8/lib-dynload/grp.cpython-38-x86_
64-linux-gnu.so'),
PosixPath('/root/django-core-test/first_django/test_view/__init__.py'),
PosixPath('/root/.pyenv/versions/3.8.6/envs/django2-core-test/lib/python3.8/site-pac
kages/django/contrib/messages/api.py'),
PosixPath('/root/.pyenv/versions/3.8.6/lib/python3.8/codecs.py')]
```

从上面的代码可以看到，watched_files()方法返回的监听文件除 first_django 项目本身的文件外，还有虚拟环境下的许多代码文件。有兴趣的读者不妨在启动 first_django 项目后（使用 python manage.py runserver 命令）修改上面的代码，看启动的服务是否能检测到对应文件变动并重新加载。下面从这 513 个文件中提取非虚拟环境下的 Python 内置模块代码，操作如下：

```
>>> fs = list(files)
>>> file_list = [ str(fs[i]) for i in range(len(fs)) if '3.8.6' not in str(fs[i]) ]
>>> file_list
['/root/django-core-test/first_django/manage.py',
'/root/django-core-test/first_django/test_view/__init__.py',
'/root/django-core-test/first_django/test_view/models.py',
'/root/django-core-test/first_django/test_view/admin.py',
'/root/django-core-test/first_django/first_django/__init__.py',
'/root/django-core-test/first_django/book_sales/__init__.py',
'/root/django-core-test/first_django/book_sales/models.py',
'/root/django-core-test/first_django/shell_test/__init__.py',
'/root/django-core-test/first_django/shell_test/models.py',
'/root/django-core-test/first_django/book_sales/admin.py',
```

```
'/root/django-core-test/first_django/shell_test/admin.py',
'/root/django-core-test/first_django/first_django/settings.py']
>>> len(file_list)
12
```

从上面的代码可以看到，first_django 项目下的大部分代码文件都被 Django 监控了，但是还缺少两个 urls.py 文件，即 first_django/urls.py 文件和 test_views/urls.py 文件。是 Django 框架不会监听相关项目下的 urls.py 文件吗？当然不是。修改虚拟环境中 Django 框架的源码，在 StatReloader 类的 snapshot_files()方法中打印监听的所有文件，提取包含 urls.py 的文件：

```
# 源码位置：django/utils/autoreload.py
# ……

class StatReloader(BaseReloader):
   # ……

   def snapshot_files(self):
      # watched_files()方法可能会产生重复路径
      seen_files = set()
      for file in self.watched_files():
         # 去除 Django 内部 urls.py 文件的干扰，在 Django 源码内一共有 2 个 urls.py 文件
         if str(file).endswith('urls.py') and '.pyenv' not in str(file):
            print('找到监控 urls.py 文件：{}'.format(file))
         # ……

   # ……
```

启动 first_django 项目，得到的打印结果如下（这里省略了一些不必要的打印结果）：

```
# 修改对应的源码文件，添加前文的打印语句
(django2-core-test) [root@master first_django]# vim
~/.pyenv/versions/django2-core-test/lib/python3.8/site-packages/django/utils/autorel
oad.py
(django2-core-test) [root@master first_django]# python manage.py runserver
192.168.26.110:8000
Watching for file changes with StatReloader
Performing system checks...

System check identified no issues (0 silenced).
找到监控 urls.py 文件：/root/django-core-test/first_django/first_django/urls.py
找到监控 urls.py 文件：/root/django-core-test/first_django/test_view/urls.py

You have 17 unapplied migration(s). Your project may not work properly until you apply
the migrations for app(s): admin, auth, contenttypes, sessions.
Run 'python manage.py migrate' to apply them.

April 16, 2021 - 19:48:47
Django version 2.2.16, using settings 'first_django.settings'
```

```
Starting development server at http://192.168.26.110:8000/
Quit the server with CONTROL-C.
找到监控 urls.py 文件：/root/django-core-test/first_django/first_django/urls.py
找到监控 urls.py 文件：/root/django-core-test/first_django/test_view/urls.py
# 每秒都会调用 snapshot_files()方法，所以会不停地输出找到监控 urls.py 文件，忽略下面的内容
# ……
```

可以看到，当使用命令运行 first_django 项目时，就能监控到 first_django 项目下的所有 urls.py 文件，然而在 Django 的 shell 模式下，却监控不到 first_django 项目中的 urls.py 文件，这是为什么呢？只需查看一下 iter_all_python_module_files()方法的源码即可：

```
# 源码位置：django/utils/autoreload.py
# ……

def iter_all_python_module_files():
    modules_view = sorted(list(sys.modules.items()), key=lambda i: i[0])
    modules = tuple(m[1] for m in modules_view if not isinstance(m[1], weakref.ProxyTypes))
    return iter_modules_and_files(modules, frozenset(_error_files))

# ……
```

sys.modules 返回的是所有导入的模块名称，没有相应的 urls.py 文件，说明这些 urls 模块并没有导入当前系统中，因此只需手工导入这些 urls 模块即可。由于在 first_django/urls.py 文件中通过 include 语句导入了 test_views.url 模块，所以这里只需导入 first_django.urls 模块即可：

```
(django2-core-test) [root@master first_django]# python manage.py shell
Python 3.8.6 (default, Oct 18 2020, 15:33:08)
[GCC 4.8.5 20150623 (Red Hat 4.8.5-39)] on linux
Type "help", "copyright", "credits" or "license" for more information.
(InteractiveConsole)
>>> from django.utils.autoreload import iter_all_python_module_files
>>> from first_django import urls
>>> files = iter_all_python_module_files()
>>> fs = list(files)
>>> file_list = [ str(fs[i]) for i in range(len(fs)) if str(fs[i]).endswith('urls.py')
and '.pyenv' not in str(fs[i]) ]
>>> file_list
['/root/django-core-test/first_django/test_view/urls.py',
'/root/django-core-test/first_django/first_django/urls.py']
>>>
```

这样 first_django 项目下的所有 urls.py 文件就都能被监控到了。继续回到 run()和 run_loop()方法，从 run_loop()方法的循环语句可以看到，自动重载器会不停地循环调用内部的 tick()方法。根据前面的分析可知，StatReloader 类实现的 tick()方法会每秒扫描一次所有的监控文件，对比其修改时间，如果文件有变动，就发送文件改变信号，通知 Django 项目重新加载服务。

继续回到 run_with_reloader()方法，当使用 get_reloader()方法获得一个重载对象后，会接着调用 start_django()函数。start_django()函数的功能是启动 Django 项目，具体的实现源码如下：

```
# 源码位置：django/utils/autoreload.py
# ……

def start_django(reloader, main_func, *args, **kwargs):
    ensure_echo_on()
    # 检查错误
    main_func = check_errors(main_func)
    # 注意，这里传入的 main_func 是 runserver.py 文件的 Command 类中定义的 inner_run()方法
    django_main_thread = threading.Thread(target=main_func, args=args, kwargs=kwargs,
                                          name='django-main-thread')
    django_main_thread.setDaemon(True)     # 设置为后台线程
    django_main_thread.start()                    # 启动线程

    while not reloader.should_stop:
        # StatReloader 对象的 should_stop 其实是线程的 event 事件标识。
        # 如果该标识没有设置，继续执行
        try:
            reloader.run(django_main_thread)
        except WatchmanUnavailable as ex:
            # 如果是 watchman 服务或者其他情况导致的异常，则使用 StatReloader 对象进行重载操作
            reloader = StatReloader()
            logger.error('Error connecting to Watchman: %s', ex)
            logger.info('Watching for file changes with %s', reloader.__class__.__name__)

# ……
```

注意，这里传入的 reloader 和 main_func 参数，前者是前面分析过的 StatReloader 对象，后者是在 runserver.py 文件的 Command 类中定义的 inner_run()方法。在上述源码中，首先设置一个线程对象 django_main_thread；然后在后台启动该线程；接着进入 while 循环，调用 StatReloader 对象的 run()方法，定时循环检查所有受监控的文件。一旦文件有变动，就立即发送信号并重新加载 Django 服务，这便是 Django 服务实现自动重载的全过程。

从上面的分析可知，Django 项目实现自动重载的核心代码在 django/utils/autoreload.py 文件中，自动的过程依赖于每秒定时对所有指定的文件进行监控，一旦文件有修改，就通过信号机制通知 Django 服务线程进行重新加载。

## 8.2 日志配置

在 Django 的源码中可以看到很多 logger = logging.getLogger('...')语句，同时在 Django 的源码中有非常多的 logger.debug()、logger.error()等语句用于打印 Django 框架的执行日志。为什么在启动

first_django 项目后，只能在控制台中看到相应的请求日志且格式十分简陋？如何将这些日志信息保存到日志文件中？如何调整打印的日志格式？下面给出一个简单的日志配置示例并完成相关测试。

## 8.2.1 日志配置实战

（1）修改 settings.py 文件，添加 LOGGING 配置：

```
(django2-core-test) [root@master first_django]# cat first_django/settings.py
# ……

# 新建日志目录
log_dirpath = os.path.join(BASE_DIR, "logs")
if not os.path.isdir(log_dirpath):
   os.makedirs(log_dirpath)

# 配置日志
LOGGING = {
   'version': 1,
   'disable_existing_loggers': False,  # 如果设置为 True，就等于把所有默认的 logger 全部删掉
   # 配置打印日志格式 formatters
   'formatters': {
      # 新的控制台日志格式
      'new_console': {
         'format': '[{asctime}] [{levelname}] {module} {message}',
         'style': '{'
      },
      # 保存在文件中的日志格式
      'file_format': {
         'format': '%(asctime)s [%(threadName)s:%(thread)d] [%(name)s:%(lineno)d]
[%(module)s:%(funcName)s] [%(levelname)s]- %(message)s'
      }
   },
   # 配置日志的 handlers
   'handlers': {
      'console': {
         'level': 'INFO',
         'class': 'logging.StreamHandler',
         # 选择日志打印格式
         'formatter': 'new_console'
      },
      'file_handler': {
         'level': 'INFO',
         # 打印日志到文件中
         'class':'logging.handlers.RotatingFileHandler',
         # 指定日志文件名
         'filename': os.path.join(BASE_DIR, "logs", 'first_django.log'),
```

```
            'maxBytes':1024 * 1024 * 5,      # 日志最大文件
            'backupCount': 3,                # 备份数
            'formatter':'file_format',       # 选择日志格式
        }
    },
    # 配置日志管理器
    'loggers': {
        'django': {
            'handlers': ['console'],   # 选择该 logger 对应的 handlers，即支持多种日志打印方式
            'level': 'INFO',
        },
        'django.request': {
            # 对应 logger = logging.getLogger('django.request')得到的日志管理器会将日志内容
            # 同时打印到控制台和文件中
            'handlers': ['console', 'file_handler'],
            'level': 'INFO',
        }
    }
}
```

（2）调整 test_view/views.py 文件。对 6.1 节中实验 2 的视图类进行调整，删除在 HelloView 类中定义的 delete()方法。

（3）测试。首先，在没有加上日志配置之前，控制台打印请求日志的结果如下：

```
# 先用"""注释掉添加的 LOGGING 配置
(django2-core-test)[root@master first_django]# python manage.py runserver
192.168.26.110:8000

# ……
[17/Apr/2021 01:49:38] "GET /test_view/hello/ HTTP/1.1" 200 13
[17/Apr/2021 01:50:21] "GET /test_view/hello_class/ HTTP/1.1" 200 11
```

在加上 LOGGING 配置后，控制台打印请求日志的结果如下：

```
(django2-core-test)[root@master first_django]# python manage.py runserver
192.168.26.110:8000

# ……
[2021-04-17 01:52:29,674] [INFO] basehttp "GET /test_view/hello/ HTTP/1.1" 200 13
[2021-04-17 01:52:51,268] [WARNING] base Method Not Allowed (DELETE):
/test_view/hello_class/
[2021-04-17 01:52:51,268] [WARNING] base Method Not Allowed (DELETE):
/test_view/hello_class/
[2021-04-17 01:52:51,269] [WARNING] log Method Not Allowed: /test_view/hello_class/
[2021-04-17 01:52:51,269] [WARNING] log Method Not Allowed: /test_view/hello_class/
[2021-04-17 01:52:51,269] [WARNING] basehttp "DELETE /test_view/hello_class/ HTTP/1.1"
405 0
^C
```

```
(django2-core-test) [root@master first_django]# cat logs/first_django.log
2021-04-17 01:52:51,268 [Thread-2:140342888232704] [django.request:100]
[base:http_method_not_allowed] [WARNING]- Method Not Allowed (DELETE):
/test_view/hello_class/
2021-04-17 01:52:51,269 [Thread-2:140342888232704] [django.request:222]
[log:log_response] [WARNING]- Method Not Allowed: /test_view/hello_class/
```

以上均是通过 curl 命令来模拟 HTTP 请求的，其中，第二次是发送 DELETE 请求。从上面的代码可以看到，在加上 LOGGING 配置后，控制台的请求日志格式已经发生了改变，和 new_console 指定的格式一致，说明上述配置已经生效。此外，当请求发生“Method Not Allowed”错误时，通过源码可以看到，相关日志信息会使用名称为“django.request”的 logger 打印，源码如下：

```
# 源码位置：django/views/generic/base.py
# ……

logger = logging.getLogger('django.request')

class View:
    # ……

    def http_method_not_allowed(self, request, *args, **kwargs):
        logger.warning(
            'Method Not Allowed (%s): %s', request.method, request.path,
            extra={'status_code': 405, 'request': request}
        )
        return HttpResponseNotAllowed(self._allowed_methods())

    # ……
```

可以看到，第二次请求不仅在 console 上打印了告警日志，还在 first_django.log 文件中写入了请求信息。

### 8.2.2 日志配置的源码追踪

接下来探索 Django 内部是如何读取 LOGGING 配置并实现上述日志打印效果的。搜索日志配置的入口函数非常简单，由于 Django 一定会读取 settings.py 文件中的 LOGGING 配置，因此可以直接在 VSCode 中搜索 Django 源码中出现的有关 settings.LOGGING 的代码，结果如图 8-1 所示。

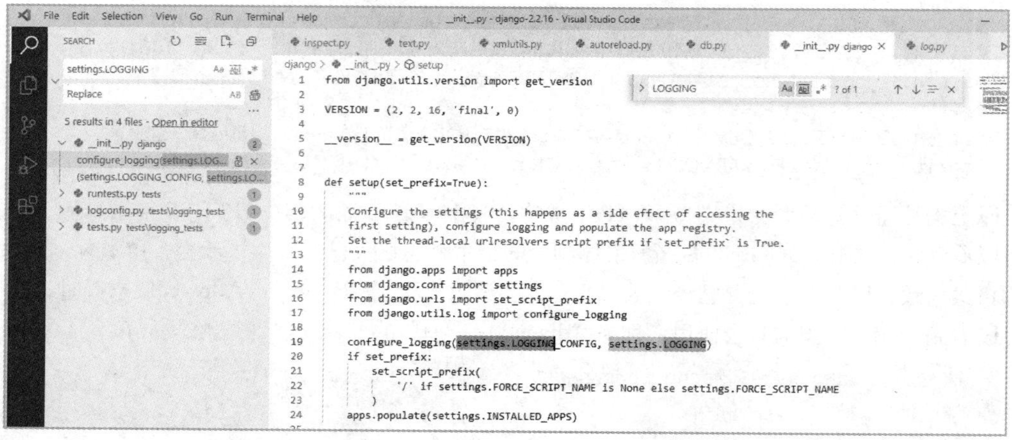

图 8-1

test 目录下的文件可以直接忽略，此时只剩下 django/__init__.py 文件，在该文件中，对应读取日志配置信息的代码如下：

```
# 源码位置：django/__init__.py
# ……

def setup(set_prefix=True):
    # ……

    # 配置日志文件
    configure_logging(settings.LOGGING_CONFIG, settings.LOGGING)

    # ……
```

configure_logging()函数是在 django/utils/log.py 文件中实现的，源码如下：

```
# 源码位置：django/utils/log.py
# ……

DEFAULT_LOGGING = {
    'version': 1,
    'disable_existing_loggers': False,
    'filters': {
        'require_debug_false': {
            '()': 'django.utils.log.RequireDebugFalse',
        },
        'require_debug_true': {  # 控制台过滤器
            '()': 'django.utils.log.RequireDebugTrue',
        },
    },
    'formatters': {
```

```
        'django.server': {
            '()': 'django.utils.log.ServerFormatter',
            'format': '[{server_time}] {message}',
            'style': '{',
        }
    },
    'handlers': {
        'console': {
            'level': 'INFO',
            'filters': ['require_debug_true'],
            'class': 'logging.StreamHandler',
        },
        'django.server': {
            'level': 'INFO',
            'class': 'logging.StreamHandler',
            'formatter': 'django.server',
        },
        'mail_admins': {
            'level': 'ERROR',
            'filters': ['require_debug_false'],
            'class': 'django.utils.log.AdminEmailHandler'
        }
    },
    'loggers': {
        'django': {
            'handlers': ['console', 'mail_admins'],
            'level': 'INFO',
        },
        'django.server': {
            'handlers': ['django.server'],
            'level': 'INFO',
            'propagate': False,
        },
    }
}

# ……

def configure_logging(logging_config, logging_settings):
    if logging_config:
        # First find the logging configuration function
        logging_config_func = import_string(logging_config)

        logging.config.dictConfig(DEFAULT_LOGGING)

        # then invoke it with the logging settings
        if logging_settings:
            logging_config_func(logging_settings)
```

```
# ……
```

可以看到，在 Django 内部有一个默认的日志配置，即 DEFAULT_LOGGING。它的写法和 settings.py 文件中的 LOGGING 配置基本一致，这就是未作任何日志配置时，项目内部默认使用的日志配置。从这个日志配置中可以看出，控制台输出的日志中有一个过滤器：require_debug_true。该过滤器的实现源码如下：

```
# 源码位置：django/utils/log.py
# ……

class RequireDebugTrue(logging.Filter):
    def filter(self, record):
        return settings.DEBUG

# ……
```

当 settings.py 文件中的 DEBUG 参数被设置为 False 时，所有使用了 console 的 logger 将不会打印任何信息。当 DEBUG 参数被设置为 True 时，handlers 中的 console 所对应的过滤器插件会一直返回 True，表示对应的 logger 操作已生效，这便是 settings.DEBUG 控制是否打印日志详情的原理。

再来看 configure_logging() 函数，它有两个参数：logging_config 和 logging_settings。从 django/__init__.py 文件中的 setup()函数的源码可知，logging_config 是在 settings.py 文件中设置的 LOGGING_CONFIG 值，而 logging_settings 是前面设置的 LOGGING 配置。由于前面没有在 settings.py 文件中设置 LOGGING_CONFIG 变量的值，因此直接搜索默认的全局配置文件 django/conf/global_settings.py，即可看到相关变量的定义，源码如下：

```
# 源码位置：django/conf/global_settings.py
# ……

LOGGING_CONFIG = 'logging.config.dictConfig'

# ……
```

可以看到，LOGGING_CONFIG 变量的值是一个 Python 内置的模块路径。函数 configure_logging() 先导入该模块，接着调用 logging.config.dictConfig（默认情况下和前面同一个模块）导入默认的日志配置。如果在 settings.py 文件中设置了 LOGGING 配置，就会导入 LOGGING_CONFIG 变量表示的模块，并设置新的日志配置（覆盖之前导入的默认配置）。整个过程很简单，就是调用 Python 内置的 logging 模块。除此之外，LOGGING 配置的规则也来自该 logging 模块。

## 8.3 时间解析

该部分代码主要涉及 django/utils 目录下的 datetime_safe.py、dateformat.py 和 dateparse.py 这 3 个文件。

### 8.3.1 datetime_safe.py 文件

datetime_safe.py 文件中的代码如下：

```
# 源码位置：django/utils/datetime_safe.py
import re
import time as ttime
# 注意，这里只是重命名了一些模块
from datetime import (
    date as real_date, datetime as real_datetime, time as real_time,
)

#与 datetime.date 模块相比，django.utils.datetime_safe.date 类调整了格式化时间的方法
class date(real_date):
    def strftime(self, fmt):
        return strftime(self, fmt)

# 同理，下面定义的 datetime 类在 datetime.datetime 类的基础上调整了格式化时间的方法
class datetime(real_datetime):
    def strftime(self, fmt):
        return strftime(self, fmt)

    @classmethod
    def combine(cls, date, time):
        # 和 datetime.datetime 类中的 combine()方法的功能几乎一致，只是去掉了抛出异常的语句
        return cls(date.year, date.month, date.day,
                   time.hour, time.minute, time.second,
                   time.microsecond, time.tzinfo)

    def date(self):   # 和 datetime.datetime 类中的 date()方法的功能几乎一致，只返回日期部分
        return date(self.year, self.month, self.day)

# 这里定义的 time 类和 datetime.time 类是一样的
class time(real_time):
    pass


def new_date(d):
    # 根据 datetime.date 对象生成安全的 date
    return date(d.year, d.month, d.day)
```

```
def new_datetime(d):
    """
    根据 datetime.date 对象或者 datetime.datetime 对象，生成一个安全的 datetime
    """
    kw = [d.year, d.month, d.day]  # 先获取年月日值
    if isinstance(d, real_datetime):
        # 如果是 datetime.datetime 模块类型，则继续获取时、分、秒等信息
        kw.extend([d.hour, d.minute, d.second, d.microsecond, d.tzinfo])
    return datetime(*kw)

# 非法格式的正则表达式。很明显，如果在字符串中有%s 或者%y，就会被匹配到
_illegal_formatting = re.compile(r"((^|[^%])(%%)*%[sy])")


def _findall(text, substr):
    # Also finds overlaps
    sites = []
    i = 0
    while True:
        i = text.find(substr, i)  # 从 i 位置开始查找匹配的子串
        if i == -1:               # 如果找不到，则直接退出循环
            break
        sites.append(i)       # 添加匹配的位置
        i += 1                # 从下一个位置开始
    return sites


def strftime(dt, fmt):
    if dt.year >= 1000:
        # 如果年份大于 1000，则直接调用 dt 对应的父类的 strftime()方法格式化时间
        return super(type(dt), dt).strftime(fmt)
    # 检查格式化字符串是否正确
    illegal_formatting = _illegal_formatting.search(fmt)
    if illegal_formatting:
        raise TypeError("strftime of dates before 1000 does not handle " +
illegal_formatting.group(0))

    # 下面处理 1000 年之前的时间格式，略复杂。直接看该方法的输出即可
    year = dt.year
    # 对于每一个非闰年，提前 6 年进入 28 年重复周期
    delta = 2000 - year
    off = 6 * (delta // 100 + delta // 400)
    year = year + off

    # 移到 2000 年附近
    year = year + ((2000 - year) // 28) * 28
    timetuple = dt.timetuple()
```

```
    s1 = ttime.strftime(fmt, (year,) + timetuple[1:])
    sites1 = _findall(s1, str(year))

    s2 = ttime.strftime(fmt, (year + 28,) + timetuple[1:])
    sites2 = _findall(s2, str(year + 28))

    sites = []
    for site in sites1:
        if site in sites2:
            sites.append(site)

    s = s1
    syear = "%04d" % (dt.year,)
    for site in sites:
        s = s[:site] + syear + s[site + 4:]
    return s
```

上面的代码比较简单，下面通过一个测试案例帮助读者理解上面的代码，特别是这里定义的 date、datetime 类与标准模块中的 date、datetime 类的区别。示例代码如下：

```
>>> from datetime import date as real_date
>>> from django.utils.datetime_safe import date
>>> date(2021, 4, 24).strftime('%Y%m%d')
'20210424'
>>> date(921, 4, 24).strftime('%Y%m%d')   # 新的格式化方法，把年份按 4 位补齐
'09210424'
>>> real_date(921, 4, 24).strftime('%Y%m%d')
'9210424'
# 对于自定义的 date 对象，会检查是否有%y 和%s，如果出现%y 和%s，则抛出异常
>>> date(921, 4, 24).strftime('%y')
Traceback (most recent call last):
  File "<console>", line 1, in <module>
  File
"/root/.pyenv/versions/django2-core-test/lib/python3.8/site-packages/django/utils/da
tetime_safe.py", line 19, in strftime
    return strftime(self, fmt)
  File
"/root/.pyenv/versions/django2-core-test/lib/python3.8/site-packages/django/utils/da
tetime_safe.py", line 78, in strftime
    raise TypeError("strftime of dates before 1000 does not handle " +
illegal_formatting.group(0))
TypeError: strftime of dates before 1000 does not handle %y
>>> real_date(921, 4, 24).strftime('%y')  # datetime.date 对象会对%y 匹配年份的后两位
'21'
>>> _findall('xtestescctestx', 'test')
[1, 10]
>>> from django.utils.datetime_safe import new_date, new_datetime
>>> d = real_date(2019, 12, 13)
```

```
>>> new_date(d)
date(2019, 12, 13)
>>> type(new_date(d))
<class 'django.utils.datetime_safe.date'>
>>> new_datetime(d)
datetime(2019, 12, 13, 0, 0)
>>> type(new_datetime(d))
<class 'django.utils.datetime_safe.datetime'>
>>> dt = real_datetime(2021, 4, 21, 17, 28, 59)
>>> new_datetime(dt)
datetime(2021, 4, 21, 17, 28, 59)
>>> from django.utils.datetime_safe import strftime
>>> d1 = date(1921, 8, 24)
>>> format(d1, "%Y-%m-%d")
'1921-08-24'
>>> d2 = date(111, 7, 24)
>>> format(d2, "%Y-%m-%d")
'0111-07-24'
>>> d3 = date(50, 7, 24)
>>> format(d3, "%Y-%m-%d")
'0050-07-24'
```

## 8.3.2 dateformat.py 文件

在 dateformat.py 文件的注释中有一些示例语句，下面我们参考示例语句完成实战演示，以了解该文件中定义的相关类与函数的功能，最后深入这些类与函数的源码进行解读。操作示例如下：

```
>>> import datetime
>>> from django.utils.dateformat import DateFormat, TimeFormat
>>> from django.utils.dateformat import format, time_format
>>> d = datetime.datetime.now()  #得到一个 datetime.datetime 类型
>>> d
datetime.datetime(2021, 4, 25, 3, 42, 18, 487294)
>>> df = DateFormat(d)            # 日期格式化对象
>>> df.format('b c d D E F I j L m M n N o r S t U w W y Y z')
# format()方法，每个字母对应一个含义
'apr 2021-04-25T03:42:18.487294 25 Sun April April 0 25 False 04 Apr 4 April 2021 Sun,
25 Apr 2021 03:42:18 +0800 th 30 1619293338 0 16 21 2021 115'
>>> dt = TimeFormat(d)            # 时间格式化对象
>>> dt.format('a A e f g G h H i O P s T u Z')  # format()方法，每个字母对应一个含义
'a.m. AM  3:42 3 3 03 03 42 +0800 3:42 a.m. 18 CST 487294 28800'
>>> format(d, 'jS F Y H:i')
'25th April 2021 03:42'
>>> time_format(d, 'A G H:i:s')
'AM 3 03:42:18'
```

上面的代码清晰地展示了在 dateformat.py 文件中定义的两个类（DateFormat、TimeFormat）与

两个函数（format()、time_format()）的功能，即主要对日期进行格式化。下面看 DateFormat 和 TimeFormat 这两个类的实现源码，如下：

```
# 源码位置：django/utils/dateformat.py
# ……

def format(value, format_string):
    # 日期格式化的便捷函数
    df = DateFormat(value)
    return df.format(format_string)


def time_format(value, format_string):
    # 时间格式化的便捷函数
    tf = TimeFormat(value)
    return tf.format(format_string)
```

上面的源码非常清晰，format()或 time_format()函数对 DateFormat 或 TimeFormat 类的格式化操作进行了简单封装。首先传入一个日期对象（如 datetime.datetime 对象），然后把该对象作为参数用于实例化 DateFormat 或 TimeFormat 类，得到一个 DateFormat 或 TimeFormat 对象，最后调用该对象的 format()方法返回参数指定格式的结果。前面操作语句的核心是 DateFormat 或 TimeFormat 类中定义的 format()方法，继续看 DateFormat 类的实现源码，如下：

```
# 源码位置：django/utils/dateformat.py
# ……

re_formatchars = re.compile(r'(?<!\\)([aAbBcdDeEfFgGhHiIjlLmMnNoOPrsStTUuwWyYzZ])')
re_escaped = re.compile(r'\\(.)')

class Formatter:
    def format(self, formatstr):
        pieces = []
        for i, piece in enumerate(re_formatchars.split(str(formatstr))):
            if i % 2:
                if type(self.data) is datetime.date and hasattr(TimeFormat, piece):
                    # 抛出异常
                    # ……
                pieces.append(str(getattr(self, piece)()))
            elif piece:
                pieces.append(re_escaped.sub(r'\1', piece))
        return ''.join(pieces)

class TimeFormat(Formatter):
    # ……

class DateFormat(TimeFormat):
    # ……
```

从上面的源码可以看到，DateFormat 和 TimeFormat 类的 format()方法均来自 Formatter 类中定义的 format()方法。为了理解上面源码的含义，下面拆解并执行 format()方法中的源码，如下：

```
>>> from django.utils.dateformat import re_formatchars, re_escaped, DateFormat
>>> from datetime import datetime
>>> d = datetime(2021, 4, 21, 14, 23, 17)
>>> df = DateFormat(d)
>>> formatstr = 'jS F Y  H:i'  #该字符串的格式化结果
>>> pieces = []
>>> piece_list = re_formatchars.split(str(formatstr))  # 根据正则表达式切割字符串
>>> piece_list
['', 'j', '', 'S', ' ', 'F', ' ', 'Y', '  ', 'H', ':', 'i', '']
>>> i = 0; piece = ''
#当 i%2 == 0 时，执行 pieces.append(re_escaped.sub(r'\1', piece))语句
>>> pieces.append(re_escaped.sub(r'\1', piece))
>>> pieces
['']
>>> i = 1; piece = 'j'  # 当 i%2==1 时，执行 pieces.append(str(getattr(df, piece)()))语句
>>> pieces.append(str(getattr(df, piece)()))
>>> pieces
['', '21']
>>> i = 2; piece = ''
>>> pieces.append(re_escaped.sub(r'\1', piece))
>>> pieces
['', '21', '']
>>> i = 3; piece = 'S'
>>> pieces.append(str(getattr(df, piece)()))  # 核心是 getattr()方法
>>> pieces
['', '21', '', 'st']
>>> i = 4; piece = ' '
>>> pieces.append(re_escaped.sub(r'\1', piece))
>>> pieces
['', '21', '', 'st', ' ']
>>> i = 5; piece = 'F'
>>> pieces.append(str(getattr(df, piece)()))
>>> pieces
['', '21', '', 'st', ' ', 'April']
# 后续过程不再演示
# ……
```

从上面的源码可以看到，当对时间按照指定格式进行格式化时，最核心的部分是切割该格式的字符串，得到列表中偶数索引位置（i%2==0）的值为空字符串或者只包含空白字符的字符串，而奇数索引位置（i%2==1）的值为相应字符，比如 j、S 等。对于这些字符，在 DateFormat（或 TimeFormat）类中会定义同名函数（比如字符 j 对应 j()函数），并且通过 getattr()方法获取该同名函数。对于字符 j 和 S 而言，在 DateFormat 类中定义的 j()函数和 S()函数的实现源码如下：

```
# 源码位置：django/utils/dateformat.py
# ……

class DateFormat(TimeFormat):
    # ……

    def j(self):
        "Day of the month without leading zeros; i.e. '1' to '31'"
        return self.data.day

    # ……

    def S(self):
        "English ordinal suffix for the day of the month, 2 characters; i.e. 'st', 'nd',
'rd' or 'th'"
        if self.data.day in (11, 12, 13):  # 特殊情况
            return 'th'
        last = self.data.day % 10
        if last == 1:
            return 'st'
        if last == 2:
            return 'nd'
        if last == 3:
            return 'rd'
        return 'th'

# ……
```

看到上述类方法后，就能理解表示时间格式的字符串中字符 j 和 S 所表示的时间含义了。因此，对于 DateFormat 类所支持的格式字符串，其中的单字符均由 DateFormat 类中对应的单字符方法进行处理（TimeFormat 类也是一样）。由于这些单字符函数的源码都非常简单，因此不再赘述，当有需要时，直接查看相应方法的源码即可。

### 8.3.3　dateparse.py 文件

在 dateparse.py 文件中，主要定义了一些将时间字符串转换成 datetime 对象的函数，而且这些函数均依赖编写好的正则表达式。parse_date()函数的源码如下：

```
# 源码位置：django/utils/dateparse.py
# ……

date_re = re.compile(
    r'(?P<year>\d{4})-(?P<month>\d{1,2})-(?P<day>\d{1,2})$'
)

# ……
```

```
def parse_date(value):
    """解析一个字符串，并返回一个 datetime.date 对象"""
    match = date_re.match(value)
    if match:
        kw = {k: int(v) for k, v in match.groupdict().items()}
        return datetime.date(**kw)

# ……
```

正则表达式 date_re 匹配 2021-04-11 这样的日期格式的字符串，并可以从中得到年、月、日，最后根据匹配的结果得到一个 datetime.date 对象，操作示例如下：

```
>>> from django.utils.dateparse import parse_date
>>> value = '2021-04-11'          # 日期字符串
>>> parse_date(value)
datetime.date(2021, 4, 11)
>>> value = '2021-0411'           # 一个不符合要求的日期字符串
>>> parse_date(value) is None     #parse_date()函数得到的结果为 None
True
```

parse_time()函数依赖正则表达式 time_re，源码如下：

```
# 源码位置：django/utils/dateparse.py
# ……

time_re = re.compile(   # 正则表达式
    r'(?P<hour>\d{1,2}):(?P<minute>\d{1,2})'
    r'(?::(?P<second>\d{1,2})(?:\.(?P<microsecond>\d{1,6})\d{0,6})?)?'
)

# ……

def parse_time(value):
    match = time_re.match(value)
    if match:
        kw = match.groupdict()
        kw['microsecond'] = kw['microsecond'] and kw['microsecond'].ljust(6, '0')
        kw = {k: int(v) for k, v in kw.items() if v is not None}
        return datetime.time(**kw)

# ……
```

parse_time()函数根据正则表达式 time_re 去解析 value，从中得到时、分、秒、毫秒、微秒等信息，最后根据这些值返回一个 datetime.time 对象。parse_time()函数的操作示例如下：

```
>>> from django.utils.dateparse import parse_time
>>> value = '13:24:17'       # 时间字符串
>>> parse_time(value)
```

```
datetime.time(13, 24, 17)
>>> value = '13:24:17.234'   # 带微秒的时间字符串
>>> parse_time(value)
datetime.time(13, 24, 17, 234000)
>>> value = '113:24:17.234'  # 不正确的时间字符串
>>> parse_time(value)
>>> parse_time(value) is None
True
```

接下来解析 parse_datetime()函数，该函数依赖正则表达式 datetime_re，源码如下：

```
# 源码位置：django/utils/dateparse.py
# ……

datetime_re = re.compile(
    r'(?P<year>\d{4})-(?P<month>\d{1,2})-(?P<day>\d{1,2})'
    r'[T ](?P<hour>\d{1,2}):(?P<minute>\d{1,2})'
    r'(?::(?P<second>\d{1,2})(?:\.(?P<microsecond>\d{1,6})\d{0,6})?)?'
    r'(?P<tzinfo>Z|[+-]\d{2}(?::?\d{2})?)?$'
)

# ……

def parse_datetime(value):
    match = datetime_re.match(value)
    if match:
        kw = match.groupdict()
        # 对于微秒需要补 0 对齐，占 6 个字符位置，即“23”要变成“000023”
        kw['microsecond'] = kw['microsecond'] and kw['microsecond'].ljust(6, '0')
        tzinfo = kw.pop('tzinfo')  # 匹配的 tzinfo
        if tzinfo == 'Z':
            tzinfo = utc
        elif tzinfo is not None:
            offset_mins = int(tzinfo[-2:]) if len(tzinfo) > 3 else 0
            offset = 60 * int(tzinfo[1:3]) + offset_mins
            if tzinfo[0] == '-':
                offset = -offset
            tzinfo = get_fixed_timezone(offset)
        kw = {k: int(v) for k, v in kw.items() if v is not None}
        kw['tzinfo'] = tzinfo
        return datetime.datetime(**kw)

# ……
```

调用该函数的示例如下：

```
>>> from django.utils.dateparse import parse_datetime
>>> parse_datetime('2021-4-25 23:11:20')
datetime.datetime(2021, 4, 25, 23, 11, 20)
```

parse_duration()函数是一个处理日期时间差的函数，该函数依赖的正则表达式有点复杂，需要仔细分析。在 parse_duration()函数中，根据日期时间差字符串可以解析出天、小时、分、秒、微秒的差值，然后根据这些差值得到一个 datetime.timedelta 对象并返回。parse_duration()函数涉及的正则表达式及源码如下：

```
# 源码位置：django/utils/dateparse.py
# ……

standard_duration_re = re.compile(        # 支持匹配标准时间差的字符串格式
    r'^'
    r'(?:(?P<days>-?\d+) (days?, )?)?'
    # 匹配 “5 days, ” “-5 days, ” “1 day, ” 这样的字符串，0 或 1 个
    r'((?:(?P<hours>-?\d+):)(?=\d+:\d+))?'
    r'(?:(?P<minutes>-?\d+):)?'        # 匹配 “10: ” “10: ” 这样的字符串，0 或 1 个
    r'(?P<seconds>-?\d+)'              # 匹配 “1000 ” “-1000 ” 这样的字符串，必须有
    r'(?:\.(?P<microseconds>\d{1,6})\d{0,6})?'
    r'$'
)

iso8601_duration_re = re.compile(
    r'^(?P<sign>[-+]?)'
    r'P'
    r'(?:(?P<days>\d+(.\d+)?)D)?'
    r'(?:T'
    r'(?:(?P<hours>\d+(.\d+)?)H)?'
    r'(?:(?P<minutes>\d+(.\d+)?)M)?'
    r'(?:(?P<seconds>\d+(.\d+)?)S)?'
    r')?'
    r'$'
)

postgres_interval_re = re.compile(
    r'^'
    r'(?:(?P<days>-?\d+) (days? ?))?'
    r'(?:(?P<sign>[-+])?'
    r'(?P<hours>\d+):'
    r'(?P<minutes>\d\d):'
    r'(?P<seconds>\d\d)'
    r'(?:\.(?P<microseconds>\d{1,6}))?'
    r')?$'
)

# ……

def parse_duration(value):
    match = (
        standard_duration_re.match(value) or
```

```
        iso8601_duration_re.match(value) or     # 支持 ISO 8601 标准时间格式
        postgres_interval_re.match(value)       # 支持 PostgreSQL 表示的日期时间格式
    )
    if match:
        kw = match.groupdict()
        # 根据解析出的天数得到一个 datetime.timedelta 对象
        days = datetime.timedelta(float(kw.pop('days', 0) or 0))
        sign = -1 if kw.pop('sign', '+') == '-' else 1
        if kw.get('microseconds'):
            kw['microseconds'] = kw['microseconds'].ljust(6, '0')
        if kw.get('seconds') and kw.get('microseconds') and
kw['seconds'].startswith('-'):
            kw['microseconds'] = '-' + kw['microseconds']
        kw = {k: float(v) for k, v in kw.items() if v is not None}
        return days + sign * datetime.timedelta(**kw)
```

在 parse_duration()函数中有 3 种时间格式，最后返回一个 datetime.timedelta 对象。上述源码的核心就是理解支持的 3 种正则表达式所匹配的字符串，对 standard_duration_re 的操作示例如下：

```
>>> from django.utils.dateparse import parse_duration
>>> from django.utils.dateparse import standard_duration_re
>>> value = '6 days, 10:20:1000'
>>> m = standard_duration_re.match(value)
>>> m
<re.Match object; span=(0, 18), match='6 days, 10:20:1000'>
>>> m.group('days')
'6'
>>> m.group('hours')
'10'
>>> m.group('minutes')
'20'
>>> m.group('seconds')
'1000'
>>> parse_duration(value)
datetime.timedelta(days=6, seconds=38200)
>>> 10 * 3600 + 20 * 60 + 1000
38200
```

对其他正则表达式的匹配测试，读者可以按照上面的方式自行完成。

## 8.4 文本处理

该部分代码主要涉及 utils 目录下的 text.py 文件和 html.py 文件。先看 text.py 文件中的源码，里面定义了非常多的处理字符串的函数，还有一个 Truncator 类。限于篇幅，本章仅介绍一些常用的函数。

## 8.4.1 text.py 文件中的 capfirst()函数和 wrap()函数

首先看在 text.py 文件中定义的 capfirst()函数和 wrap()函数，以及一些常规的正则表达式，内容如下：

```
# 源码位置：django/utils/text.py
# ……

@keep_lazy_text
def capfirst(x):
    """将字符串的首字母变为大写"""
    return x and str(x)[0].upper() + str(x)[1:] # 如果 x 不为空，返回 and 后面的结果

# Set up regular expressions
re_words = re.compile(r'<[^>]+?>|([^<>\s]+)', re.S)    # 匹配单词
re_chars = re.compile(r'<[^>]+?>|(.)', re.S)           # 匹配字符
# \s 匹配空白符，\S 匹配所有非空白符
re_tag = re.compile(r'<(/)?(\S+?)(?:(\s*/)|\s.*?)?>', re.S)
re_newlines = re.compile(r'\r\n|\r') # 匹配'\r\n'或者'\r'
re_camel_case = re.compile(r'(((?<=[a-z])[A-Z])|([A-Z](?![A-Z]|$)))') # 匹配驼峰变量名


@keep_lazy_text
def wrap(text, width):
    def _generator():
        for line in text.splitlines(True):  # True keeps trailing linebreaks
            max_width = min((line.endswith('\n') and width + 1 or width), width)
            # 确保每行的长度在大于 max_width 后停止
            while len(line) > max_width:
                space = line[:max_width + 1].rfind(' ') + 1
                if space == 0:
                    space = line.find(' ') + 1
                    if space == 0:
                        yield line
                        line = ''
                        break
                yield '%s\n' % line[:space - 1]
                line = line[space:]
                max_width = min((line.endswith('\n') and width + 1 or width), width)
            if line:
                yield line
    return ''.join(_generator())
```

capfirst()函数比较容易理解，而后面的正则表达式及 wrap()函数则略微复杂，下面通过示例理解上面的源码，操作如下：

```
>>> from django.utils import text
>>> text.capfirst('xxxx')  # 返回首字母大写的字符串
```

```
'xxxx'
>>> text.re_words.findall('This is a test!')
['This', 'is', 'a', 'test!']
# re_words 中的表达式只能被[^<>\s]+选中，所以不会匹配<xxx>中的字符串 xxx
>>> text.re_words.findall('This is a <test>!')
['This', 'is', 'a', '', '!']
# <[^>]+?>匹配<xxx>中的 xxx，但是没有加上括号，所以这里的字符不会被匹配
>>> text.re_chars.findall('This is a <test>!')
['T', 'h', 'i', 's', ' ', 'i', 's', ' ', 'a', ' ', '', '!']
>>> text.re_tag.search('This <is> a test!')     # 匹配标签，<>中的单词匹配(\S+?)表达式
<re.Match object; span=(5, 9), match='<is>'>
>>> text.re_tag.search('This <is  > a test!')
# <>中的字符串有空格也可以，由(?:(\s*/)|\s.*?)?匹配
<re.Match object; span=(5, 11), match='<is  >'>
>>> text.re_newlines.findall('This is \r\n a \rtest!')
['\r\n', '\r']
# 在对每个单词进行分割后，当多个单词长度在 width 内时，可以把这些单词合并在一行内，
#例如这里的 is 和 a
>>> print(text.wrap('This is a text. Hello, world', 4))
This
is a
text.
Hello,
world
#这里的 This、is 和 a 被分在一行，它们的总长度之和小于 10
>>> print(text.wrap('This is a text. Hello, world', 10))
This is a
text.
Hello,
world
```

接下来讲解在 text.py 文件中定义的一个用于截断字符串的类 Truncator。首先看该类的源码，内容如下：

```
# 源码位置：django/utils/text.py
# ……

class Truncator(SimpleLazyObject):
    """
    通过字符或者单词截断文本的类
    """
    def __init__(self, text):
        # 注意，通过查看父类可知，text 被保存在属性__wrapped 中，
        # __wrapped 属性值由父类的_setup()方法设置
        super().__init__(lambda: str(text))

    def add_truncation_text(self, text, truncate=None):
        if truncate is None:  # 如果没有设置截断的文本，使用默认值
```

```
        truncate = pgettext(
            'String to return when truncating text',
            '%(truncated_text)s…')
    if '%(truncated_text)s' in truncate:
        return truncate % {'truncated_text': text}
    if text.endswith(truncate):
        # 如果已经以截断文本结尾，则直接返回
        return text
    # 返回在 text 后添加截断文本的字符串
    return '%s%s' % (text, truncate)

def chars(self, num, truncate=None, html=False):
    self._setup()  # 查看父类，可知这里将在初始化中输入的 str(text)赋给了属性值_wrapped
    length = int(num)
    text = unicodedata.normalize('NFC', self._wrapped)

    # Calculate the length to truncate to (max length - end_text length)
    truncate_len = length
    for char in self.add_truncation_text('', truncate):
        if not unicodedata.combining(char):
            truncate_len -= 1
            if truncate_len == 0:
                break
    if html:
        return self._truncate_html(length, truncate, text, truncate_len, False)
    return self._text_chars(length, truncate, text, truncate_len)

def _text_chars(self, length, truncate, text, truncate_len):
    """在某个长度的字符后截断文本"""
    s_len = 0
    end_index = None
    for i, char in enumerate(text):
        if unicodedata.combining(char):
            # 不要考虑组合字符，会增加字符长度
            continue
        s_len += 1
        if end_index is None and s_len > truncate_len:
            end_index = i
        if s_len > length:
            # 返回截断的字符串
            return self.add_truncation_text(text[:end_index or 0],
                                            truncate)

    # 由于不需要截断，所以直接返回原始字符串
    return text

def words(self, num, truncate=None, html=False):
    self._setup()
```

```
        length = int(num)
        if html:
            return self._truncate_html(length, truncate, self._wrapped, length, True)
        return self._text_words(length, truncate)

    def _text_words(self, length, truncate):
        words = self._wrapped.split()
        if len(words) > length:
            words = words[:length]
            return self.add_truncation_text(' '.join(words), truncate)
        return ' '.join(words)

    # ……

# ……
```

Truncator 类中的大部分方法都比较简单，下面通过示例理解该类及该类中的一些方法，如下：

```
>>> from django.utils import text
>>> tc = text.Truncator('This is a long long story')
# 该方法只在文本后添加后缀字符串，默认为'…'
>>> tc.add_truncation_text('xxxxxxxx')
'xxxxxxxx…'
>>> tc.add_truncation_text('xxxxxxxx', '#...')
'xxxxxxxx#...'
>>> tc.add_truncation_text('xxxxxxxx', '$')
'xxxxxxxx$'
# chars()方法的功能是按某个长度截断文本，原文本会在执行 self._setup()语句后被保存在属性_wrapped 中
>>> tc._wrapped
<object object at 0x7f3982e910e0>
>>> tc.chars(8, '#...')
'This#...'
>>> len(tc.chars(8, '#...'))
8
>>> tc._wrapped
'This is a long long story'
>>> tc._text_words(3, '#...')    # 按单词数进行截断
'This is a#...'
>>> tc.words('3', '#...')        # 这里的单词数可以是字符串
'This is a#...'
# 通过源码可知，在切割单词后，最终按单词数合并会去掉多余的空格，每个单词之间只有单个空格
>>> tc = text.Truncator('This     is a long     long story')
>>> tc.words('3', '#...')
'This is a#...'
>>> tc._text_chars(3, '#...', 'xxxxxxxxxxxx', 1)
'x#...'
# 当 truncate_len<=length 时，truncate_len 为截取字符串的长度
>>> tc._text_chars(3, '#...', 'xxxxxxxxxxxx', 2)
```

```
'xx#...'
>>> tc._text_chars(3, '#...', 'xxxxxxxxxxxx', 3)
'xxx#...'
>>> tc._text_chars(3, '#...', 'xxxxxxxxxxxx', 4)
'#...'
```

通过上面的操作示例，相信读者可以迅速理解一些方法的功能。比如_text_chars()方法。当该方法中的参数 truncate_len<=length 时，truncate_len 为从头截取 text 字符串的长度，当该值大于 length 时，只返回 truncate。

接下来定义的一些函数都非常简单，我们直接调用这些函数来理解它们的功能，示例如下：

```
>>> from django.utils import text
# 通过观察其源码可知，空格将被替换成下画线，$、!等特殊字符将被移除
>>> text.get_valid_filename('$x!   xxxxx')
'x___xxxxx'
# 当列表中的元素个数为 0 时返回空字符串，当只有 1 个元素时返回该元素
# 对于多个元素，前面用逗号分隔，最后两个元素用指定连接符连接，默认为 or
>>> text.get_text_list(['a', 'b', 'c', 'd'])
'a, b, c or d'
>>> text.get_text_list(['a', 'b', 'c'], 'and')
'a, b and c'
>>> text.get_text_list(['a'], 'and')
'a'
>>> text.get_text_list([])
''
#该函数使用正则表达式将“\r\n”或者“\r”字符串转换成“\n”
>>> text.normalize_newlines('xxxxx\r\nsssssssr\nxxx\rx')
'xxxxx\nsssssssr\nxxx\nx'
#该函数通过映射表将字母转成对应的数字
>>> text.phone2numeric('aaXbssdd')
'22927733'

# StreamingBuffer()方法继承自 BytesIO()方法，并且重写了 read()方法
>>> sb = text.StreamingBuffer()
>>> sb.write(b'hello')
5
>>> sb.read()
b'hello'
>>> sb.write(b'x')
1
>>> sb.read()
b'x'
>>> sb.getvalue()
b''
# 演示 BytesIO()方法的用法
>>> from io import BytesIO
>>> bio = BytesIO()
```

```
>>> bio.write(b'hello')
5
>>> bio.read()
b''
>>> bio.getvalue()
b'hello'
>>> bio.write(b'x')
1
>>> bio.getvalue()
b'hellox'
>>> bio.read()
b''
# 按字符串中出现的大写字母分割字符串，同时把所有字母转成小写
>>> text.camel_case_to_spaces('testTestHexxxxXyz')
'test test hexxxx xyz'
>>> text.camel_case_to_spaces('XestZZ TestHexxxxXyz')
'xest z z  test hexxxx xyz'
>>> text.camel_case_to_spaces('XestTestHexxxxXyz')
'xest test hexxxx xyz'
>>> text.camel_case_to_spaces('Xest TestHexxxxXyz')
'xest  test hexxxx xyz'
```

对于 text.py 文件中的其他函数，如 compress_string()、smart_split()等，本书不再一一演示。所有的函数都可以按照上面的方式，要么研究函数源码掌握其功能，要么对着源码完成实际测试，通过结果分析出其功能，之后再学习其实现源码。

## 8.4.2　html.py 文件中的代码

下面看 html.py 文件中的代码，首先看在 django/utils/safestring.py 文件中定义的一些安全的文本类（如 SafeData、SafeBytes、SafeText 等）以及标记安全的函数 mark_safe()，如下：

```
# 源码位置：djangp/utils/safestring.py

from django.utils.functional import wraps


class SafeData:
   def __html__(self): # 定义了魔法函数__html_()，返回该对象本身
      return self


# 该类继承自 bytes 和 SafeData
class SafeBytes(bytes, SafeData):
   def __add__(self, rhs):                  # 自定义魔法函数__add__()
      t = super().__add__(rhs)              # 使用父类加法
      if isinstance(rhs, SafeText):         # 如果 rhs 是 SafeText 实例，则返回 SafeText 对象
```

```
            return SafeText(t)
        elif isinstance(rhs, SafeBytes):    # 如果 rhs 是 SafeBytes 实例，则返回 SafeBytes 对象
            return SafeBytes(t)
        return t                            # 直接返回父类调用魔法函数__add__()的结果


class SafeText(str, SafeData):
    def __add__(self, rhs):             # 逻辑类似 SafeBytes
        t = super().__add__(rhs)
        if isinstance(rhs, SafeData):
            return SafeText(t)
        return t

    def __str__(self):
        return self


SafeString = SafeText                   # SafeString 就是 SafeText，或者说起了一个别名


def _safety_decorator(safety_marker, func):   # 定义一个装饰器
    @wraps(func)
    def wrapped(*args, **kwargs):
        return safety_marker(func(*args, **kwargs))
    return wrapped


def mark_safe(s):
    if hasattr(s, '__html__'):     # 如果定义了魔法函数__html__()的属性，则直接返回
        return
    if callable(s):                # 如果 s 是可调用的，则调用装饰器函数处理
        return _safety_decorator(mark_safe, s)
    return SafeText(s)             # 返回一个 SafeText 对象
```

在上述代码中定义的 SafeData、SafeBytes、SafeText 和 SafeString 都非常简单，SafeData 只是简单实现了魔法函数__html__ (),而 SafeBytes、SafeText 则分别继承了 bytes 和 str,同时继承了 SafeData 类，并在此基础上重新实现了魔法函数__add__()。

以下是在 html.py 文件中定义的两个简单转义函数 escape()和 escapejs(),用于对 HTML 文本或者 JavaScript 中的一些字符串进行转换，它们的源码如下：

```
# 源码位置：djangp/utils/html.py
# ……

_html_escapes = {
    ord('&'): '&amp;',
    ord('<'): '&lt;',
```

```
    ord('>'): '&gt;',
    ord('"'): '&quot;',
    ord("'"): '&#39;',
}


@keep_lazy(str, SafeText)
def escape(text):
    return mark_safe(str(text).translate(_html_escapes)) # 对 text 中出现的&、<、>等进行转义


_js_escapes = {
    ord('\\'): '\\u005C',
    ord('\''): '\\u0027',
    ord('"'): '\\u0022',
    ord('>'): '\\u003E',
    ord('<'): '\\u003C',
    ord('&'): '\\u0026',
    ord('='): '\\u003D',
    ord('-'): '\\u002D',
    ord(';'): '\\u003B',
    ord('`'): '\\u0060',
    ord('\u2028'): '\\u2028',
    ord('\u2029'): '\\u2029'
}

# 跳过每一个值小于 32 的 ASCII 字符
_js_escapes.update((ord('%c' % z), '\\u%04X' % z) for z in range(32))


@keep_lazy(str, SafeText)
def escapejs(value):
    # 对 value 中出现的&、=等符号进行转义
    return mark_safe(str(value).translate(_js_escapes))

# ……
```

通过查看源码即可明确这两个转义函数的功能，下面在交互模式下测试这两个转义函数，示例如下：

```
>>> from django.utils import html
>>> html.escape('<p>标题</p>')
'&lt;p&gt;标题&lt;/p&gt;'
>>> html.escapejs('console.log("hello, world")')
'console.log(\\u0022hello, world\\u0022)'
```

在 html.py 文件中还定义了一个用于处理换行符的 linebreaks()函数，该函数会将输入字符串中的\r\n 或'\n 转成<br>，同时将文本中的每行（要求有 2 个及以上的换行符）用<p>和</p>包围起来，形成 HTML 中的段落文本。linebreaks()函数的源码如下：

```
# 源码位置：django/utils/html.py
# ……

@keep_lazy_text
def linebreaks(value, autoescape=False):
    """将新的行转成<p>和<br>"""
    value = normalize_newlines(value)      # 去掉字符串中的\r
    paras = re.split('\n{2,}', str(value))     # 对含 2 个\n 以上的文本进行切割
    if autoescape:
        paras = ['<p>%s</p>' % escape(p).replace('\n', '<br>') for p in paras]
    else:
        paras = ['<p>%s</p>' % p.replace('\n', '<br>') for p in paras]
    return '\n\n'.join(paras)                  # 按照\n\n 补回来

# ……

@keep_lazy_text
def normalize_newlines(text):
    return re_newlines.sub('\n', str(text))    # 将\r\n 和\r 全部替换成\n

# ……
```

在交互模式下测试 linebreaks()函数，示例如下：

```
>>> from django.utils import html
>>> html.normalize_newlines('xxxxx\r\n\nsssss\r\nkkkkkk\r')
'xxxxx\n\nsssss\nkkkkkk\n'
>>> html.linebreaks('xxxxx\r\n\nsssss\r\nkkkkkk\r')
'<p>xxxxx</p>\n\n<p>sssss<br>kkkkkk<br></p>'
```

在 html.py 文件中还定义了一个用于解析 HTML 文本的类：MLStripper。MLStripper 类继承自 Python 内置的 HTMLParser 类，并重新实现了 HTMLParser 类中的若干方法，源码如下：

```
# 源码位置：django/utils/html.py
# ……

class MLStripper(HTMLParser):
    def __init__(self):
        super().__init__(convert_charrefs=False)
        self.reset()
        self.fed = []

    def handle_data(self, d):
        # 处理数据，标签之间的文本。比如<p>百度</p>文本，d 即为提取的“百度”字符串
```

```
        self.fed.append(d)

    def handle_entityref(self, name):
        # 处理以&开头、以分号结尾的字符串，将得到的中间的字符串赋给 name 变量
        # 例如对于字符串“&hello; ”，name 变量为 hello
        self.fed.append('&%s;' % name)

    def handle_charref(self, name):
        # 处理以&#开头、以分号结尾的字符串，将得到的中间的字符串赋给 name 变量
        # 例如对于字符串“&#hello; ”，name 变量为 hello
        self.fed.append('&#%s;' % name)

    def get_data(self):
        return ''.join(self.fed)
# ……
```

HTMLParser 类中的 handle_data()方法的作用是处理标签中的文本。比如对于“<p>百度</p>”字符串，可以先调用 HTMLParser 类的 feed()方法在内部扫描字符串，得到一个文本，即 “百度”字符串；再通过 HTMLParser 类的 handle_data()方法处理得到的文本。

MLStripper 类中的 handle_data()方法会将文本加入 fed 属性中。handle_entityref()方法和 handle_charref()方法的作用就是处理文本中分别以“&”和“&#”开头的字符串（在 Python 内部使用正则表达式匹配）。例如，对于字符串“<span>&hello</span>”，当调用 MLStripper 类的 feed()方法时，该类会从头扫描输入的字符串，当匹配到“&hello”后，会调用 MLStripper 类的 handle_entityref()方法进行处理。根据 handle_entityref()方法的源码可知，MLStripper 类同样会对匹配到的“hello”字符串进行扩展，在得到字符串“&hello; ”后再追加到 fed 属性中。下面简单演示在 HTMLParser 内部实现的一些关键源码：

```
# 源码位置：lib/html/parser.py
# ……

entityref = re.compile('&([a-zA-Z][-.a-zA-Z0-9]*)[^a-zA-Z0-9]')
charref = re.compile('&#(?:[0-9]+|[xX][0-9a-fA-F]+)[^0-9a-fA-F]')

# ……

class HTMLParser(_markupbase.ParserBase):
    # ……

    def feed(self, data):
        # 添加的数据
        self.rawdata = self.rawdata + data
        self.goahead(0)  # 从头开始解析
```

```
# ……

def goahead(self, end):
    rawdata = self.rawdata
    i = 0
    n = len(rawdata)
    while i < n:
        # 得到本次开始扫描字符串的位置 i
        # ……

        if i == n: break  # 已到达最后，直接退出循环
        startswith = rawdata.startswith
        if startswith('<', i):
            # 处理标签中的文本
            # ……
        elif startswith("&#", i):
            # 当前位置开始的字符串以“&#”开头，然后使用正则表达式进行匹配
            match = charref.match(rawdata, i)
            if match:
                name = match.group()[2:-1]    # 去掉开头的“&#”及最后的“;”，得到 name
                self.handle_charref(name)    # 调用 handle_charref()方法进行处理
                k = match.end()
                if not startswith(';', k-1):
                    k = k - 1
                i = self.updatepos(i, k)         # 更新下一次扫描位置
                continue
            else:
                # 没有匹配到，直接跳出循环
                if ";" in rawdata[i:]:  # bail by consuming &#
                    self.handle_data(rawdata[i:i+2])
                    i = self.updatepos(i, i+2)
                break
        elif startswith('&', i):    # 匹配以“&”开头的字符串
            match = entityref.match(rawdata, i)   # 使用正则表达式进行匹配
            if match:
                name = match.group(1)                  # 匹配得到的 name 变量
                self.handle_entityref(name)        # 调用 handle_entityref()方法处理
                k = match.end()  # 更新扫描位置，然后跳回循环从最开始处继续处理
                if not startswith(';', k-1):
                    k = k - 1
                i = self.updatepos(i, k)
                continue
            # 没有匹配到 entityref 表达式的处理
            # ……
        else:
            assert 0, "interesting.search() lied"
    # 最后一些额外处理
    # ……
```

```
    # ……
```

在看完 HTMLParser 类中关于解析文本的源码及相关的注释后，对于 handle_entityref()、handle_charref()方法的作用及相关参数是否有了更清楚地认识呢？最后，我们在控制台中演示对 MLStripper 类的使用，以便理解该类的功能及相关方法的作用，示例如下：

```
>>> from django.utils.html import MLStripper
>>> mls = MLStripper()
>>> mls.feed('<p><a href="http://www.baidu.com">百度一下
(&#913;)</a><span>&Alpha;</span></p>')
>>> mls.fed
['百度一下(', '&#913;', ')', '&Alpha;']
>>> mls.get_data()
'百度一下(&#913;)&Alpha;'
```

html.py 文件中的 smart_urlquote()方法会对 URL 中包含的一些特殊字符，如空格、中文等进行编码处理，类似于在浏览器上输入该 URL 后的结果：

```
>>> html.smart_urlquote("http://192.168.26.110:8000/test_view/a=xxx&b=xyz &k='中文'")
"http://192.168.26.110:8000/test_view/a=xxx&b=xyz%20&k='%E4%B8%AD%E6%96%87'"
```

还有一个比较重要的函数 urlize()，它会搜索输入字符串中的 URL 子串，并将其转换成可单击的 HTML 文本，即用<a></a>元素包围该子串，示例如下：

```
>>> html.urlize('<p>你好啊，单击测试 http://192.168.26.110:8000/test_view/</p>')
'<p>你好啊，单击测试 <a
href="http://192.168.26.110:8000/test_view/">http://192.168.26.110:8000/test_view/</
a></p>'
>>> html.urlize('<p>你好啊，单击测试 http://192.168.26.110:8000/test_view/</p>',
trim_url_limit=5)
'<p>你好啊，单击测试 <a href="http://192.168.26.110:8000/test_view/">http…</a></p>'
# 设置 nofollow=True，将在生成的元素<a>中添加一个 rel 属性，指定值为 nofollow
>>> html.urlize('<p>你好啊，单击测试 http://192.168.26.110:8000/test_view/</p>',
trim_url_limit=5, nofollow=True)
'<p>你好啊，单击测试 <a href="http://192.168.26.110:8000/test_view/"
rel="nofollow">http…</a></p>'
>>> html.urlize('<p>你好啊，单击测试 http://192.168.26.110:8000/test_view/</p>',
trim_url_limit=5, nofollow=True, autoescape=True)
'&lt;p&gt;你好啊，单击测试 <a href="http://192.168.26.110:8000/test_view/"
rel="nofollow">http…</a>&lt;/p&gt;'
```

在 urlize()函数中共有 4 个参数：text、trim_url_limit、nofollow 和 autoescape，它们的含义如下：

◎ text：待转换的文本，必须输入。

◎ trim_url_limit：限制生成链接元素的文本，即<a></a>中的文本，默认会显示全部匹配到的 URL 文本。

◎ nofollow：当把该参数设置为 True 时，将在生成的<a>元素中添加 rel 属性，并设置其值为 nofollow。

◎ autoescape：当把该参数设置为 True 时，将自动转义链接文本和 URLs。

至此，对 html.py 文件中的部分重要函数的介绍就结束了，下面介绍 utils 目录下的一些代码文件。

# 8.5 其他的类与函数

本节介绍 utils 目录下的 archive.py 文件在 archive.py 文件中定义了多个用于处理压缩文件的类，包括处理 zip 压缩包的 ZipArchive 类和处理 tar 压缩包的 TarArchive 类。首先讲解在 archive.py 文中定义的压缩基类——BaseArchive，源码如下：

```
# 源码位置：django/utils/archive.py
# ……

class BaseArchive:
   @staticmethod
   def _copy_permissions(mode, filename):
      # 如果设置的 mode 有对其他人的读权限，那么就对指定文件设置该 mode 权限
      if mode & stat.S_IROTH:
         os.chmod(filename, mode)

   def split_leading_dir(self, path):
      path = str(path)  #
      path = path.lstrip('/').lstrip('\\')  # 去掉开头的/或者\\
      if '/' in path and (('\\' in path and path.find('/') < path.find('\\')) or '\\' 
not in path):
         #字符串“root/test/hosts”将被分割成['root', 'test/hosts']
         return path.split('/', 1)
      elif '\\' in path:
         return path.split('\\', 1)
      else:
         # 如果在 path 中没有/或者\\，则直接返回二元组(path, '')
         return path, ''

   def has_leading_dir(self, paths):
      """
      如果所有的 paths 路径都是同一个目录开头，则返回 True，否则返回 False
      """
      common_prefix = None
      for path in paths:
         # 获取前缀目录
         prefix, rest = self.split_leading_dir(path)
```

```
            if not prefix: # 如果没有prefix，则直接返回 False
                return False
            elif common_prefix is None:
                # 如果是第一次进入循环，则直接将得到的主目录名赋给 common_prefix 变量
                common_prefix = prefix
            elif prefix != common_prefix:
                # 如果本次得到的主目录与公共主目录不匹配，则直接返回 False
                return False
        #有公共的主目录，返回 True
        return True

    def extract(self):
        raise NotImplementedError('...')

    def list(self):
        raise NotImplementedError('...')

# ……
```

在 BaseArchive 类中定义了三个基础方法和两个待子类实现的方法。下面演示三个基础方法的使用，示例如下：

```
>>> import os, stat
>>> oct(os.stat('/root/hosts').st_mode)[-3:] # 查看文件权限
'644'
>>> os.chmod("/root/hosts", stat.S_IREAD|stat.S_IWRITE)  # 重新修正权限
>>> oct(os.stat('/root/hosts').st_mode)[-3:]
'600'
>>> os.stat('/root/hosts').st_mode & stat.S_IROTH        # 不包含其他人的读权限
0
>>> os.chmod("/root/hosts", stat.S_IREAD|stat.S_IWRITE|stat.S_IROTH)
# 设置其他人的读权限
>>> oct(os.stat('/root/hosts').st_mode)[-3:]
'604'
>>> os.stat('/root/hosts').st_mode & stat.S_IROTH   # 取&之后，结果为 4，即为 000 000 100
4
>>> from django.utils.archive import BaseArchive
>>> ba = BaseArchive()
>>> ba.split_leading_dir('/root')  # 得到的第 1 个元素为主目录
('root', '')
>>> ba.split_leading_dir('/root/hosts')
['root', 'hosts']
>>> ba.split_leading_dir('/root/test/hosts')
['root', 'test/hosts']
>>> ba.has_leading_dir(['/root', '/root/hosts', '/root/test/hosts'])
True
# 最后一个路径的主目录为 etc，与前面的 root 不符，所以返回 False
```

```
>>> ba.has_leading_dir(['/root', '/root/hosts', '/etc/hosts'])
False
```

Django 在 BaseArchive 类的基础上实现了两个子类 TarArchive 和 ZipArchive，它们分别用于处理两种不同的压缩文件，这两个子类的核心正是基于 Python 内置的 tarfile 模块和 zipfile 模块。TarArchive 类和 ZipArchive 类的源码如下：

```
# 源码位置：django/utils/archive.py
# ……

class TarArchive(BaseArchive):

    def __init__(self, file):
        self._archive = tarfile.open(file)  # 调用 tarfile 模块的 open()方法打开 tar 压缩文件

    def list(self, *args, **kwargs):
        # 调用能打开 tar 包的 list()函数，显示压缩包中的文件列表
        self._archive.list(*args, **kwargs)

    def extract(self, to_path):
        members = self._archive.getmembers()  # 获取 tar 包中的所有文件
        leading = self.has_leading_dir(x.name for x in members)  # 获取所有文件的主目录
        for member in members:
            name = member.name
            if leading:
                name = self.split_leading_dir(name)[1]
            # 解压缩文件的完整路径
            filename = os.path.join(to_path, name)
            if member.isdir():   # 如果列表元素为目录，则只需创建该目录即可
                if filename and not os.path.exists(filename):
                    os.makedirs(filename)
            else:
                try:
                    # 提取单个文件
                    extracted = self._archive.extractfile(member)
                except (KeyError, AttributeError) as exc:
                    # 打印异常
                    # ……
                else:
                    dirname = os.path.dirname(filename)
                    if dirname and not os.path.exists(dirname):
                        os.makedirs(dirname)
                    # 保存文件内容和权限信息到新的路径上
                    with open(filename, 'wb') as outfile:
                        shutil.copyfileobj(extracted, outfile)  # 第 1 个参数为文件内容
                        self._copy_permissions(member.mode, filename)
                finally:
                    if extracted:
```

```
                extracted.close()

    def close(self):
        self._archive.close()


class ZipArchive(BaseArchive):

    def __init__(self, file):
        self._archive = zipfile.ZipFile(file)

    def list(self, *args, **kwargs):
        self._archive.printdir(*args, **kwargs)

    def extract(self, to_path):
        namelist = self._archive.namelist()
        leading = self.has_leading_dir(namelist)
        for name in namelist:
            # 读取压缩的文件内容
            data = self._archive.read(name)
            info = self._archive.getinfo(name)
            if leading:
                name = self.split_leading_dir(name)[1]
            filename = os.path.join(to_path, name)
            dirname = os.path.dirname(filename)
            if dirname and not os.path.exists(dirname):
                os.makedirs(dirname)
            if filename.endswith(('/', '\\')):   #如果不存在目录，只需创建即可
                # A directory
                if not os.path.exists(filename):
                    os.makedirs(filename)
            else:
                # 保存文件内容到新的路径上，同时复制权限信息
                with open(filename, 'wb') as outfile:
                    outfile.write(data)
                mode = info.external_attr >> 16
                self._copy_permissions(mode, filename)

    def close(self):
        self._archive.close()

# ……
```

上面的代码比较简单，只需掌握 Python 的几个内置模块（tarfile、zipfile、shutil 等）即可。接下来在 Python 交互模式下进行实战演示。在演示之前，需要建立几个压缩包作为实验的目标文件：

```
(django2-core-test) [root@master ~]# tree /root/test
/root/test
```

```
├── django
│   ├── basic_check.sh
│   └── bird6-1.6.8-1.el7.x86_64.rpm
├── django_pm0lQkb
└── test.txt

1 directory, 4 files
[root@master test]# cat test/test.txt
# 这是一个普通文件
[root@master ~]#  zip -r django_test.zip test   # 压缩成 zip 包
  adding: test/ (stored 0%)
  adding: test/django/ (stored 0%)
  adding: test/django/basic_check.sh (deflated 52%)
  adding: test/django/bird6-1.6.8-1.el7.x86_64.rpm (deflated 2%)
  adding: test/django_pm0lQkb (deflated 2%)
  adding: test/test.txt (stored 0%)
[root@master ~]# tar -cvf django_test.tar test  # 压缩成 tar 包
test/
test/django/
test/django/basic_check.sh
test/django/bird6-1.6.8-1.el7.x86_64.rpm
test/django_pm0lQkb
test/test.txt
[root@master ~]# ls -l django_test.tar django_test.zip
-rw-r--r-- 1 root root 624640 Apr 28 16:19 django_test.tar
-rw-r--r-- 1 root root 608260 Apr 28 16:18 django_test.zip
```

在交互模式下测试 tarfile 模块和 zipfile 模块的常用方法，示例如下：

```
>>> import tarfile
>>> archive = tarfile.open('/root/django_test.tar')
>>> archive.list()
?rwxr-xr-x root/root          0 2021-04-28 16:16:59 test/
?rwxr-xr-x root/root          0 2021-04-02 13:43:40 test/django/
?rw-r--r-- root/root       1271 2021-04-02 12:41:48 test/django/basic_check.sh
?rw-r--r-- root/root     307964 2021-04-02 13:43:40
test/django/bird6-1.6.8-1.el7.x86_64.rpm
?rw-r--r-- root/root     307964 2021-04-02 13:19:26 test/django_pm0lQkb
?rw-r--r-- root/root         27 2021-04-28 16:05:28 test/test.txt
>>> members = archive.getmembers()
>>> members
[<TarInfo 'test' at 0x7f2296785f40>, <TarInfo 'test/django' at 0x7f2296785dc0>, <TarInfo
'test/django/basic_check.sh' at 0x7f2296785880>, <TarInfo
'test/django/bird6-1.6.8-1.el7.x86_64.rpm' at 0x7f2296785040>, <TarInfo
'test/django_pm0lQkb' at 0x7f2296785e80>, <TarInfo 'test/test.txt' at 0x7f2296761040>]
>>> for member in members:
...     print(member.name)
...
```

```
test
test/django
test/django/basic_check.sh
test/django/bird6-1.6.8-1.el7.x86_64.rpm
test/django_pm0lQkb
test/test.txt
>>> extracted = archive.extractfile(members[-1]) # 抽取 test/test.txt 文件的内容
>>> extracted.read().decode('utf-8')   # read()读取的是字节，需要转成字符串
'# 这是一个普通文件\n'
>>> import zipfile
>>> archive = zipfile.ZipFile('/root/django_test.zip')
>>> archive.printdir()
File Name                                             Modified             Size
test/                                          2021-04-28 16:17:00          0
test/django/                                   2021-04-02 13:43:40          0
test/django/basic_check.sh                     2021-04-02 12:41:48       1271
test/django/bird6-1.6.8-1.el7.x86_64.rpm       2021-04-02 13:43:40     307964
test/django_pm0lQkb                            2021-04-02 13:19:26     307964
test/test.txt                                  2021-04-28 16:05:28         27
>>> namelist = archive.namelist()
>>> namelist
['test/', 'test/django/', 'test/django/basic_check.sh',
'test/django/bird6-1.6.8-1.el7.x86_64.rpm', 'test/django_pm0lQkb', 'test/test.txt']
>>> data = archive.read(namelist[-1])
>>> info = archive.getinfo(namelist[-1])
>>> data
b'#
\xe8\xbf\x99\xe4\xb8\xaa\xe4\xb8\x80\xe4\xb8\xaa\xe6\x99\xae\xe9\x80\x9a\xe6\x96\x87
\xe4\xbb\xb6\n'
>>> data.decode('utf-8')
'# 这是一个普通文件\n'
>>> info
<ZipInfo filename='test/test.txt' filemode='-rw-r--r--' file_size=27>
>>> archive.close()
>>>
```

接下来演示 TarArchive 类和 ZipArchive 类的使用，示例如下：

```
(django2-core-test) [root@master first_django]# ls /root/test_archive_tar
ls: cannot access /root/test_archive_tar: No such file or directory
(django2-core-test) [root@master first_django]# ls /root/test_archive_zip
ls: cannot access /root/test_archive_zip: No such file or directory
(django2-core-test) [root@master first_django]# python manage.py shell
Python 3.8.6 (default, Oct 18 2020, 15:33:08)
[GCC 4.8.5 20150623 (Red Hat 4.8.5-39)] on linux
Type "help", "copyright", "credits" or "license" for more information.
(InteractiveConsole)
>>> from django.utils.archive import TarArchive
```

```
>>> ta = TarArchive('/root/django_test.tar')
>>> ta.extract('/root/test_archive_tar')
>>> ta.close()
>>> from django.utils.archive import ZipArchive
>>> za = ZipArchive('/root/django_test.zip')
>>> za.extract('/root/test_archive_zip')
>>> za.close()
>>> exit()
# 退出后查看解压缩后的目录与文件是否存在
(django2-core-test) [root@master first_django]# ls /root/test_archive_tar
django  django_pm0lQkb  test.txt
(django2-core-test) [root@master first_django]# ls /root/test_archive_zip
django  django_pm0lQkb  test.txt
```

最后，为了不让开发者自行选择压缩的类，Django 准备了一个通用的压缩类 Archive。Archive 类将直接根据文件的后缀名来选择压缩类，比如路径为/root/django_test.zip 的文件，在提取后缀“zip”后，可知对应的压缩类为 ZipArchive，源码如下：

```
# 源码位置：django/utils/archive.py
# ……

class Archive:
    def __init__(self, file):
        # 根据输入的 file 值得到压缩对象
        self._archive = self._archive_cls(file)(file)

    @staticmethod
    def _archive_cls(file):
        cls = None
        if isinstance(file, str):
            # 输入的字符串被认为是文件名
            filename = file
        else:
            try:
                filename = file.name
            except AttributeError:
                raise UnrecognizedArchiveFormat(
                    "File object not a recognized archive format.")
        # 切割字符串，得到后缀 tail_ext
        base, tail_ext = os.path.splitext(filename.lower())
        # 从字典中选择对应的 Archive 类
        cls = extension_map.get(tail_ext)
        if not cls:
            base, ext = os.path.splitext(base)
            cls = extension_map.get(ext)
        if not cls:
            raise UnrecognizedArchiveFormat(
                "Path not a recognized archive format: %s" % filename)
```

```
        return cls

# ……

extension_map = {
    '.tar': TarArchive,
    '.tar.bz2': TarArchive,
    '.tar.gz': TarArchive,
    '.tgz': TarArchive,
    '.tz2': TarArchive,
    '.zip': ZipArchive,
}
```

上面的代码非常明确：提取后缀，然后到字典中读取相应配置好的压缩类进行实例化。

限于篇幅，对 utils 目录下的其他源码不再介绍。一些复杂的源码，只需慢慢拆解，慢慢分析，最后都能理解清楚。

## 8.6　小结

本章介绍了 Django 源码中 utils 目录下的部分源码，重点剖析了 Django 的自动重载、日志配置、时间解析及文本处理等功能所涉及的源码，为读者自行分析 utils 目录下的其他源码做好了铺垫工作。至此，对 Django 2.2.16 的源码剖析就结束了，虽然仍有很多细节没有介绍，但是本书已经给出了一个学习 Django 源码的良好方式：从 Django 使用案例出发，抽丝剥茧，剖析该案例背后的源码。天下没有不散的筵席，期待我们青山不改，江湖再会。

# 反侵权盗版声明